D1754543

Sie Chin Tjong

Polymer Composites with Carbonaceous Nanofillers

Related Titles

Tjong, S. C.

Carbon Nanotube Reinforced Composites

Metal and Ceramic Matrices

2009
ISBN: 978-3-527-40892-4

Thomas, S., Stephen, R.

Rubber Nanocomposites

Preparation, Properties and Applications

2010
ISBN: 978-0-470-82345-3

Hierold, C. (ed.)

Carbon Nanotube Devices

Properties, Modeling, Integration and Applications

2008
ISBN: 978-3-527-31720-2

Haley, M. M., Tykwinski, R. R. (eds.)

Carbon-Rich Compounds

From Molecules to Materials

2006
ISBN: 978-3-527-31224-5

Sperling, L. H.

Introduction to Physical Polymer Science

2006
ISBN: 978-0-471-70606-9

Reich, S., Thomsen, C., Maultzsch, J.

Carbon Nanotubes

Basic Concepts and Physical Properties

2004
ISBN: 978-3-527-40386-8

Sie Chin Tjong

Polymer Composites with Carbonaceous Nanofillers

Properties and Applications

WILEY-VCH

WILEY-VCH Verlag GmbH & Co. KGaA

The Author

Prof. Sie Chin Tjong
City University of Hong Kong
Department of Physics and
Materials Science
Hongkong, PR China

■ All books published by **Wiley-VCH** are carefully produced. Nevertheless, authors, editors, and publisher do not warrant the information contained in these books, including this book, to be free of errors. Readers are advised to keep in mind that statements, data, illustrations, procedural details or other items may inadvertently be inaccurate.

Library of Congress Card No.: applied for

British Library Cataloguing-in-Publication Data
A catalogue record for this book is available from the British Library.

Bibliographic information published by the Deutsche Nationalbibliothek
The Deutsche Nationalbibliothek lists this publication in the Deutsche Nationalbibliografie; detailed bibliographic data are available on the Internet at <http://dnb.d-nb.de>.

© 2012 Wiley-VCH Verlag & Co. KGaA, Boschstr. 12, 69469 Weinheim, Germany

All rights reserved (including those of translation into other languages). No part of this book may be reproduced in any form – by photoprinting, microfilm, or any other means – nor transmitted or translated into a machine language without written permission from the publishers. Registered names, trademarks, etc. used in this book, even when not specifically marked as such, are not to be considered unprotected by law.

Composition Laserwords Private Ltd., Chennai, India
Printing and Binding Markono Print Media Pte Ltd, Singapore
Cover Design   Adam-Design, Weinheim

Print ISBN: 978-3-527-41080-4
ePDF ISBN: 978-3-527-64875-7
ePub ISBN: 978-3-527-64874-0
mobi ISBN: 978-3-527-64873-3
oBook ISBN: 978-3-527-64872-6

Printed in Singapore
Printed on acid-free paper

Contents

Preface *XI*
Abbreviations *XIII*

1	**Introduction** *1*	
1.1	Graphene-Based Nanomaterials *1*	
1.1.1	Graphite Intercalation Compound *3*	
1.1.2	Graphene Oxide *4*	
1.2	Carbon Nanotubes *10*	
1.2.1	Synthesis of Carbon Nanotubes *11*	
1.2.1.1	Physical Vapor Deposition *11*	
1.2.1.2	Chemical Vapor Deposition *13*	
1.2.2	Purification of Carbon Nanotubes *19*	
1.2.2.1	Chemical Techniques *19*	
1.2.2.2	Physical Techniques *21*	
1.2.3	Characterization of Purified Carbon Nanotubes *23*	
1.3	Carbon Nanofibers (CNFs) *28*	
1.4	Physical Properties of Graphene *30*	
1.4.1	Mechanical Behavior *30*	
1.4.2	Electrical Behavior *31*	
1.4.3	Thermal Behavior *32*	
1.5	Properties of Carbon Nanotubes *35*	
1.5.1	Mechanical Behavior *35*	
1.5.1.1	Theoretical Prediction *35*	
1.5.1.2	Experimental Measurement *36*	
1.5.1.3	Flexibility of Carbon Nanotubes *37*	
1.5.2	Electrical Behavior *38*	
1.5.3	Thermal Behavior *39*	
1.6	Properties of Carbon Nanofibers *40*	
1.7	Current Availability of Carbonaceous Nanomaterials *41*	
1.8	Multifunctional Composite Materials *41*	
1.8.1	Composites with Carbon Black Nanoparticles *44*	
1.8.2	Composites with Graphene Oxide and Graphite Nanoplatelet Fillers *45*	

1.8.3	Composites with Carbon Nanotubes	46
	Nomenclature 46	
	References 46	
2	**Preparation of Polymer Nanocomposites** **53**	
2.1	Overview 53	
2.2	Dispersion of Nanofillers 54	
2.2.1	Surface Modification of Graphene Oxide 54	
2.2.2	Surface Modification of Graphene Nanoplatelet and Expanded Graphite 58	
2.2.3	Functionalization of CNTs and CNFs 59	
2.2.3.1	Covalent Functionalization 60	
2.2.3.2	NonCovalent Functionalization 62	
2.3	Solution Mixing 66	
2.3.1	Nanocomposites with Graphene-Like Fillers 67	
2.3.2	Nanocomposites with EG and GNP Fillers 68	
2.3.3	Nanocomposites with CNT and CNF Fillers 76	
2.4	Melt Mixing 78	
2.4.1	Nanocomposites with Graphene-Like Fillers 78	
2.4.2	Nanocomposites with EG and GNP Fillers 80	
2.4.3	Nanocomposites with CNT and CNF Fillers 81	
2.5	*In situ* Polymerization 84	
2.5.1	Nanocomposites with Graphene-Like Fillers 84	
2.5.2	Nanocomposites with EG and GNP Fillers 89	
2.5.3	Nanocomposites with CNT and CNF Fillers 91	
2.6	Patent Processes 98	
	References 98	
3	**Thermal Properties of Polymer Nanocomposites** **103**	
3.1	Crystallization 103	
3.2	Characterization Techniques for Crystallization 104	
3.2.1	Dynamic Mechanical Analysis 104	
3.2.2	Differential Scanning Calorimetry 105	
3.2.3	Nanocomposites with Graphene Nanofillers 107	
3.2.4	Nanocomposites with Carbon Nanotubes 111	
3.3	Thermal Stability 120	
3.3.1	Thermogravimetry Analysis 121	
3.3.2	Linear Thermal Expansion 124	
3.3.3	Heat Deflection Temperature 125	
3.4	Thermal Conductivity 127	
3.4.1	Composites with CNTs 127	
3.4.1.1	Thermal Interface Resistance 127	
3.4.1.2	Dispersion and Functionalization of Carbon Nanotubes 130	
3.4.2	Composites with GNP and Graphene Nanofillers 133	

Nomenclature 139
References 139

4 Mechanical Properties of Polymer Nanocomposites 143
4.1 Background 143
4.2 General Mechanical Behavior 144
4.3 Fracture Toughness 146
4.4 Strengthening and Toughening Mechanisms 148
4.4.1 Interfacial Shear Stress 148
4.4.2 Interfacial Interaction 150
4.4.3 Micromechanical Modeling 151
4.4.4 Toughening Mechanism 152
4.5 Nanocomposites with Graphene Fillers 153
4.5.1 Thermoplastic Matrix 153
4.5.2 Thermosetting Matrix 163
4.5.3 Elastomeric Matrix 168
4.6 Nanocomposites with EG and GNP Fillers 169
4.7 Nanocomposites with CNT and CNF Fillers 172
4.7.1 Thermoplastic Matrix 172
4.7.2 Thermosetting Matrix 180
4.8 Composites with Hybrid Fillers 186
Nomenclature 189
References 190

5 Electrical Properties of Polymer Nanocomposites 193
5.1 Background 193
5.2 Percolation Concentration 193
5.2.1 Theoretical Modeling 193
5.2.2 Excluded Volume 196
5.3 Electrical Conductivity and Permittivity 197
5.3.1 Graphene-Filled Polymer Composites 201
5.3.1.1 Thermoplastic Matrices 201
5.3.1.2 Thermosetting Matrices 207
5.3.1.3 Elastomeric Matrices 208
5.3.2 EG- and GNP-Filled Polymers 210
5.3.2.1 Thermoplastic Matrices 210
5.3.2.2 Thermosetting Matrices 213
5.3.3 CNT- and CNF-Filled Polymer Composites 213
5.3.3.1 Thermoplastic Matrices 213
5.3.3.2 Thermosetting Matrices 222
5.3.3.3 Elastomeric Matrices 228
5.4 Current-Voltage Relationship 229
5.5 Positive Temperature Coefficient Effect 234
5.6 Hybrid Nanocomposites 238
References 243

6		**Carbonaceous Nanomaterials and Polymer Nanocomposites for Fuel Cell Applications** *247*
6.1		Overview *247*
6.2		Polymer Exchange Membrane Fuel Cell *248*
6.3		Conventional Bipolar Plates *249*
6.4		Polymer Nanocomposite Bipolar Plates *252*
6.4.1		CNT-Filled Composite Plates *252*
6.4.2		Graphene-Sheet-Filled Composite Plates *254*
6.5		Electrical Characteristics of Nanocomposite Bipolar Plates *259*
6.5.1		Polymer/Graphite/CNT Hybrid Plates *259*
6.5.2		Polymer/EG and Polymer/GNP Composite Plates *262*
6.5.3		Polymer/CB Composite Plates *264*
6.5.4		Fuel Cell Performance *266*
6.5.4.1		Voltage Loss *266*
6.5.4.2		Polarization Curves *267*
6.6		Mechanical Properties of Nanocomposite Bipolar Plates *269*
6.6.1		Hybrid Composite Bipolar Plates *269*
6.6.2		Polymer/EG Composites *272*
6.7		Electrocatalyst Supports *274*
6.7.1		Carbon-Nanotube-Supported Platinum Electrocatalysts *274*
6.7.2		Graphene-Supported Platinum Electrocatalysts *279*
		Nomenclature *282*
		References *282*
7		**Polymer Nanocomposites for Biomedical Applications** *285*
7.1		Overview *285*
7.2		Bone Implants *286*
7.3		Biocompatibility of Carbon Nanotubes *290*
7.3.1		Potential Health Hazards *290*
7.3.2		Cell Viability Assays *296*
7.3.3		Tissue Cell Responses *301*
7.4		CNT/Polymer Nanocomposites for Load-Bearing Implants *306*
7.4.1		Mechanical Properties *307*
7.4.1.1		MWNT/PE Nanocomposites *307*
7.4.1.2		MWNT/PEEK Nanocomposites *310*
7.5		CNT/Polymer Nanocomposite Scaffolds *310*
7.5.1		Types and Structures of Scaffolds *311*
7.5.2		Electrospinning: Principle and Applications *312*
7.6		Nervous System Remedial Applications *320*
7.7		Biocompatibility of Graphene Oxide and Its Nanocomposites *323*
		References *326*
8		**Polymer Nanocomposites for Electromagnetic Interference (EMI) Shielding** *331*
8.1		Introduction *331*

8.2	EMI Shielding Efficiency	*332*
8.3	Nanocomposites with Graphene Fillers	*335*
8.4	Nanocomposites with GNPs	*335*
8.5	Nanocomposites with CNTs and CNFs	*338*
8.6	Foamed Nanocomposites for EMI Applications	*343*
8.6.1	CNT/Polymer Foamed Nanocomposites	*343*
8.6.2	Graphene/Polymer Foamed Nanocomposites	*347*
	Nomenclature *349*	
	References *349*	
9	**Polymer Nanocomposites for Sensor Applications**	*351*
9.1	Introduction *351*	
9.2	Pressure/Strain Sensors	*352*
9.2.1	Piezoresistivity *352*	
9.2.2	Nanocomposites with GNPs	*355*
9.2.3	Nanocomposites with CNTs	*356*
9.3	Gas and Humidity Sensors	*359*
9.3.1	Gas Sensitivity *359*	
9.4	Organic Vapor Sensors	*361*
9.4.1	Nanocomposites with CBs	*361*
9.4.1.1	Nanocomposites with CNTs and CNFs	*364*
9.4.2	Humidity Sensors *371*	
9.4.2.1	Graphene Oxide Sensors	*372*
9.4.2.2	Nanocomposite Sensors with CNT Fillers	*375*
	Nomenclature *377*	
	References *377*	

Index *381*

Preface

Carbon nanotubes and graphene sheets exhibit unique and extraordinary electrical, mechanical, and thermal properties rendering them attractive fillers for reinforcing polymers to form functional and structural composite materials of high performance. The performance of the polymer nanocomposites relies on the inherent properties of carbonaceous nanofillers, and on optimizing the dispersion, interfacial interaction, and nanoscale exfoliation of those fillers within the polymer matrix. Designing smart polymer nanocomposite materials with the appropriate processing-structure-property relationships for biomedical, electronic, electromagnetic interference shielding, and chemical sensing as well as structural engineering applications is challenging. In recent years, one-dimensional carbon nanotubes have been incorporated into various types of polymeric materials for achieving these purposes. However, the high cost, tedious purification and high tendency of agglomeration of carbon nanotubes hurdle the development of nanotube/polymer composites in engineering applications. The recent successful synthesis of two-dimensional graphene layers from graphite oxide via chemical and thermal reduction techniques has sparked enormous interest in their properties, functions, and applications. The low cost and ease of fabrication of graphene offer tremendous opportunities for chemists and materials scientists to explore and develop novel graphene/polymer nanocomposites with excellent biological, mechanical, and physical properties. This book focuses exclusively on the latest research related to the synthesis and property characterization of one- and two dimensional carbonaceous nanomaterials and their polymer nanocomposites, and addresses potential applications of these materials to bipolar plates of fuel cells, electrocatalysts, human orthopedic implants and scaffolds, electromagnetic interference shielding materials, and gas-, pressure- and temperature sensors. This book serves as a valuable and informative reference to scientists, engineers, medical technologists, and practitioners engaged in the teaching, research, development, and use of functional polymer composites with carbonaceous nanofillers.

Sie Chin Tjong
CEng CSci FIMMM
City University of Hong Kong

Abbreviations

AC	alternating current
AFM	atomic force microsopy
AIBN	2,2′-azobisisobutyronitrile
ARTP	atom transfer radical polymerization
BIB	α-bromoisobutyryl bromide
BP	benzoyl peroxide
CB	carbon black
CF	carbon fiber
CMG	chemically modified graphene
CNF	carbon nanofiber
CNT	carbon nanotube
CS	chitosan
CTAB	hexadecyltrimethylammonium bromide
CTE	coefficient of thermal expansion
CVD	chemical vapor deposition
DBP	dibutyl phthalate
DC	direct current
DGEBA	diglycidyl ether bisphenol-A
DENT	double-edge-notched tension
DMA	dynamic mechanical analysis
DMAc	N,N dimethylacetamide
DMF	dimethylformamide
DMSO	dimethyl sulfoxide
DSC	differential scanning calorimetry
DWNT	double-walled carbon nanotube
EG	expanded graphite
ECM	extracellular matrix
ECSA	electrochemically active surface area
EDS	energy dispersive spectroscopy
EMI	electromagnetic interference
EVA	ethylene vinyl acetate
EWF	essential work of fracture
FGS	functional graphene sheet

FMWNT	functionalized multiwalled carbon nanotube
GIC	graphite intercalation compound
GDL	gas diffusion layer
GNP	graphite nanoplatelet
GO	graphene oxide
HA	hydroxyapatite
HDPE	high-density polyethylene
HDT	heat deflection temperature
HEK	human epidermal keratinocyte
HiPCo	High-pressure carbon oxide disproportionation
HOPG	highly oriented pyrolytic graphite
HOR	hydrogen oxidation reaction
iGO	isocyanate-treated graphene oxide
LDPE	low-density polyethylene
LEFM	linear elastic fracture mechanics
LLDPE	linear low-density polyethylene
MA-g-PP	maleic anhydride-grafted polypropylene
MD	molecular dynamics
MEA	membrane electrode assembly
MMT	montmorillonite
MTS	3-(4,5-dimethylthiazol-2-yl)-5(3-carboxymethonyphenol)-2-(4-sulfophenyl)-2H-tetrazolium
MTT	3-(4,5-dimethylthiazol-2-yl)-2,5-diphenyltetrazolium bromide
MWNT	multiwalled carbon nanotube
ORR	oxygen reduction reaction
PA	polyamide
PAA	poly(acrylic acid); polyallylamine; polyamic acid
PC	polycarbonate
PCL	polycaprolactone
PDMS	poly(dimethyl siloxane)
PE	polyethylene
PECVD	plasma-enhanced chemical vapor deposition
PEDOT	poly(3,4-ethylenedioxythiophene)
PEEK	poly(etheretherketone)
PEMFC	proton exchange membrane fuel cell
PEN	poly(ethylene-2,6-naphthalate)
PEO	poly(ethylene oxide)
PET	polyethylene terephthalate
PGMA	poly(glycidyl methacrylate)
PI	polyimide
PLA	polylactic acid
PmPV	poly(m-phenylene vinylene)
PMMA	poly(methyl methacrylate)
PS	polystyrene
PSF	polysulfone

PTC	positive temperature coefficient
PTT	polytrimethylene terephthalate
PU	polyurethane
PVA	poly(vinyl alcohol)
PVC	polyvinyl chloride
PVD	physical vapor deposition
PVDF	polyvinylidene fluoride
PVP	polyvinyl pyrrolidone
P3HT	poly(3-hexylthiophene)
rGO	reduced graphene oxide
RBM	radial breathing mode
SAED	selected-area electron diffraction
SAN	styrene-acrylonitrile
SBR	styrene-butadiene rubber
SDBS	sodium dodecylbenzenesulfonate
SDS	sodium dodecyl surfate
SE	shielding efficiency
SEM	scanning electron microscopy
SENB	single-edge-notched bending
SGF	short carbon fiber
SIP	surface-initiated polymerization
sPS	syndiotactic polystyrene
SR	silicone rubber
SWNT	single-walled carbon nanotube
TEA	triethylaminc
TEGO	thermally expanded graphene oxide
TEM	transmission electron microscopy
TETA	triethylenetetramine
T_g	glass-transition temperature
TGA	thermogravimetric analysis
THF	tetrahydofuran
TLP	tissue culture plate
TPU	thermoplastic polyurethane
TRG	thermally reduced graphene
VGCNF	vapor-grown carbon nanofiber
VLS	vapor-liquid-solid
WST-1	2-(4-iodophenyl)-3-(4-nitrophenyl)-5-(2,4-disulfophenyl)-2H-tetrazolium
XRD	X-ray diffraction

1
Introduction

1.1
Graphene-Based Nanomaterials

Carbon exists in many forms including buckyballs, diamond, nanotubes, and graphite. It is naturally abundant as coal and natural graphite. Two-dimensional (2D) graphene, a new class of carbon nanostructure, has attracted tremendous attention in recent years since the successful isolation of graphene by micromechanical cleavage of highly oriented pyrolytic graphite (HOPG) [1, 2]. Graphene is a single atomic layer of sp^2 hybridized carbon atoms covalently bonded in a honeycomb lattice. It is a building block for carbon materials of different dimensionalities, including 0D buckyballs, 1D nanotubes, and 3D graphite (Figure 1.1). It shows great potential for technological applications in several areas such as electronics, optoelectronics, nanocomposites, sensors, batteries, and so on [3–7]. Graphene sheets stack together to form graphite with an interlayer spacing of 0.34 nm, showing strong in-plane bonding but weak van der Waals interaction between layers. By virtue of this layered structure, large efforts have been tempted to exfoliate graphite into individual atomic layers. It is difficult to obtain a fully separated sheet layer of graphene because freestanding atomic layer is widely considered to be thermodynamically unstable. A lack of an effective approach to exfoliate graphite into individual, pure graphene sheet in large quantities remains a major obstacle to exploiting its full potential applications.

In 2004, Geim and coworkers of the Manchester University (United Kingdom) prepared single layer of graphene using the cohesive tape method through repeated peeling of graphite and deposited onto a Si/SiO_2 substrate [1, 2]. This is often referred to as a *scotch tape* or *drawing method*. Optical microscopy was initially used to distinguish individual graphene layers followed by their identification in an atomic force microscope (AFM). Geim and Novoselov received the Nobel Prize in Physics for 2010 for their pioneering work in the fabrication and physical characterization of graphene. Such novel preparation of graphene has opened up a new era in nanotechnology and materials science and prompted much excitement in these fields. This technique can only produce low-yield, high-purity graphene for research purposes, and insufficient for practical applications. Moreover, it is hard to control the number of layers for peeled off pieces.

Polymer Composites with Carbonaceous Nanofillers: Properties and Applications, First Edition. Sie Chin Tjong.
© 2012 Wiley-VCH Verlag GmbH & Co. KGaA. Published 2012 by Wiley-VCH Verlag GmbH & Co. KGaA.

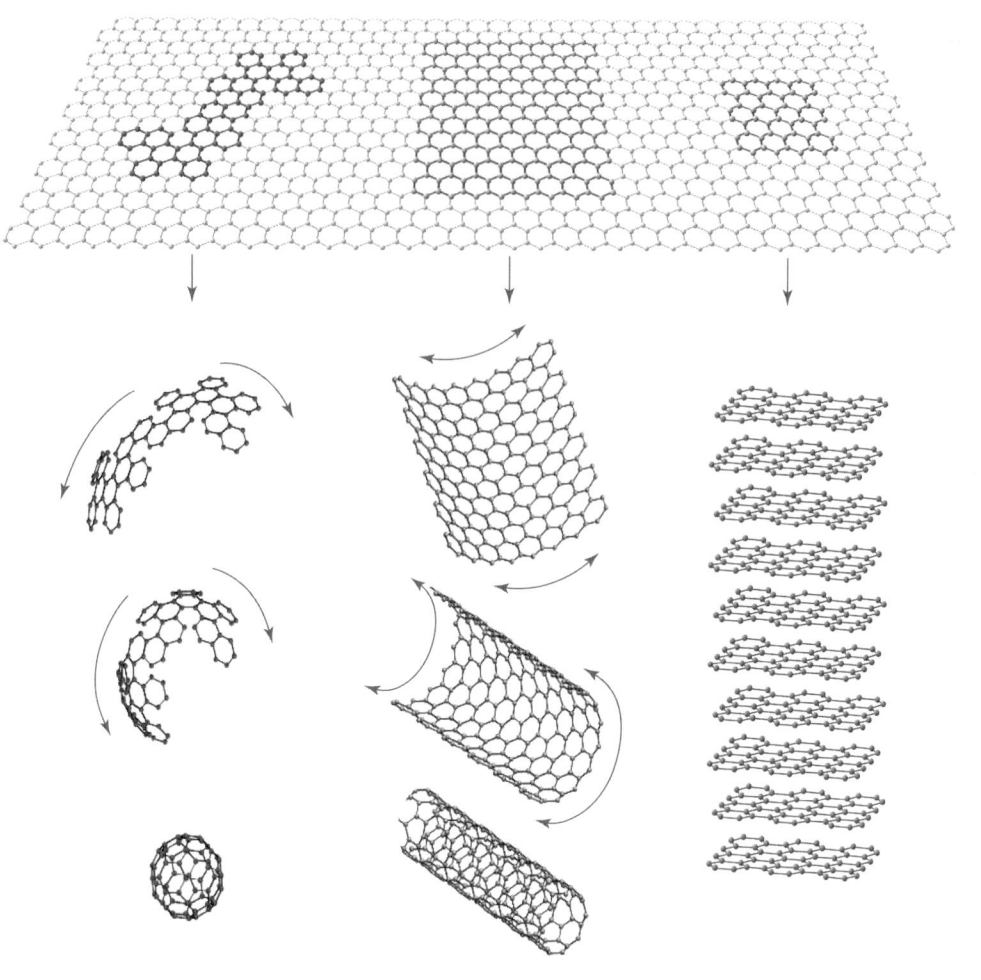

Figure 1.1 Graphene is a 2D building material for carbon materials of different dimensionalities. It can be wrapped up into 0D buckyballs, rolled into 1D nanotubes, or stacked into 3D graphite. (Source: Reproduced with permission from Ref. [3], Nature Publishing Group (2007).)

As an alternative, graphene can be grown directly on solid substrates using two different approaches. The first involves graphitization of single-crystal silicon carbide substrate through thermal desorption of silicon in ultrahigh vacuum at high temperatures (circa above 1300 °C). Consequently, excess carbon is left behind on the surface. The carbon-enriched surface then undergoes reorganization and graphitization to form graphene under proper control sublimation conditions. This process yields epitaxial graphene with dimensions dependant on the size of SiC substrate [8, 9]. The shortcomings of this process are the use of high processing temperature, the formation of atomic scale defects in the graphene lattice and the difficulty of achieving large graphite domains with uniform thickness. The second approach involves epitaxial growth of graphene on metal carbide (e.g., TaC, TiC) or metallic substrates (e.g., Ni, Cu) via chemical vapor deposition (CVD) of hydrocarbons at high temperatures. This is commonly followed by chemical etching and transfer printing to arbitrary substrates [10–14]. For example, Kim et al. [11] prepared patterned graphene film on thin nickel layer using a gas mixture of CH_4, H_2, and Ar, followed by transferring the printing film onto target substrates. The growth of graphene on nickel with higher carbon solubility (>0.1 at%) occurs by the diffusion of the carbon species into the metal surface before segregating and precipitating to the surface on fast cooling. Ni can dissolve more carbon atoms and thus it is difficult to obtain uniform graphene films due to precipitation of extra C during fast cooling. In contrast, the graphene growth on low carbon solubility Cu substrates occurs by means of surface adsorption process [13]. CVD graphene generally exhibits lower electron mobility than mechanically exfoliated graphene because of its higher concentration of point defects, smaller grain sizes, and residual impurities from the transfer or growth processes [14]. The transfer-printing process is also difficult to scale up for industrial applications. Accordingly, wet chemical processing through oxidation of graphite into graphene oxide (GO) followed by reduction appears to be a cost-effective method for mass-producing graphenelike materials.

1.1.1
Graphite Intercalation Compound

Apparently, high-yield production processes for graphene sheets are necessary for practical applications as conductive films and nanofillers for composite materials. Hence, chemical conversion from graphite offers significant advantages over physical approaches and the CVD process for preparing graphene for large-scale applications. This approach converts natural graphite into graphite intercalation compound (GIC) by reacting with electron-donor agents such as alkali metals and electron-acceptor agents such as halogens and acids [15]. Because of its layered structure, acid molecules and alkali metal can penetrate within the gallery spaces of graphite. The layers of graphite interact with the guest molecules through charge transfer process. For example, potassium can be inserted into graphite galleries to yield both first stage and higher stages of intercalation. Stage implies the number of graphite host layers divided by the number of guest layers that occur periodically in the galleries. In the case where every carbon layer in graphite is intercalated, a stage I compound forms, while intercalating on average every other layer yields a stage II compound [16a]. The first-stage intercalation compound, KC_8, has a larger d-spacing (0.541 nm) compared to that of graphite. The second-stage compound, KC_{24}, and the third-stage material, KC_{36}, have a spacing of 0.872 and 1.2 nm, respectively (Figure 1.2). KC_8 generally forms by heating graphite with potassium under vacuum at 200 °C [16b]. The KC_8 compound then reacts with ethanol to yield potassium ethoxide and hydrogen gas, which aid in separating the graphitic sheets

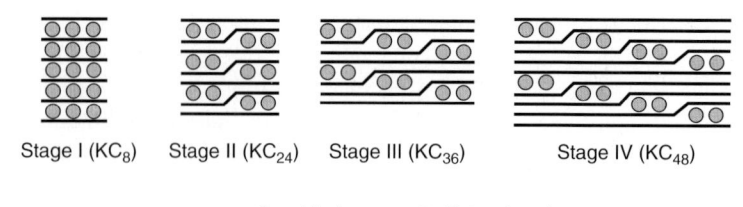

Stage I (KC$_8$) Stage II (KC$_{24}$) Stage III (KC$_{36}$) Stage IV (KC$_{48}$)

— Graphite layer   ○ Potassium ion

Figure 1.2 Schematic diagram of graphite host layers intercalated with different numbers of potassium guest layers. (Source: Reproduced with permission from [16a], Elsevier (2007).)

to form exfoliated graphite. The reaction takes place as follows:

$$KC_8 + CH_3CH_2OH \rightarrow 8C + KOCH_2CH_3 + \tfrac{1}{2}H_2 \tag{1.1}$$

Apart from potassium, the intercalate species normally employed for forming GICs include sulfuric acid, perchloric acid, and selenic acid. In the former case, sulfuric species intercalates into the gallery spaces of graphite in the presence of oxidizing nitric acid. The chemical reaction that takes place between graphite and concentrated sulfuric acid is given by Chen et al. [17a,b]

$$n(\text{graphite}) + nH_2SO_4 + n/2[O] \rightarrow n[\text{graphite} \cdot HSO_4] + n/2\,H_2O \tag{1.2}$$

where O is the oxidant and graphite · HSO$_4$ is the GIC.

Expanded graphite (EG) is an industrial term for exfoliated graphite obtained from sulfuric acid-based GIC precursor [18]. Rapid heating or microwave irradiation causes a large expansion of the graphite flakes along their c-axis to produce EGs of ∼50–400 nm thickness. In general, microwave heating is more effective to exfoliate GIC than conventional thermal treatment because of its high-energy density and fast heating process. Microwave heating vaporizes the acids within the layers of graphite, producing a significant and rapid expansion of the graphite gallery [19]. Figure 1.3a,b shows the low- and high-magnification scanning electron micrographs showing typical porous, vermicular, or wormlike morphology of EGs. The EGs can be further exfoliated to graphite nanoplatelets (GNPs) of 1–15 μm in diameter and a thickness of <10 nm under sonication in the solvents [17b, 20]. The morphology of individual GNP particle is shown in Figure 1.4a,b. The term *nanoplatelet* describes the formation of several layers of graphene rather than a single graphene layer.

1.1.2
Graphene Oxide

Graphite oxide can be obtained by reacting graphite with strong oxidizers such as sulfuric acid, nitric acid, potassium chlorate, and potassium permanganate. The typical Staudenmaier [21] oxidation method involves a mixture of sulfuric acid, nitric acid and potassium chlorate. The Hummers process involves chemical oxidation of graphite with KMnO$_4$ and NaNO$_3$ in concentrated H$_2$SO$_4$ [22]. Till

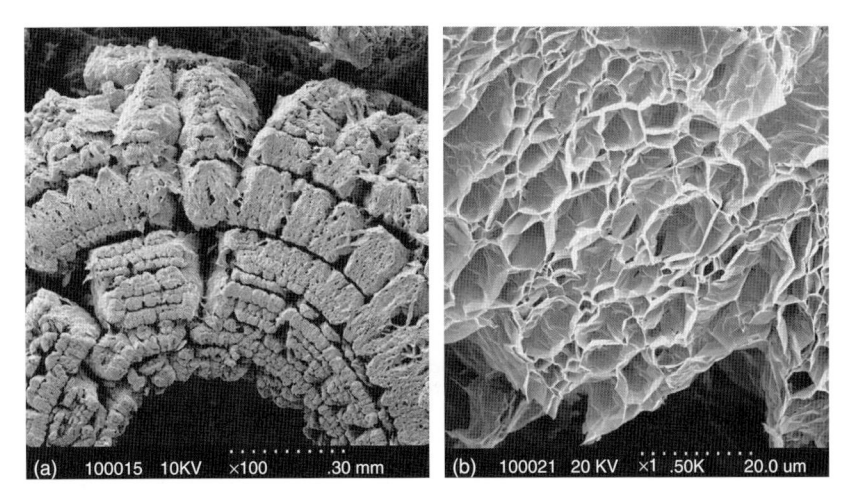

Figure 1.3 (a) SEM image of EGs showing wormlike morphology. (b) High-magnification SEM image showing pores intersperse with EG platelets. (Source: Reproduced with permission from Ref. [18], Elsevier (2005).)

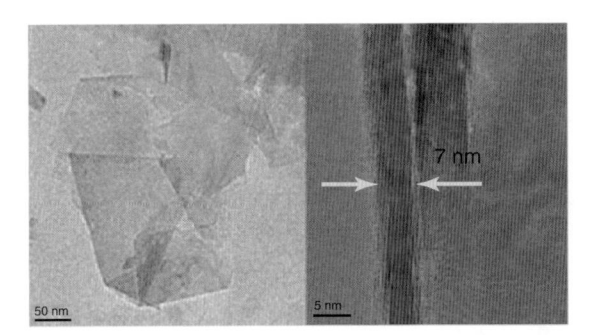

Figure 1.4 TEM images of (a) surface and (b) cross-sectional view of an individual graphite nanoplatelet particle. (Source: Reproduced with permission from Ref. [20], Elsevier (2011).)

present, this is the most commonly used process. Graphite oxide is decorated with hydrophilic oxygenated graphene sheets bearing oxygen functional groups on their basal planes and edges [23, 24]. In other words, functional groups such as epoxide, hydroxyl, carbonyl, and carboxyl groups are formed in the basal planes (Figure 1.5). Thus graphite oxide exhibits an increased interlayer spacing from original 3.4 Å of graphite to 6.0–10 Å nm depending on the water content [25]. Such functional groups make graphite oxide hydrophilic and weaken the van der Waals forces between layers. Thus graphite oxides can be dispersed in aqueous media readily to form colloidal suspensions [26]. This facilitates exfoliation of graphite oxide into GO sheets via sonication [27]. Figure 1.6 is an AFM image of GO exfoliated in water via sonication showing the presence of sheets with uniform thickness of ∼1 nm. This thickness is somewhat larger than the theoretical value of 0.34 nm found in

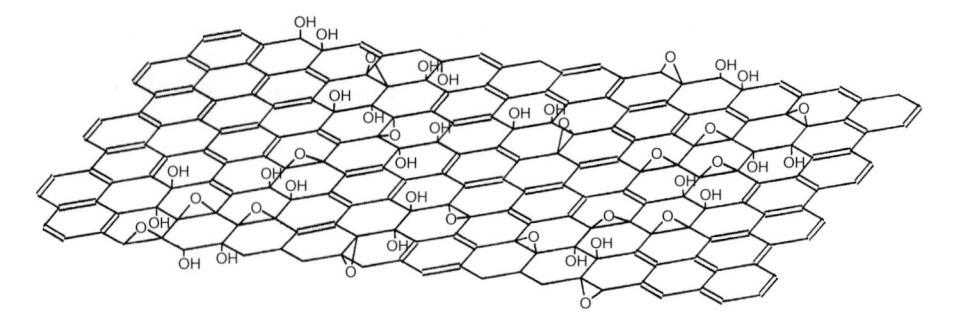

Figure 1.5 Chemical structure of graphite oxide. Carboxylic groups at the edges are not shown. (Source: Reproduced with permission from [24a], Elsevier (1998).)

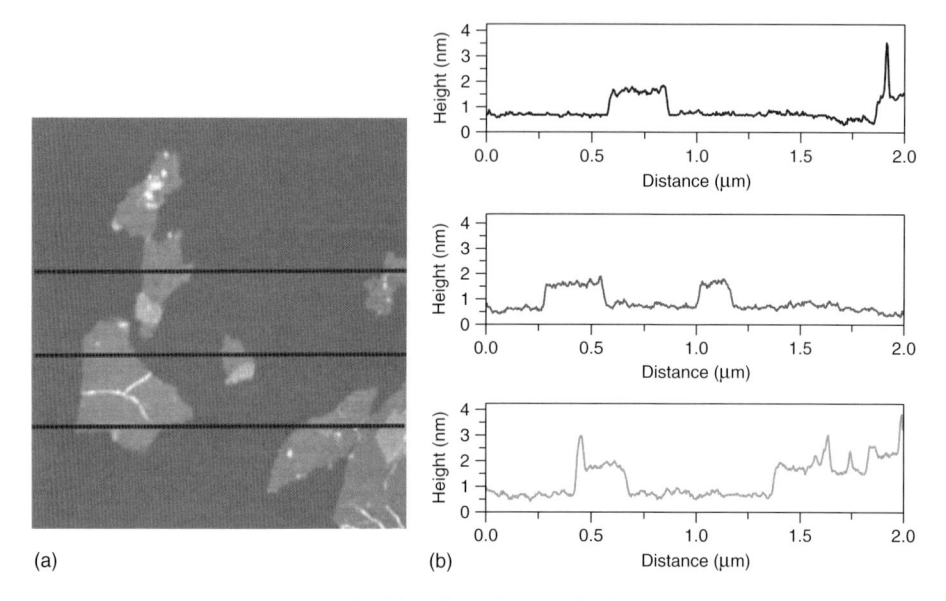

Figure 1.6 (a) AFM image of exfoliated graphene oxide sheets. The sheets are ∼1 nm thick. The horizontal lines in the image indicate the sections (in order from top to bottom) corresponding to the height profiles shown (b). (Source: Reproduced with permission from Ref. [27], Elsevier (2007).)

graphite. This is attributed to the presence of covalently bound oxygen in the GO. It appears that large-scale production of graphene sheets can be achieved through chemical oxidation and exfoliation of graphite flakes in the liquid phase, and the subsequent deoxygenation reduction owing to its simplicity, reliability, and low material cost.

GO is electrically insulating because the functional groups distort intrinsic network of the sp^2 carbon–carbon bonds in the graphene sheets. To recover electrical conductivity, chemical reducing agents such as hydrazine and its derivatives have been used to eliminate oxygen functionalities. For example, Ruoff and coworkers added hydrazine hydrate directly to aqueous dispersions of GO to remove epoxide complexes, producing reduced graphene oxide (rGO), and often referred to as chemically modified graphene (CMG) [25, 28]. A possible reaction pathway for epoxide reduction is given by Stankovich et al. [27]

$$\text{(epoxide)} + H_2N-NH_2 \longrightarrow \text{(intermediate with HO, NH, H}_2\text{N)} \xrightarrow{-H_2O} \text{(intermediate with N-NH}_2\text{)} \xrightarrow{-N_2H_2} \text{(reduced product)} \quad (1.3)$$

It is noted that hydrazine (N_2H_4) is highly toxic and the treatment causes the formation of unsaturated and conjugated carbon atoms, which in turn degrades electrical conductivity. Residual carbonyl and carboxyl groups still can be found in the C1s X-ray photoelectron spectroscopy (XPS) spectrum because of incomplete chemical reduction by hydrazine (Figure 1.7). Further, C–N groups are also incorporated during chemical reduction. The residual oxygen forms sp^3 bonds with carbon atoms in the basal plane such that the carbon sp^2 bonding fraction in fully reduced GO is ~0.8 [29]. Very recently, Shin et al. [30] reported that sodium borohydride ($NaBH_4$) is more effective to remove oxygen moieties in GO than hydrazine. Nevertheless, rGO shows promise for technological applications since it can be processed in liquid phase in large quantities, thus facilitating the fabrication of thin films and composites using low-cost solution processing techniques [28, 31]. Figure 1.8 outlines the process scheme for fabricating rGO-based films for polymer composite and graphene-related electronics applications.

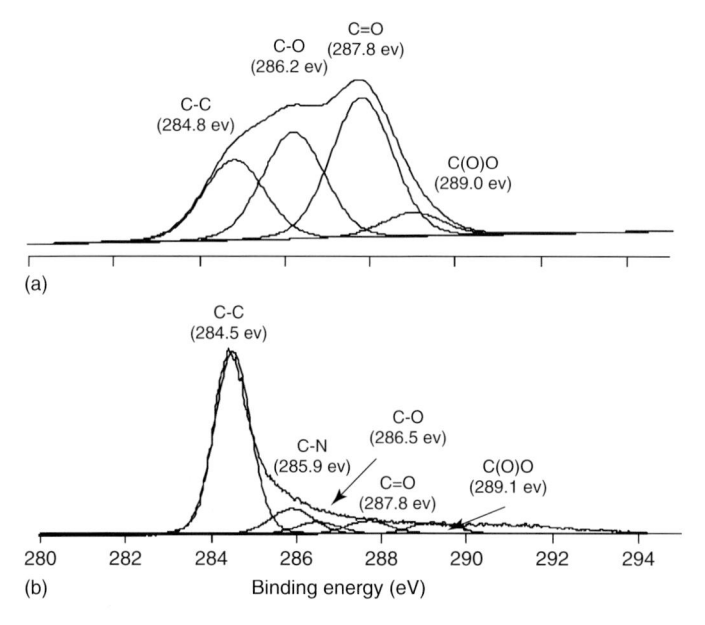

Figure 1.7 C1s spectra of (a) GO and (b) reduced GO. (Source: Reproduced with permission from Ref. [27], Elsevier (2007).)

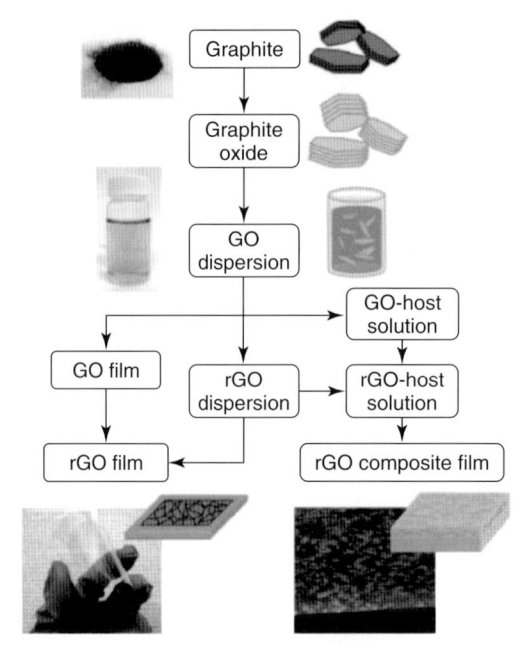

Figure 1.8 Process scheme for fabricating rGO-based thin films. (Source: Reproduced with permission from Ref. [28], Wiley-VCH (2010).)

In addition to chemical reduction, large quantities of graphene can be obtained by reducing GO thermally. This involves rapidly heating GO in an inert atmosphere to form thermally reduced graphene oxide (TRG) or thermally expanded graphene oxide (TEGO). Aksay and coworkers [32a,b] reduced GO by rapid heating ($>2000\,°C\,min^{-1}$) to $1050\,°C$, resulting in the evolution of carbon dioxide due to the decomposition of hydroxyl and epoxide groups. The evolved gas pressure

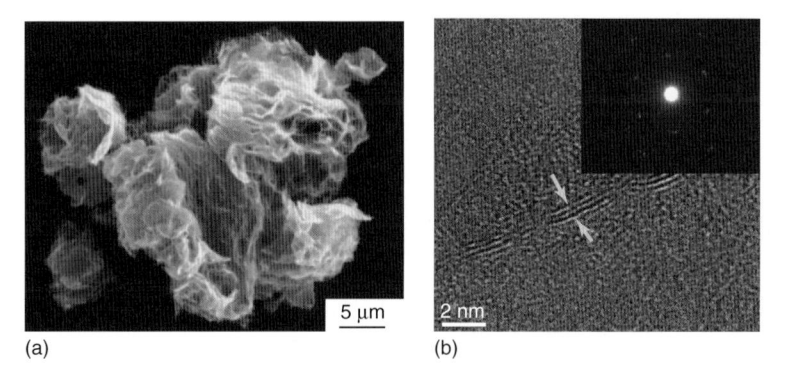

Figure 1.9 (a) SEM image of dry FGS powder. (Source: Reproduced with permission from Ref. [32b], the American Chemical Society (2007).) (b) HRTEM image of TEGO. The inset is selected-electron diffraction pattern of TEGO. (Source: Reproduced with permission from Ref. [33], the American Chemical Society (2009).)

then increases, forcing the sheets apart and producing exfoliation of graphene sheets. TEGO is also known as functionalized graphene sheet (FGS) having a wrinkled morphology (Figure 1.9a). Some functional groups are still retained despite high-temperature annealing. High-resolution transmission electron microscopy (HRTEM) image of TEGO reveals the presence of approximately three to four individual graphene layers within the platelet [33] (Figure 1.9b). Erickson et al. [34] investigated the local chemical structures rGO and TEGO films using a transmission electron microscopy (TEM) corrected with monochromatic aberration. GO was produced using a modified Hummers method and drop cast into TEM carbon grids. GO-containing grids were reduced in a hydrazine atmosphere and then slowly heated to 550 °C under flowing nitrogen to form TEGO. The TEM image of GO obviously shows the presence of the oxidized areas (A and B) and graphene region (C) (Figure 1.10). The TEM image of TEGO reveals a high amount of

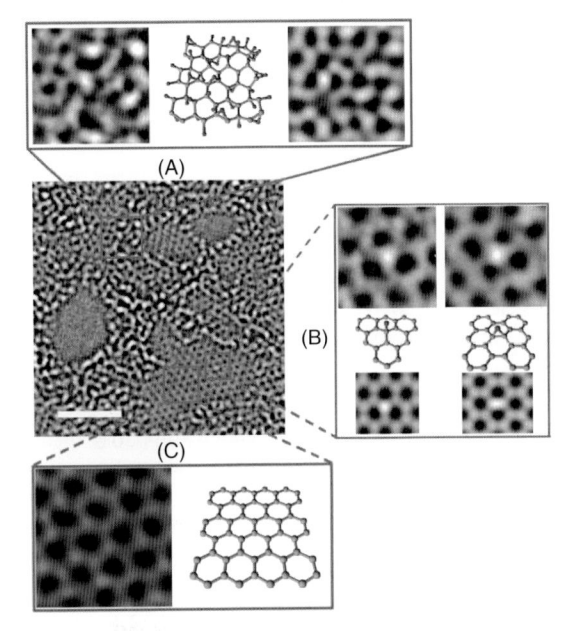

Figure 1.10 Aberration-corrected TEM image of a single sheet of suspended GO. The scale bar is 2 nm. Expansion (A) shows, from left to right, a 1 nm² enlarged oxidized region of the material, then a proposed possible atomic structure of this region with carbon atoms in gray and oxygen atoms in dark gray, and finally the average of a simulated TEM image of the proposed structure and a simulated TEM image of another structure where the position of oxidative functionalities has been changed. Expansion (B) focuses on the white spot on the graphitic region. This spot moved along the graphitic region but stayed stationary for three frames (6 s) at a hydroxyl position (left portion of expansion (B)) and for seven frames (14 s) at a (1,2) epoxy position (right portion of expansion (B)). The ball-and-stick figures below the microscopy images represent the proposed atomic structure for such functionalities. Expansion (C) shows a 1 nm² graphitic portion from the exit plane wave reconstruction of a focal series of GO and the atomic structure of this region. (Source: Reproduced with permission from Ref. [34], Wiley-VCH (2010).)

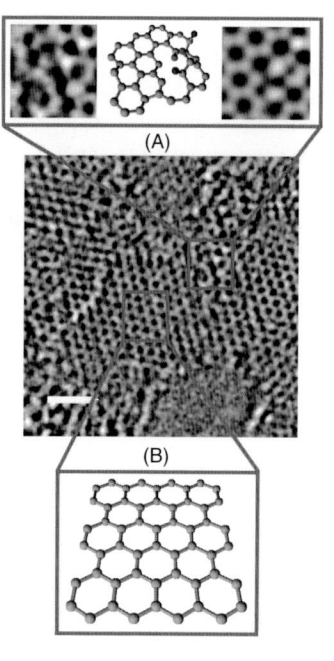

Figure 1.11 Aberration-corrected TEM image of a monolayer of TEGO. The scale bar is 1 nm. Expansion (A) shows, from left to right, an enlarged region of the micrograph, then a proposed possible structure for the region and finally a simulated TEM image for this proposed structure. Expansion (B) shows the structure of a graphitic region. (Source: Reproduced with permission from Ref. [34], Wiley-VCH (2010).)

restored graphene area (B) with little oxidized area (A) associated with oxygenated functional groups (Figure 1.11).

1.2
Carbon Nanotubes

Since the documented discovery of carbon nanotubes (CNTs) by Iijima in 1991, the properties and potential applications of CNTs have attracted considerable interests among scientific and technological communities. This is because one-dimensional CNTs exhibit large aspect ratio, exceptionally high tensile strength and modulus, as well as excellent electrical and thermal conductivity. Single-walled carbon nanotube (SWNT) consists of a single graphene layer rolled up into a seamless cylinder. Thus CNT is also composed of a network with the sp^2 carbon–carbon bonds. The nanotube structure is typically characterized by a chiral vector ($\vec{C}_h$) defined by the relation: $\vec{C}_h = n\vec{a}_1 + m\vec{a}2$ where $\vec{a}_1$ and $\vec{a}_2$ are the graphene lattice vectors and n and m are integers. The chiral vector determines the direction along which the graphene sheets are rolled up to form nanotubes (Figure 1.12). The electronic properties of SWNTs are highly sensitive to the variations in diameter and the indices of

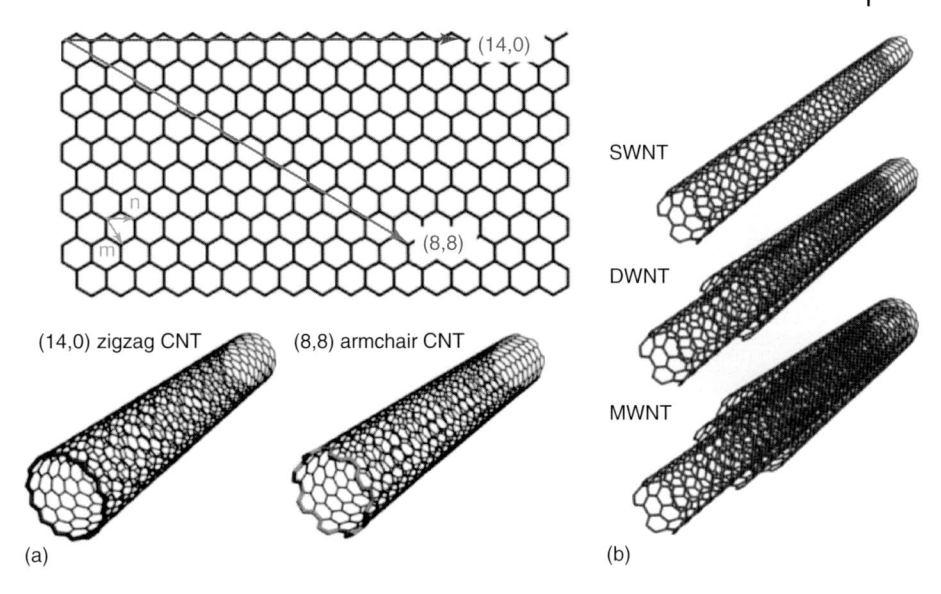

Figure 1.12 Structural variety of CNTs. (a) Orientation of the carbon network in armchair (*n,n*) and zigzag (*n*,0) CNTs. (b) Single-, double-, and multiwalled CNTs. (Source: Reproduced with permission from Ref. [36], the American Chemical Society (2011).)

their chiral vector, n and m [35, 36]. Armchair nanotubes ($n = m$) generally show metallic conducting behavior, while zigzag ($m = 0$) or chiral ($n \neq m$) CNTs are semiconducting. Double-walled and multiwalled carbon nanotubes (MWNTs) are composed of two and several graphene layers wrapped onto concentric cylinders with an interlayer spacing of 0.34 nm. The diameters of CNTs range from a few for SWNTs to several nanometers for MWNTs.

1.2.1
Synthesis of Carbon Nanotubes

1.2.1.1 Physical Vapor Deposition

CNTs have been synthesized by a variety of physical vapor deposition (PVD) and CVD processes. Both processes have their advantages and disadvantages for synthesizing nanotubes. PVD can be classified into direct current (DC) arc discharge and the laser ablation. Those involve condensation of hot carbon vapor generated by evaporating solid graphite. In the former method, an arc is formed between two high-purity graphite electrodes under a protective atmosphere of inert gases. The carbon vapor then condenses on the cold cathode forming a cigarlike deposit with a hard outer shell and a softer inner core [37]. CNTs are then deposited in the weblike soot attached on the chamber walls or electrodes [38]. This technique generally favors the formation of CNTs with a higher degree of crystallinity and structural integrity because of the high temperature of arc plasma. However, by-products such as amorphous carbon and other carbonaceous species are also generated. The quality and yield of nanotubes depend on the processing

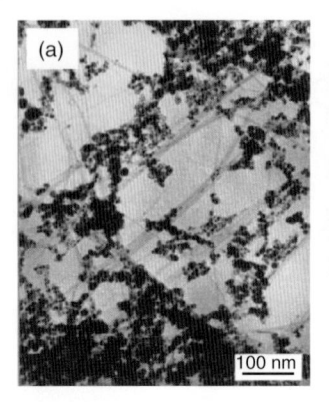

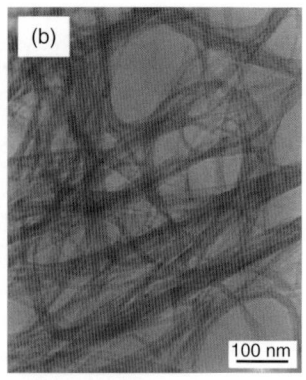

Figure 1.13 TEM images of (a) arc-grown SWNTs in H_2 and (b) purified SWNTs. (Source: Reproduced with permission from Ref. [41], Elsevier (2008).)

conditions such as efficient cooling of cathode, the gap between electrodes, reaction chamber pressure, uniformity of the plasma arc, plasma temperature, and so on. [39]. Comparedto laser ablation, electric arc discharge technique is cheaper and easier to implement but has lower output yield.

SWNTs can also be synthesized by the arc discharge evaporation of a carbon electrode with the aid of transition metal catalyst in hydrogen-containing environments such as H_2–inert gas (Ne, Ar, Kr, and Xe) or H_2–N_2 [40]. The as-prepared SWNT soot generally contains a large amount of impurities including transition metal catalysts, amorphous carbon, and carbonaceous particle impurities, rendering purification of the arc-SWNTs a big challenge [41] (Figure 1.13). Ando and coworkers [42, 43] demonstrated that the arc discharge of pure graphite in pure hydrogen results in the formation of MWNTs of high crystallinity in the cathode deposit.

Laser ablation refers to removal of substantial amount of material from the target by an intense laser pulse. In the process, a graphite target placed inside a tube furnace is irradiated with a focused laser beam. A stream of inert gas is admitted into the specimen chamber for carrying vaporized species downstream to a cold finger [44] (Figure 1.14). The Nd:YAG and CO_2 laser sources operated in a continuous wave mode are typically used for generating carbon vapor species. The quality and yield of CNTs can be manipulated by several experimental parameters including the composition of the target, gas atmosphere, pressure and its flow rate, as well as the laser energy, peak power, and repetition rate [38, 45–48]. SWNTs synthesized from this process often assemble into bundles because of the van der Waals attractions between them [49]. Figure 1.15a,b shows the respective TEM images of SWNTs produced by vaporizing a graphite target containing Pt, Rh, and Re catalysts at 1450 °C in nitrogen or helium gas atmospheres with a laser source [48]. SWNT bundles and amorphous carbon can be readily seen in these micrographs. However, the amount of amorphous carbon is relatively higher in the SWNTs synthesized in helium.

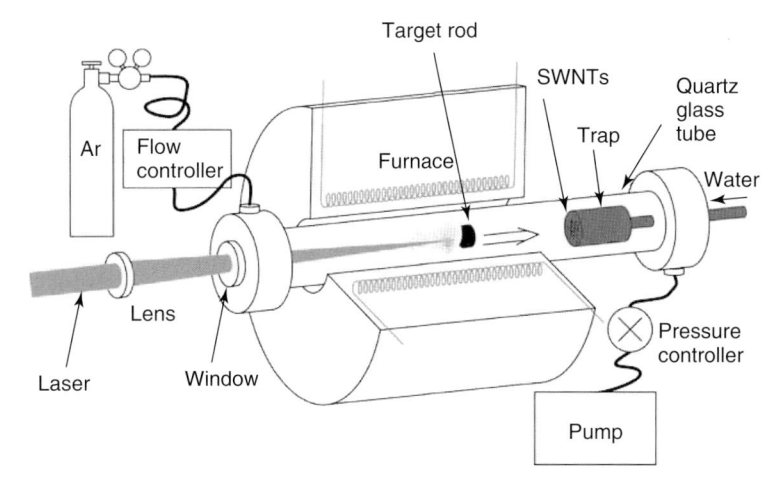

Figure 1.14 Laser ablation facility for synthesizing CNTs. (Source: Reproduced with permission from Ref. [44], Elsevier (2004).)

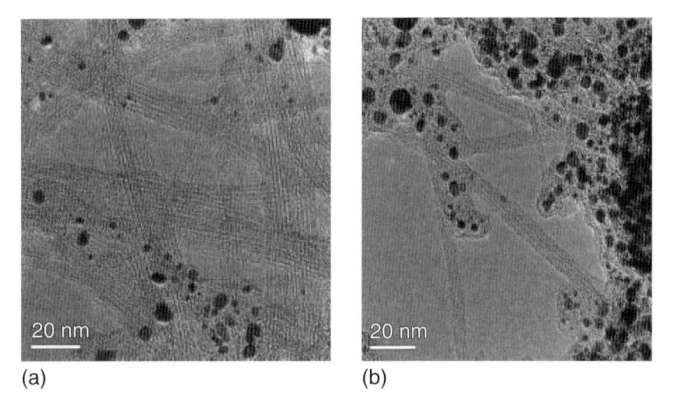

Figure 1.15 TEM images of SWNTs synthesized from a graphite target containing Pt, Rh, and Re catalysts by a Nd:YAG laser irradiation at 1450 °C in (a) nitrogen and (b) helium. (Source: Reproduced with permission from Ref. [48], the American Chemical Society (2007).)

1.2.1.2 Chemical Vapor Deposition

CVD method is a versatile and effective technique for massive synthesis of CNTs at low production cost using a wide variety of hydrocarbon gases such as methane, ethylene, propylene, acetylene, and so on. It allows the manufacture of CNTs into various forms including thin films, aligned, or entangled tubes as well as free-standing nanotubes. The process involves catalytic dissociation of hydrocarbon gases over metal nanoparticle catalysts in a high-temperature reactor. Comparing with the arc-grown CNTs, CVD-synthesized nanotubes generally have higher density of structural defects. The types of CNTs produced depend mainly on the synthesis temperatures. MWNTs are generally formed at ∼600–850 °C, while SWNTs are produced at higher temperatures of 900–1200 °C [44]. CVD can be classified into two categories, that is, thermal- and plasma-enhanced processes depending on the heating sources employed. Thermal CVD decomposes hydrocarbon gases using thermal energy. The hydrocarbon precursor is usually diluted with H_2, normally acting as carrier gases [50, 51].

The catalytic decomposition of hydrocarbons can be further enhanced by using plasmas generated from DC, hot filament aided with DC, microwave, radio frequency (rf), electron cyclotron resonance (ECR), and inductively coupled plasma sources. These sources ionize the gas precursors, producing plasmas, electrons, ions, and excited radical species. The setup of DC plasma reactor is relatively simple and consists of a couple of electrodes with one grounded and the other connecting to a power supply. The negative DC bias applied to the cathode causes the decomposition of hydrocarbon gases. To enhance deposition efficiency, a metallic wire (e.g., tungsten) is added to the system as resistively heated hot filament [52]. This is known as the *DC-plasma*-enhanced hot filament CVD [53], typically used for depositing diamond and diamondlike films. However, the DC plasma is less effective in producing reactive species since the majority of energy consumed in DC plasmas is lost in accelerating ions in the sheath, and this leads to a substantial substrate heating [52]. The plasma instability of DC reactors has led to the adoption of high-frequency plasma reactors in the semiconductor industry. In high-frequency plasma reactors, the gas molecules are activated by electron impact. For example, a microwave source operated at a frequency at 2.45 GHz at 1.5–2 kW oscillates electrons effectively, leading to an increase in their density (Figure 1.16a). These electrons then collide with the feed gas to form radicals and ions. In general, the ECR source is capable of producing higher fluxes of lowenergy ions than other sources (Figure 1.16b). It is well known that electrons travel in a circular path with a cyclotron frequency under the influence of a magnetic field. The cyclotron frequency is proportional to the strength of magnetic field. Electrons that move in a circular path in a magnetic field can absorb energy from an AC electric field provided that the frequency of the field matches the cyclotron frequency. In the ECR reactor, microwaves are introduced into a volume of the reactor at a frequency matches closely with ECR. The absorbed energy then increases the velocity of electrons, leading to enhanced ionization of the feed gas. Thus ECR source has the advantages of obtaining high dissociation levels of the precursor gas and high uniformity of plasma energy distribution.

Plasma-enhanced chemical vapor deposition (PECVD) allows nanotubes to be synthesized at lower temperatures [54–56]. It also offers another advantage by forming vertically aligned nanotubes over a large area with superior uniformity in diameter and length by means of the electric field present in the plasma sheath [57–60]. PECVD process is more complicated than thermal CVD since several plasma parameters such as ion density, ion energy, radical species, radical densities, and applied substrate bias are involved [61]. Since the dissociation of hydrocarbon feedstock creates many reactive radicals, leading to the deposition of amorphous carbon on the substrate. It is necessary to dilute the hydrocarbon with hydrogen, argon, or ammonia. The chemistry of gas precursors is considered of primarily

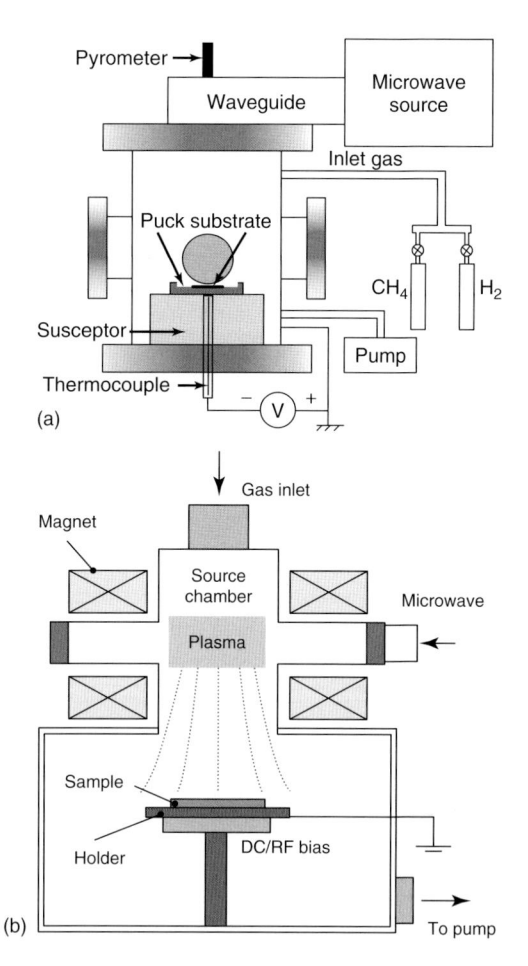

Figure 1.16 Schematic diagrams of (a) microwave PECVD and (b) ECR PECVD reactors. (Source: Reproduced with permission from Refs. [54] and [55], Elsevier (2006) and (2007), respectively.)

importance for growing aligned nanotubes. The hydrogen gas generally assists the growth of aligned nanotubes by etching the substrate. Atomic hydrogen generated within the plasma can remove amorphous carbon deposit on the substrate [62]. Figure 1.17a,b shows the respective low- and high-magnification scanning electron micrographs of aligned CNT film grown on a cobalt-catalyzed Si substrate using microwave PECVD [60]. However, Wong et al. [63] demonstrated that ammonia gas is even more effective to form aligned CNTs from microwave PECVD process. This is because ammonia can inhibit the formation of amorphous carbon during the initial synthesis stage [64].

The growth of CNTs during CVD process involves two main step sequences, that is, dissolution of the gas precursors into carbon atoms on the metal catalyst surface and surface diffusion of carbon atoms to the growth site to form the nanotube.

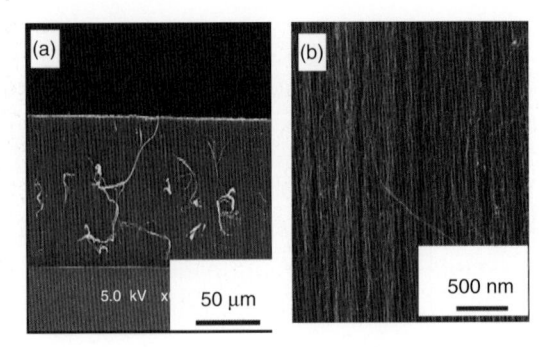

Figure 1.17 (a) Low- and (b) high-magnification scanning electron micrographs of aligned CNT film grown on a Co-catalyzed Si substrate using mixed CH_4-H_2 gases. (Source: Reproduced with permission from Ref. [60], Elsevier (2007).)

At an earlier stage of the process, catalytic dissociation of hydrocarbon molecules occurs near metal nanoparticles. Carbon atoms then absorb and diffuse on the metal surfaces, forming liquid alloy droplets. The droplets act as preferential sites for further adsorption of the carbon atoms, forming metal carbide clusters [65]. When the clusters reach supersaturation, nucleation, and preferential growth of 1D carbon nanostructure takes place [66]. This growth behavior is commonly known as the vapor-liquid-solid (VLS) mechanism [67]. Further, the growth can initiate either below or above the metal catalyst, regarding as the "base" or "tip" growth models (Figure 1.18). The former growth mode occurs assuming the presence of a strong metal catalyst-substrate interaction. Thus the nanotube grows up in 1D manner with the catalyst particle pinned on the substrate surface. In the case of a weak catalyst-substrate interaction, the catalyst particle is lifted up by the growing nanotube and encapsulated at the nanotube tip eventually. There is a speculation

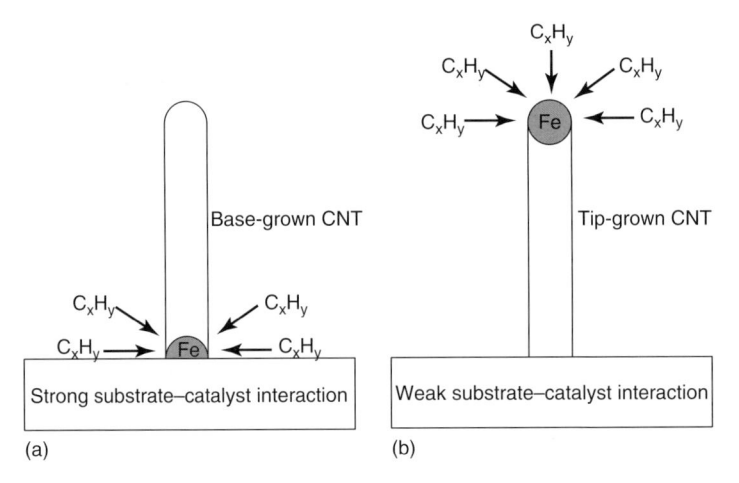

Figure 1.18 (a,b) "Base" and "tip" growth models for CNTs. (Source: Reproduced with permission from Ref. [44], Elsevier (2004).)

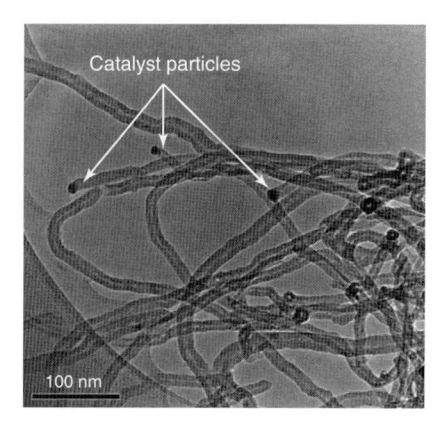

Figure 1.19 TEM image of CVD MWNTs synthesized using Ni-Al catalyst at 550 °C. (Source: Reproduced with permission from Ref. [68], Elsevier (2008).)

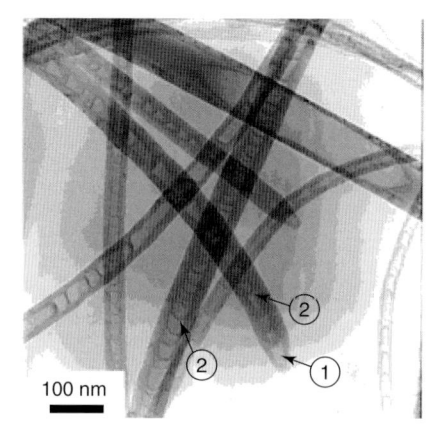

Figure 1.20 TEM image of CNTs grown at 850 °C. CNTs exhibit a bamboolike structure. The arrow 1 indicates the closed tip with no encapsulated Fe particle. The arrow 2 corresponds to the compartment layers whose curvature is directed to the tip. (Source: Reproduced with permission from Ref. [69], the American Chemical Society (2001).)

that MWNTs follow the "top" growth mode, while SWNTs adopt the "base" growth fashion [61]. Figure 1.19 is a TEM image showing the presence of nickel-aluminum composite catalysts at the tips of CVD MWNTs [68].

Park and Lee [69] synthesized CNTs using thermal CVD on a Fe-catalyzed silica substrate using C_2H_2 gas from 550 to 950 °C. They reported that CNTs grown at 950, 850, 750, and 600 °C exhibit a bamboolike morphology. There are no encapsulated Fe particles at the closed tips (Figure 1.20). The CNTs grown at 550 °C possess encapsulated Fe particles at the closed tips. Accordingly, they proposed propose a base growth model for the bambootype CNTs (Figure 1.21a–e). Carbons produced from the decomposition of C_2H_2 adsorb on the catalytic metal particle. They then diffuse on the metal particle surface to form the graphitic sheet cap (Figure 1.21a).

As the cap grows up from the catalytic particle, a closed tip with a hollow tube is produced (Figure 1.18b). The compartment graphite sheets are then formed on the inner surface of the catalyst due to carbon accumulation as a result of bulk diffusion (Figure 1.21c). The growth process continues progressively, forming a bambootype CNT (Figure 1.21d,e). Similarly, Chen *et al.* also reported that the CNTs deposited on Ni-catalyzed Si substrate via hot-filament CVD also follow the base growth mode. Figure 1.22a clearly shows the presence of encapsulated Ni particles at the nanotube end-substrate interface. The nanotubes also exhibit a bamboolike morphology. A high-resolution transmission electron micrograph reveals that the Ni particle exhibits a conical morphology (Figure 1.22b). The bamboolike feature of the nanotubes is also apparent.

Apart from the hydrocarbon feedstock, carbon monoxide is an alternative carbon source gas for synthesizing CNTs. The research group at Rice University developed and commercialized high-pressure carbon oxide disproportionation (HiPCo) process for synthesizing SWNTs using carbon monoxide and iron pentacarbonyl ($Fe(CO)_5$) catalyst [71]. This process enables the production of a relatively high yield of SWNTs. The nanotubes are synthesized in a flowing CO reactor at temperatures of 800–1200 °C under high pressures of 1–10 atm. On heating, the $Fe(CO)_5$ decomposes into Fe atoms that condense into larger nanoparticles. SWNTs then nucleate and grow on these particles in the gas phase via CO disproportionation (decomposition into C and CO_2) reaction

$$CO + CO \rightarrow CO_2 + C \; (SWNT) \tag{1.4}$$

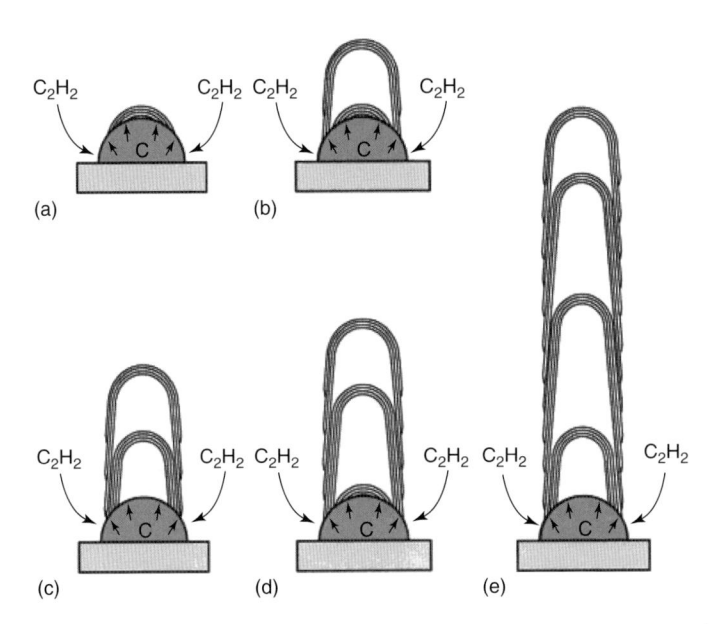

Figure 1.21 Schematic diagrams of a base growth model. (Source: Reproduced with permission from Ref. [69], the American Chemical Society (2001).)

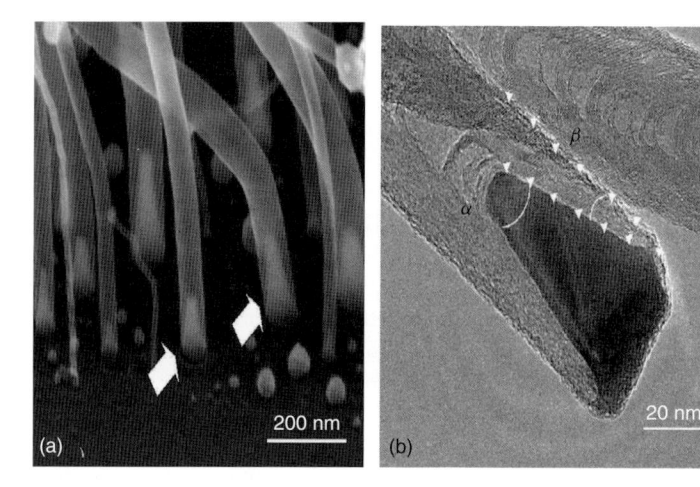

Figure 1.22 (a) SEM and (b) HRTEM images of CNTs grown on the Ni film/silicon substrate in a C_2H_4/NH_3 atmosphere using hot-filament CVD process. Conical-shape nickel nanoparticles are marked with arrows in (a). (Source: Reproduced with permission from Ref. [70], Elsevier (2004).)

1.2.2
Purification of Carbon Nanotubes

The as-synthesized CNTs generally contain a large amount of impurities including metal catalyst particles, amorphous carbon, fullerenes, and multishell graphitic carbon. For example, arc discharge nanotubes possess a considerable content of amorphous carbon, fullerenes, and carbon enclosing metal catalyst particles. The SWNTs synthesized from the CVD process contain metallic catalyst and amorphous carbon. These impurities affect electrical, mechanical, and biological properties of the CNTs markedly. Thus they must be removed before their practical applications in biomedical engineering and industrial sectors. Because of the diversity of the nanotube synthesis techniques, the as-prepared CNTs may have different morphologies, structures, and impurity levels. Thus the purification techniques must be properly tailored and selected to obtain CNTs with desired purity. It is a challenging task to effectively purify CNTs without damaging their structures. The purification routes can be classified into chemical- and physical-based techniques. The former includes gas- and liquid-phase oxidation, while the latter includes filtration, centrifugation, high-temperature annealing, and chromatography. The most widely adopted purification process is the liquid-phase oxidation because of their simplicity and capability for the removal of metal impurities to a certain concentration level. However, such process often results in structural damages to the nanotubes because of their vulnerability to chemical oxidation.

1.2.2.1 Chemical Techniques
Gas-phase oxidation is the simplest technique for purifying CNTs by removing carbonaceous impurities. It is ineffective to remove metal catalyst particles [41].

Accordingly, this method is suitable for purifying arc-grown MWNTs containing no metal catalysts. CNTs can be oxidized in air, pure oxygen, or chlorine atmosphere at a temperature range of 300–600 °C. The oxidants breach the carbon shell and then oxidize metal catalysts into metal oxides. As the oxidation proceeds, the volume of nanotube increases, and metal oxides crack open the carbon shell surroundings accordingly [72]. Carbonaceous impurities are oxidized at a faster rate than the nanotube material via selective oxidation [73]. The main disadvantage of thermal oxidation is that the process can burn off more than 95% of the nanotube material. This leads to an extremely low purification yield [74]. Alternatively, microwave heating has been found to be an effective method to purify arc-grown SWNTs because of its short processing time. The microwave induces rapid local heating of the catalyst particles, causing both the oxidation and rupture of the carbon layer surrounding metal catalyst particles [74].

In general, thermal oxidation is ineffective for the purification of SWNTs having large tube curvatures and metallic impurities. Additional step procedure is necessary for removing metal catalysts. Smalley and coworkers [75] developed the oxidation and deactivation of metal oxides for purifying raw HiPCo SWNTs. The metal oxides formed in the SWNTs by oxygen oxidation were deactivated into metal fluorides through reacting with $C_2H_2F_4$, SF_4, and other fluorine-containing gases [75]. As a result, the iron content was significantly decreased from ∼30 to ∼1 wt% with ∼70% SWNT yield. This purification process is mainly designed for the HiPCo SWNTs with predominant Fe particles. The toxicity of the reagents is another issue that must be considered from using this process.

Liquid-phase oxidation involves the use of oxidizing agents such as concentrated HNO_3, mixed HNO_3/H_2SO_4 (1 : 3 by volume), HCl, $KMnO_4$, $HClO_4$, and H_2O_2 for purifying the nanotubes, followed by the filtering and drying procedures. Strong oxidants such as $KMnO_4$ and $HClO_4$ are mainly used for the purification of MWNTs with higher resistance to oxidation. Nitric acid is found to be fairly effective in removing metal catalysts and amorphous carbon in the arc-grown SWNTs [76]. In general, a mixed HNO_3/H_2SO_4 (1 : 3 by volume) solution is more effective than concentrated nitric acid in removing impurities [77]. Thus this solution is widely used for the liquid-phase oxidation of CNTs today. As nitric acid is a mild reducing agent, a prolonged oxidation time of up to 50 h is needed to eliminate metallic impurities of the arc-grown SWNTs to a level below 0.2% [76] (Table 1.1). This produces significant wall damages, length reduction, and losses of the nanotube materials. In this regard, a two-step (e.g., gas-phase thermal oxidation followed by dipping in acid) process is adopted for further eliminating metallic component of the arc-grown SWNTs. For a combined thermal and acid oxidation treatment, the acid can easily dissolve the metal oxides formed from the gaseous oxidation. Table 1.1 compares metal content of the arc-grown SWNTs containing Ni and Y catalysts before and after purifying by nitric acid, air oxidation/HNO_3 reflux, and microwave heating/HCl reflux treatments. Apparently, an initial oxidation of the SWNTs in air at 400 °C for 30 min can reduce the refluxing time in a nitric acid to 6 h to yield a residual metal content <1%. In contrast, microwave heating in air at 500 °C for 20 min followed by refluxing in HCl for 6 h can further remove residual

Table 1.1 Residual metal contents in Arc-Grown SWNTs containing Ni and Y impurities purified by different approaches.

Sample	Purification treatments	Metal content (wt%)	Comments (based on high-resolution transmission electron microscopy)
SWNTs	As-prepared	~35	SWNTs and metal residue covered by amorphous and multishell carbon
	Reflux HNO_3 acid at 130 °C for 25 h	~4	Significant wall damages and shorten tubes, losses of SWNTs
	Reflux HNO_3 acid at 130 °C for 50 h	<0.2	Significant wall damages and shorten tubes, losses of SWNTs
	Oxidation in air at 400 °C for 30 min, then reflux HNO_3 for 6 h	<1	No amorphous carbon, wall damages, losses of SWNTs
	Microwave heating in air at 500 °C for 10 min, reflux HCl for 6 h	~1.5	Some amorphous carbon on the walls, metal residue covered by multishell carbon
	Microwave heating in air at 500 °C for 20 min, reflux HCl for 6 h	<0.2	Slight wall damages due to the metals attached to the tube's walls

Source: Reproduced with permission from Ref. [76], the American Chemical Society (2002).

metal catalyst content to a level below 0.2%. Hou et al. [41] used a reverse sequence strategy, that is, an initial sonication in ethanol followed by air oxidation to purify hydrogen arc-grown SWNTs. Sonication is recognized as one of the effective ways to eliminate amorphous impurities in CNTs using suitable solvents and acid [78, 79]. Using such a two-step procedure, a purity of about 96% with a 41% SWNT yield can be achieved (Figure 1.23b). However, sonication can also induce structural defects in CNTs including buckling and bending [80].

It is noteworthy that the acid treatment induces structural defects and shortens the lengths of both the MWNTs and SWNTs (Table 1.1). The treatment also disrupts and opens the ends of CNTs, thus introducing oxygenated functional groups (hydroxyl and carboxyl) on the nanotube surfaces. These functional groups degrade electrical conductivity of the nanotubes markedly. However, these are beneficial in improving the dispersion of nanotubes in the polymer matrix.

1.2.2.2 Physical Techniques

Physical-based purification is an alternative route to reduce the damage of CNTs caused by chemical oxidation. It can retain the intrinsic structure of CNTs that is highly desirable for scientific research purposes and technological applications, particularly for the device and sensor applications. Physical techniques are mainly based on the dispersion of nanotubes in a stable colloidal suspension by the size separation through filtration, centrifugation, or chromatography. CNTs generally exhibit poor dispersibility in polar media such as water and organic solvents. SWNTs often agglomerate into bundles and ropes [49]; thus surfactants are widely used to stabilize the nanotube suspensions in water. Insoluble CNTs of large aspect ratios are extracted from the suspensions through filtration. Bonard et al. [82] dispersed pristine MWNTs in water with sodium dodecyl sulfate (SDS), forming a stable colloidal suspension. The suspension was filtered through a funnel large enough to allow the insertion of an ultrasonic probe. In order to increase the separation yield, successive filtrations were carried out until the attainment of desired purity. Smalley and coworkers employed both the filtration and the microfiltration to purify SWNTs produced by pulsed laser ablation [83]. The process separates carbon nanospheres, metal nanoparticles, polyaromatic carbons and fullerenes from the SWNTs. Purity of SWNTs in excess of 90 wt% can be achieved. The disadvantage of this technique is the use of a number of successive filtration steps to achieve desired purity.

Centrifugation is based on the gravity effects to separate carbon nanoparticles, amorphous carbon from the nanotube suspension. Thus it involves the use of centrifugal force for sedimentation of the suspensions. Amorphous carbon can be removed from the nitric-acid-treated SWNTs by low-speed centrifugation (2000g) in an acid solution of pH 2, leaving the SWNTs in the sediment in an acid [84]. Carbon nanoparticles in the SWNTs can be eliminated by high-speed centrifugation (20 000g) of several cycles in a neutral pH. This leads to the sedimentation of carbon nanoparticles, leaving the SWNTs suspended in aqueous media [85]. In another study, centrifugation in the SDS aqueous solutions is found to be effective for the removal of carbonaceous impurities in the SWNTs [86]. In general, filtration and centrifugation can only remove carbonaceous impurities and ineffective for the elimination of metallic impurities.

High-temperature annealing is the only one physical technique that does not involve the use of colloidal solutions. This process can effectively remove residual metal impurities in the CNTs when compared with other purification techniques such as acid treatment, filtration, and centrifugation. As recognized, a trace amount of metallic impurities in the CNTs can induce inflammation and cell apoptosis in mammals. These impurities must be completely removed for biomedical engineering applications. The annealing process involves heat treatment of nanotubes at high temperatures ($>1400\,^\circ$C) under inert atmosphere or high vacuum [81, 87–90]. Andrews et al. [81] annealed MWNTs at temperatures between 1600 and 3000 $^\circ$C. They reported that heat treatment of MWNTs at temperatures above 1800 $^\circ$C is very effective to remove residual metals (Figure 1.23a,b). Furthermore, high-temperature annealing can induce graphitization, resulting in the appearance of high-order diffraction spots such as (004) and (006) reflections in the selected-area electron diffraction (SAED) pattern. High-temperature annealing converts disordered structure of the walls of MWNTs into a more perfect graphitic structure (Figure 1.24a–f). In general, high-temperature heating can induce structural changes in the CNTs [88]. A twofold increase in the diameter of SWNTs is observed by heating at 1500 $^\circ$C under argon and hydrogen atmospheres [89]. Similarly, Yudasaka et al. [90] reported that the diameters of HiPCo SWNTs can be increased via heat treatments at 1000–2000 $^\circ$C.

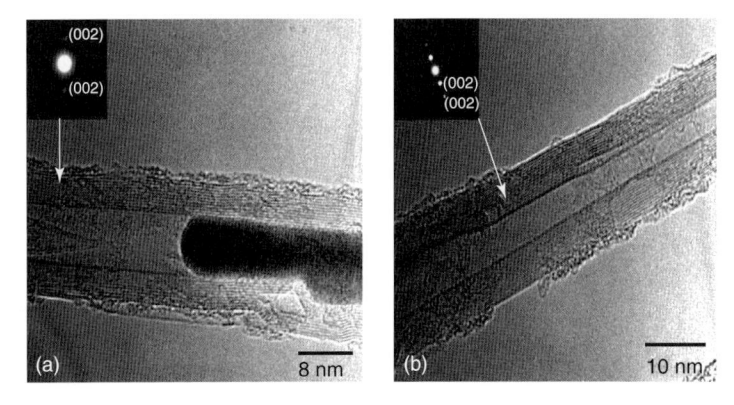

Figure 1.23 TEM images and selected-electron diffraction patterns of (a) CVD MWNT and (b) MWNT annealed at 2250 °C. (Source: Reproduced with permission from Ref. [81], Elsevier (2001).)

1.2.3
Characterization of Purified Carbon Nanotubes

The final quality and purity of purified CNTs can be examined by several analytical techniques including scanning electron microscopy (SEM), TEM, energy dispersive spectroscopy (EDS), Raman spectroscopy, AFM, X-ray diffraction

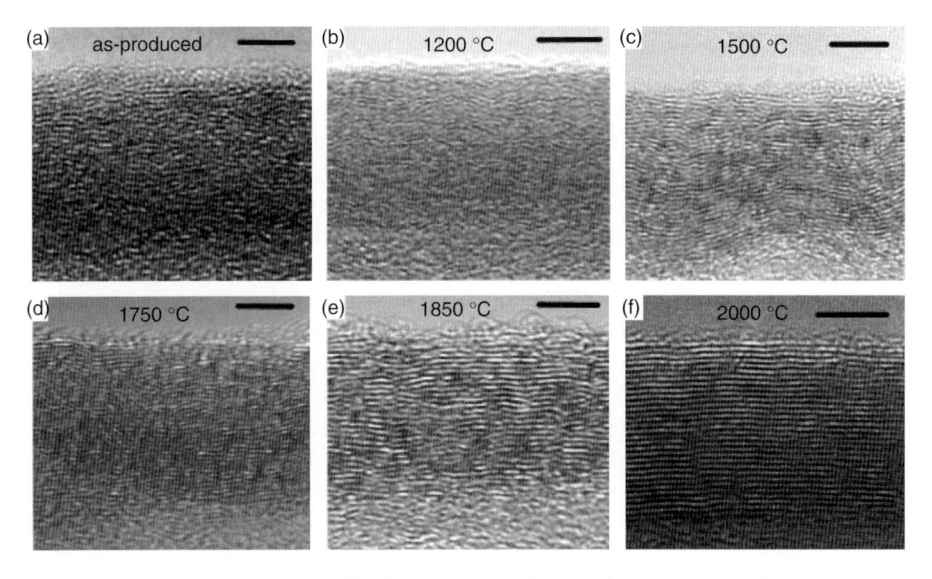

Figure 1.24 TEM images of the walls of (a) noncatalytic CVD-grown MWNTs in a porous alumina template and those annealed at (b) 1200 °C, (c) 1500 °C, (d) 1750 °C, (e) 1850 °C, and (f) 2000 °C. The initial, disordered carbon structure is converted into a more perfect graphitic structure with increasing annealing temperature. Scale bar is 5 nm. (Source: Reproduced with permission from Ref. [87], the American Chemical Society (2006).)

(XRD), ultraviolet-visible-near infrared (UV-vis-NIR), absorption spectroscopy, NIR spectroscopy, and thermogravimetric analysis (TGA). These have been recommended by the International Team in Nanosafety (TITNT) and NASA-Johnson Space Center (United States) as the characterization techniques for both the pristine and purified SWNTs [91, 92]. In general, SAED and TEM offer useful information relating the chemical identity of impurities of both the pristine and purified nanotubes. However, these techniques are unsuitable for quantitative evaluation of the purity of CNTs. Further, the sample preparation procedures for TEM examination are tedious and time consuming. The EDS attached to either SEM or TEM can provide characteristic X-ray peaks of various elements, particularly useful for qualitative assessment of the metallic impurities. The EDS/SEM is routinely used in the research institutions and industrial laboratories world wide for qualitative analysis of inorganic materials. The EDS/SEM can provide quantitative analysis if suitable standards are available. However, it is difficult to analyze and quantify light elements in the EDS spectra because of the absorption of their X-ray radiation by the window material (beryllium) of the spectrometer. This effect can be minimized by using an SEM-windowless EDS facility.

At present, there appears to be a need of reliable techniques for fast and accurate assessment of the nanotube purity. In this regard, Raman spectroscopy is particularly useful for qualitative analysis of purified nanotubes by examining vibrational frequency responses of different carbon species. The position, width and relative intensity of Raman peaks of various carbonaceous species (e.g., amorphous carbon, fullerene, diamond, and SWNT) are related to their sp^3 and sp^2 configurations [93–95]. Raman spectra of SWNTs are well characterized for the radial breathing mode (RBM) at 150–200 cm^{-1} and the tangential G-band at $\sim$1550–1605 cm^{-1} [94]. The tangential-mode G-band involves out-of-phase intralayer displacement in the graphene structure of the nanotubes. It is a measure of the presence of ordered carbon. Figure 1.25 shows the Raman profile of the SWNT. A band located at $\sim$1350 cm^{-1} is attributed to the disorder-induced band (D-band) and related to the presence of nanoparticles and amorphous carbon. Furthermore, a second-order mode at $\sim$2600 cm^{-1} is referred to the D^* mode, commonly known as the G' *mode*. The RBM is associated with the collective in-phase radial displacement of carbon atoms, and only can be found in the SWNTs. It gives direct information for the tube diameter since its frequency is inversely proportional to the tube diameter. The D-peak is absence in graphite and only found in the presence of disorder defects. The D- and G-peak ratio characterizes the disorder degree of the materials studied. Thus the purity level of SWNTs can be qualitatively determined from the D/G intensity ratio and the width of the D-band [95, 96]. The full width at half maximum (FWHM) of the D-band for the carbonaceous impurities is much broader than that of the SWNT. An asymmetry on the right-hand side of G-band is a peak characteristic of MWNTs and normally appears at 1620 cm^{-1} [97].

XRD is a nondestructive and useful material characterization technique for identifying the crystal structure, interlayer spacing, structural strain, and crystal orientation. It is a global characterization technique since all its data are averaged over the whole regions of the specimen. On the contrary, TEM-SAED can provide

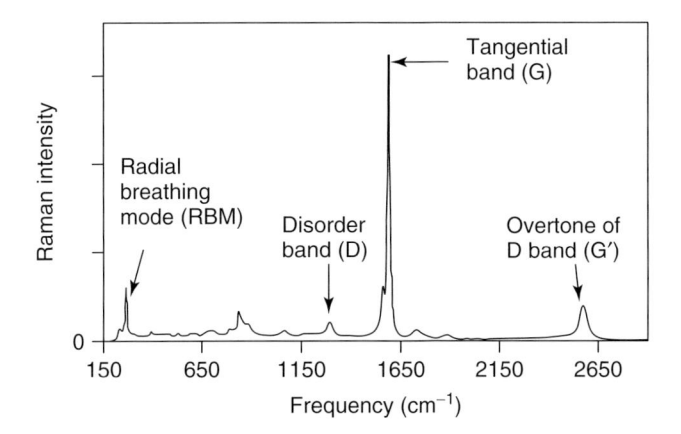

Figure 1.25 Raman profile of SWNT showing the presence of radial breathing mode and tangential G-band.

morphological and structural information at local levels. Thus a combination of these techniques can provide further insights into the structure and morphology of nanotubes after purification. Very recently, Vigolo et al. [98] used Raman, TEM, XRD, and infrared spectroscopy to assess the purity of arc-grown SWNTs containing Ni and Y catalyst nanoparticles (denoted as PS0). Three progressive purification steps were adopted: (i) air oxidation at 365 °C for 90 min (denoted as PS1), (ii) acid treatment in a 6N HCl solution for 24 h under reflux (denoted as PS2), and (iii) high-temperature annealing in vacuum (1100 and 1400 °C) for 1 h (denoted as PS3a and PS3b, respectively). Figure 1.26 shows Raman spectra of the specimens after each step of the purification procedures. The D-band range is given in the inset. The D-band becomes sharper after initial and second steps of purification and undergoes a slight upshift with a decrease in intensity. These result from the removal of carbonaceous impurities. The D-band intensifies after high-temperature annealing. Figure shows their corresponding XRD patterns. The reflections associated with the Ni nanoparticles can be readily seen in the pattern of pristine nanotube. A strong background at small wave vector $Q(=4\pi \sin\theta)/\lambda)$ below 1 Å^{-1} is apparent. This is associated with the (10) reflection of the SWNT bundles featuring as the nanoparticles in this background. A faint peak at $\sim$1.8 Å^{-1} reveals the presence of some graphitic particles. After air oxidation, NiO peaks derived from nickel oxidation are clearly visible. But weak reflections from nickel still remain. Most of the NiO Å^{-1} particles are removed after the HCl treatment. The (10) reflection and other (hk) reflections of the SWNT bundles are quite obvious at this stage. The intense peak at 1.8 Å^{-1} is assigned to the graphitic particles. The (hk) reflections from SWNT bundles become more intense after high-temperature annealing. The TEM images of the specimens after each step of the purification procedure are shown in Figure 1.28a–d. A large quantity of impurities including metal nanoparticles coated with amorphous carbon and graphitized carbon together with SWNT bundles can be seen (Figure 1.28a). The number of catalyst particles

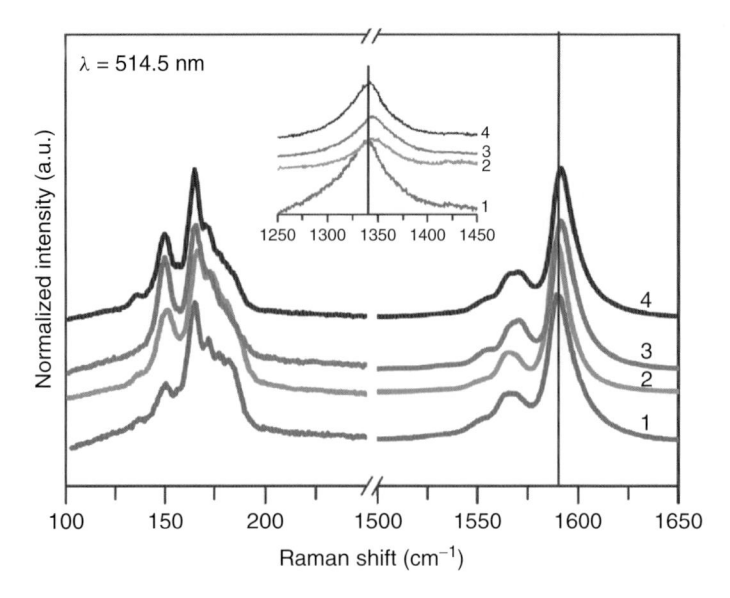

Figure 1.26 Raman spectra of arc-grown SWNT (PS0) after each step of the purification procedure. (1) PS0, (2) PS1, (3) PS2, and (4) PS3a. (Source: Reproduced with permission from Ref. [98], Elsevier (2010).)

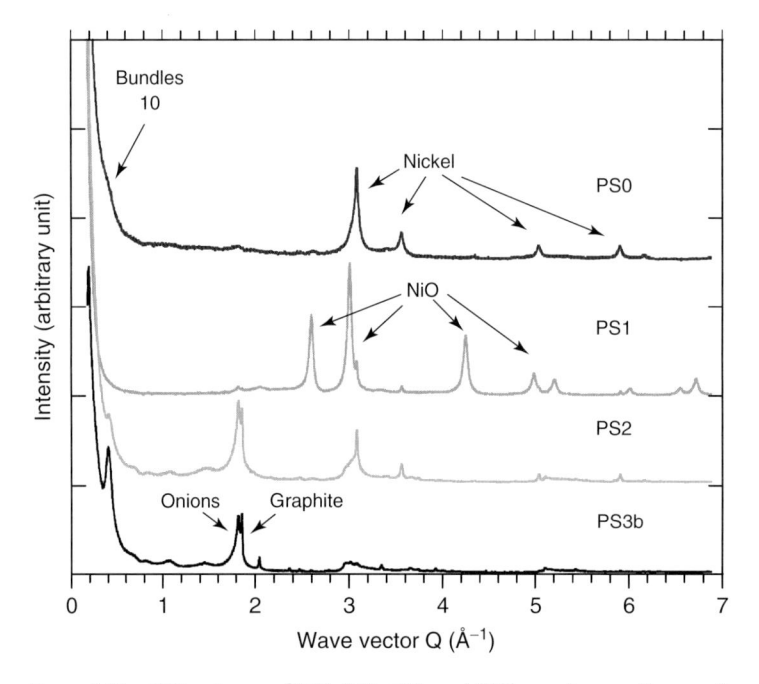

Figure 1.27 XRD patterns of PS0, PS1, PS2, and PS3b specimens. (Source: Reproduced with permission from Ref. [98], Elsevier (2010).)

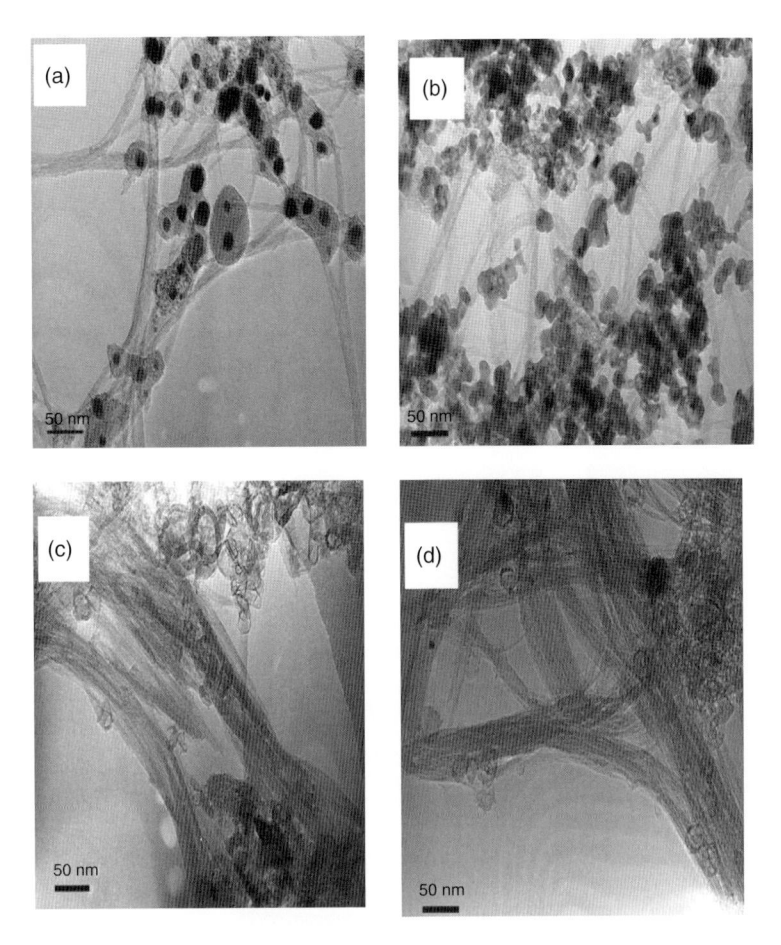

Figure 1.28 TEM images of (a) PS0, (b) PS1, (c) PS2, and (d) PS3a. (Source: Reproduced with permission from Ref. [98], Elsevier (2010).)

reduces significantly after acid treatment. Numerous empty carbon cells derived from the removal of catalyst particles are visible (Figure 1.28c).

TGA is a simple quantitative method for measuring the specimen mass in relation to changes in temperatures using a thermobalance. During the tests, the temperature is raised and the weight of the sample is recorded continuously, which permits the monitoring of weight losses with time. It can be performed in air, oxygen, or inert gas (or vacuum). TGA with air/oxygen atmosphere allows the determination of temperatures at which nanotubes and impurities oxidize. Since amorphous carbon, carbon nanoparticle, multishell carbon, fullerene and SWNT have different affinity toward oxygen, thus those impurities and CNTs oxidize at different temperatures as expected [99, 100]. Other quantitative method such as UV-vis-NIR spectroscopy has also been employed for purity evaluation of SWNTs [101, 102].

1.3
Carbon Nanofibers (CNFs)

Vapor-grown carbon nanofibers (VGCNFs) have received considerable attention in recent years because of their relatively low cost when compared to CNTs, especially SWNTs. Thus VGCNFs are an attractive alternative for CNTs for research purposes and technological applications. VGCNFs are also widely known as carbon nanofibers (CNFs). A CNF consists of stacked curved graphene layers forming cones or cups with an angle alpha with respect to the longitudinal axis of the fiber [103–107]. The stacked cone graphene sheets yield the so-called "herringbone" morphology, while the stacked-cup structure is referred to as a *bamboo feature*

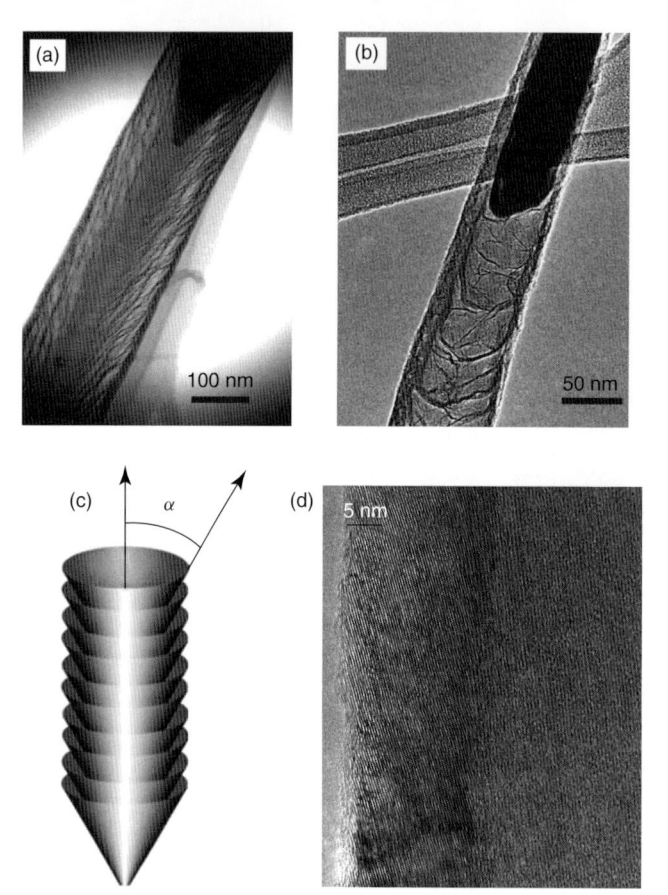

Figure 1.29 (a) STEM (scanning transmission electron microscopy) micrograph of a herringbone nanofiber and (b) TEM image of a bambootype CNF. (c) Schematic illustration of stacked cone (herringbone) nanofiber. (Source: Reproduced with permission from Ref. [103], American Institute of Physics (2005).) (d) TEM image of the VGCNF wall with canted graphene planes comprising of stacked-cup morphology. (Source: Reproduced with permission from Ref. [105], Elsevier (2007).)

(Figure 1.29a–d). The diameters of CNFs range from 50 to 500 nm. CNFs generally have more crystalline defects, rendering them exhibit poorer mechanical properties than MWNTs. Graphitization of MWNTs by high-temperature heat treatment can improve their electrical conductivity [87]. Analogously, graphitization can improve the perfection of grapheme planes in the walls of the VGCNFs. Tibbetts and coworkers [108] reported that heat treatment of CNFs at temperatures above 1500 °C results in significant rearrangement of the core morphology. Heat treatment at temperatures of 1800–3000 °C reduces the structural disorder and increases the graphitic content of the fiber [109, 110]. In other words, increasing graphitization temperature leads to a reduction of the turbostratic layer and an increase of the order in the graphene planes [106].

CNFs were produced commercially by the Applied Sciences, Inc. using high-temperature decomposition of natural gas in the presence of the $Fe(CO)_5$ catalyst (Figure 1.30). When the catalyst particles are properly dispersed and activated with sulfur, VGCNFs are abundantly produced in a reactor at 1100 °C [105]. This manufacturer has produced Pyrograf® III nanofibers of two types (PR-19 and PR-24) in four different grades, that is, as-grown (AS), pyrolytically stripped (PS), LHT and HHT [111]. Pyrograf®-III is available in diameters ranging from 70 to 200 nm and a length estimated to be 50–100 μm. The diameters of PR-19 and PR-24 fibers are about 150 and 100 nm, respectively. The AS fibers have polyaromatic hydrocarbons on their surfaces. The PS-grade fiber is pyrolytically stripped at 600 °C to remove polyaromatic hydrocarbons from the fiber surface. The LHT-grade nanofiber is treated at 1500 °C in order to carbonize and chemically vaporize deposited carbon present on the fiber surface, and the HHT-grade nanofiber is treated to temperatures up to 3000 °C, leading to graphitization of the fiber. On the other hand, vertically aligned CNFs designed for electronic device applications can be synthesized by means of PECVD process [103].

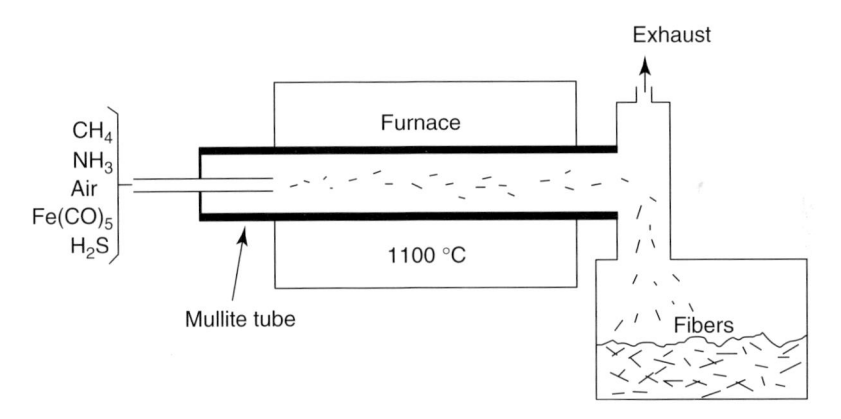

Figure 1.30 Apparatus for manufacturing VGCNFs. (Source: Reproduced with permission from Ref. [105], Elsevier (2007).)

1.4
Physical Properties of Graphene

1.4.1
Mechanical Behavior

The mechanical behavior of graphene can be determined from both theoretical modeling and experimental measurement. The numerical simulation methods commonly used for predicting mechanical behaviors of graphene are equivalent-continuum, quantum mechanical (QM) and atomistic modeling. Equivalent-continuum modeling is based on the well-established principles of elasticity of shells, beams, and rods. This method serves as a bridge between computational chemistry and solid mechanics by substituting discrete molecular structures with equivalent-continuum models [112]. For example, a beam undergoes stretching; bending and torsional deformation in structural mechanics can be used to mimic bond stretching, angle bending, and dihedral stretching of covalently bonded carbon atoms. Using this approach, Sakhaee-Pour [113] determined the elastic properties of a single-layered graphite sheet and reported that its Young-modulus (E) is around 1 TPa, regardless of the chirality. The QM calculations are performed using the density functional theory. The *ab initio* QM calculation for the phonon spectra of graphene as a function of uniaxial tension generates $E = 1.05$ TPa [114].

Atomistic simulation can reveal the physical nature of graphene mechanical deformation at atomistic scale using relevant interatomic potential energy models. In this regard, molecular dynamics (MD) is typically used to simulate the physical motion of atoms and molecules numerically. The atoms are allowed to interact for a period of time, giving rise to the movement and force field among them. The trajectories of atoms in the system under consideration can be estimated from the Newtonian equation of motion. The potential energy of the system is a function of the positions of atoms and thus can be described in terms of force field functions and parameters. Several interatomic potentials have been developed for use with a select group of materials including Lennard-Jones, Morse and Tersoff-Brenner potentials [115]. Lennard-Jones and Morse potentials introduced in the 1920s are empirical isotopic pair potentials and particularly suitable for atoms with no valence electrons. The Tersoff-Brenner potential is widely used for simulations in the carbon-based materials such as diamond, graphene, and nanotubes.

The mechanical deformation of graphene can be predicted by MD simulations using appropriate interatomic potential models under tensile/shear loadings. Thus the Young's modulus and shear modulus of graphene can be determined accordingly. In general, there exists two kinds of atomistic interactions between the carbon atoms, that is, covalently bonded and nonbonded. The potential energy of graphene is derived from covalently bonded atom interactions such as bond stretching, bond angle bonding, dihedral angle torsion and inversion, as well as nonbonded interaction (van der Waals force) characterized by Lennard-Jones potential [116]. Tsai and Tu employed MD simulations for predicting mechanical properties of a single graphite layer using these assumptions and obtained a stiffness value of 0.912 TPa. Jiang et al. [117] used MD to obtain thermal vibrations of graphite and then evaluate the Young's modulus from the thermal mean-square vibration amplitudes. They indicated that the Young's modulus increases from 0.95 to 1.1 TPa as the temperature increases from 100 to 500 K. In recent years, coupled quantum mechanical/molecular mechanical (QM/MM) calculations have been for mechanical modeling for the graphene sheet [118, 119]. Kim and coworkers [119] determined $E = 1.086$ TPa for armchair graphene and $E = 1.05$ TPa for zigzag graphene using the QM/MM simulations.

The experimental measurements of mechanical properties of pure graphene are known to be difficult and tedious. Lee et al. [120] obtained a Young's modulus of 1.02 TPa and intrinsic tensile strength (σ) of 130 GPa for a defect-free graphene monolayer via nanoindentation using an AFM. Using a similar method, the E values of graphene bilayer and trilayer were determined to be 1.04 and 0.98 TPa, while the σ values to be 126 and 101 GPa, respectively [121]. Gomez-Navarro et al. [122] obtained $E = 0.25$ TPa for freely suspended rGO monolayer, being a quarter of that of pristine graphene. The Young's modulus was determined through a tip-induced deformation by indenting an AFM tip at the center of the suspended layer.

The graphene paper produced by dispersing GO sheets in water followed by filtration and air drying, displaying $E = 25.6$ GPa and σ of 81.9 MPa. These GO colloids modified with either MCl_2 salts (M = Mg or Ca) followed by filtration and air drying, showing Young's modulus in the range of 24.6–28.2 GPa and ultimate tensile stress in the range of 87.9–125.8 MPa [123]. The Mg^{2+} or Ca^{2+} ions can bind oxygen functional groups on the basal planes and the carboxylate groups on the edges of the GO sheets, thereby improving their mechanical properties. Table 1.2 summarizes experimental tensile properties of graphene and chemically rGO specimens. Commercial exfoliated GNP comprising multiple stacked graphene sheets with a trade name of "xGnP" (XG Sciences) is reported to have tensile modulus and strength of $\sim$1 TPa and $\sim$10–20 GPa, respectively [124].

1.4.2
Electrical Behavior

Graphene exhibits excellent electron mobility in excess of 200 000 $cm^2\ V^{-1}\ s^{-1}$ [125, 126]. GO is electrically insulating; hence, additional chemical reduction and thermal annealing treatments are needed for eliminating oxygen functional groups. These treatments partially restore its electrical conductivity. The conductivity of rGO generally ranges from 0.05 to 500 $S\ cm^{-1}$ depending on the degree of deoxygenation. The residual oxygen forms sp^3 bonds with carbon atoms in the basal plane of rGO. The sp^3 bonds disrupt charge transport in the sp^2 carbon clusters, thereby reducing electrical conductivity of rGO. Mattevi et al. studied the effect of sp^2 carbon fractions on the conductivity of GO treated by chemical reduction and thermal annealing [29]. The conductivity can be tailored over about 12 orders of magnitude by increasing sp^2 graphene domain fractions (Figure 1.31). At

Table 1.2 Tensile properties of graphene and its derivatives.

Research group	Material	Facility	Young's modulus (GPa)	Tensile strength (GPa)	Year
Lee et al. [120]	Graphene monolayer	AFM	1.02×10^3	130	2008
Lee et al. [121]	Graphene bilayer	AFM	1.04×10^3	126	2009
Lee et al. [121]	Graphene trilayer	AFM	0.98×10^3	101	2009
Gomez-Navarro et al. [122]	rGO monolayer	AFM	250	–	2008
Park et al. [123]	GO paper	DMA at 35 °C	25.6	81.9×10^{-3}	2008
Park et al. [123]	Mg-modified GO paper	DMA at 35 °C	24.6	87.9×10^{-3}	2008
Park et al. [123]	Rinsed Mg-modified GO paper	DMA at 35 °C	27.9	80.6×10^{-3}	2008
Park et al. [123]	Ca-modified GO paper	DMA at 35 °C	21.5	75.4×10^{-3}	2008
Park et al. [123]	Rinsed Ca-modified GO paper	DMA at 35 °C	28.1	125.8×10^{-3}	2008

DMA, dynamic mechanical analyzer.

low sp^2 fractions, rGO contains a large number of oxidized domains; thus hopping or tunneling transport among the sp^2 clusters is expected. By increasing the sp^2 fraction to more than 0.9, the conductivity of rGO or TRG sheets approaches that of graphene.

Chen et al. [127] fabricated freestanding graphene paper by vacuum filtration of rGO suspensions through a commercial membrane filter, followed by air drying and peeling from the filter membrane. The as-produced paper had electrical conductivity of about 30 S cm^{-1} but increased to 118 and 351 S cm^{-1} by heat treating at 220 and 500 °C, respectively. Ruoff and coworkers [128] reported that the electrical conductivity of freestanding rGO paper is 6.87×10^2 S m^{-1}. Very recently, Schwamb et al. [129] indicated that the electrical conductivity of TEGO flake is in the range of 6.2×10^2 to 6.2×10^3 S m^{-1}.

1.4.3
Thermal Behavior

Recently, Baladin and coworkers [130, 131] achieved very high thermal conductivity (κ) of 4840–5300 Wm^{-1}K^{-1} for a single-layer graphene (SLG) produced by mechanical cleavage of bulk graphite. During the measurements, a single graphene layer suspended a trench in Si/SiO$_2$ substrate was irradiated with a laser beam of 488 nm wavelength. The heat was transferred laterally toward the sinks made

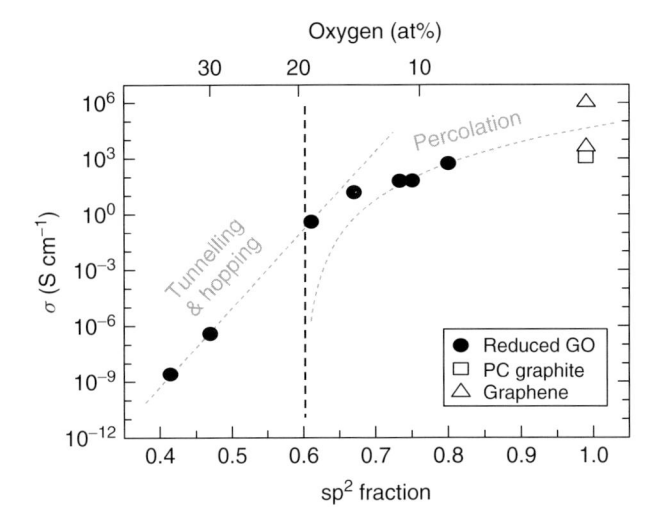

Figure 1.31 Conductivity of thermally reduced GO versus sp² carbon fraction. The vertical dashed line indicates the percolation threshold at sp² fraction of ~0.6. Fitting of the experimental data reveals two different regimes for electrical transport with sp² fraction. Tunneling and/or hopping (straight dashed line) mechanisms dominate at sp² fractions below 0.6, while percolation amongst the sp² clusters dominates above the percolation threshold. The 100% sp² materials are polycrystalline (PC) graphite and graphene. (Source: Reproduced with permission from Ref. [29], Wiley-VCH (2009).)

from graphitic layers (Figure 1.32). The amount of power dissipated in graphene and corresponding temperature rise were determined from the spectral position and integrated intensity in the graphene G-peak using confocal micro-Raman spectroscopy. SLG is commonly supported on a dielectric substrate for electronic device applications. Thus fully contact with a substrate could affect the thermal transport property of graphene. Ruoff and coworkers [132] made thermal measurements on an SLG supported on amorphous SiO_2 substrate, reporting a κ value of 600 W m^{-1} K^{-1} near room temperature. The fall in the κ value of graphene supported on SiO_2 is due to the scattering by substrate phonons and impurities. Schwamb et al. [129] determined the κ value of thermally reduced GO flake to be 0.14–2.87 W m^{-1} K^{-1}. Very recently, Xiang and Drzal [20] fabricated freestanding GNP paper by vacuum filtration of GNP suspension using an ANODISC membrane (0.2 μm, Whatman) (Figure 1.33). After filtration, the paper was suction dried and placed in a vacuum oven at 60 °C overnight before peeling the GNP paper off the membrane. The "as-made" freestanding paper was annealed at a furnace at 340 °C for 2 h and termed as *annealed* paper. The annealed paper was finally cold pressed at 100 psi at room temperature for 1 h and identified as 'annealed and cold pressed' (Figure 1.34a,b). The paper has the advantages of ease of handling and storage in readiness for its incorporation into polymers to form strong and conducting nanocomposites for industrial applications. The in-plane thermal conductivities of

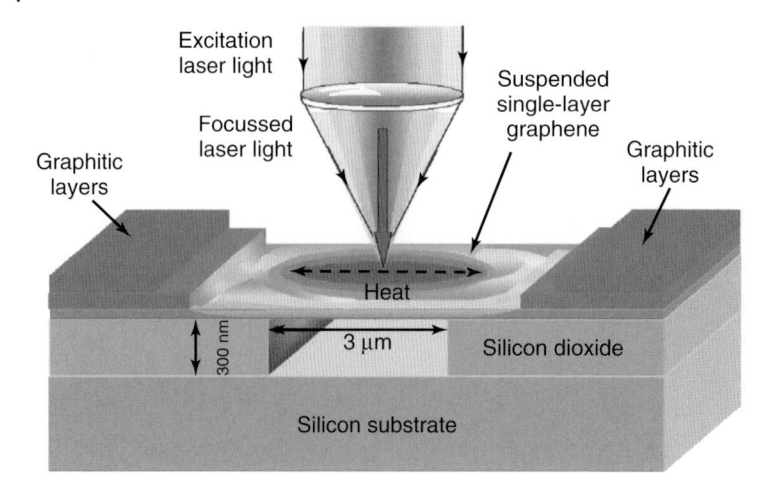

Figure 1.32 Schematic diagram showing the excitation laser beam focused on a graphene layer suspended across a trench. Laser beam creates a local hot spot and generates a heat wave inside SLG propagating toward heat sinks. (Source: Reproduced with permission from Ref. [130], the American Chemical Society (2008).)

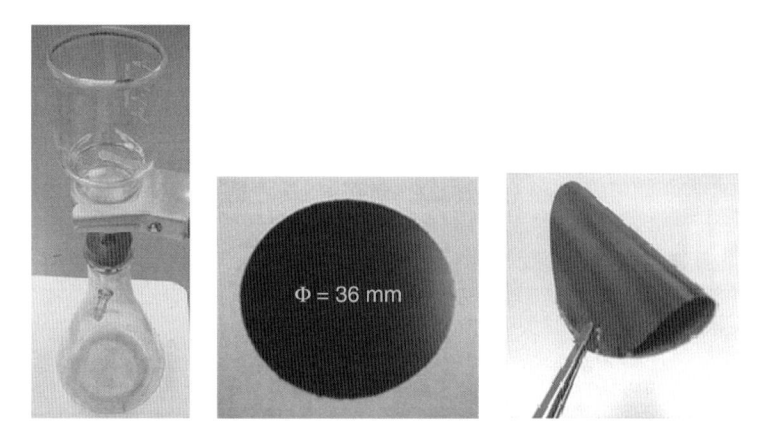

Figure 1.33 Photographs of the filtration device, a sample of GNP paper peeled off the ANODISC filter paper and a mechanical flexible GNP paper. (Source: Reproduced with permission from Ref. [20], Elsevier (2011).)

the as-prepared, annealed (340 °C for 2 h), and annealed and cold-pressed GNP papers are 98, 107, and 178 W m^{-1} K^{-1}, respectively. Cold pressing improves the thermal contact by eliminating most of the pores in the paper, resulting in highest thermal conductivity. The through-plane thermal conductivities of the as-prepared, annealed (340 °C for 2 h), and annealed and cold-pressed GNP papers are 1.43, 1.71, and 1.28 W m^{-1} K^{-1}, respectively.

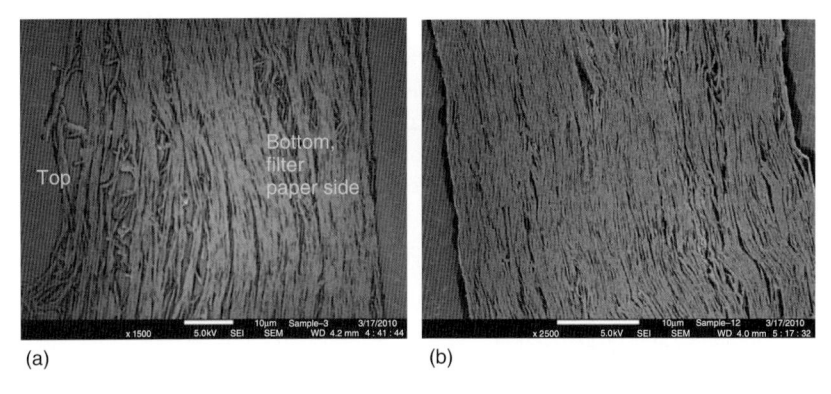

Figure 1.34 Scanning electron micrographs of cross section of the (a) as-made GNP paper and (b) annealed and cold-pressed GNP paper. (Source: Reproduced with permission from Ref. [20], Elsevier (2011).)

1.5
Properties of Carbon Nanotubes

1.5.1
Mechanical Behavior

1.5.1.1 Theoretical Prediction

Basic understanding of the mechanical properties of CNTs, such as the Young's modulus, strength and fracture strain, is essential for the development of their technological applications. The Young's modulus of a material is generally related to the chemical bonding of its constituent atoms. The interaction forces between atoms are typically represented in the form of interatomic potential energy models. Many theoretical and experimental investigations have been carried out to determine the Young's modulus of CNTs. Theoretical studies show that the Young's modulus varies greatly from 0.6 to 5.5 TPa [133–138]. A large variation in predicted mechanical stiffness arises from the different methods used for the simulations, different CNT chiralities or lengths, different potentials used to define the C–C bond in the plane of the graphene sheets, and the presence of intrinsic defects.

Yakobson *et al.* [133] employed MD simulations to determine the Young's modulus of SWNTs under axial compression. The Tersoff-Brenner potential was used for characterizing the C–C interactions. They obtained a Young's modulus of 5.5 TPa. The simulation also predicts that SWNTs can sustain a large strain of 40% without damaging its graphitic structure. Robertson *et al.* [134] adopted empirical Brenner potential to examine the energetics and elastic properties of CNTs with radii <9 Å. The Young's modulus of CNTs was close to that of in-plane graphene, that is, ∼1.06 TPa. Yao and Lordi [135] employed MD simulations with the universal force field and reported a stiffness value of ∼1 TPa for SWNTs. Furthermore, the Young's modulus was found to increase significantly with decreasing tube diameter and increase slightly with decreasing tube helicity. Hernandez *et al.* [136] employed a tight-binding model to calculate the Young's modulus of SWNTs. The stiffness was dependant on the diameter and chirality of the CNTs, ranging from 1.22 TPa for the (10,0) and (6,6) tubes to 1.26 TPa for the (20,0) tube. Haskins et al. [137] also used a tight-binding MDs to predict the Young's modulus of nanotubes. For a pristine (5,5) nanotube, the stiffness was calculated to be ∼1.1 TPa. Very recently, Qin et al. used Lennard-Jones potential for the MD simulations. The predicted modulus of the (5,5) and (9,0) SWNTs ranges from 0.6 to 0.7 TPa [138]. Apparently, most MD simulations produce a stiffness value of around 1 TPa. The predicted Young's modulus of around 1 TPa is acceptable for the SWNT since it is close to that of the graphene sheet.

1.5.1.2 Experimental Measurement

Experimental measurements of the Young's modulus of CNTs are mostly conducted in electron microscopes and AFMs [139–144]. Treacy et al. [139] determined the Young's modulus of individual MWNTs (arc-grown) anchored in a TEM from their amplitudes of thermal vibrations. They obtained an average value of 1.8 TPa with a large scatter in the data ranging from 0.4 to 4.15 TPa. A similar method was adopted by Krishnan et al. [140] for measuring the stiffness of MWNTs synthesized by the laser ablation process, producing a value of 1.25 TPa. Considering CNT acts as a cantilever beam under bending in an AFM. The cantilever deflection and force applied can be measured accordingly. Using this technique, Wong et al. [141] obtained a stiffness value of 1.28 TPa for arc-grown MWNTs. Salvetat et al. [142] obtained the Young's modulus values of both the arc- and CVD-grown MWNTs to be 0.81 and 0.027 TPa, respectively. Arc-grown MWNTs exhibit higher Young's modulus because of their more perfect crystalline structure. Yu et al. [143] used a tensile testing tool inside an SEM to study the tensile behavior of MWNTs. An *in situ* AFM tip was employed to assist the manipulation, attachment, and stretching of MWNTs. They reported the Young's modulus values ranging from 0.29 to 0.95 TPa and a strain at break of up to 12%. The failure of the MWNT under tensile load can be described in term of the "sword-in sheath" type mode, in which the outermost shell fails first followed by the sliding and pulling out of internal shells from the outer shell of the MWNT [144]. Yu et al. [145] then used a similar method for tensile loading of SWNT ropes and reported Young's modulus values of 0.32–47 TPa with a mean value of 1 TPa. Demczyk et al. [146] *in situ* tensile loaded arc-grown MWNT in a TEM, obtaining a Young's modulus of 0.9 TPa and a tensile strength of 150 GPa.

Raman spectroscopy has been used to extract the mechanical properties of CNTs embedded in a polymer resin [147, 148]. For example, Lourie and Wagner [147] employed micro-Raman spectroscopy to monitor cooling-induced compressive deformation of CNTs embedded in an epoxy matrix. The thermal stress induced can cause a shift in the G'-band. Cooper et al. [148] used a self-made four-point bending rig and placed on the Raman microscope stage to induce mechanical deformation for the CNT/epoxy specimens. From the shift in the G'-band as a result of tensile deformation on the specimen surfaces, the effective moduli of arc-/laser-grown SWNTs and arc-grown MWNTs dispersed in the composites are ∼1 and 0.3 TPa, respectively. Table 1.3 summarizes experimental values of

Table 1.3 Experimental young's modulus and tensile strength of CNTs determined using electron microscopic and raman spectroscopic facilities.

Research group	Nanotube type	Young's modulus (TPa)	Tensile strength (GPa)	Year
Treacy et al. [139]	MWNT	1.8	–	1996
Wong et al. [141]	MWNT	1.28	–	1997
Krishnan et al. [140]	SWNT	1.25	–	1998
Salvetat et al. [142]	MWNT	0.81	–	1999
Salvetat et al. [142]	MWNT	0.027	–	1999
Yu et al. [143]	MWNT	0.27–0.95	11–63	2000
Yu et al. [145]	SWNT rope	0.32–1.47	13–52	2000
Cooper et al. [148]	SWNT	1	–	2001
Cooper et al. [148]		0.3	–	2001
Demczyk et al. [146]		0.9	150	2002

MWNTs and SWNTs determined from different techniques. A large variation in experimental values of Young's modulus of CNTs is observed. This is attributed to the difficulty in manipulating and measuring individual nanotubes of nanometer dimension [149]. Furthermore, the CNTs listed in this table were synthesized from either PVD or CVD techniques, having different nanotube types, impurity contents, lengths, and diameters.

1.5.1.3 Flexibility of Carbon Nanotubes

Iijima et al. [150] employed the Tersoff-Brenner potential for simulating the bending behavior of SWNTs of different diameters and helicity. Their results showed that the nanotubes are extremely flexible on the application of large deformation strains. The hexagonal network of CNTs is resilient and can resist bond breaking up to a very high strain values. The network can be preserved even by bending up to $\sim 110°$, forming a kink consequently. The kink enables the nanotubes to relax elastically. Furthermore, HRTEM observation also provides a substantial evidence for the formation of a single kink or multiple kinks in the CNTs on bending (Figure 1.35a–d). Falvo et al. [151] employed AFM to observe bending features of MWNTs up to a 42% strain value. They found that the nanotubes exhibit very high flexibility and possess the ability to buckle elastically through large angles without inducing visible defects. This is an interesting and important feature of CNTs, demonstrating their good ductile characteristics. The fracture tensile strain predicted by MD simulations could reach up to 30% [152], or even higher depending on simulated temperatures and interatomic potentials employed [153]. The tensile fracture strain of SWNTs at room temperature is 15% or higher and could reach up to 280% at temperatures $\geq 2000\,°C$ [154].

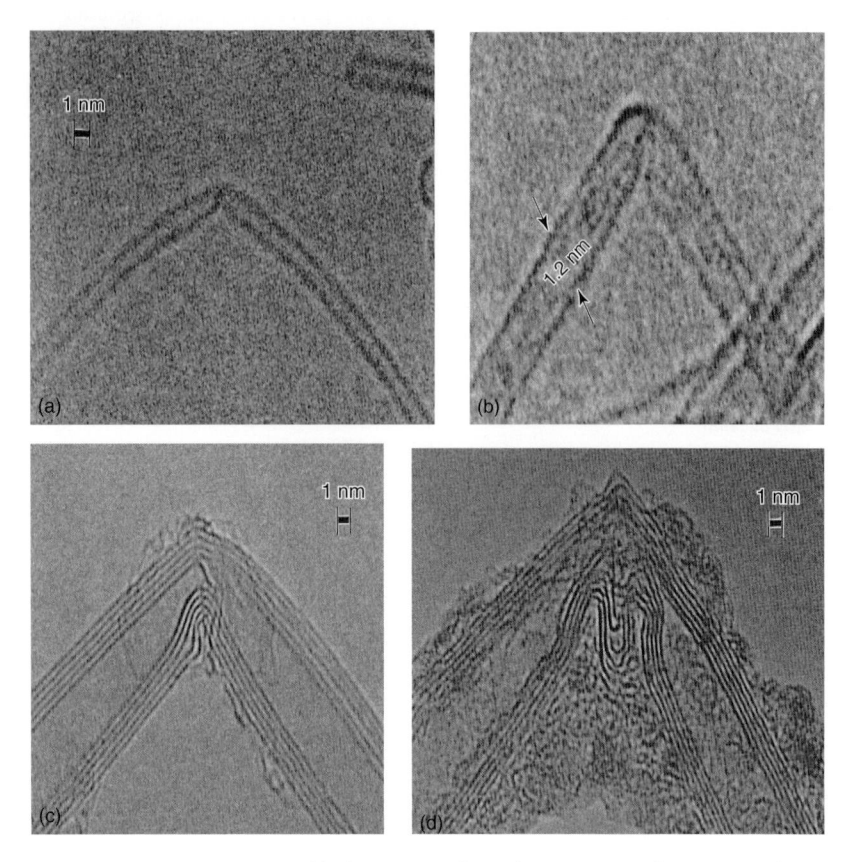

Figure 1.35 HRTEM images of kink structures formed in SWNTs and MWNTs under mechanical bending. A single kink in the middle of SWNTs with diameters of (a) 0.8 and (b) 1.2 nm. MWNTs with (c) single-kink and (d) two-kink complex. (Source: Reproduced with permission from Ref. [150], The American Institute of Physics (1996).)

1.5.2
Electrical Behavior

The sp^2 carbon bonding in the CNTs gives their π electrons very large mobility. The electrical conductivity of CNTs varies from metallic to semiconducting depending on their chirality and diameter. Armchair SWNTs ($n = m$) possess a large carrier concentration, rendering them metallic in nature. In contrast, zigzag and other chiral nanotubes exhibit semiconducting to semimetallic behavior since they possess medium to small band gaps between the valence and conduction bands. The electrical conductivity of SWNT ropes is reported in the range of 1×10^4 to 3×10^4 S cm^{-1} at 300 K [155, 156]. The agglomeration of nanotubes into ropes may degrade the conductivity of SWNTs. The individual tubes of an MWNT may have different chiralities. Accordingly, a single MWNT may have both metallic and semiconducting layers. The electrical conductivity of carbon arc-grown MWNTs determined from the four-point technique ranging from ~1.25 to 2×10^5 S cm^{-1}, corresponding to semiconducting to metallic behavior [157]. High-temperature annealing can enhance electrical conductivity of CNTs synthesized by CVD. The conductivity of MWNTs produced by noncatalytic CVD in a porous alumina template increases from about 1×10^4 to 2×10^5 S m^{-1} after annealing at 2000 °C. This is due to the perfection in the structure of nanotubes walls by high-temperature annealing [87].

Electrons in one-dimensional CNTs are believed to conduct ballistically. In this regard, electrons experience no resistance in the CNTs. An ideal metallic SWNT is predicted to have a quantum conductance (Go) of 2 Go defined by the following equation [158]

$$\text{Go} = \frac{2e^2}{h} = (12.9\,\text{k}\Omega)^{-1} \tag{1.5}$$

where e is electronic charge and h the Plank's constant. Frank et al. determined the electrical conductance of MWNT using a scanning probe microscope attached with a nanotube fiber. They found that the nanotube conducts current ballistically with quantized conductance behavior. The MWNT conductance jumped by the increments of one unit of Go [159].

1.5.3
Thermal Behavior

The transport of thermal energy in CNTs is known to occur via phonon conduction mode. The phonon thermal conductivity can be determined from $\kappa = C_p v \lambda$ where C_p is the specific heat, v the speed of sound and λ the mean free path of phonon. The mean free path is the average distance travels by a phonon between successive scatterings and estimated to be greater than 1 µm [160]. Thus thermal conductivity of CNTs depends mainly on the mean free path of phonons and temperature [161]. When SWNTs have comparable or shorter length than the free path of phonons, they can be considered as ballistic phonon conductors at low temperatures with four quanta of thermal conductance (Goth) given by Schwab et al. [162]

$$\text{Go}^{\text{th}} = \pi^2 k_B^2 T / 3h \tag{1.6}$$

where k_B is a Boltzmann constant. The factor of T in thermal quantum conductance implies its linear dependance on temperature. In the case of electronic conduction, the corresponding quantity is the electron charge, e, and the electrical quantum conductance given by Eq. (1.5) is temperature independent. The ballistic thermal conductivity of CNTs increases linearly with temperature at low-temperature regime (~1–50 K) [161]. As the temperature increases, both acoustic and optical phonons involve in the transport, hence reducing the phonon mean free path. At higher temperatures, inelastic Umklapp scattering becomes dominant. It is noted that the thermal conductivity of nanotubes is also affected by the structural defects and impurities induced in the tubes during synthesis.

Theoretical MD simulation prediction exhibits an unusually high value of thermal conductivity, that is, 6600 W m^{-1} K^{-1} at room temperature for the (10,10)

Table 1.4 Room temperature thermal conductivity of carbonaceous materials.

Research group/company	Material	Thermal conductivity (W m^{-1} K^{-1})	Method	Specimen testing condition
Baladin et al. [130]	SLG	~4840–5300	Optical	Individual, suspended
Seol et al. [132]	SLG	600	Electrical	Supported on a substrate
Schwamb et al. [129]	TEGO	0.14–2.87	Electrical	Individual, suspended
Pop et al. [167]	SWNT	~3500	Electrical	Individual, suspended
Hone et al. [164]	SWNT	1750–5800	Thermocouples	Bundles
Kim et al. [165]	MWNT	>3000	Electrical	Individual, suspended
Fuji et al. [166]	MWNT	2069	Electrical	Individual, suspended

SWNT [160]. Osman and Srivastava determined thermal conductivity of SWNTs at 100–500 K using MD simulations with the Tersoff-Brenner potential for C–C interactions [163]. The temperature dependance of the thermal conductivity of nanotubes exhibits a peak behavior before falling off at higher temperatures. This is due to the onset of anharmonic Umklapp phonon scattering. The armchair (5,5) SWNT displays a κ value of ~2250 W m^{-1} K^{-1} at room temperature (Table 1.4).

Pop et al. [167] determined the thermal properties of a suspended SWNT of length 2.6 µm and diameter 1.7 nm at 300–800 K by means of electrical measurements. The κ value is about 3500 W m^{-1} K^{-1} at room temperature. Hone et al. [164] determined thermal conductivity of SWNT ropes using typical thermocouple measurements, obtaining κ value of 1750–5800 W m^{-1} K^{-1} at room temperature. Fuji et al. [166] used a suspended sample-attached T-type nanosensor and reported that the thermal conductivity of a single MWNT at room temperature increases with decreasing tube diameter. The conductivity of an MWNT with an outer diameter of 9.8 nm is 2069 W m^{-1} K^{-1} and decreases to about 500 W m^{-1} K^{-1} for a tube of 28.2 nm. Kim et al. [165] ed a suspended microdevice and found that the thermal conductivity of an MWNT is more than 3000 W m^{-1} K^{-1} at room temperature. Yi et al. [168] stripped CVD-grown MWNT bundles from their aligned arrays and obtained κ value of 25 W m^{-1} K^{-1} using a self-heating 3ω method. The large difference between the thermal conductivity of a single tube and MWNT bundles can be attributed to high thermal contact resistance (Kapitza resistance), intertube Umklapp scattering and the presence of defects [168, 169].

1.6
Properties of Carbon Nanofibers

VGCNFs have larger diameters but lower crystallinity than the MWNTs. The mechanical properties of VGCNFs are generally poorer than the MWNTs. The tensile properties of VGCNFs depend greatly on the graphitization heat treatment. Very recently, Ozkan et al. [170] reported that the Young's modulus and tensile strength of Pyrograf® III PR-24 nanofiber of PS grade are 180 GPa and tensile strength of 2.9 GPa, while those of the HHT-grade are 245 and 2.35 GPa, respectively. Thus graphitization treatment increased the nanofiber elastic modulus by 35% because of the reduction of the outermost turbostratic layer and the increase of the order in the graphene planes. The thermal conductivity of VGCNF is reported to be 1950 W m^{-1} K^{-1} by the manufacturer [111].

1.7
Current Availability of Carbonaceous Nanomaterials

The demands for carbonaceous nanomaterials are ever increasing today because of their attractive applications in biomedical and technological sectors. Extensive research studies have been carried out in worldwide research institutions and industrial laboratories on the issues relating to the improvement of the synthesis techniques and physical/mechanical properties of carbon nanostuctures. Several patents have been granted recently by the US Patent and Trademark Office, and European Patent office for innovative synthesis processes of carbon nanomaterials (Tables 1.5 and 1.6). These offices granted exclusive rights for the researchers for their novel synthesis of carbonaceous materials. These patents form the basis for successful commercialization of carbonaceous nanomaterials. Further, global industrial manufacturers also produce carbon nanostructures of high qualities to meet the demands of academic research and technological application. The current availability of some carbonaceous nanomaterials is listed in Table 1.7.

1.8
Multifunctional Composite Materials

Conventional polymer composites have been extensively used as structural materials for industrial applications because of their light weight, high mechanical strength and excellent corrosion resistance. A typical example is carbon fiber/epoxy composites showing high-end applications in the aerospace and military sectors. They have successfully replaced most metallic components for structural engineering applications. However, the limits of improving properties of microcomposites have been reached because the need of using high filler volume contents for achieving required physical and mechanical properties. Large-volume fractions of fillers with micrometer dimensions have adverse effects on the processing and mechanical properties of resulting polymer composites.

Recent advances in nanotechnology have provided great opportunities for developing advanced techniques to synthesize novel nanomaterials with functional properties. These include nanoceramics (e.g., clay platelets, alumina, silica, silicon carbide, titania, zinc oxide, etc.) and carbonaceous nanomaterials. Nanomaterials can be added in very low loading levels to polymers for enhancing their properties. The physical and mechanical properties of polymer nanocomposites

Table 1.5 Patent processes for manufacturing graphenelike sheets and graphite nanoplatelets.

Patent number and year	Patent title	Inventor	Assignee
US 7658901 (9 February 2010)	Thermally exfoliated graphite oxide	Prud'Homme, R.K., Aksay, I.A., Adamson, D., Abdala, A.	Princeton University (USA)
US 7771824 (10 August 2010)	Bridged graphite oxide materials	Herrera-Alonso, M., McAllister, M.J., Aksay, I.A., Prud'Homme, R.K.	Princeton University (USA)
US 7785557 (31 August 2010)	Method of producing a graphene film as transparent and electrically conducting material	Gruner, G., Hecht, D., Hu, L.B.	Unidym, Inc. (USA)
US 7550529 (23 June 2009)	Expanded graphite and products produced therefrom	Drzal, L.T., Fukushima, H.	Michigan State University (USA)
US 7785492 (31 August 2010)	Mass production of nano-scaled platelets and products	Jang, B.Z., Zhamu, A., Guo, J.S.	Nanotech Instruments, Inc. (USA)
US 7799309 (21 September 2010)	Area weight uniformity flexible graphite sheet material	Reynolds III, R.A., Greinke, R.A.	Graftech International Holdings, Inc. (USA)
US 7824651 (2 November 2010)	Method of producing exfoliated graphite, flexible graphite and nano-scaled graphene platelets	Zhamu, A., Shi, J.J., Guo, J.S., Jang, B.Z.	Nanotech Instruments, Inc. (USA)

differ substantially from those of traditional composite counterparts. For example, clay silicates have been incorporated into polymers to improve their mechanical strength, flame retardancy and gas barrier resistance because of their abundancy and low cost [171, 172]. To achieve these desired properties, clay silicate must delaminate into individual nanoplatelets in the polymer matrix to produce exfoliated clay/polymer composites. However, hydrophilic clay silicates show poor dispersion in the polymer matrix because of their high tendency to agglomerate into large clusters. To increase their compatibility with the polymers, clay silicate surfaces are commonly modified with organic surfactants such as $R - NH_3^+$, yielding the so-called organoclays. The use of organomodifiers increases the production cost of polymer nanocomposites. Alkylammonium surfactants generally exhibit low thermal stability since they decompose into organic constituents during melt processing at high temperatures [173]. This case becomes even more serious for

Table 1.6 Patent processes for producing carbon nanotubes.

Patent number and year	Patent title	Inventor	Assignee
US 6413487 (2 July 2002)	Method and apparatus for producing carbon nanotubes	Resasco, D.E, Kitiyanan, B., Alvarez, W.E. Balzano, L.	University of Oklahoma (USA)
US 6962892 (8 Nov 2005)	Metallic catalytic particle for producing single-walled carbon nanotubes	Resasco, D.E, Kitiyanan, B., Harwell, J.H., Alvarez, W.E	University of Oklahoma (USA)
EP 1401763 B1 (30 Mar 2005)	Catalyst supports and carbon nanotubes produced thereon	Nagy, J., Nagaraju, N., Willems, I., Fonseca, A.	Facultes Universitaires Notre-Dame de la Paix (Belgium)
EP 1413551 B1 (21 July 2010)	Methods of manufacturing multiwalled carbon nanotubes	Sakurabayashi, Y., Kondo, T., Yamazawa, Y., Suzuki, Y., Monthioux, M. LeLay, M.	Toyota Jidosha Kabushiki Kaisha (Japan)
EP 1968889 B1 (10 Aug 2011)	Method for synthesis of carbon nanotubes	Bordere, S., Gaillard, P., Baddour, C.	Arkema France (France)
US 7993620 (9 Aug 2011)	Systems and methods for the formation and harvesting of nanofibrous materials	Lashmore, D.S., Brown, J.J., Chaffee, J.K., Resnicoff, B. Antoinette, P.	Nanocomp Technologies, Inc. (USA)
US 8057777 (15 Nov 2011)	Systems and methods for controlling chirality of nanotubes	Lashmore, D.S., Lombard, C.	Nanocomp Technologies, Inc. (USA)

melt compounding of high-performance polymers such as polyetherimide of high melting point. Furthermore, the lack of electrical conductivity of clay platelets also limits their applications in functional electronic devices and sensors. In this regard, carbonaceous nanomaterials of very high aspect ratio, extraordinary high mechanical strength and stiffness, and superior electrical and thermal conductivity are excellent fillers for forming polymer nanocomposites with functional properties. The incorporation of these nanofillers into polymers improves their physical, mechanical, and thermal properties markedly. Such polymer nanocomposites show attractive applications as structural materials for bipolar plates of fuel cells, load-bearing materials for orthopedic implants, electromagnetic interference shielding materials, and functional materials for switching devices, chemical vapor sensors, and transducers.

Table 1.7 Manufacturers of carbonaceous nanomaterials.

Material	Manufacturer	Trade name	Web site information
Carbon nanotube	Carbon Solutions, Inc., USA	AP-SWNT; P2-SWNT P3-SWNT; P9-SWNT	http://www.carbonsolution.com/
Carbon nanotube	Carbon Nanotechnologies, Inc., USA	HiPCo	http://www.cnanotech.com/
Carbon nanotube	Nanoshel LLC, USA	Nanoshel™	http://www.nanoshel.com/
Carbon nanotube	Nanocyl S.A, Belgium	Nanocyl™ NC 7000 (MWNT)	http://www.nanocyl.com/
Carbon nanotube	Timesnano Company, China	TNM5, TNM7 (MWNT) TNS (SWNT)	http://www.timesnano.com/
Carbon nanotube yarn	Nanocomp Technologies, Inc., USA	C-Tex™	http://www.nanocomptech.com/
Carbon nanotube	Hanwha Nanotech Company, South Korea	ASA-100F (SWNT) CM-95 (MWNT)	http://www.hanwhananotech.com/
Carbon nanofiber	Applied Sciences, Inc., USA	Pyrograf® III	http://www.apsci.com/
Graphite nanoplatelet	XG Sciences, Inc., USA	xGnP	http://www.xgsciences.com/
Thermally reduced graphene oxide	Vorbeck Materials, Inc., USA	Vor-x™	http://www.vorbeck.com/

1.8.1
Composites with Carbon Black Nanoparticles

Carbon blacks (CBs) are colloidal, amorphous carbon particles with a distorted graphite structure [174, 175]. CBs are generally used as the reinforcements and conductive fillers for improving dimensional stability and imparting conductive behavior for rubbers and plastic products [176]. Furthermore, CB-filled polymer composites also find applications in electronic packaging and thermistor devices because of their cost advantage over other carbon-based fillers. In general, CBs can be classified into five types according to their manufacturing processes, that is, furnace black, thermal black, lamp black, channel black, and acetylene black. Among these, furnace black is the most commonly used and produced by thermal decomposition of aromatic oils in furnace reactors. This process yields CBs with diameters ranging from 10 to 100 nm and surface area from 25 to 1500 m² g⁻¹. Thermal black is manufactured from natural gas feedstock in the furnace. The particle size, structure, and surface chemistry are three important characteristics of CBs for using as conductive fillers [177]. The CB morphology composes of prime particles fused into primary aggregates. The aggregate size, shape and the number of particles per aggregate determine the structure of CB. CBs with high surface area, high degree of porosity and elongated aggregates composing of prime particles with extensive branching (high-structure) can impart electrical conductivity at lower loadings in the polymer composites. However, such CBs are more expensive to produce. On the other hand, CBs with primary aggregates consist of relatively few prime particles, terming as the low structure, often yield large percolation threshold in the polymer composites. The structure of CBs can be evaluated from dibutyl phthalate (DBP) absorption test as described in the ASTM D2414-09a [178]. This test measures the amount of oil that can be absorbed by the CBs. A higher DBP value implies a higher structure of CB or vice versa. Balberg [179] indicated that the electrical conductivity of CB/polymer composites depends greatly on the CB structures. The percolation threshold of CB/PE (polyethylene) composites filled with high-structure CBs having a DBP value of 350 cm³/100 g is 10 vol% while that of low-structure CBs with a DBP value of 43 cm³/100 g is 39 vol%. The large filler contents in low-structure CB composites can cause high viscosity during processing and degrade mechanical performance of the composites markedly.

1.8.2
Composites with Graphene Oxide and Graphite Nanoplatelet Fillers

As aforementioned, rGO can be produced in large quantities via chemical reduction of GO colloidal dispersion. GO and rGO are inexpensive, versatile building blocks for carbon-based materials. GO can serve either as freestanding papers and thin films to form components of field-effect transistors, while rGO films can act as efficient electrical conductors for the sensors and electrodes in solar cells [180]. Functionalized rGO can be readily dispersed in a wide variety of polymers, this opens up tremendous opportunities for chemists and materials scientists to synthesize novel polymer nanocomposites with multifunctional properties at relatively low cost. The addition of low loadings of rGO and TRG into polymers can improve their electrical, mechanical, and thermal properties as well as dimensional stability significantly. Such nanocomposites find potential applications as electrode materials for batteries and ultracapacitors, hydrogen storage, electromagnetic interference shielding materials, automotive body panel, and so on. GNPs with diameters of 1–15 μm and thickness of <10 nm consist of a stack of multiple graphene sheets (Figure 1.14a,b). They possess smaller aspect ratio (width/thickness) than rGOs and TRGs as expected. Nevertheless, the incorporation of GNPs into polymers can produce mechanically strong and highly conductive polymer nanocomposites for the construction of bipolar plates in fuel cells and the electrodes for supercapacitors [181].

1.8.3
Composites with Carbon Nanotubes

CNT/polymer composites are characterized by high-strength, good electrical conductivity, very good thermal resistance and dimensional stability, and enhanced optical properties. The reinforcing effect of nanotubes in polymers renders them effective load-bearing materials for structural components of terrestrial, space and other engineering applications as well as human orthopedic implants. The excellent electrical conductivity of CNTs makes them suitable fillers for polymer nanocomposites that find useful applications in electronic and optoelectronic devices, coatings, electromagnetic interference shielding materials, actuators, chemical vapor sensors, and so on [182–186]. The most challenging issues for fabricating CNT-polymer nanocomposites are the high cost of CNTs, long processing time of purified nanotubes, and poor dispersion of nanotubes in the polymer matrix because of their high tendency for clustering.

Nomenclature

E	Young's modulus
e	Electronic charge
Go	Quantum electrical conductance
Goth	Quantum thermal conductance
h	Plank's constant
k_B	Boltzmann constant

References

1. Novoselov, K.S., Geim, A.K., Morozov, S., Jiang, D., Zhang, Y., Dubonos, S.V., Grigorieva, I.V., and Firsov, A.A. (2004) *Science*, **306**, 666–669.
2. Novoselov, K.S., Jiang, D., Schedin, F., Booth, T.J., Khotkevich, V.V., Morozov, S.V., and Geim, A.K. (2005) *Proc. Natl. Acad. Sci. U.S.A.*, **102**, 10451–10453.
3. Geim, A.K. and Novoselov, K.S. (2007) *Nat. Mater.*, **6**, 183–191.
4. Stankovich, S., Dikin, D.A., Dommett, G.H., Kohlhaas, K.M., Zimney, E.J., Stach, E.A., Piner, R.D., Nguyen, S.T., and Ruoff, R.S. (2006) *Nature*, **442**, 282–285.
5. Kim, H., Abdala, A.A., and Macosko, C.W. (2010) *Macromolecules*, **43**, 6515–6530.
6. Soldano, C., Mahmood, A., and Dujardin, E. (2010) *Carbon*, **48**, 2127–2150.
7. Lian, P., Zhu, X., Liang, X., Li, Z., Yang, W., and Wang, H. (2010) *Electrochem. Acta*, **55**, 3909–3914.
8. Berger, C., Song, Z.M., Li, X.B., Ogbazghi, A.Y., Feng, R., Dai, Z., Marchenkov, A.N., Conrad, E.H., First, P.N., and de Heer, W.A. (2004) *J. Phys. Chem. B*, **108**, 19912–19916.
9. Berger, C., Song, Z.M., Li, X.B., Wu, X.S., Brown, N., Naud, C., Mayo, D., Li, T.B., Hass, J., Marchenkov, A.N., Conrad, E.H., First, P.N., and de Heer, W.A. (2006) *Science*, **312**, 1191–1196.
10. Reina, F., Jia, X., Ho, H., Nezich, D., Son, H., Bulovich, V. Dresselhaus, M.S., and Kong, M.J. (2009) *Nano Lett.*, **9**, 30–35.

11. Kim, S.J., Zhao, Y., Jang, H., Lee, S.Y., Kim, J.M., Kim, K.S., Ahn, J.H., Kim, P., Choi, J.Y., and Hong, B.H. (2009) *Nature*, **457**, 706–710.
12. Li, X., Cai, W., An, J., Kim, S., Nah, J., Yang, D., Piner, R., Velamakanni, A., Jung, I., Tutuc, E., Barnejee, S.K., Colombo, L., and Ruoff, R.S. (2009) *Science*, **324**, 1312–1314.
13. Li, X.S., Cai, W.W., Colombo, L., and Ruoff, R.S. (2009) *Nano Lett.*, **9**, 4268–4272.
14. Zhao, L., Rim, K.T., Zhou, H., He, R., Heinz, T.F., Pinczuk, A., Flyn, G.W., and Pasupathy, A.N. (2011) *Solid State Commun.*, **151**, 509–513.
15. Shioyama, H. (2000) *Synth. Met.*, **114**, 1–15.
16. (a) Kim, H., Hahn, H.T., Viculis, L.M., Giljie, S., and Kaner, R.B. (2007) *Carbon*, **45**, 1578–1582; (b) Viculis, L.M., Mack, J.J., Mayer, O.M., Hahn, H.T., and Kaner, R.B. (2005) *J. Mater. Chem.*, **15**, 974–978.
17. (a) Chen, G.H., Wu, D.J., Weng, W.G., and Yan, W.L. (2001) *J. Appl. Polym. Sci.*, **82**, 2506–2513; (b) Chen, G.H., Weng, W., Wu, D., Wu, C., Lu, J., Wang, P., and Chen, X. (2004) *Carbon*, **42**, 753–759.
18. Celzard, A., Mareche, J.F., and Furdin, G. (2005) *Prog. Mater. Sci.*, **50**, 93–179.
19. Fukushima, H. (2003) Graphite reinforcements in polymer nanocomposites. PhD thesis. Michigan State University, East Lansing, MI.
20. Xiang, J.L. and Drzal, L.T. (2011) *Carbon*, **49**, 773–778.
21. Staudenmaier, L. (1898) *Ber. Dtsch. Chem. Ges.*, **31**, 1481–1487.
22. Hummers, W.S. and Offeman, R.E. (1958) *J. Am. Chem. Soc.*, **80**, 1339–1341.
23. Hontoria-Lucas, C., Lopez-Peinado, A.J., Lopez-Gonzalez, J., Rojas-Cervantes, M.L., and Martin-Aranda, R.M. (1995) *Carbon*, **33**, 1585–1592.
24. (a) He, H.Y., Klinowski, J., Forster, M., and Lerf, A. (1998) *Chem. Phys. Lett.*, **287**, 53–56; (b) Lerf, A., He, H.Y., Forster, M., and Klinowski, J. (1998) *J. Phys. Chem. B*, **102**, 4477–4482.
25. Park, S. and Ruoff, R.S. (2009) *Nat. Nanotechnol.*, **4**, 217–224.
26. Paredes, J.I., Villar-Rodil, S., Martinez-Alonso, A., and Tascon, J.M. (2008) *Langmuir*, **24**, 10560–10564.
27. Stankovich, S., Dikin, D.A., Piner, R.D., Kohlhaas, K.A., Kleinhammes, A., Jia, Y., Wu, Y., Nguyen, S.T., and Ruoff, R.S. (2007) *Carbon*, **45**, 1558–1565.
28. Eda, G. and Chhowalla, M. (2010) *Adv. Mater.*, **22**, 2392–2415.
29. Mattevi, C., Eda, G., Agnoli, S., Miller, S., Mkhoyan, K.A., Celik, O., Mastrogiovanni, D., Granozzi, C., Garfunkel, E., and Chhowalla, M. (2009) *Adv. Funct. Mater.*, **19**, 2577–2583.
30. Shin, H.J., Kim, K.K., Benayad, A., Yoon, S.M., Park, H.K., Jung, I.S., Jin, M.H., Jeong, H.K., Kim, J.M., Choi, J.Y., and Lee, Y.H. (2009) *Adv. Funct. Mater.*, **19**, 1987–1992.
31. Eda, G., Fanchini, G., and Chhowalla, M. (2008) *Nat. Nanotechnol.*, **3**, 270–274.
32. (a) Schniepp, H.C., Li, J.L., McAllister, M.J., Sai, H., Herrera-Alonso, M., Adamson, D.H., Car, R., Prud'homme, R.K., Saville, D.A., and Aksay, I.A. (2006) *J. Phys. Chem. B*, **110**, 8535–8539; (b) McAllister, M.J., Li, J.L., Adamson, D.H., Schniepp, H.C., Abdala, A.A., Liu, J., Herrera-Alonso, M., Milius, D.L., Car, R., Prud'homme, R.K., and Aksay, I.A. (2007) *Chem. Mater.*, **19**, 4396–4404.
33. Rafiee, M.A., Rafiee, J., Wang, Z., Song, H., Yu, Z.Z., and Koratkar, N. (2009) *ACS Nano*, **3**, 3884–3890.
34. Erickson, K., Erni, R., Lee, Z.H., Alem, N., Gannett, W., and Zettl, A. (2010) *Adv. Mater.*, **22**, 4467–4472.
35. Odom, T.W., Huang, J.L., Kim, P., and Lieber, C.M. (2000) *J. Phys. Chem. B*, **104**, 2794–2809.
36. Schnorr, J.M. and Swager, T.M. (2011) *Chem. Mater.*, **23**, 646–657.
37. Ebbesen, T.W. (1994) *Annu. Rev. Mater. Sci.*, **24**, 235–264.
38. Bystrzejewski, M., Rummeli, M.H., lange, H., Huczko, A., Baranowski, P., Gemming, T., and Pichler, T. (2008) *J. Nanosci. Nanotechnol.*, **8**, 6178–6186.

39. Ebbesen, T.W. and Ajayan, P.M. (1992) *Nature*, **358**, 220–222.
40. Zhao, X., Ohkohchi, M., Inoue, S., Suzuki, T., Kadoya, T., and Ando, Y. (2006) *Diamond Relat. Mater.*, **15**, 1098–1102.
41. Hou, P.X., Liu, C., and Cheng, H.M. (2008) *Carbon*, **46**, 2003–2025.
42. Zhao, X., Liu, Y., Inoue, S., Suzuki, T., Jones, R.O., and Ando, Y. (2004) *Phys. Rev. Lett.*, **92**, 125502.
43. Ando, Y. (2010) *J. Nanosci. Nanotechnol.*, **10**, 3726–3738.
44. Ando, Y., Zhao, X., Sugai, T., and Kumar, M. (2004) *Mater. Today*, **7**, 22–29.
45. Yudasaka, M., Yamada, R., Sensui, N., Wilkins, T., Ichihashi, T., and Iijima, S. (1999) *J. Phys. Chem. B*, **103**, 6224–6229.
46. Kusaba, M. and Tsunawaki, Y. (2006) *Thin Solid Films*, **506–507**, 255–258.
47. Muñoz, E., Maser, W.K., Benito, A.M., Martínez, M.T., de la Fuente, G.F., Righi, A., Anglaret, E., and Sauvajol, J.L. (2001) *Synth. Met.*, **121**, 1193–1194.
48. Rümmeli, M.H., Kramberger, C., Loeffler, M., Jost, O., Bystrzejewski, M., Gruneis, A., Gemming, T., Pompe, W., Buechner, B., and Pichler, T. (2007) *J. Phys. Chem. B*, **111**, 8234–8241.
49. Thess, A., Lee, R., Nikolaev, P., Dai, H., Petit, P., Robert, J., Xu, C., Lee, Y.H., Kim, S.G., Rinzler, A.G., Colbert, D.T., Scuseria, G.E., Tomanek, D., Fischer, J.E., and Smalley, R.E. (1996) *Science*, **273**, 483–487.
50. Yao, Y., Falk, L.K., Morjan, R.E., Nerushev, O.A., and Campbell, E.E. (2007) *Carbon*, **45**, 2065–2071.
51. Nessim, G.D., Seita, M., Plata, D.L., O'Brien, K.P., Hart, A.J., Reddy, C.M., Gschwend, P.M., and Thompson, C.V. (2011) *Carbon*, **49**, 804–810.
52. Bruden, B.A., Cassell, A.M., Ye, Q., and Meyyappan, M. (2003) *J. Appl. Phys.*, **94**, 4070–4078.
53. Jung, K.H., Shin, Y.S., Boo, J.H., Kim, Y.J., and Hong, B.Y. (2006) *Thin Solid Films*, **501**, 238–242.
54. Maschmann, M.R., Amama, P.B., Goyal, A., Iqbal, Z., and Fisher, T.S. (2006) *Carbon*, **44**, 10–18.
55. Fang, T.H., Chang, W.J., Lu, D.M., and Lien, W.C. (2007) *Appl. Surf. Sci.*, **253**, 8749–8753.
56. Garg, R.K., Kim, S.S., Hash, D.B., Gore, J.P., and Fisher, T.S. (2008) *J. Nanosci. Nanotechnol.*, **8**, 3068–3076.
57. Eres, G., Puretzky, A.A., Geohegan, D.B., and Cui, H. (2004) *Appl. Phys. Lett.*, **84**, 1759–1761.
58. Maschmann, M.R., Amama, P.B., Goyal, A., Iqbal, Z., and Fisher, T.S. (2006) *Carbon*, **44**, 2758–2763.
59. Hayashi, Y., Fukumura, T., Odani, K., Matsuba, T., and Utsonomiya, R. (2010) *Thin Solid Films*, **518**, 3506–3508.
60. Hiramatsu, M., Deguchi, T., Nagao, H., and Hori, M. (2007) *Diamond Relat. Mater.*, **16**, 1126–1130.
61. Levchenko, I., Ostrikov, K., and Keidar, M. (2008) *J. Nanosci. Nanotechnol.*, **8**, 6112–6122.
62. Meyyappan, M., Delzeit, L., Cassell, A., and Hash, D. (2003) *Plasma Sources Sci. Technol.*, **12**, 205–216.
63. Wong, W.K., Lee, C.S., and Lee, S.T. (2006) *J. Appl. Phys.*, **97**, 1871354 (6 pp).
64. Choi, K.S., Cho, Y.S., Hong, S.Y., Park, J.B., and Kim, D.J. (2001) *J. Eur. Ceram. Soc.*, **21**, 2095–2098.
65. Harutyunyan, A.R., Tokune, T., and Mora, E. (2005) *Appl. Phys. Lett.*, **87**, 051919 (3 pp).
66. Hu, J., Wang, O., and Lieber, C.M. (1999) *Acc. Chem. Res.*, **32**, 435–445.
67. Wu, Y. and Yang, P. (2001) *J. Am. Chem. Soc.*, **123**, 3165–3166.
68. Li, H., Shi, C., Du, X., He, C., Li, J., and Zhao, N. (2008) *Mater. Lett.*, **62**, 1472–1475.
69. Lee, C.J. and Park, J. (2001) *J. Phys. Chem. B*, **105**, 2365–2368.
70. Chen, X., Wang, R.M., Xu, J., and Yu, D.P. (2004) *Micron*, **35**, 455–460.
71. Nikolaev, P., Bronikowski, M.J., Bradley, R.K., Rohmund, F., Colbert, D.T., Smith, K.A., and Smalley, R.E. (1999) *Chem. Phys. Lett.*, **313**, 91–97.
72. Park, Y.S., Choi, Y.C., Kim, K.S., Chung, D.C., Bae, D.J., An, K.H., Lim, S.C., Zhu, X.Y., and Lee, Y.H. (2001) *Carbon*, **39**, 655–661.

73. Chiang, I.W., Brinson, B.E., Huang, A.Y., Willis, P.A., Bronikowski, M.J., Margrave, J.L., Smalley, R.E., and Hauge, R.H. (2001) *J. Phys. Chem. B*, **105**, 8297–8301.
74. Ebbesen, T.W., Ajayan, P.M., Hiura, H., and Tanigaki, K. (1994) *Nature*, **367**, 319–319.
75. Xu, Y.Q., Peng, H.Q., Hauge, R.H., and Smalley, R.E. (2005) *Nano Lett.*, **5**, 163–168.
76. Harutyunyan, A.R., Pradhan, B.K., Chang, J., Chen, G., and Eklund, P.C. (2002) *J. Phys. Chem. B*, **106**, 8671–8675.
77. Li, Y., Zhang, X.B., Luo, J.H., Huang, W.Z., Cheng, J.P., Luo, Z.Q., Li, T., Xu, G.L., Ke, X.X., Li, L., and Geise, H.J. (2004) *Nanotechnology*, **15**, 1645–1649.
78. Nepal, D., Kim, D.S., and Geckeler, K.E. (2005) *Carbon*, **43**, 660–662.
79. Ko, F.H., Lee, C.Y., Ko, C.J., and Chu, T.C. (2005) *Carbon*, **43**, 727–733.
80. Lu, K., Lago, R.M., Chen, Y.K., Green, M.L., Harris, P.J., and Tsang, S.C. (1996) *Carbon*, **34**, 814–816.
81. Andrews, R., Jacques, D., Qian, D., and Dickey, E.C. (2001) *Carbon*, **39**, 1681–1687.
82. Bonard, J.M., Stora, T., Salvetat, J.P., Maier, F., Stockli, T., Duschl, C., Forro, L., de Heer, W.A., and Chatelain, A. (1997) *Adv. Mater.*, **9**, 827–831.
83. Bandow, S., Rao, A.M., Williams, K.A., Thess, A., Smalley, R.E., and Eklund, P.C. (1999) *J. Phys. Chem. B*, **101**, 8839–8842.
84. Hu, H., Yu, A., Kim, E., Zhao, B., Itkis, M.E., Bekyarova, E., and Haddon, R.C. (2005) *J. Phys. Chem. B*, **109**, 11520–11524.
85. Yu, A., Bekyarova, E., Itkis, M.E., Fakhrutdinov, D., Webster, R., and Haddon, R.C. (2006) *J. Am. Chem. Soc.*, **128**, 9902–9908.
86. Vichchulada, P., Shim, J., and Lay, M.D. (2008) *J. Phys. Chem. C*, **112**, 19186–19192.
87. Mattia, D., Rossi, M.P., Kim, B.M., Korneva, G., Bau, H.H., and Gogotsi, Y. (2006) *J. Phys. Chem. B*, **110**, 9850–9855.
88. Yudasaka, M., Ichihashi, T., Kasuya, D., Kataura, H., and Iijima, S. (2003) *Carbon*, **41**, 1273–1280.
89. Nikolaev, P., Thess, A., Rinzler, A.G., Colbert, D.T., and Smalley, R.E. (1997) *Chem. Phys. Lett.*, **266**, 422.
90. Yudasaka, M., Kataura, H., Ichihashi, T., Qin, L.C., Kar, S., and Iijima, S. (2001) *Nano Lett.*, **1**, 487–489.
91. Arepalli, S., Nilolaev, P., Gorelik, O., Hadjiev, V.G., Holmes, W., Files, B., and Yowell, L. (2004) *Carbon*, **42**, 1783–1791.
92. Schweinberger, F.F., and Plath, A.M. (2011) *J. Phys. Conf. Ser.*, **304**, 012087 (10 pp).
93. Ferrari, A.C. and Robertson, J. (2000) *Phys. Rev. B*, **61**, 14095–140096.
94. Dresselhaus, M.S., Dresselhaus, G., Jorio, A., and Souza Filho, A.G. (2002) *Carbon*, **40**, 2043–2061.
95. Dillon, A.C., Parilla, P.A., Alleman, J.L., Gennett, T., Jones, K.M., and Heben, M.J. (2005) *Chem. Phys. Lett.*, **401**, 522–528.
96. Dillon, A.C., Yudasaka, M., and Dresselhaus, M.S. (2004) *J. Nanosci. Nanotechnol.*, **4**, 691–703.
97. Rao, A.M., Jorio, A., Pimenta, M.A., Dantas, M.S., Saito, R., Dresselhaus, G., and Dresselhaus, M.S. (2000) *Phys. Rev. Lett.*, **84**, 1820–1823.
98. Vigolo, B., Herold, C., Mareche, J.F., Ghanbaja, J., Gulas, M., Normand, F., Almairac, R., Alvarez, L., and Bantignies, J.L. (2010) *Carbon*, **48**, 949–963.
99. Wang, Y., Wu, J., and Fei, F. (2003) *Carbon*, **34**, 814–816.
100. Boccaleri, E., Arrais, A., Frache, A., Gianelli, W., Fino, P., and Camino, G. (2006) *Mater. Sci. Eng. B*, **131**, 72–82.
101. Ryabenko, A.G., Dorofeeva, T.V., and Zverera, G.I. (2004) *Carbon*, **42**, 1523–1535.
102. Itkis, M.E., Perea, D.E., Jung, R., Niyoki, S., and Haddon, R.C. (2005) *J. Am. Chem. Soc.*, **27**, 3439–3448.
103. Melechko, A.V., Merkulov, V.I., McKnight, T.E., Guillorn, M.A., Klein, K.L., Lowndes, D.H., and Simpson, M.L. (2005) *J. Appl. Phys.*, **97**, 41301 (39 pp)

104. Klein, K.L., Melechko, A.V., McKnight, T.E., Retterer, S.T., Rack, P.D., Fowlkes, J.D., Joy, D.C., and Simpson, M.L. (2008) *J. Appl. Phys.*, **103**, 061301 (6 pp).
105. Tibbets, G.G., Lake, M.L., Strong, K.L., and Rice, B.P. (2007) *Compos. Sci. Technol.*, **67**, 1709–1718.
106. Memon, M.O., Haillot, S., and Lafdi, K. (2011) *Carbon*, **49**, 3820–3828.
107. Miyagawa, H., Rich, M.J., and Drzal, L.T. (2006) *Thermochim. Acta*, **442**, 67–73.
108. Howe, J.Y., Tibbetts, G.G., Kwag, C., and Lake, M.L. (2001) *J. Mater. Res.*, **21**, 2646–2652.
109. Zhu, H., Li, X., Ci, L., Xu, C., Wu, D., Mao, Z. et al. (2003) *Mater. Chem. Phys.*, **78** (3), 670–675.
110. Endo, M., Kim, Y.A., Hayashi, T., Yanagisawa, T., Muramatsu, H., Ezaka, M., Terrones, H., Terrones, M., and Dresselhaus, M.S. (2003) *Carbon*, **41**, 1941–1947.
111. http://www.apsci.com/.
112. Odegard, G.M., Gates, T.S., Nicholson, L.M., and Wise, K.E. (2002) *Compos. Sci. Technol.*, **62**, 1869–1880.
113. Sakhaee-Pour A. (2009) *Solid State Commun.*, **149**, 91–95.
114. Liu, F., Ming, P.B., and Li, J. (2007) *Phys. Rev. B*, **76**, 064120 (7 pp).
115. Lu, Q. and Bhattacharya, B. (2005) *Eng. Fract. Mech.*, **72**, 2037–2071.
116. Tsai, J.L. and Tu, J.F. (2010) *Mater. Des.*, **31**, 194–199.
117. Jiang, J.W., Wang, J.S., and Li, B.W. (2009) *Phys. Rev. B*, **80**, 113405 (4 pp).
118. Khare, R., Mielke, S.L., Paci, J.T., Zhang, S., Ballarini, R., Schatz, G.C., and Belytschko, T. (2007) *Phys. Rev. B*, **75**, 075412 (12 pp).
119. Zheng, Q., Geng, Y., Wang, S., Li, Z., and Kim, J.K. (2010) *Carbon*, **48**, 4315–4322.
120. Lee, C.G., Wei, X.D., Kysar, J.W., and Hone, J. (2008) *Science*, **321**, 385–388.
121. Lee, C.G., Wei, X.D., Li, Q.Y., Carpick, R., Kysar, J.W., and Hone, J. (2009) *Phys. Status Solidi B*, **242**, 2562–2567.
122. Gomez-Navarro, C., Burghard, M., and Kern, K. (2008) *Nano Lett.*, **8**, 2045–2049.
123. Park, S.J., Lee, K.S., Bozoklu, G., Cai, W.W., Nguyen, S.B., and Ruoff, R.S. (2008) *ACS Nano*, **2**, 572–578.
124. http://www.xgsciences.com/.
125. Bolotin, K.I., Sikes, I.J., Jiang, Z., Klima, M., Fudenberg, G., Hone, J., Kim, P., and Stormer, H.L. (2008) *Solid State Commun.*, **146**, 351–355.
126. Du, X., Skachko, I., Barker, A., and Andrei, E.Y. (2008) *Nat. Nanotechnol.*, **3**, 491–495.
127. Chen, H., Muller, M.B., Gilmore, K.J., Wallace, G.D., and Li, D. (2008) *Adv. Mater.*, **20**, 3557–3561.
128. Park, S.J., An, J.H., Piner, R.D., Jung, I., Yang, D.X., Velamakanni, A., Nguyen, S.B., and Ruoff, R.S. (2008) *Chem. Mater.*, **20**, 6592–6594.
129. Schwamb, T., Burg, B.R., Schirmer, N.C., and Poulikakos, D. (2009) *Nanotechnology*, **20**, 405704 (5 pp).
130. Baladin, A.A., Ghosh, S., Bao, W., Calizo, I., Teweldebrhan, D., Miao, F., and Lau, C.N. (2008) *Nano Lett.*, **8**, 902–907.
131. Ghosh, S., Calizo, I., Teweldebrhan, D., Pokatilov, E.P., Nika, D.L., Baladin, A.A., Bao, W., Miao, F., and Lau, C.N. (2008) *Appl. Phys. Lett.*, **92**, 151911 (3 pp).
132. Seol, J.H., Jo, I.S., Moore, A.L., Lindsay, L., Aitken, Z.H., Pettes, M.T., Li, X.S., Yao, Z., Broido, D., Mingo, N., Ruoff, R.S., and Shi, L. (2010) *Science*, **328**, 213–216.
133. Yakobson, B.I., Brabec, C.J., and Bernholc, J. (1996) *Phys. Rev. Lett.*, **76**, 2511–2514.
134. Robertson, D.H., Brenner, D.W., and Mintmire, J.W. (1992) *Phys. Rev. B*, **45**, 12592–12595.
135. Yao, N. and Lordi, V. (1998) *J. Appl. Phys.*, **84**, 1939–1943.
136. Hernandez, E., Goze, C., Bernier, P., and Rubio, A. (1998) *Phys. Rev. Lett.*, **80**, 4502–4505.
137. Haskins, R.W., Maier, R.S., Ebeling, R.M., Marsh, C.P., Majure, D.L., Bednar, A.J., Welch, C.R., and Barker, B.C. (2007) *J. Chem. Phys.*, **127**, 074708 (10 pp).
138. Qin, Z., Qin, Q.H., and Feng, X.Q. (2008) *Phys. Lett. A*, **372**, 6661–6666.

139. Treacy, M.M., Ebbesen, T.W., and Gibson, J.W. (1996) *Nature*, **38**, 678–680.
140. Krishnan, A., Dujardin, E., and Ebbesen, T.W. (1998) *Phys. Rev. B*, **58**, 14013–14019.
141. Wong, E.W., Sheehan, P.E., and Lieber, C. (1997) *Science*, **277**, 1971–1975.
142. Salvetat, J.P., Kulik, A.J., Bonard, J.M., Briggs, A.D., Stockli, T., Metenier, K., Bonnamy, S., Beguin, F., Burnham, N.A., and Forro, L. (1999) *Adv. Mater.*, **11**, 61–65.
143. Yu, M.F., Lourie, O., Dyer, M.J., Kelly, T.F., and Ruoff, S. (2000) *Science*, **287**, 637–641.
144. Yu, M.F., Yakobson, B.I., and Rodney, S. (2000) *J. Phys. Chem. B*, **104**, 8764–8767.
145. Yu, M.F., Files, B.S., Arepalli, S., and Ruoff, S. (2000) *Phys. Rev. Lett.*, **84**, 5552–5555.
146. Demczyk, B.G., Wang, Y.M., Cumings, J., Hetman, M., Han, W., Zettl, A., and Ritchie, R.O. (2002) *Mater. Sci. Eng. A*, **334**, 173–178.
147. Lourie, O. and Wagner, H.D. (1998) *J. Mater. Res.*, **13**, 2418–2422.
148. Cooper, C.A., Young, R.J., and Harsall, M. (2001) *Composites Part A*, **32**, 401–411.
149. Yu, M.F. (2004) *J. Eng. Mater. Technol.*, **126**, 271–278.
150. Iijima, S., Brabec, C., Maiti, A., and Bernholc, J. (1996) *J. Chem. Phys.*, **104**, 2089–2092.
151. Falvo, M.R., Clary, G.J., Taylor, R.M., Chi, V., Brooks, J.P., Wasburn, S., and Superfine, B. (1997) *Nature*, **389**, 582–584.
152. Yakobson, B.I., Campbell, M.P., Brabec, C.J., and Bernholc, J. (1997) *Comput. Mater. Sci.*, **8**, 341–348.
153. Dereli, G. and Ozdogan, C. (2003) *Phys. Rev. B*, **67**, 0354161–0354166.
154. Huang, J.Y., Chen, S., Wang, Z.Q., Kempa, K., Wang, Y.M., Jo, S.H., Chen, G., Dresselhaus, M.S., and Ren, Z.F. (2006) *Nature*, **439**, 281–281.
155. Fischer, J.E., Dai, H., Thess, A., Lee, R., Hajani, N.M., Dehaas, D.L., and Smalley, R.E. (1997) *Phys. Rev. B*, **55**, R4921–R4924.
156. Bockrath, M., Cobden, D.H., McEun, P.L., Chopra, N.G., Zettl, A., Thess, A., and Smalley, R.E. (1997) *Science*, **275**, 1922–1925.
157. Ebbesen, T.W., Lezec, H.J., Hiura, H., Bennett, J.W., Ghaemi, H.F., and Thio, T. (1996) *Nature*, **382**, 54–56.
158. Sanvito, S., Kwon, Y.K., Tomanek, D., and Lambert, C.L. (2000) *Phys. Rev. Lett.*, **84**, 1974–1977.
159. Frank, S., Poncharal, P., Wang, Z.I., and de Heer, W.A. (1998) *Science*, **280**, 1744–1746.
160. Berber, S., Kwon, Y.K., and Tománek, D. (2000) *Phys. Rev. Lett.*, **84**, 4613–4616.
161. Osman, M.A. and Srivastave, D. (2001) *Nanotechnology*, **12**, 21–24.
162. Schwab, K., Henriksen, E.A., Worlock, J.M., and Roukes, M.L. (2000) *Nature*, **404**, 974–976.
163. Mingo, N. and Broido, D.A. (2005) *Phys. Rev. Lett.*, **95**, 096105 (4 pp).
164. Hone, J., Whitney, M., Piskoti, C., and Zettl, A. (1999) *Phys. Rev. B*, **59**, R2514–R2516.
165. Kim, P., Shi, L., Majumdar, A., and McEuen, P.L. (2001) *Phys. Rev. Lett.*, **87**, 215502 (4 pp).
166. Fuji, M., Zhang, X., Xie, H., Ago, H., Takahashi, K., Ikuta, T., Abe, H., and Shimizu, T. (2005) *Phys. Rev. Lett.*, **95**, 065502 (4 pp).
167. Pop, E., Mann, D., Wang, Q., Goodson, K., and Dai, H. (2006) *Nano Lett.*, **6**, 96–199.
168. Yi, W., Lu, L., Zhang, D.L., Pan, Z.W., and Xie, S.S. (1999) *Phys. Rev. B*, **59**, R9015–R9018.
169. Prasher, R. (2008) *Phys. Rev. B*, **77**, 075424 (11 pp).
170. Ozkan, T., Naraghi, M., and Chasiotis, I. (2010) *Carbon*, **48**, 239–244.
171. Tjong, S.C., Meng, Y.Z., and Hay, A.S. (2002) *Chem. Mater.*, **14**, 44–51.
172. Tjong, S.C. and Mai, Y.W. (2010) *Physical Properties and Applications of Polymer Nanocomposites*, Woodhead Publishing, Oxford.
173. Vanderhart, D.L., Asano, A., and Gilman, J.W. (2001) *Chem. Mater.*, **13**, 3796–3809.
174. Bourrat, X. (1993) *Carbon*, **31**, 288–300.

175. Donnet, J.B. (1994) *Carbon*, **32**, 1305–1310.
176. *http://carbon-black.org/*.
177. Huang, J.C. (2002) *Adv. Polym. Technol.*, **21**, 299–313.
178. ASTM (2009) D2414-09a. *Standard Test Method for Carbon Black: Oil Absorption Number (OAN)*, American Society for Testing and Materials, West Conshohocken, PA.
179. Balberg, I. (2002) *Carbon*, **40**, 139–143.
180. Compton, O.C. and Nguyen, S.B. (2010) *Small*, **6**, 711–723.
181. Zhamu, A. and Jang, B.Z. (2011) Process for producing nano-scaled graphene platelet nanocomposites electrodes for supercapacitors. U.S. patent 7875219.
182. Park, C., Watson, K.A., Ounaies, Z., Connell, J.W., Smith, J.G., and Harrison, J.S. (2009) Electrically conductive, optically transparent polymer/carbon nanotube composites and process for preparation thereof. U.S. patent 7588699.
183. Harmont, J.P. and Clayton, L.M. (2008) Polymer/carbon nanotube composites, methods of use and methods of synthesis thereof. U.S. patent 7399794.
184. Zhao, Y.P., Chen, Y.C., Zhang, X.C., Raravikar, N.R., Ajayan, P.M., Lu, T.M., Wang, G.C., and Schadler, L.S. (2004) Ultrafast all-optical switch using carbon nanotube polymer composites. U.S. patent 6782154.
185. Terai, M., Hasegawa, M., Okusawa, M., Otani, S., and Oshima, C. (1998) *Appl. Surf. Sci.*, **130–132**, 876–882.
186. Geohegan, D.B., Ivanov, I.N., Puretzky, A.A., Jesse, S., Hu, B., Garrett, M., and Zhao, B. (2011) Transparent conductive nanocomposites. U.S. patent 7923922.

2
Preparation of Polymer Nanocomposites

2.1
Overview

Carbonaceous nanofillers with large surface areas and aspect ratios generally disperse poorly in organic solvents and polymers. Pristine carbon nanotubes (CNTs) have a high tendency for forming bundles or ropes because of strong van der Waals interactions among them. This leads to weak interfacial interactions between the nanotubes and the polymer matrix. As a result, CNTs can be pulled out easily during mechanical deformation. The mechanical and physical properties of polymer composites filled with carbonaceous nanofillers rely on the attainment of homogeneous dispersion of those fillers in the polymer matrix and the establishment of strong interfacial filler–matrix interaction. The dispersion of carbonaceous fillers in solvents and polymers can be improved by means of mechanical mixing and chemical functionalization. The former route includes ultrasonication and high-shear mixing. Those techniques can induce structural damage to carbonaceous nanofillers, especially one-dimensional CNTs and carbon nanofibers (CNFs) by reducing their aspect ratios markedly. Therefore, chemical functionalization of nanofillers is found to be more effective to improve their dispersion in organic solvents and polymers homogeneously. This, in turn, enhances interfacial interactions between the polymer matrix and the nanofillers. In general, chemical modification of carbonaceous nanofillers can be achieved by covalent attachment of functional groups onto the filler surfaces, or through noncovalent adsorption or wrapping of functional moieties. With the progress and experience in surface modification of nanofillers and materials processing techniques, novel polymer nanocomposites with excellent mechanical strength and biocompatibility can be fabricated accordingly.

2.2
Dispersion of Nanofillers

2.2.1
Surface Modification of Graphene Oxide

As discussed in Chapter 1, graphene oxide (GO) is an attractive filler for reinforcing polymer composites because of its low cost, ease of preparation and potential scalability. The GO precursor contains hydroxyl and epoxide functional groups on their basal planes, as well as carbonyl and carboxyl groups at its sheet edges (Figure 1.5). These functional groups make GO hydrophilic, allowing its exfoliation in water easily. Therefore, GO can form nanocomposites readily with water-soluble polymers such as poly(vinyl alcohol) (PVA) and poly(ethylene oxide) (PEO) [1–5]. However, hydrophylic GO is poorly exfoliated in nonaqueous solvents, for example polar aprotic solvents such as N,N-dimethylformamide (DMF), dimethyl sulfoxide (DMSO) and tetrahydrofuran (THF). Such solvents cannot penetrate the interlayer spaces of GO, thus preventing its exfoliation. Therefore, surface modification of GOs is needed for exfoliating in aprotic solvents and for reacting with nonpolar polymers such as polyethylene (PE) and polypropylene (PP).

GO can be chemically functionalized with long alkyl chain molecules, isocyanates and polyallylamine to establish covalent linkages. Alkylation is the simplest way to make GOs compatible with organic solvents [6–8]. For example, acidic functional groups of GO can react with long chain octadecylamine (ODA) [$CH_3(CH_2)_{17}NH_2$] to yield graphene sheets with a large solubility in organic solvents [6]. A typical reaction between GO and ODA in the presence of DMF and N,N-dicyclohexylcarbodiimide (DCC) at 90 °C under a nitrogen atmosphere is given by Cao et al. [7]:

$$\text{Graphene}-\text{C}(=\text{O})\text{OH} + NH_2(CH_2)_{17}CH_3 \xrightarrow[N_2, 90\,°C, \text{Reflux}]{\text{DMF, DCC}} \text{Graphene}-\text{C}(=\text{O})NH(CH_2)_{17}CH_3 \quad (2.1)$$

Accordingly, functionalized-GO-containing amide groups can disperse in nonpolar xylene, a solvent commonly used to dissolve nonpolar polymers. GO can also react with isocyanate to yield isocyanate-treated graphene oxide (iGO) [9]. In this case, the edge carboxyl and surface hydroxyl groups of GO react with isocyanate to form amides and carbamate esters, respectively (Figure 2.1). As a result, iGO can form stable colloidal suspensions in polar aprotic solvents such as DMF, DMSO, N-methylpyrrolidone (NMP), and hexamethylphosphoramide (HMPA) [9]. These dispersions enable GO sheets to be readily mixed with polymers, facilitating the formation of graphene-polymer composites. The addition of polyallylamine (PAA) to an aqueous suspension of GO sheets under sonication can also produce a homogeneous aqueous colloidal suspension of chemically cross-linked GO sheets

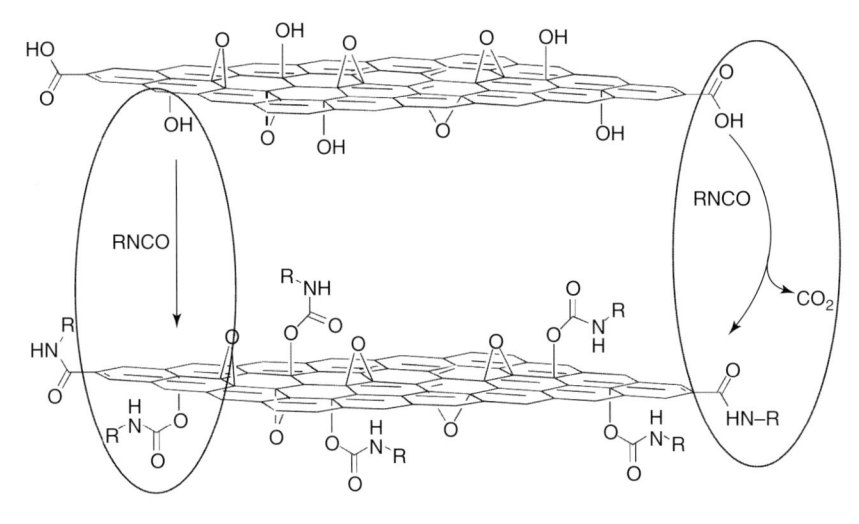

Figure 2.1 Organic isocyanates react with hydroxyl (left oval) and carboxyl groups (right oval) of graphene oxide sheets to form carbamate and amide functionalities, respectively. (Reproduced with permission from Ref. [9], Elsevier (2006).)

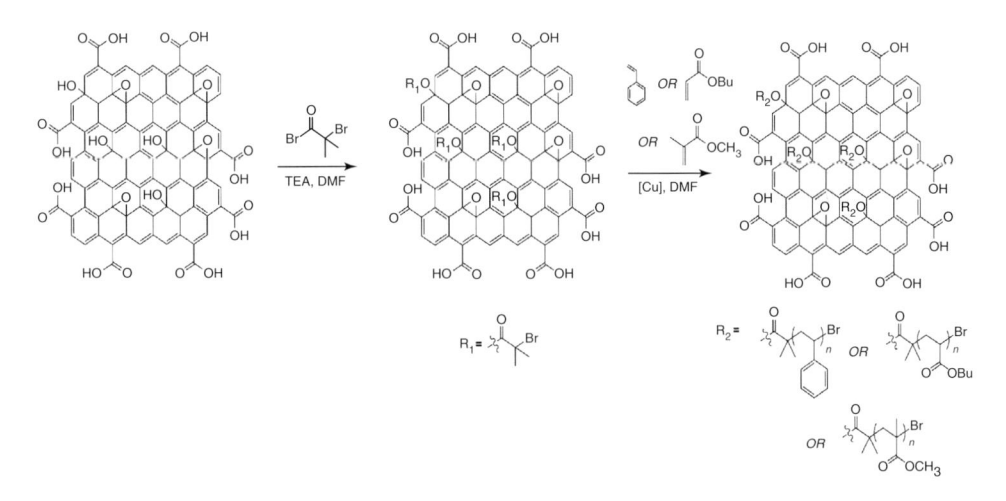

Figure 2.2 Synthesis of functionalized graphene oxide via attachment of an initiator, α-bromoisobutyryl bromide. This is followed by the polymerization of styrene, butyl acrylate and methyl methacrylate using ATRP. (Reproduced with permission from Ref. [11], Wiley-VCH (2010).)

[10]. More recently, Ruoff and coworkers [11] attached polymer brushes to the GO using a surface initiator, that is α-bromoisobutyryl bromide (BIB) in the presence of triethylamine (TEA) and DMF (Figure 2.2). This strategy allows surface functionalized GO to react with polymer monomers such as styrene, butyl acrylate and methyl methacrylate via atom transfer radical polymerization (ATRP).

In the case of thermally exfoliated GO, McAllister et al. [12] reported that thermally reduced graphenes (TRGs) disperse well in several solvents such as NMP, DMF, 1,2-dichlorobenzene (DCB), nitromethane or THF under sonication (Chapter 1, [32b]). The presence of oxygen functionalities in TRG allows it to react with water-soluble PEO [13] and polar polymers such as poly(methyl methacrylate) (PMMA), poly(acrylonitrile) and poly(acrylic acid) [14]. This favors the formation of nanocomposites having good dispersion of graphene nanosheet fillers and enhanced filler–matrix interactions.

Noncovalent functionalization of GO sheets has attracted considerable attention in recent years [15–20]. Choi et al. achieved stable dispersions of reduced graphene oxide (rGO) in various organic solvents through noncovalent functionalization with amine-terminated polystyrene (PS-NH$_2$). GO sheets can also be functionalized noncovalently using a water-soluble pyrene derivative (1-pyrenebutyrate) through the π–π interactions [16]. The pyrene derivative has a strong affinity for reacting with the basal plane of graphene to form a stable π–stacking bond [17]. Very recently, Ma's group [19] functionalized thermally expanded graphene oxide (TEGO) noncovalently through π–π stacking of pyrene (Py) with a functional segmented polymer chain. In the process, Py-Br initiator was first synthesized by adding 2-BIB drop-wise to a stirring mixture of 1-pyrenemethanol and TEA in THF, and subsequent filtration and drying treatments. The functional segmented poly(glycidyl methacrylate)-containing pyrene groups (Py-PGMA) was then prepared using ATRP (Figure 2.3). Finally, Py-PGMA was dissolved in acetone, followed by adding TEGO (weight ratio of 1 : 1) under sonication. Py-PGMA was physically absorbed on TEGO through the π–π stacking (Figure 2.4). Liu et al. modified chemically modified graphene (CMG) sheets noncovalently with poly(styrene-co-butadiene-co-styrene)

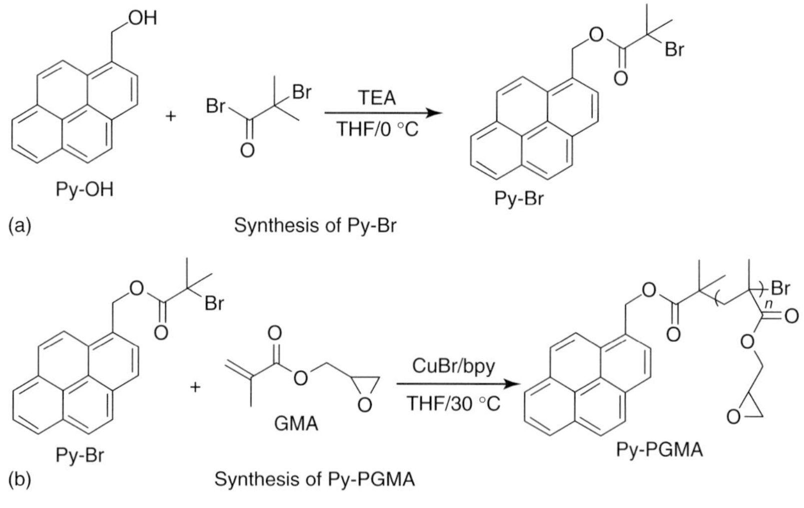

Figure 2.3 Synthesis of (a) Py-Br initiator and (b) Py-PGMA polymer. (Reproduced with permission from Ref. [19], Elsevier (2011).)

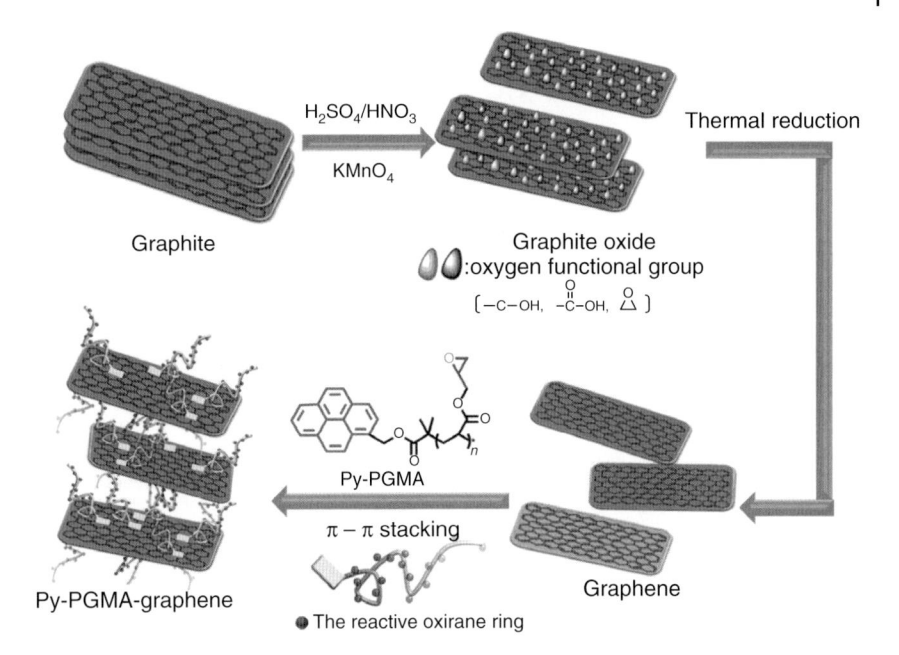

Figure 2.4 Schematic diagram showing the preparation of TEGO and its functionalization. (Reproduced with permission from Ref. [19], Elsevier (2011).)

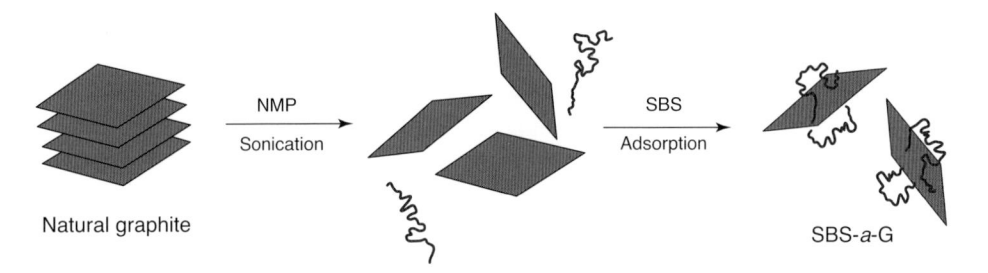

Figure 2.5 Scheme for noncovalent functionalization of exfoliated graphene sheets with SBS. (Reproduced with permission from Ref. [20], Elsevier (2011).)

(SBS) in which the graphene sheets are stabilized by SBS through π–π stacking with the PS chains (Figure 2.5) [20]. In general, GO sheets tend to re-aggregate after chemical reduction with hydrazine derivatives. Surfactants or macromolecular additions can effectively prevent re-aggregation of graphene sheets. Lomeda et al. [18] obtained surfactant-wrapped GO sheets by dispersing them in sodium dodecylbenzenesulfonate (SDBS) surfactant followed by hydranize reduction at a pH value of 10, 80 °C for 24 h. The surfactant-wrapped dispersion of rGO was then functionalized with diazonium salts (Figure 2.6). The resulting functionalized nanosheets disperse readily in polar aprotic solvents, permitting their incorporation into different polymer matrices.

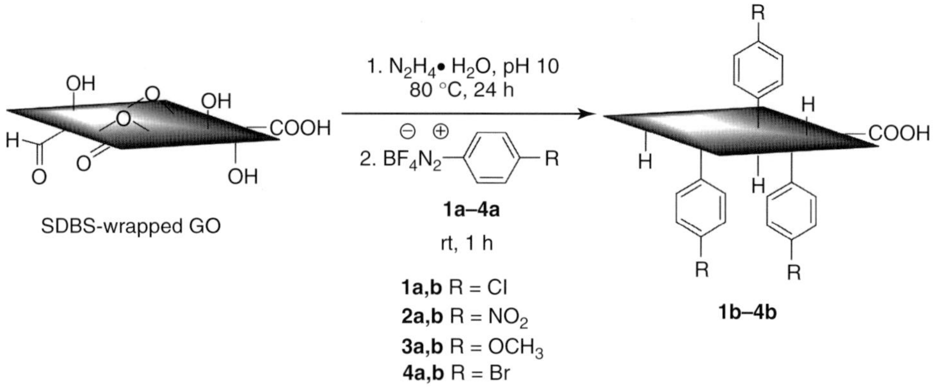

Figure 2.6 Wrapping GO with SDBS, and functionalization of intermediate SDBS-wrapped. rGO with diazonium salts. (Reproduced with permission from Ref. [18], The American Chemical Society (2008).)

2.2.2
Surface Modification of Graphene Nanoplatelet and Expanded Graphite

Because of their inherent chemical inertness, graphite nanoplatelets (GNPs) have relatively low compatibility with polymers. Hydrophobic GNPs disperse poorly in an aqueous solution and show a high tendency for forming agglomerates. In this regard, the additions of pristine GNPs to polymers always lead to low mechanical strength due to weak filler–matrix interactions. To tackle these problems, pristine GNPs are subjected to photo-oxidative treatment using ultraviolet (UV) light and ozone gas. The oxidation treatment introduces carbonyl, carboxyl and hydroxyl groups on the GNP surfaces on absorption of the UV energy [21]. This process is well known for inducing oxygen functionalities in conventional carbon fibers. The advantages of UV/ozone treatment are its simplicity, low cost and ease of operation at atmospheric pressure without using a vacuum system [22]. In the ultraviolet/ozone (UV/O$_3$) treatment, the mercury grid lamp emits two strong spectral lines at wavelengths of 185.9 and 253.7 nm. The oxidative reactions of carbon fiber take place as follows [22, 23]:

$$O_2 + h\nu \text{ (185.9 nm)} \longrightarrow 2[O] \tag{2.2}$$

$$[O] + O_2 \longrightarrow O_3 \tag{2.3}$$

$$O_3 \xrightarrow{h\nu (\lambda = 253.7 \text{ nm})} [O] + O_2 \tag{2.4}$$

$$C_S + [O]/O_3 \xrightarrow{h\nu (\lambda = 253.7 \text{ nm})} CO_x \tag{2.5}$$

where [O] represents atomic oxygen, s stands for surface and near surface, and x is the number of oxygen atoms in the functional group. By forming oxygen

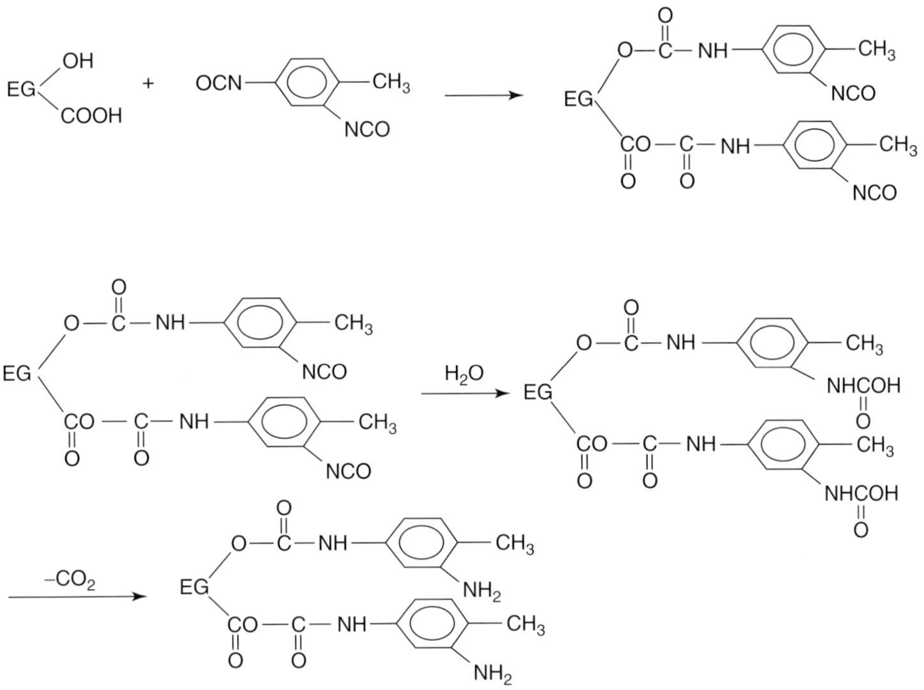

Figure 2.7 Reaction scheme between EG and isocyanate. (Reproduced with permission from Ref. [25], Elsevier (2004).)

functionalities in pristine GNPs, amino groups can be covalently functionalized to the GNPs through the addition of triethylenetetramine (TETA) [24].

Expanded graphite (EG) obtained by heating sulfuric-acid-intercalated graphite possesses oxygen functionalities on their surfaces. Amine moieties can be covalently bonded to the EG by adding toluene-2,4-diisocyanate and DMSO. The reactions between hydroxyl and carboxylic groups of the EG with isocyanate are shown in Figure 2.7 [25]. Noncovalent functionalization is closely related to the physical adsorption of surfactants and charge polymers onto the GNP surfaces. This treatment does not induce structural damage and disruption of inherent π-network of the GNPs, thereby preserving the integrity of nanoplatelets. A variety of surfactants and polyelectrolytes have been used to improve the dispersion of GNPs and their interfacial interactions with the polymer matrix [26].

2.2.3
Functionalization of CNTs and CNFs

In general, the hexagonal lattice of the CNTs is not perfect, and contains many defects that are induced during the nanotube synthesis, purification, and mechanical deformation [27]. These defects include pentagon–heptagon (5–7) pairs, commonly known as Stone–Wales (S-W) defects, and vacancies [28]. The five-

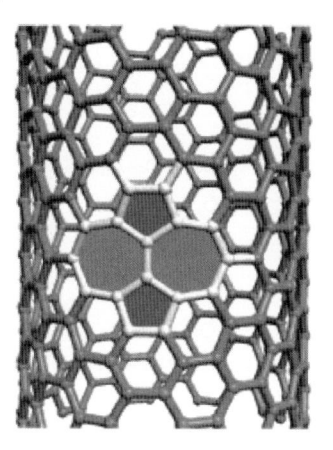

Figure 2.8 Formation of (5,7,7,5) pairs by bond rotation of SWNT. (Reproduced with permission from Ref. [27], Elsevier (2010).)

or seven-membered rings in the carbon lattice produces a bend in the tube [29]. The S-W defects are created by rotating a C–C bond by 90° within four adjacent hexagonal rings into two pentagons and two heptagons (Figure 2.8). The vacancies are generated by the removal of a single atom from the surface of the nanotube sidewall. Purification of a single-walled carbon nanotube (SWNT) involves the removal of its metal catalysts and amorphous carbon on immersion in strong oxidizing acids. This oxidative treatment introduces carboxyl groups at the nanotube open end (Figure 2.9). Moreover, the introduction of the carboxyl groups at the nanotube sidewall creates holes behind. Covalent functionalization tends to disrupt the π-electron conjugation of the CNTs, leading to deterioration in their electrical and thermal properties. The strategies for chemically functionalizing SWNTs generally fall into five categories including (i) covalent sidewall functionalization, (ii) covalent functionalization at defect sites, (iii) noncovalent surfactant adsorption, (iv) noncovalent polymer wrapping and (v) incorporation of guest molecules into the nanotube (Figure 2.10).

2.2.3.1 Covalent Functionalization

Acid oxidative treatment of CNTs results in the opening of the tube tip and shortening of the tube length. The attachment of functional groups to the surface of CNTs generally occurs at the nanotube tips and surface defect sites created by the oxidative treatment. In the latter case, the functional groups anchor at defective carbon sites in which the hybridization changing from sp^2 to sp^3. This leads to a loss of the π-conjugation network. The carboxylic groups serve as the precursors for further chemical reactions with molecules to form covalent bonding with the nanotube. Thus, functionalized nanotubes become soluble in many solvents, and this in turn improves the compatibility of CNTs with polymers. The carboxylic acid groups of the SWNTs can be converted to highly reactive acyl chloride by reacting with thionyl chloride ($SOCl_2$). Thus, polymer chains terminated with amino (RNH_2) or hydroxyl (ROH) moieties can react with acyl chloride groups via amidation or esterification reactions [30–32]. In the presence of bifunctional amines, pendant functional groups can attach to other moieties such as metal

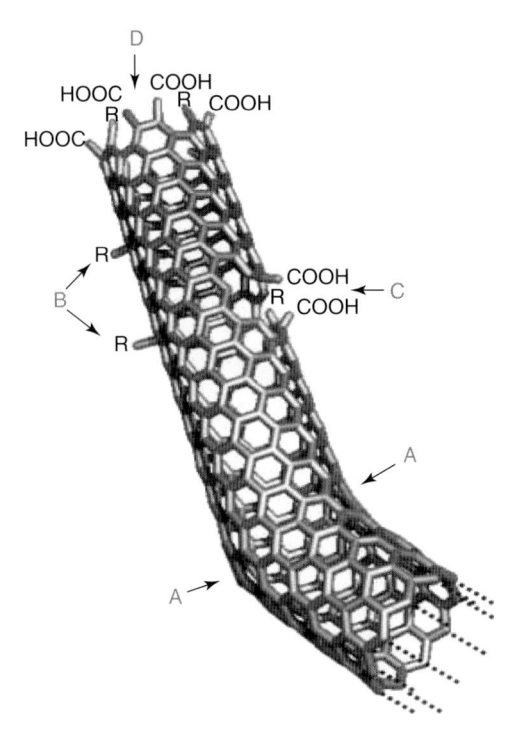

Figure 2.9 Structural defects in an SWNT. (A) Five- or seven-membered rings in the C lattice, (B) sp^3-hybridized defects (R = H and OH), (C) C framework damaged by oxidative treatment, leaving a hole lined with carboxyl groups and (D) carboxyl groups at open end of the nanotube. (Reproduced with permission from Ref. [29], Wiley-VCH (2002).)

colloids effectively (Figure 2.11) [33]. For some cases, covalent linkages can be established by zwitterionic interactions between carboxyl groups of the nanotube and amine groups of the polymer.

Covalent sidewall functionalization, which does not require previous oxidation of the nanotubes, opens up new frontiers in functionalization chemistry. These include alkylation, ozonation, cycloaddition, nucleophilic addition, fluorination and so on (Figure 2.12). The sidewall functionalization improves the solubility of SWNTs in organic solvents and processability of the SWNT-polymer nanocomposites. Alkylation is a simple surface modification method by reacting SWNTs with alkyl groups of long chain molecules. The long polymer chains assist the dissolution of bundled nanotubes in organic solvents [34]. The reaction of CNT with ozone yields a primary ozonide group that can further transform into secondary ozonide and carboxyl groups. Sidewall modifications of the nanotubes via addition reactions can by achieved using cycloaddition, nucleophilic and electrophilic additions. A typical reaction is the 1,3-dipolar cycloaddition on the CNT developed by Prato and coworkers [35, 36]. An azomethine-ylide created by the condensation of α-amino acid and aldehyde is added to the CNT, forming a pyrrolidine ring on its surface. Coleman et al. [37] functionalized SWNTs with a cyclopropane group using Bingel

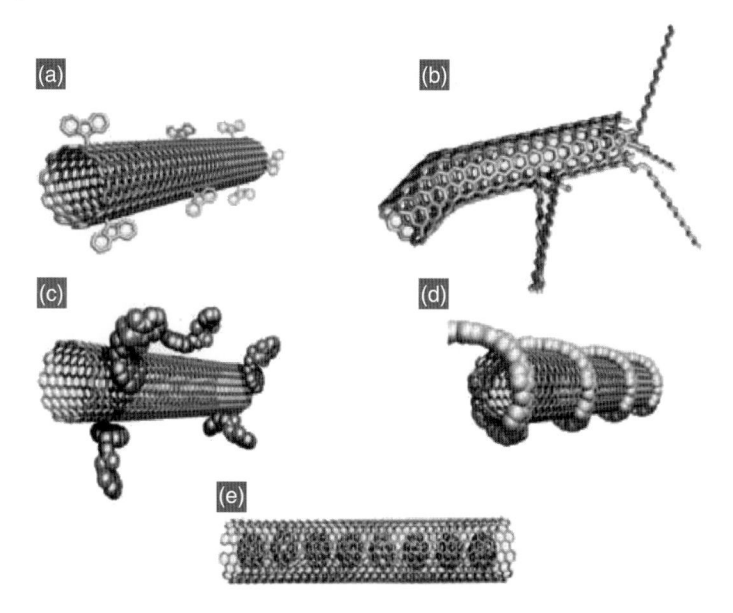

Figure 2.10 Functionalization strategies for SWNTs. (a) covalent sidewall functionalization, (b) defect-site functionalization, (c) noncovalent surfactant attachment, (d) noncovalent polymer wrapping and (e) encapsulation of guest molecules by SWNT. (Reproduced with permission from Ref. [29], Wiley-VCH (2002).)

reaction conditions. The Bingel reaction was based on a [2 + 1] cycloaddition reaction. In the process, SWNT dispersed in o-DCB was reacted with diethyl bromomalonate and 1,8-diazabicyclo[5.4.0]undecene (DBU) catalyst under stirring to give functionalized nanotubes. Similarly, cycloaddition of a reactive nitrene reagent, that is alkyl azidoformate to attack the SWNT wall can yield aziridine ring on the nanotube surface [38].

For the nucleophilic addition, a nucleophilic dipyridyl imidazolidene reacts with electrophilic SWNT to give zwitterionic 1 : 1 adducts. In this addition reaction, one negative charge is transferred from the imidazolidene to the nanotube surface, rendering CNT as a negative charge. Negatively charged nanotubes can be dispersed homogeneously in organic solvents by virtue of electrostatic repulsion [38]. Lastly, fluorination of CNTs offers the opportunity to enhance their compatibility with epoxy resins and fluoropolymers [39–41]. The fluorination treatment of CNTs generally employs reactive F_2 at high temperatures. Khare et al. [42] reported that a high degree of fluorination of SWNTs can be obtained in a very short time (a few minutes) using the microwave-generated CF_4 plasma. Further, the fluorine atoms in fluorinated CNTs can be replaced through nucleophilic substitution reactions, thus bonding the nanotube sidewalls with a variety of functional groups.

2.2.3.2 Noncovalent Functionalization

Noncovalent functionalization typically involves the attachment of weakly bonded functional groups on CNTs through physical interactions. It can be realized by the

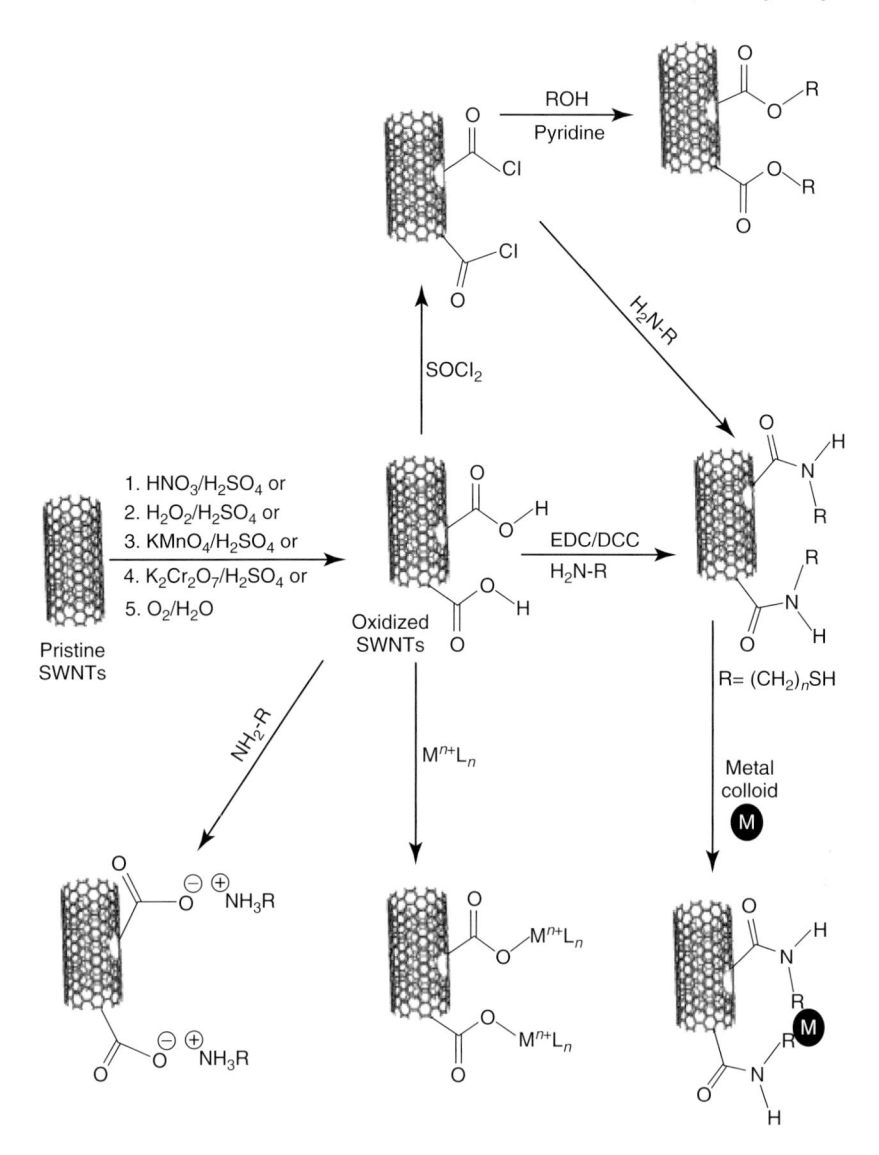

Figure 2.11 Schematic of covalent functionalization of SWNTs at ends and defect sites. (Reproduced with permission from Ref. [33], Wiley-VCH (2005).)

surfactant adsorption [42–45] and polymer wrapping [29, 46–50] on the nanotube surfaces (Figure 2.10). Noncovalent functionalization maintains the structural integrity and intrinsic property of CNTs. Moreover, physical adsorption of surfactant molecules on the nanotube surfaces reduces their surface tension, thereby preventing formation of the aggregates. Surfactants are surface-active agents consisting of hydrophilic head groups and hydrophobic hydrocarbon chain molecules. The

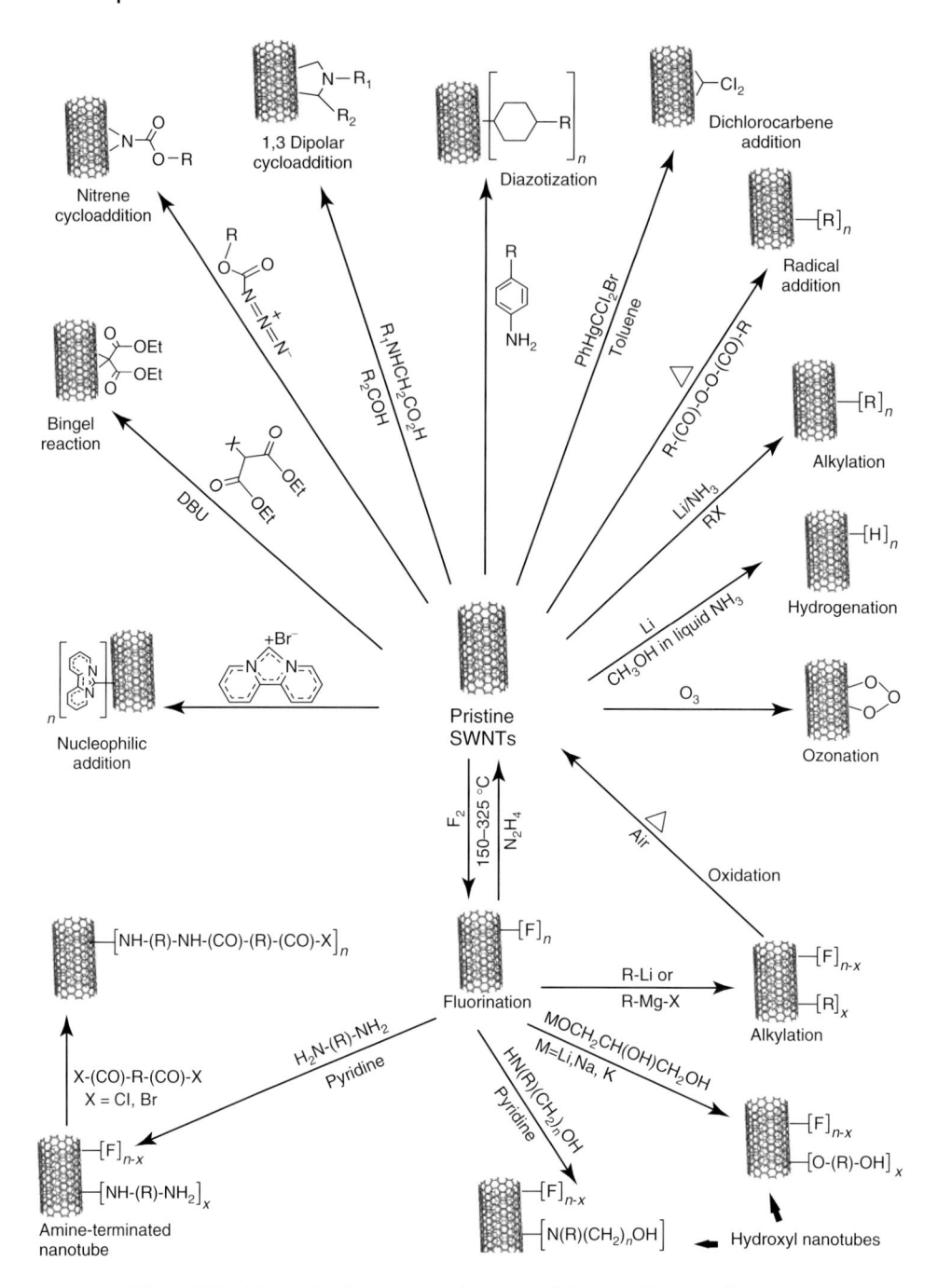

Figure 2.12 Schematic of various covalent sidewall functionalization of SWNTs. (Reproduced with permission from Ref. [33], Wiley-VCH (2005).)

SDBS

$C_{12}H_{25}$—⟨⟩—SO_3^- Na^+

SDS

$CH_3(CH_2)_{11}OSO_3^-$ Na^+

Triton X-100

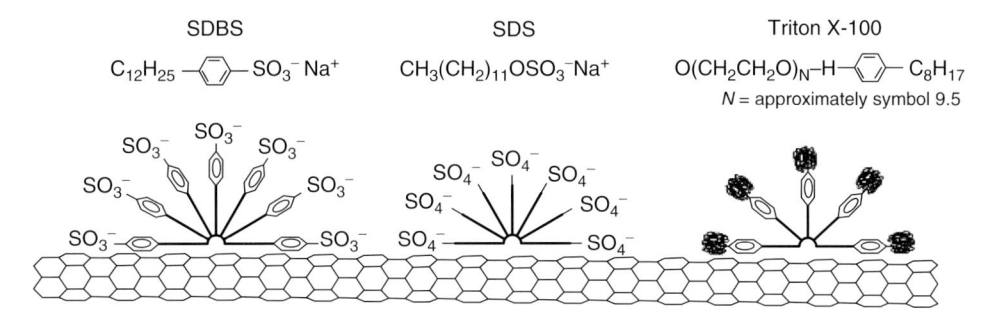

$O(CH_2CH_2O)_N$–H—⟨⟩—C_8H_{17}
N = approximately symbol 9.5

Figure 2.13 Schematic of surfactant adsorption on carbon nanotube surfaces. Tube stabilization depends on the surfactant molecules that lie on the tube surface parallel to the cylindrical axis. The alkyl chains groups of a surfactant molecule adsorb flat along the length of the tube instead of the diameter. (Reproduced with permission from Ref. [43], The American Chemical Society (2003).)

hydrophilic group of surfactants interacts with polar solvent molecules, while the hydrophobic group adsorbs onto nanotube surfaces [41]. Several ionic surfactants such as SDBS, sodium dodecyl sulfate (SDS) and hexadecyltrimethylammonium bromide (CTAB), as well as the nonionic surfactant, octyl phenol ethoxylate (Triton X-100), are good dispersants for CNTs in water [42–44]. In general, the dispersion efficiency of those surfactants depends mainly on their chemical structures such as the length of hydrophobic alkyl chains and the type of hydrophilic headgroups. In this regard, SDBS and Triton X-100 disperse CNTs in water more effectively than SDS because of the presence of benzene rings in SDBS and Triton X-100 (Figure 2.13). The SDBS exhibits the same alkyl chain length as SDS, but contains a phenyl ring in the alkyl chain. The phenyl ring of SDBS interacts preferentially with nanotubes within micelles, leading to better dispersibility in water. Furthermore, SDBS disperses better than Triton X-100 because of its sulfonate headgroup and slightly longer alkyl chain. The order of dispersibility is SDBS > Triton X-100 > SDS [42].

Polymers such as polystyrene sulfonate (PSS), polyvinyl pyrrolidone (PVP), poly(m-phenylene vinylene) (PmPV) and poly(3-hexylthiophene) (P3HT) can enhance the solubility of CNTs in water or solvents through a wrapping approach. The CNTs are wrapped helically by those polymers through the π–π interactions [50] (Figure 2.14). Consequently, the van der Walls force between the nanotubes weakens, thereby increasing dispersibility of the CNTs. Smalley and coworkers [46] reported that pristine SWNTs can be solubilized in water by wrapping them with linear PVP and PSS polymers. To achieve this, SWNTs were first dissolved in SDS such that the hydrophobic segment of the surfactant interacts with the nanotubes. The surfactant may introduce the steric hindrance and repulsive electrostatic force between the SWNTs, which overcomes the inter-tube van der Waals force. This leads to the formation of a stable suspension in the aqueous phase. PVP or PSS is then dispersed in the nanotube suspension. Very recently, Hwang

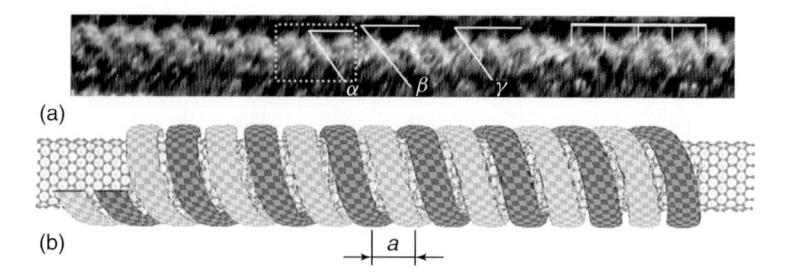

Figure 2.14 (a) Scanning tunneling microscopic image of an MWNT covered by P3HT self-organized into a coiled structure of different coiling angles, α, β and γ. (b) Schematic representation of a double helical structure with coils equally spaced at a distance a. (Reproduced with permission from Ref. [50], American Institute of Physics (2009).)

et al. [49] modified multiwalled carbon nanotubes (MWNTs) using a polymer wrapping approach by dispersing MWNTs and P3HT in chloroform under sonication. Figure 2.15a,b clearly shows the difference between the pristine and functionalized MWNTs in chloroform. Functionalized nanotubes show stable dispersion in chloroform without showing aggregation. In contrast, pristine nanotubes aggregate and precipitate readily in chloroform as expected. Further, the high-resolution transmission electron microscopy (HRTEM) image reveals that the MWNT surface is coated or wrapped by P3HT as evidenced by the presence of sulfur atoms in the energy-dispersive spectroscopy (EDS) result (Figure 2.15c,d).

2.3 Solution Mixing

Solution mixing or solvent casting is a process widely adopted to fabricate polymer nanocomposites using appropriate organic solvents. Thus it is suitable for polymers that are soluble in organic solvents. The process involves the dissolution of polymer in a suitable solvent followed by mixing with the carbonaceous nanomaterial suspension under vigorous stirring and/or sonication. During mixing, the polymer coats the surface of nanofillers effectively. The CNT/polymer composites are produced by evaporation of the solvent, or by precipitation and filtration processes. There is a high tendency for the nanofillers to form agglomerates during slow solvent evaporation. This can be prevented by reducing the evaporation time using a spin casting technique. The advantages of solution mixing are its simplicity and effectiveness for dispersing nanofillers in the polymer matrix. The disadvantages are the necessities for solvent and sonication for achieving homogeneous dispersion of nanofillers. Organic solvents are not always commercially available, and many polymers have low solubility in those solvents. The consumption of solvents in large quantities gives rise to environmental pollution issues. Furthermore, sonication is destructive to the aspect ratio of nanofillers, causing inferior mechanical strength of resulting composites.

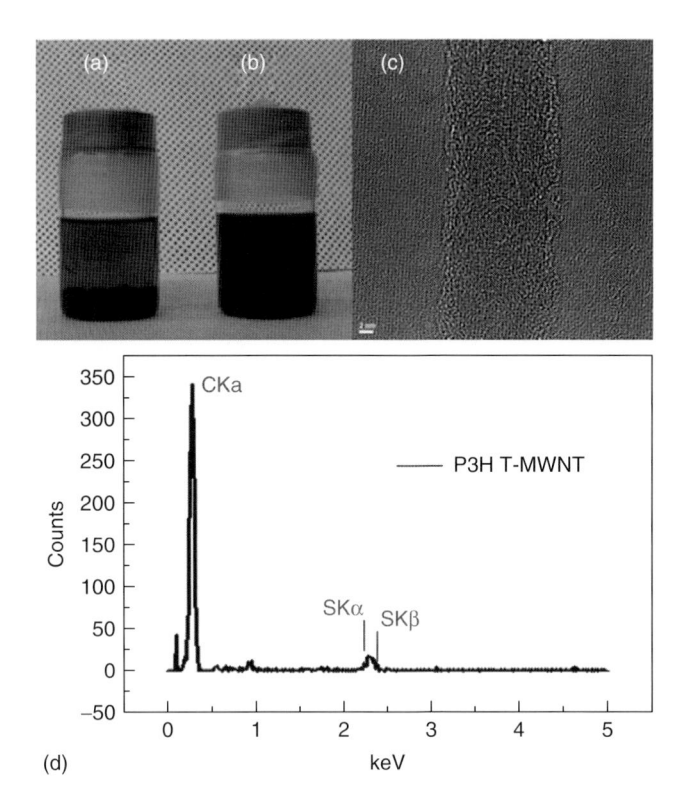

Figure 2.15 Macrographs of (a) pristine MWNT and (b) P3HT-wrapped MWNT in chloroform. (c) HRTEM micrograph of P3HT-MWNT (1 : 1 ratio); scale bar = 2 nm and (d) EDS signal of P3HT-MWNT. (Reproduced with permission from Ref. [49], Elsevier (2011).)

2.3.1
Nanocomposites with Graphene-Like Fillers

GO is readily processable from aqueous and polar solvents; it offers the advantage to synthesize polymer nanocomposites using solution mixing. Polymer nanocomposites such as graphene/PVA [1, 2], graphene/polyvinyl chloride (PVC) [51], graphene/polyurethane (PU) [52, 53], graphene/PP [7, 8], graphene/PS [54] and graphene/PMMA [14] can be prepared using the solution mixing method. This strategy can be employed to synthesize epoxy-based nanocomposites requiring the use of a hardener for curing [55, 56]. GO with oxygen functionalities can react with water-soluble polymers such as PVA and PEO readily since GO swells readily in water and other protic solvents via hydrogen bonding interactions. Accordingly, GO disperses homogeneously in those polymeric matrix. Similarly, TRG fillers with oxygen functionalities also disperse uniformly in the matrix of the 1 wt% TRG/PMMA composite [14]. In contrast, the dispersion of GO in aprotic solvents is poor; hence, the surface modification with isocyanate can assist its exfoliation in such solvents. Ruoff and coworkers [54] prepared graphene/PS nanocomposites via solution mixing iGO and PS in DMF followed by hydrazine reduction at 80 °C for 24 h. Chemical reduction is essential for restoring electrical conductivity because the composites with pristine iGO fillers are insulating. Moreover, the addition step procedure of the hydrazine reduction agent is critical for achieving homogeneous dispersion of graphene sheets. Re-stacking and re-aggregation of graphene sheets often occurs by introducing the reduction agent directly to a single GO suspension. To prevent re-stacking of graphene sheets, blending iGO with PS before the chemical reduction is recommended. In this regard, hydrazine is incorporated into a premixed iGO and PS solution. As the reduction proceeds, graphene-like sheets are coated with polymer to form nanocomposites. Such nanocomposites exhibit a low percolation threshold of 0.1 vol% at room temperature. By increasing filler loading to 1 vol%, a conductivity of $\sim 0.1\,\text{S}\,\text{m}^{-1}$ can be achieved, sufficient for electronic device applications.

Cao et al. [7] synthesized graphene/PP nanocomposites by mixing alkyl-functionalized GO or pristine GO fillers with PP. In the process, functionalized GO and bare GO were dispersed separately in xylene followed by the PP addition. Scanning electron microscopy (SEM) observation revealed that functionalized graphene sheets disperse uniformly in the PP matrix. The composite with pristine GO showed poor dispersion of fillers in the PP matrix as expected. Similarly, alkylation of GOs with tetra-*n*-octylammonium bromide (TOAB) can enhance their dispersion in the PP matrix [8]. When pristine GO or bare TRG fillers are used for reinforcing polyolefins, homogeneous dispersion of the fillers in the polyolefin matrix can be achieved provided that the polymer material is functionalized. For example, Macosko and coworkers [57] prepared TRG/linear low-density polyethylene (LLDPE) using both solution mixing and melt-blending techniques. To homogenize the dispersion of graphene sheets, nonpolar LLDPE was functionalized with cyano-(PE-CN), primary amino-(PE-NH$_2$) and secondary amino-(PE-NHEt) groups, respectively. For comparison, commercial LLDPE-copolymer-containing 24 wt% octane (Dow Chemical Co.), designated as EG-8200, and maleated EG-8200 (designated as EG-8200-MA) were used. Figure 2.16a–d shows transmission electron microscopy (TEM) images of solution mixed 1 wt% TRG/PE, 1 wt% TRG/PE-NH$_2$, 1 wt% TRG/EG-8200 and 1 wt% TRG/EG-8200-MA nanocomposites, respectively. The TRG sheets are partially exfoliated in the EG-8200 matrix with the presence of graphene aggregates. However, TRG sheets are well dispersed throughout the EG-8200-MA matrix.

2.3.2
Nanocomposites with EG and GNP Fillers

Glassy thermoplastics based on PMMA, polyetherimide (PEI) and polysulfone filled with EGs and GNPs have been prepared successfully using the solution mixing method [13, 58–61]. For the composites with PMMA matrix, GNPs are found to disperse more uniformly than EGs (Figure 2.17a,b). GNPs exhibit large lateral

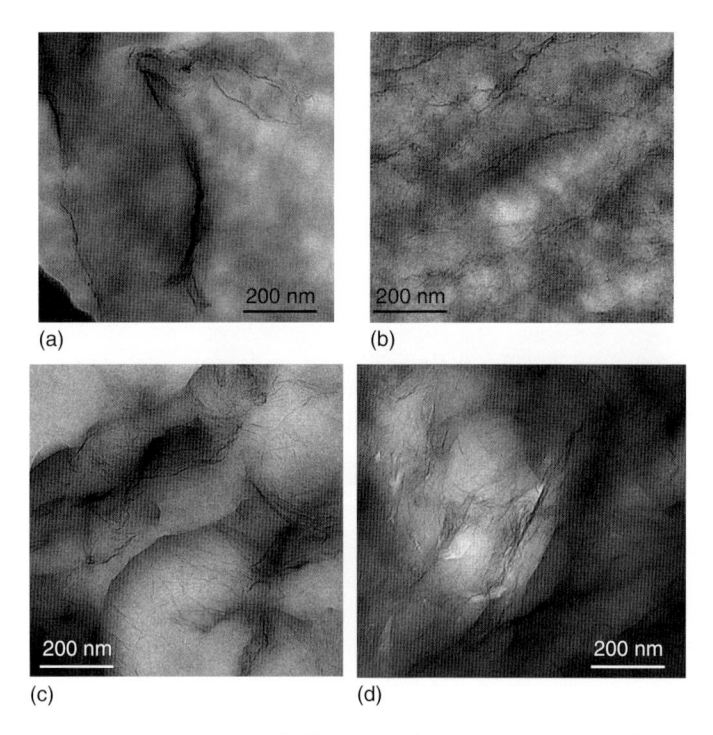

Figure 2.16 TEM images of solution mixed 1 wt% TRG in (a) PE, (b) PE-NH$_2$, (c) EG-8200 and (d) EG-8200-MA. (Reproduced with permission from Ref. [57], Elsevier (2011).)

dimension (width) and thickness of <10 nm (Figure 1.4a,b), hence having larger aspect ratio (width/thickness) than EGs. In general, EGs show poor bonding with the PMMA matrix (Figure 2.18a). Ruoff and coworkers also reported that interfacial adhesion between the EGs and the PMMA matrix is rather poor. Instead, TRG sheets disperse well and interact intimately with polar PMMA compared with the EGs. This is because oxygen functionalities in TRGs promote enhanced interaction with polar polymer matrix. EGs also possess carboxylic and hydroxyl functional groups on their surfaces. It is considered that the size and morphology of EGs affect their dispersion and interaction with PMMA. As mentioned previously, TRG sheets exhibit a wrinkled morphology because of the small thickness of the sheets (only three to four graphene layers) and the distortion generated from thermal exfoliation of GO precursor (Figure 1.9,b). The wrinkled texture appears to promote mechanical interlocking with the polymer chains, leading to strong interfacial interactions with the host polymer (Figure 2.18b).

The fabrication of GNP-filled polyvinylidene fluoride (PVDF) composites using the solution mixing process was reported by Tjong and coworkers [62, 63] very recently. PVDF is a semicrystalline thermoplastic with excellent thermal stability, unique piezoelectric and pyroelectric characteristics. Therefore, PVDF finds a variety of industrial and biomedical engineering applications because of these attractive properties. For fabricating the composites, PVDF was first dissolved in

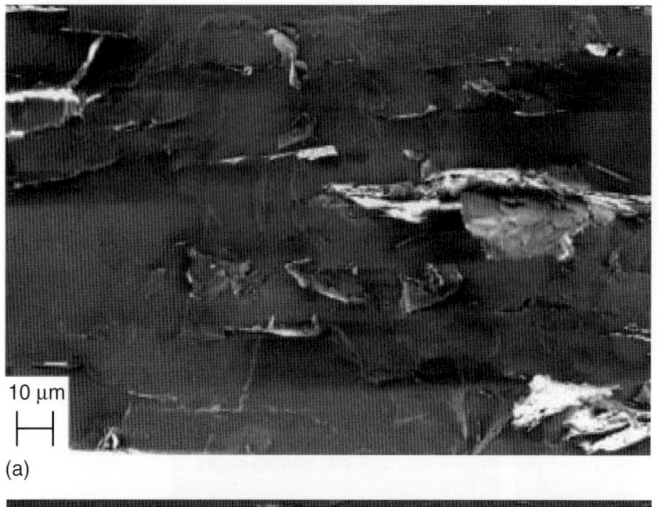

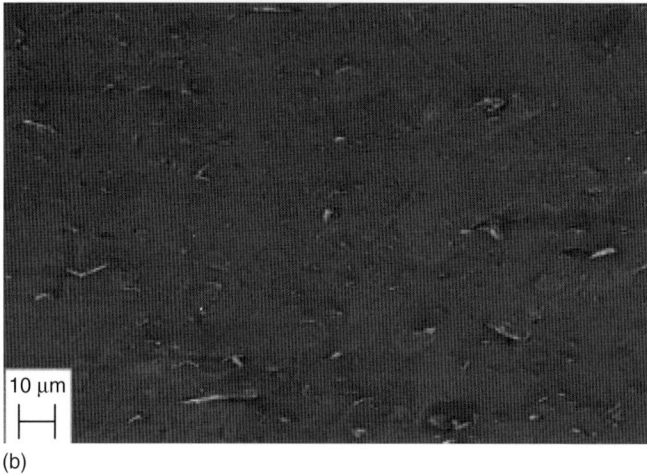

Figure 2.17 SEM images of fracture surfaces show the dispersion of (a) 2 wt% EG/PMMA and (b) 2 wt% GNP/PMMA nanocomposites. (Reproduced with permission from Ref. [59], Wiley (2007).)

a mixed solvent consisting of DMF and acetone (2 : 1(vol/vol)) under continuous stirring. Then the GNP suspension was slowly introduced into the polymer solvent and stirred continuously for 5 h. By evaporating the solvent in an oven at 70 °C overnight, the resulting composites were compression molded at 210 °C for 5 min under a pressure of ~15 MPa. Figure 2.19 shows X-ray diffraction (XRD) patterns of neat PVDF, solvent-cast 2 wt% GNP/PVDF and compression molded 2 wt% GNP/PVDF nanocomposite specimens. PVDF exhibits several polymorphs including α-, β-, and γ-forms. The β-PVDF form is of practical interest since it exhibits a high dielectric constant and piezoelectric effect. The peaks at 18.5° and 20° are characteristic reflections of nonpolar α-PVDF. The β-PVDF peak at 20.5°

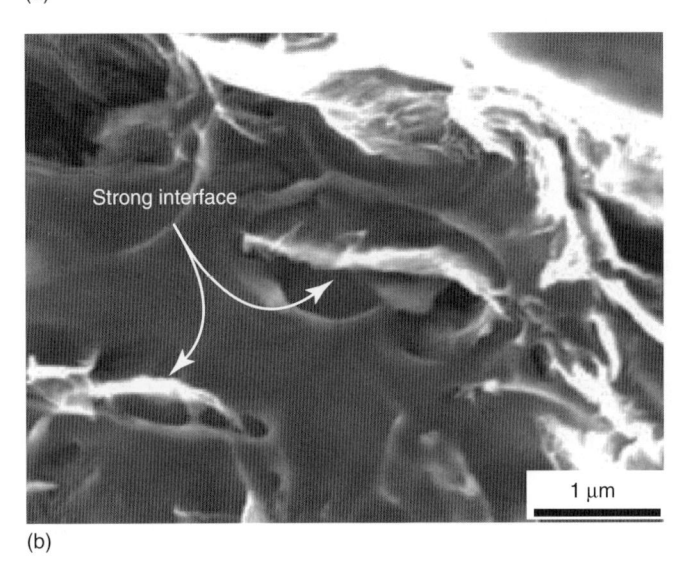

Figure 2.18 (a) SEM image of fracture surface shows the interfacial characteristic between EG and polymer for 5 wt% EG/PMMA nanocomposite. (Reproduced with permission from Ref. [59], Wiley (2007).) (b) SEM image of fracture surface of 1 wt% TRG/PMMA nanocomposite with strong interfacial interaction. (Reproduced with permission from Ref. [14], Nature Publishing Group (2008).)

is the dominant reflection of PVDF for solvent-cast 2 wt% GNP/PVDF composite. The peak at 26.6° is associated with the graphite reflection. After compression molding, all diffraction peaks intensify to a certain degree. Compression molding allows a better perfection and uniformity of PVDF polymer chains as indicated by a shift of broad β-PVDF at 20.5° into sharp α-PVDF peak at 20°. Solution mixing generally enables a better dispersion of GNPs in the PVDF matrix (Figure 2.20).

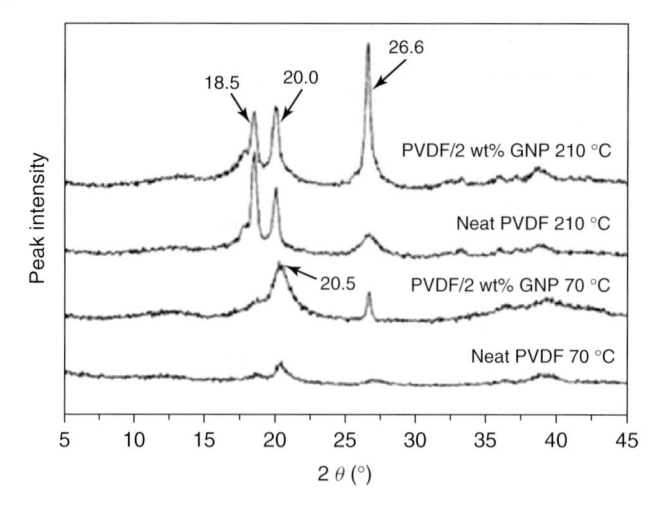

Figure 2.19 XRD traces of PVDF, cast 2 wt% GNP/PVDF and compression molded 2 wt% GNP/PVDF specimens.

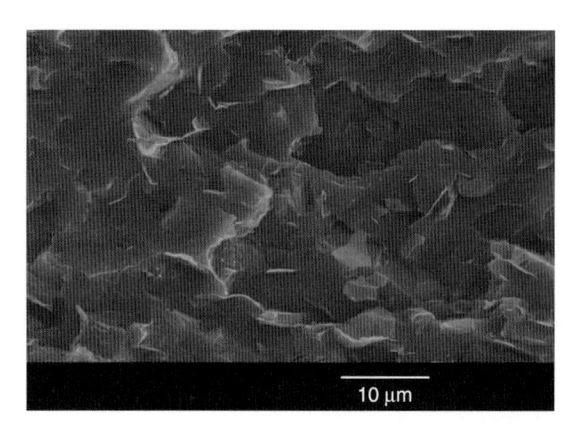

Figure 2.20 SEM image of a compression molded 2 wt% GNP/PVDF nanocomposite.

Epoxy resin has been used widely for industrial application because of its good physical and chemical properties including excellent adhesion, good chemical resistance, and lower rate of cure shrinkage compared with other thermoset resins. Some processing strategies such as direct mixing, sonication mixing, shear mixing, combined sonication and shear mixing have been employed to achieve better dispersion of EGs in diglycidyl ether bisphenol-A (DGEBA; Epon 828) [64] (Figure 2.21). Epon 828 is produced by Shell Chemical Co. from a reaction of bisphenol with epichlorohydrin to yield a resin with two epoxy functional groups per molecule. It requires heating and curing for solidification through cross-linking. Figure 2.22a–d depicts optical micrographs showing the dispersion of EGs in the epoxy matrix. Large EG particles are found to disperse nonuniformly in the epoxy matrix prepared from the direct mixing route. In contrast, other processing routes

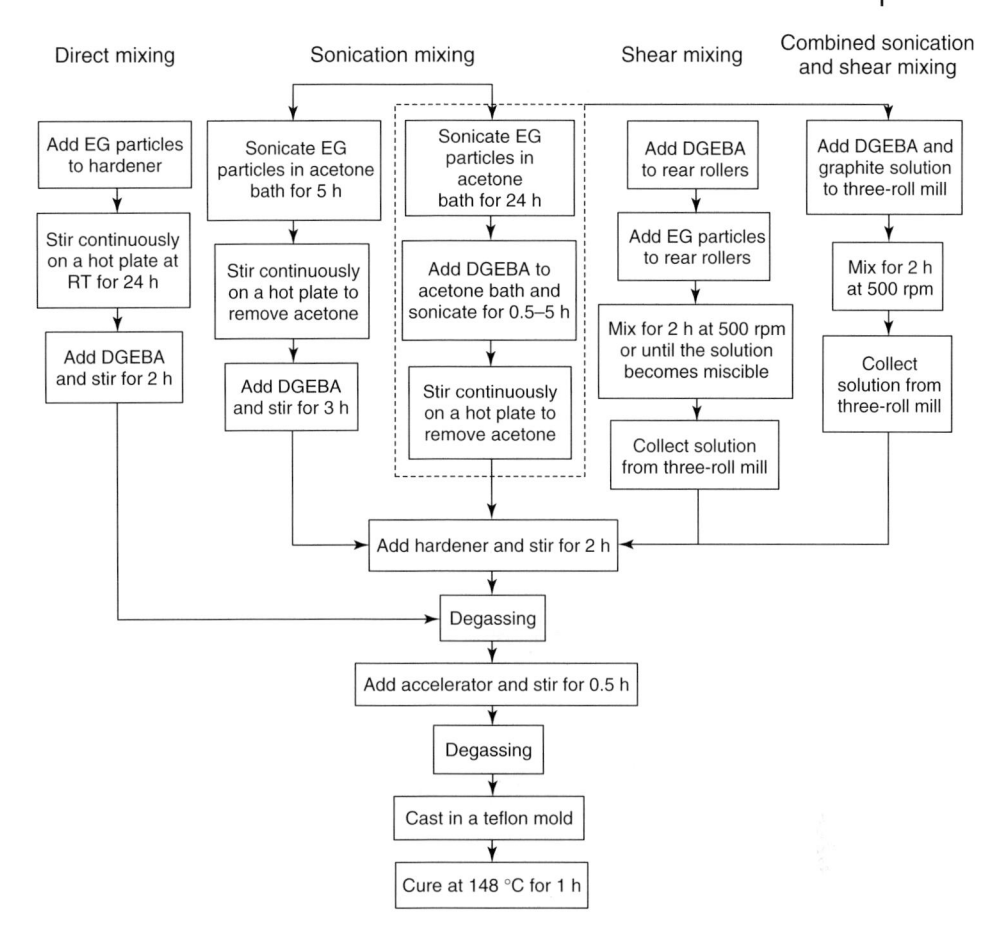

Figure 2.21 Flowchart of processing techniques of EG/epoxy nanocomposites. (Reproduced with permission from Ref. [64], Elsevier (2006).)

are more effective in dispersing EG fillers uniformly in the epoxy matrix, especially for the composites prepared by combined sonication and shear mixing.

Haddon's group employed the shear mixing strategy to fabricate the GNP/epoxy nanocomposites [65]. GNPs with a different degree of exfoliation were prepared from EGs by thermal shock exposure to temperatures of 200, 400 and 800 °C, denoting as GNP-200, GNP-400 and GNP-800, respectively. Figure 2.23a–c shows atomic force microscopy (AFM) images of the GNP-200, GNP-400 and GNP-800, respectively. It is apparent that the degree of exfoliation of GNPs depends greatly on the thermal shock exposure temperatures. AFM images reveal that the lateral dimension (L) and average thickness (t) of the GNP-200 are about 1.7 μm and 60 nm, respectively. The L and t values of GNP-400 are 1.1 μm and 25 nm, while those of GNP-800 are 0.35 μm and 1.7 nm, respectively. Therefore, the aspect ratios of GNP-200, GNP-400 and GNP-800 are ∼30, 50 and 200, respectively.

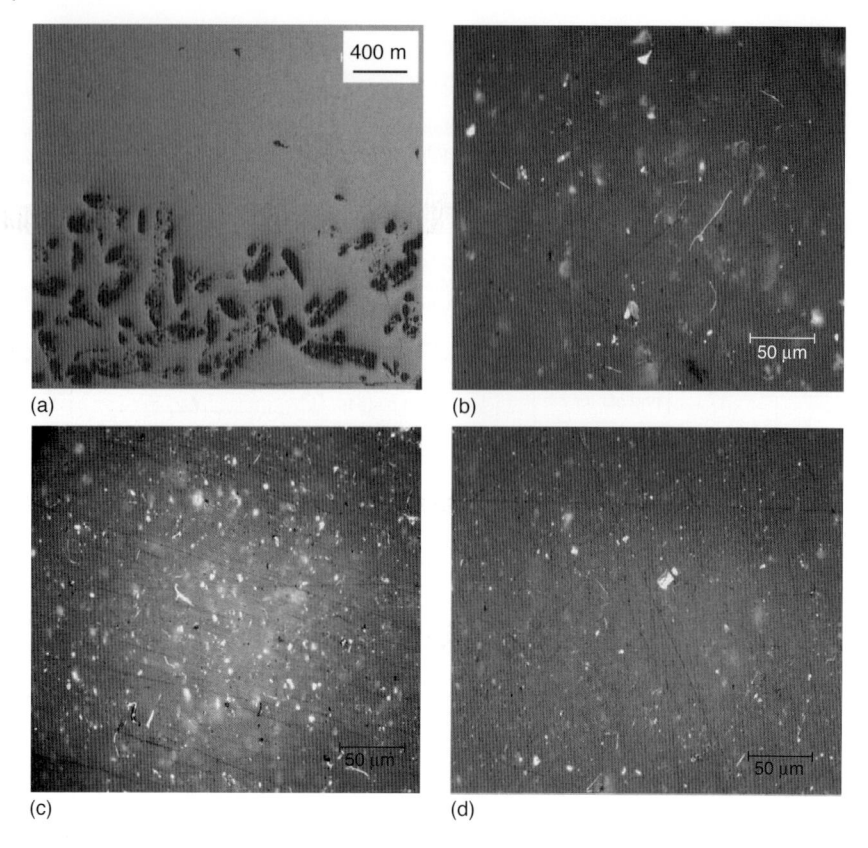

Figure 2.22 Optical micrographs show the dispersion of nanofillers in 1 wt% EG/epoxy nanocomposite prepared by (a) direct mixing, (b) sonication mixing, (c) shear mixing and (d) combined mixing. (Reproduced with permission from Ref. [64], Elsevier (2006).)

TEM micrographs of the epoxy composites filled with GNP-200, GNP-400 and GNP-800 are shown in Figure 2.23d–f, respectively. It is obvious that the number of graphene layers decreases markedly from GNP-200 to GNP-800. Complete graphite exfoliation takes place at 800 °C, leading to final separation of individual GNPs. This implies that the degree of graphite exfoliation increases with increasing thermal shock exposure temperature. As a result, the thermal conductivity performance of the GNP/epoxy nanocomposites improves with increasing the degree of exfoliation.

Very recently, Raza *et al.* [66] fabricated GNP/silicone composites using high-shear mechanical mixing. In the process, GNPs were mixed with appropriate weights of silicone using a mechanical mixer at a high speed of 2500 rpm. This was followed by the addition of the curing agent. The batch was degassed under vacuum and poured into a mold. Figure 2.24a–c shows SEM images of the fracture surfaces of 8 wt% GNP/silicone and 20 wt% GNP/silicone composites, respectively. The GNPs as indicated by the arrows are well dispersed in the silicone matrix. For the 20 wt% GNP/silicone composite, some GNPs link together with

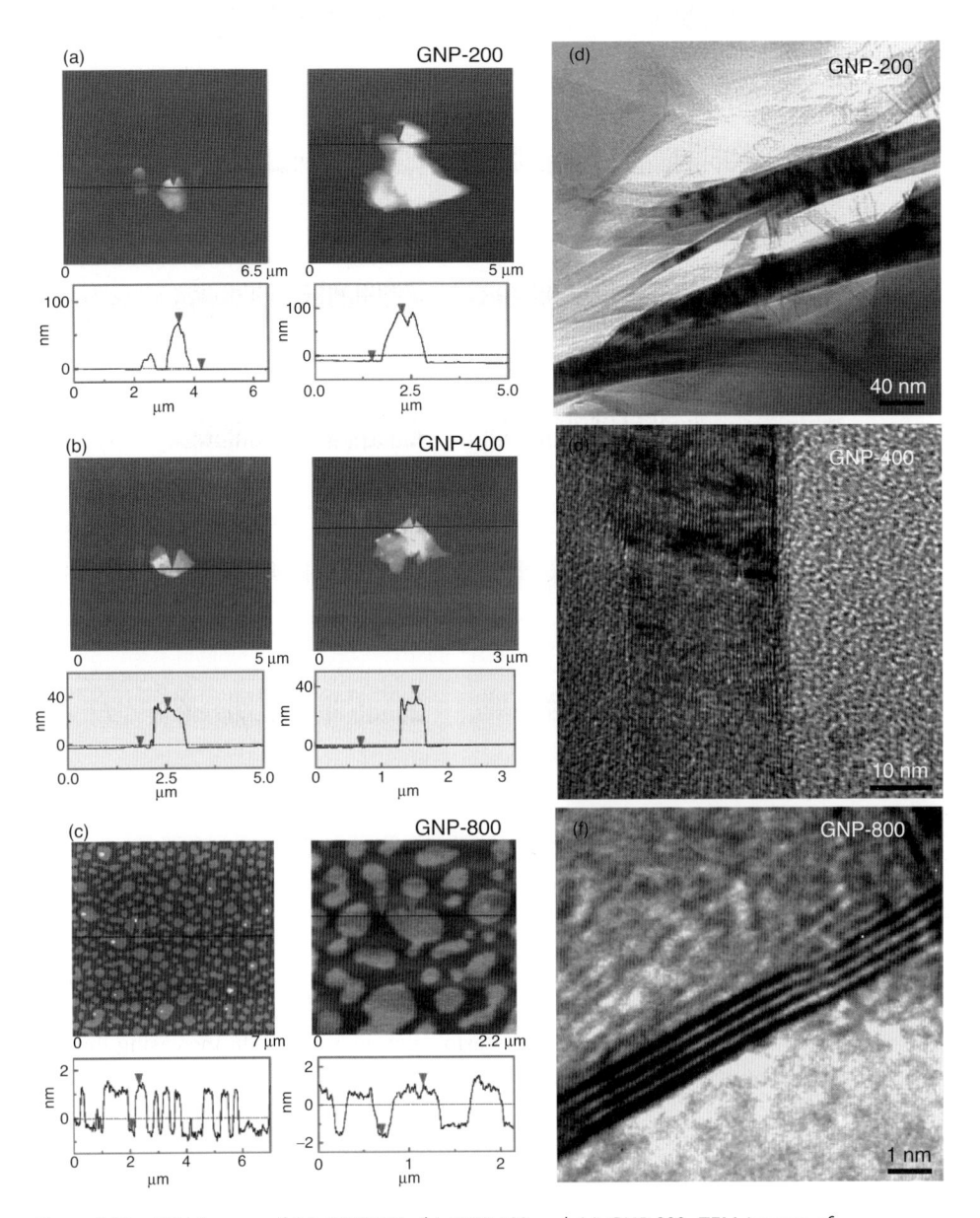

Figure 2.23 AFM images of (a) GNP-200, (b) GNP-400 and (c) GNP-800. TEM images of epoxy composites filled with (d) GNP-200, (e) GNP-400 and (f) GNP-800. (Reproduced with permission from Ref. [65], The American Chemical Society (2007).)

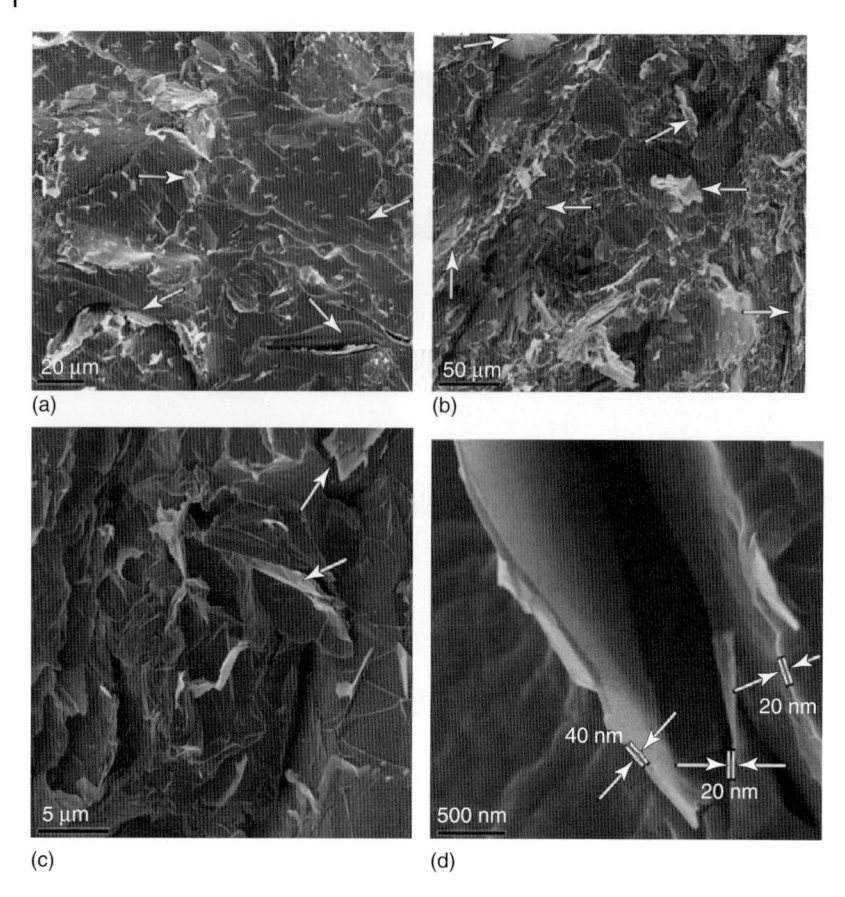

Figure 2.24 SEM images of the fracture surfaces of (a) 8 wt/% GNP/silicone and (b) 20 wt% GNP/silicone composites. (c,d) Higher magnified fractographs of 20 wt% GNP/silicone composite. (Reproduced with permission from Ref. [66], Elsevier (2011).)

one another, forming a typical conducting network. The thickness of the GNPs is estimated to be ~20–40 nm determined from the pullout sheets (Figure 2.24d).

2.3.3
Nanocomposites with CNT and CNF Fillers

Qian et al. [67] conducted an early research study of the MWNT/PS composite film. The film was prepared by mixing 1 wt%. MWNT (pristine) with PS in toluene under ultrasonication and a subsequent solvent evaporation [67]. They reported that the elastic modulus and break stress of the composite film increases markedly by adding only 1 wt% MWNT. The use of bare MWNTs also increases electrical conductivity in solvent-cast MWNT/syndiotactic PS nanocomposites [68]. In general, the solubilization of CNTs or CNFs through covalent functionalization is more effective to attain homogeneous dispersion of nanofillers in the polymer matrix.

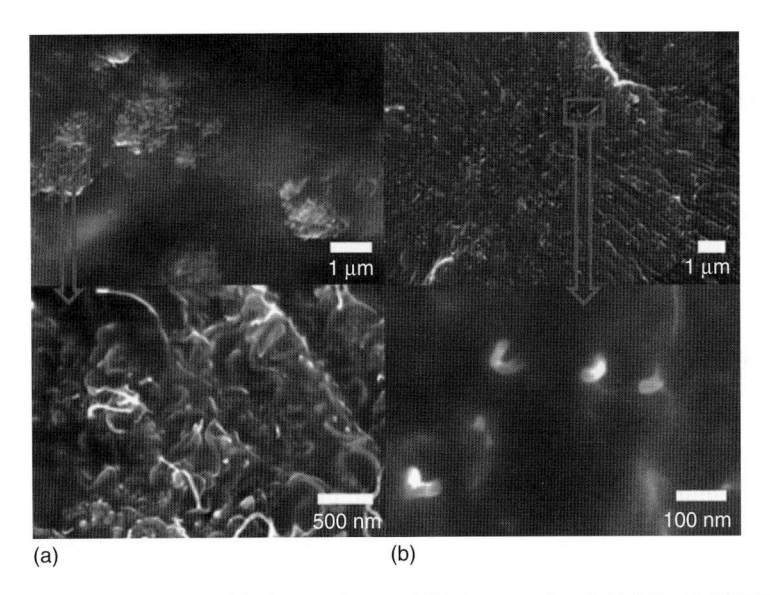

Figure 2.25 Low- and high-magnification SEM fractographs of (a) 1.5 wt% MWNT/PDMS and (b) P3HT-MWNT/PDMS nanocomposites. (Reproduced with permission from Ref. [49], Elsevier (2011).)

Thus the addition of acid-functionalized MWNTs greatly enhances mechanical properties of the PS-based nanocomposites [69]. Araujo et al. [70] functionalized CNFs through the reaction of 1,3-dipolar cycloaddition. The 1,3-dipole was formed by the condensation reaction of N-benzyloxycarbonylglycine and formaldehyde. The functionalized CNFs were then mixed with PA-6 (polyamide) via solvent casting to produce CNF/PA-6 composites. The functionalized nanofibers are distributed uniformly in the PA-6 matrix.

Very recently, Hwang et al. [49] found an improved dispersion and enhanced electrical properties of poly(dimethyl siloxane) (PDMS) in the presence of P3HT-wrapped MWNTs. Figure 2.25a,b depicts SEM micrographs showing the dispersion of pristine MWNTs and P3HT-MWNT in the PDMS matrix, respectively. Pristine nanotubes show a certain degree of agglomeration in the PDMS matrix. The P3HT-MWNTs prepared using the noncovalent polymer wrapping approach are found to disperse homogeneously in the polymer matrix.

Ajayan et al. [71] were the first to prepare aligned MWNT/epoxy composites. In the process, arc-discharged MWNTs were dispersed in ethanol under sonication, followed by mechanical mixing with epoxy resin and curing agent. The solvent was removed by evaporation, and the nanotube-epoxy mixture was poured into a mold and cured. Since then, many studies have been conducted on the fabrication, morphological and material characterization of structural and functional CNT/epoxy and CNF/epoxy nanocomposites [72–75]. As an example, Wang et al. [74] functionalized CNFs (vapor grown carbon nanofiber (VGCNF), PR-19-HT) with amine-containing pendants via acylation reaction with 4-(3-aminophenoxy)benzoic acid. The resulting H_2N-VGCNF was treated with epichlorohydrin and sodium hydroxide solution to yield N,N-diglycidyl-modified VGCNF, designating as the "epoxy-VGCNF." Subsequently, epoxy-VGCNF was dispersed in an epoxy resin with the aid of acetone and sonication, followed by pouring into water to obtain the precipitate. Comparing with pristine nanofibers, the functionalized VGCNFs disperse homogeneously in the epoxy matrix of the composites as expected.

2.4
Melt Mixing

The melt mixing process employs high temperature and intense shear forces to disperse nanofillers in molten polymer using rotating screws of an extruder or mixer. Carbonaceous nanofillers are incorporated directly into the polymer melt. This leads to a large increase in viscosity even at low loadings of filler. The process avoids the use of toxic solvents as in the case of solution mixing. The state of dispersion of nanofillers in the matrix of a composite depends on the nature of the polymer, surface modification condition of the fillers, and shear force applied, which in turn relates to the types of the compounder employed [76]. As recognized, solution mixing and *in situ* polymerization are ineffective to manufacture polymer nanocomposites in large quantities because of the environmental issue of organic solvents and the high cost of monomers. On the contrary, melt mixing is a versatile and cost-effective commercial process capable of producing a variety of polymer and composite products in different shapes and large volume scales. It is compatible with current industrial processing facilities for manufacturing polymer composites using extruders and injection molders. The process is applicable to both nonpolar and polar polymers. For a small-scale production for research purposes, internal batch mixers such as Haake (Thermo-Scientific, Co., Germany) and Brabender (C. W. Brabender Instruments, Inc., USA) are commonly used by researchers to compound polymer nanocomposites. In general, a high-shear mixing condition is essential to achieve better dispersion of nanofillers in the polymer matrix. This is often accompanied by a decrease in the aspect ratio of nanofillers, leading to property degradation of the nanocomposites.

2.4.1
Nanocomposites with Graphene-Like Fillers

During melt mixing, the polymer chains are intercalated or exfoliated into graphene-like sheets to form nanocomposites. Several graphene-filled composites with polymer matrices based on LLDPE [57], PU [53], PP [77], polycarbonate (PC) [78], polyethylene terephthalate (PET) [79], poly(ethylene-2,6-naphthalate) (PEN) [80] and PA-12 [81] have been fabricated by this process. TRG can be melt-compounded in a one-step procedure by direct feeding the desired polymer pellets and TRG in an extruder. For rGO nanofillers, a two-step procedure is generally recommended, that is combined solution mixing and melt compounding. Accordingly, solution mixing of GO with thermoplastics must be first performed

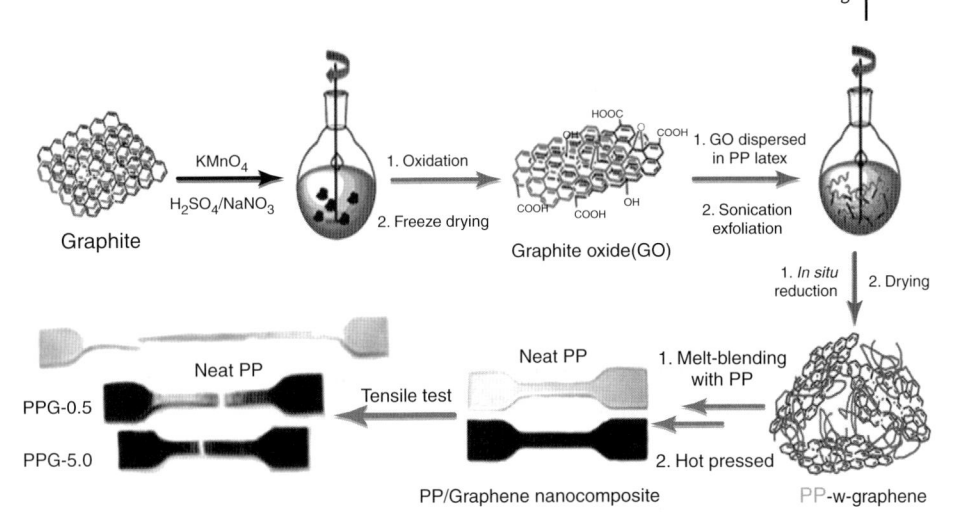

Figure 2.26 Schematic representation of process flow for fabricating graphene/PP nanocomposite and its tensile test specimens. (Reproduced with permission from Ref. [77], Elsevier (2011).)

followed by adding the hydrazine agent to avoid re-stacking of rGO sheets. On completion of the reduction reaction, the solution is removed and polymer-coated rGO is produced. The polymer-coated rGO is then melt blended with selected polymers to form nanocomposites. Low throughput of rGO sheets generally restricts its use in the melt-blending process. Figure 2.26 shows schematic diagrams for fabricating graphene/PP composites using combined solution mixing and melt compounding processes.

In general, melt compounding is less effective in dispersing graphene nanosheets uniformly in the polymer matrix compared with solvent processing because of the higher viscosity of the composites at increased filler loadings. More recently, Mahmoud [13] further treated TRGs with acetonitrile agent to reduce their residual oxygen groups. The product is termed as the foliated graphene sheets (FGSs). AFM and TEM examinations revealed that the treated FGSs consist of four to five graphene layers. Figure 2.27a,b shows TEM images of the 0.3 vol% FGS/PEO nanocomposite specimens prepared by solvent mixing and melt mixing, respectively. It is obvious that the FGS sheets are fully exfoliated into two graphene layers (inset), and dispersed homogeneously in the PEO matrix of solution-cast composite. This is because sonication treatment during solution mixing can further exfoliate FGS into thin graphene layers. When the graphene sheets are blended with PEO under sonication, the polymer solution diffuses between graphene sheets and eventually coats their surfaces. This produces highly exfoliated and uniformly dispersed graphene sheets in the polymer matrix. For melt-compounded composite, FGSs are blended with polymer pellets in a mixer, and then this mixture compression molded in a heating press. Thus, the polymer chains tend to intercalate FGS sheets rather than diffuse between the graphene layers. This leads to less uniform dispersion of FGS sheets in the polymer matrix (Figure 2.27b).

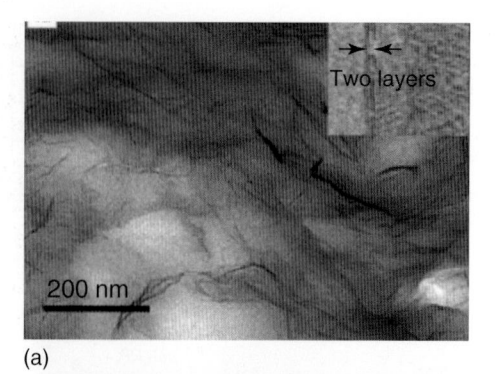

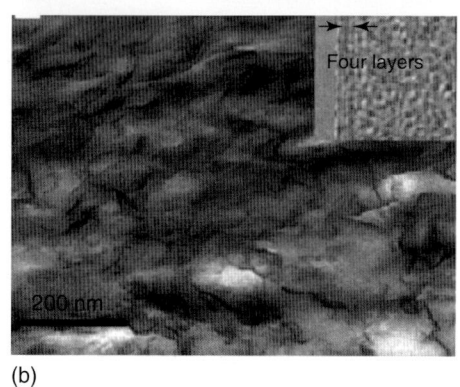

Figure 2.27 TEM micrographs of (a) solution-blended and (b) melt-blended 0.3 vol% FGS/PEO nanocomposite specimens. Insets are HRTEM images of respective specimens. (Reproduced with permission from Ref. [13], Elsevier (2011).)

For the composites with nonpolar polymer matrix, uneven distribution of graphene sheets is even more apparent. Figure 2.28a–d shows the dispersion of TRG sheets in melt-compounded 1 wt% TRG/EG-8200 and 1 wt% TRG/EG-8200-MA nanocomposites, respectively [57]. Comparing with solution-blended 1 wt% TRG/EG-8200 and 1 wt% TRG/EG-8200-MA counterparts (Figure 2.16c,d), some areas or islands highly concentrated with TRG sheets are observed in the melt-compounded specimens. The aggregated sheet islands are mostly isolated from one another, forming graphene-rich and graphene-poor phases.

2.4.2
Nanocomposites with EG and GNP Fillers

The versatility of melt compounding allows the manufacture of a variety of EG- and GNP-filled polymers including glassy thermoplastics [82, 83], semicrystalline polymers [83–88], and elastomers [89]. The state of dispersion of EG- and GNP fillers in those polymers is influenced by the type of polymer selected and chemical functionalization of the fillers.

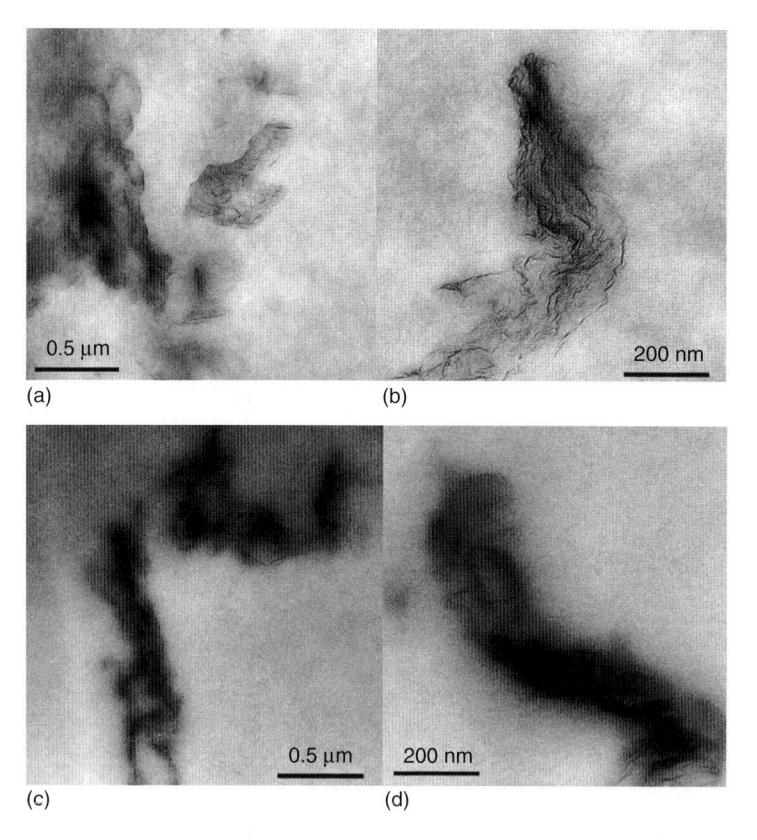

Figure 2.28 TEM images of (a,b) 1 wt% TRG/EG-8200 and (c,d) 1 wt% TRG/EG-8200-MA composites prepared by melt compounding. (Reproduced with permission from Ref. [57], Elsevier (2011).)

2.4.3
Nanocomposites with CNT and CNF Fillers

The dispersion efficiency of CNTs during melt-compounding depends greatly on the mixing energy, that is the mixing time or mixing speed employed [76, 90]. Conventional extruders generally use low-shear forces for mixing. This often leads to poor dispersion of CNTs. Figure 2.29 shows the effect of mixing energy on the dispersion and aspect ratio of MWNTs [90]. Mixing energy is the mechanical energy consumed to disperse the nanofillers during compounding. In this figure, a dispersion index of 1 defines poor dispersion of nanotubes, while a maximum value of 10 demonstrates the absence of nanotube agglomerates. Apparently, high mixing energy results in more uniform dispersion at the expense of aspect ratio of nanotubes. Higher energy input is needed with increasing nanotube content because of an increase in melt viscosity. There is a trade-off between the mixing energy and the aspect ratio of CNTs for melt-compounding CNT/polymer composites. Therefore, a balance must be reached between the dispersion and the

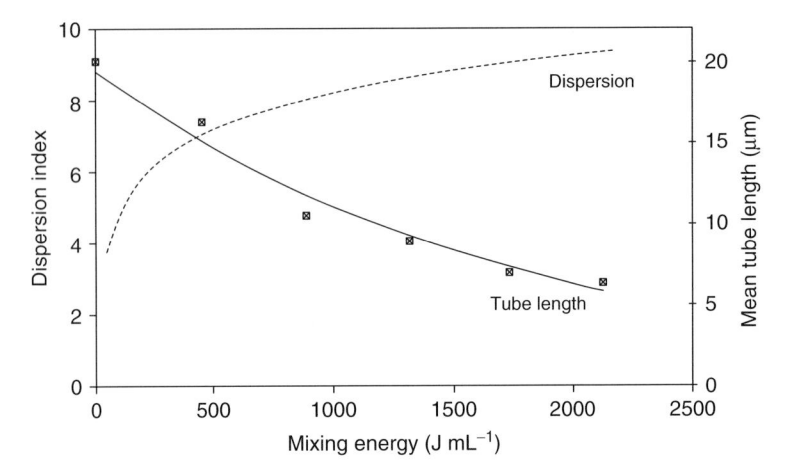

Figure 2.29 Dispersion and aspect ratio of MWNTs as a function of mixing energy. (Reproduced with permission from Ref. [90], Wiley-VCH (2002).)

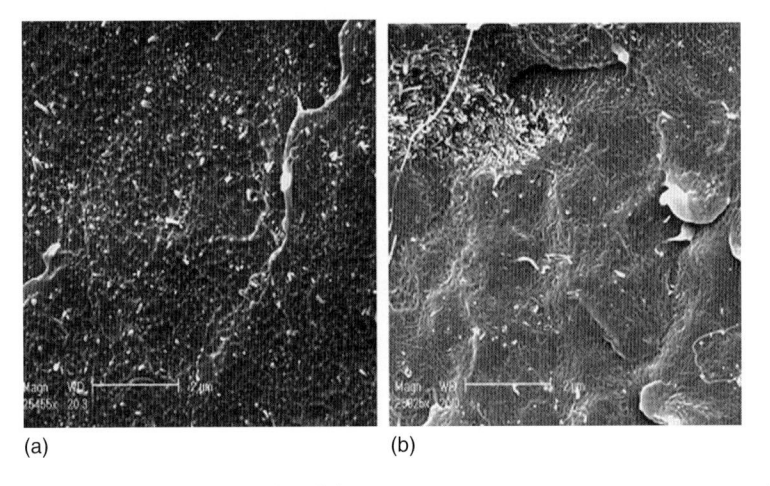

Figure 2.30 SEM fractographs of the 2 wt% MWNT/PVDF nanocomposite prepared under (a) high shearing and (b) low shearing. (Reproduced with permission from Ref. [91], Elsevier (2007).)

aspect ratio of CNTs by properly selecting appropriate mixing energy. Recently, Chen *et al.* [91] employed a tailor-made screw extruder with a screw rotation speed of 1000 rpm to disperse pristine MWNTs in PVDF, corresponding to an average shear rate of $1470\,s^{-1}$. Figure 2.30a shows an SEM image of the 2 wt% MWNT/PVDF nanocomposite. MWNTs are uniformly dispersed in the PVDF matrix of the nanocomposite. However, MWNT aggregates can be observed in the 2 wt% MWNT/PVDF nanocomposite prepared using a low-shear extruder at 100 rpm (Figure 2.30b).

Generally, good mixing and homogeneous dispersion of CNTs are frequently found in polar polymers filled with functionalized nanotubes. Zhang et al. [92] reported that nitric-acid-treated MWNTs disperse uniformly in the PA-6 matrix because of the strong interfacial interactions between the nanotubes and the PA-6 matrix during melt-compounding. For nonpolar polyolefins, a uniform dispersion of CNTs in the polymer matrix is rather difficult to achieve. McNally et al. [93] prepared MWNT/PE composites using a Haake twin screw extruder. The nanotubes dispersed mainly as aggregates of varying dimensions in addition to some independent entities in the PE matrix. As a result, there were no improvements in both yield strength and breaking stress of the MWNT/PE nanocomposites by adding MWNTs. The yield strength and breaking stress of the nanocomposites decreased with increasing filler content, indicating poor interfacial adhesion between the filler and the polymer matrix.

To improve interfacial bonding between CNTs and nonpolar polymer matrix, maleic-anhydride-grafted polymer or copolymer acting as a compatibilizer is added to the CNT/polyolefin composites. Maleic-anhydride-grafted polypropylene (MA-g-PP) is commonly used for improving dispersion of CNTs by reducing the interfacial tension between the PP matrix and nanotubes. Lee et al. [94] fabricated MWNT/PP composites in the presence of MA-g-PP or maleic-anhydride-grafted styrene-ethylene/butylene-styrene (MA-g-SEBS) compatibilizers via direct melt-mixing of composite constituents. Kumar and coworkers [95] employed a combined solution and melt processing approach to disperse and exfoliate CNTs in PP. In the process, nitric-acid-treated SWNTs and MWNTs as well as MA-g-PP were dispersed in butanol/xylene solvent mixture. The solvent-cast CNT/MA-g-PP masterbatch was then melt blended with PP in a Brabender mixer to form polymer nanocomposites. Both MWNTs and SWNTs were dispersed homogeneously in the PP matrix using this processing route.

Recently, a preformed masterbatch containing 10–20 wt% CNTs has been used increasingly for dispersing CNTs homogeneously in thermoplastics [76, 96–98]. The preformed CNT-polymer material offers an additional advantage by reducing the risk of exposure to hazardous nanotube powders. Logakis et al. [97] diluted a masterbatch containing 20 wt% MWNT/PA-6 with PA-6 using melt-compounding. They reported that the melt dilution technique can yield homogeneous dispersion of nanotubes in the PA-6 matrix (Figure 2.31a,b).

Thermoplastic polyurethane (PU of TPU) is a linear copolymer consisting of alternating hard and soft segments. The hard segment composes of alternating diisocyanate and chain extender molecules, that is low-molecular-weight diol or diamine, while the soft segment is made up from high-molecular-weight polyester or polyether diol. These segments aggregate into microdomains, resulting in a material with glassy and hard domains, as well as rubbery and soft domains. Barick and Tripathy [99] compounded acid-functionalized CNF and TPU in a Haake mixer, followed by compression molding. CNFs with a nearly straight fiber feature exhibit better processability than coiled CNTs. Figure 2.32a–d shows SEM micrographs of the composites with 1, 7, 10 and 15 wt% CNF, respectively. The nanofibers are generally well dispersed in the polymer matrix of those composites because

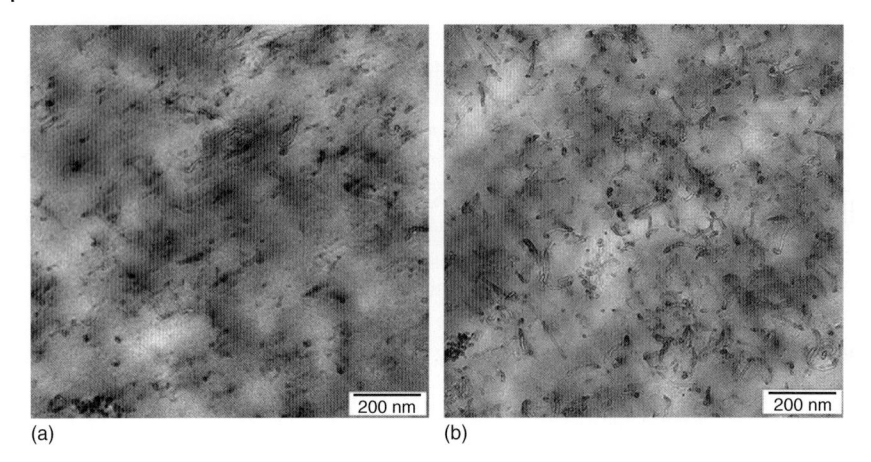

Figure 2.31 TEM micrographs of (a) 2.5 wt% MWNT/PA-6 and (b) 5 wt% MWNT/PA-6 nanocomposites fabricated using melt dilution process. Dark spots are transition metal catalysts entrapped inside MWNTs. (Reproduced with permission from Ref. [97], Elsevier (2009).)

of proper selection of mixing time for compounding. The TEM micrographs also display the homogeneous dispersion of nanofibers, but some fiber aggregates can be observed at higher filler loadings (Figure 2.33a–d).

2.5
In situ Polymerization

In situ polymerization involves an initial dispersion of nanofillers in monomers with the aid of mechanical stirring and sonication, either in the presence or absence of a solvent, followed by polymerizing monomers via free radical, open ring, emulsion or condensation reactions. This process facilitates stronger interactions between the carbonaceous nanofillers and the polymeric phase, and is hence capable of obtaining better dispersion of nanofillers in the polymer matrix. In this regard, nanocomposites prepared by *in situ* polymerization process exhibit better mechanical properties and lower percolation concentration than those fabricated by the melt-blending technique. The shortcomings of *in situ* polymerization are the high cost of monomers and the deployment of high-temperature reaction reactors for polymerizing monomers into polymer nanocomposites.

2.5.1
Nanocomposites with Graphene-Like Fillers

The polymerization process for making graphene/polymer nanocomposites involves the dispersion of GO, rGO or iGO in monomers followed by polymerizing monomers [100, 101]. For example, Ruoff and coworkers prepared

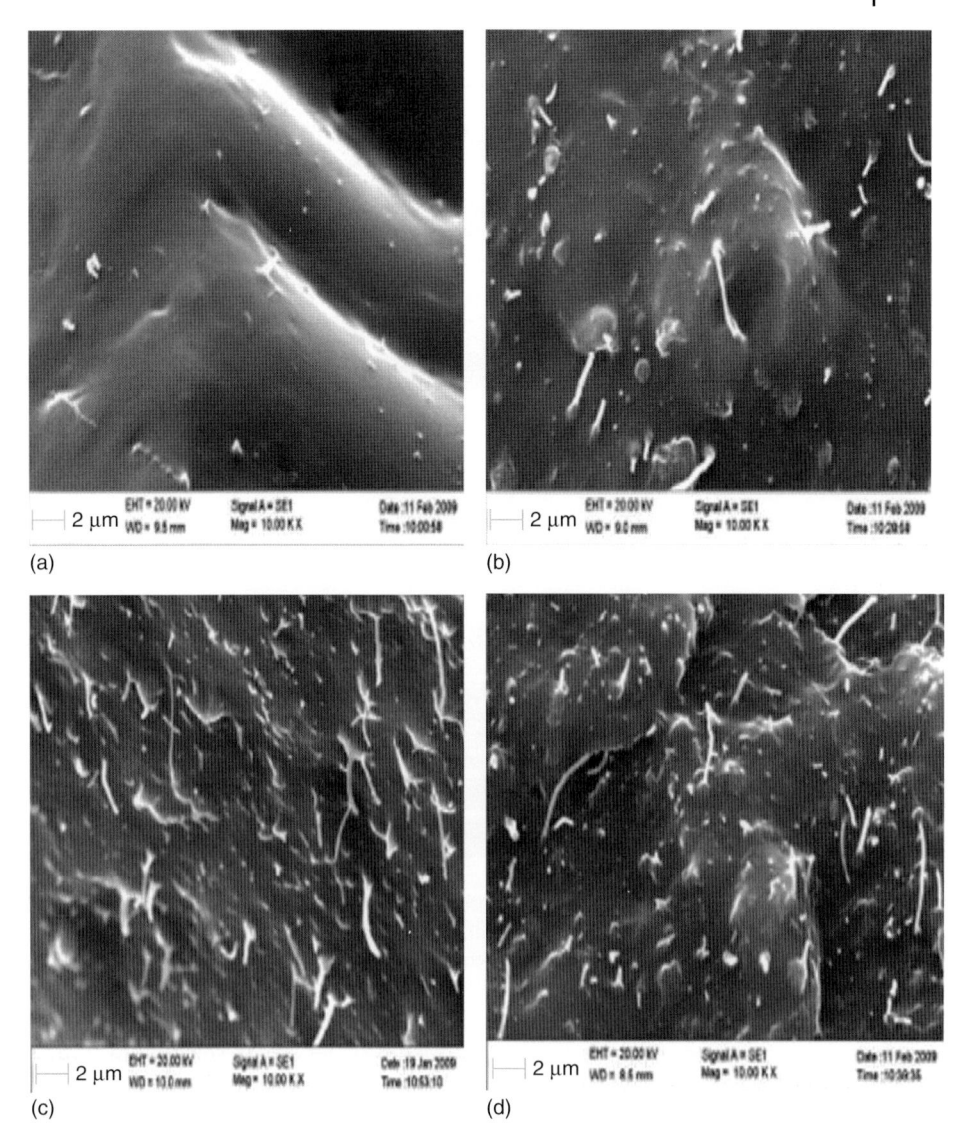

Figure 2.32 Field-emission SEM images of CNF/TPU composites with (a) 1 wt%, (b) 7 wt%, (c) 10 wt% and (d) 15 wt% CNF. (Reproduced with permission from Ref. [99], Elsevier (2010).)

graphene/PMMA nanocomposites by *in situ* polymerization using methyl methacrylate (MMA) monomer and GO [100]. In the process, GO was suspended in DMF under sonication at room temperature, and purged with nitrogen. The MMA monomer and a free radical initiator such as benzoyl peroxide (BP) were added to the suspension, and heated to 80 °C to promote polymerization. On heating, the bond of the initiator molecule was cleaved, producing radicals.

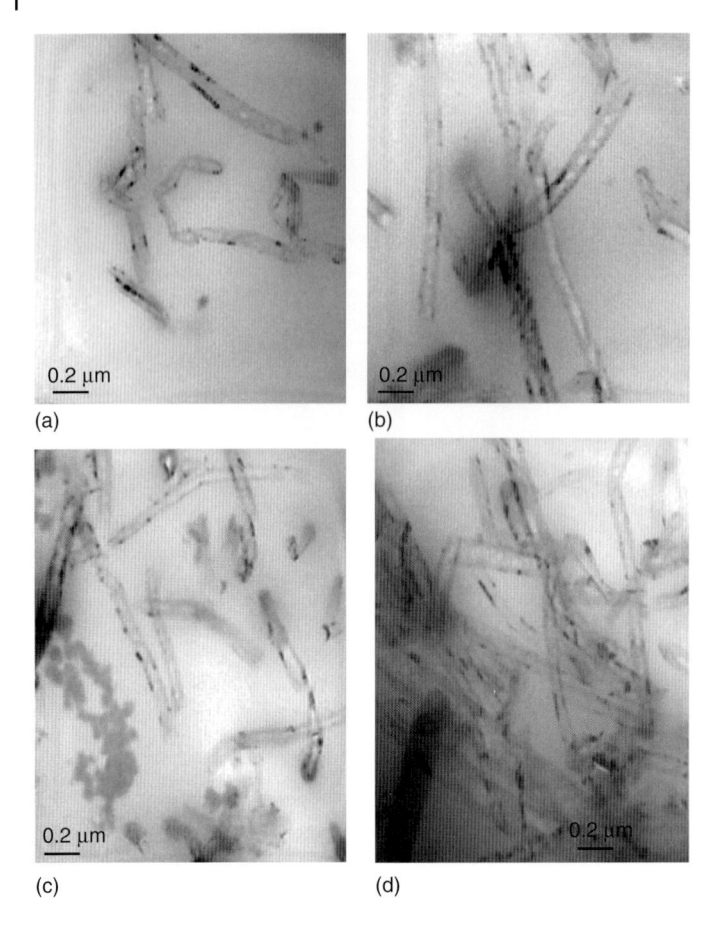

Figure 2.33 TEM images of CNF/TPU composites with (a) 1 wt%, (b) 4 wt%, (c) 7 wt% and (d) 15 wt% CNF. (Reproduced with permission from Ref. [99], Elsevier (2010).)

The radicals then attacked the monomer to initiate polymerization. After the polymerization was completed and while the polymer was in solution, hydrazine was added to reduce GO into graphene sheets.

In general, polymer chains can be attached to the GO or rGO surfaces using "grafted to" and "grafted from" approaches. The former route involves the use of a presynthesized polymer with reactive groups for reacting with GO. The latter route involves growing polymer from GO surfaces through *in situ* polymerization of monomers in the presence of a surface initiator. As mentioned earlier, polymer monomers such as styrene, butyl acrylate and methyl methacrylate can be grafted from the GO surface via ATRP as shown in Figure 2.2 [11]. The ATRP process is also commonly referred to as surface-initiated polymerization (SIP). It offers the advantages of preparing multifunctional polymer with precise manipulation over the polymer molecular weight, composition, polydispersity and end group functionality. Polymers with reactive endgroups can be grafted from the substrate surfaces, resulting in the so-called polymer brushes. The modification of surfaces with thin polymer films that are covalently bonded to the substrate is effective for tailoring surface properties such as wettability, biocompatibility and mechanical hardness. MMA monomer is widely used to form polymer brushes via ATRP [102a]. Very recently, Layek et al. [102b] used this strategy to attach MMA monomers on the GO surface (Figure 2.34). In the process, GO was dispersed in pyridine under sonication. Subsequently, α-BIB was added slowly at 0 °C. Bromoisobutyryl bromide-attached graphene (BIBG) was obtained by repeatedly washing the product with water. BIBG was then dispersed in a nitrogen-purged vessel containing DMF, $CuCl_2$ and N, N, N′,N′,N″-pentamethyl diethylene triamine (PMDETA) followed by adding MMA monomer. The reaction vessel was kept at 60 °C for 48 h. Finally,

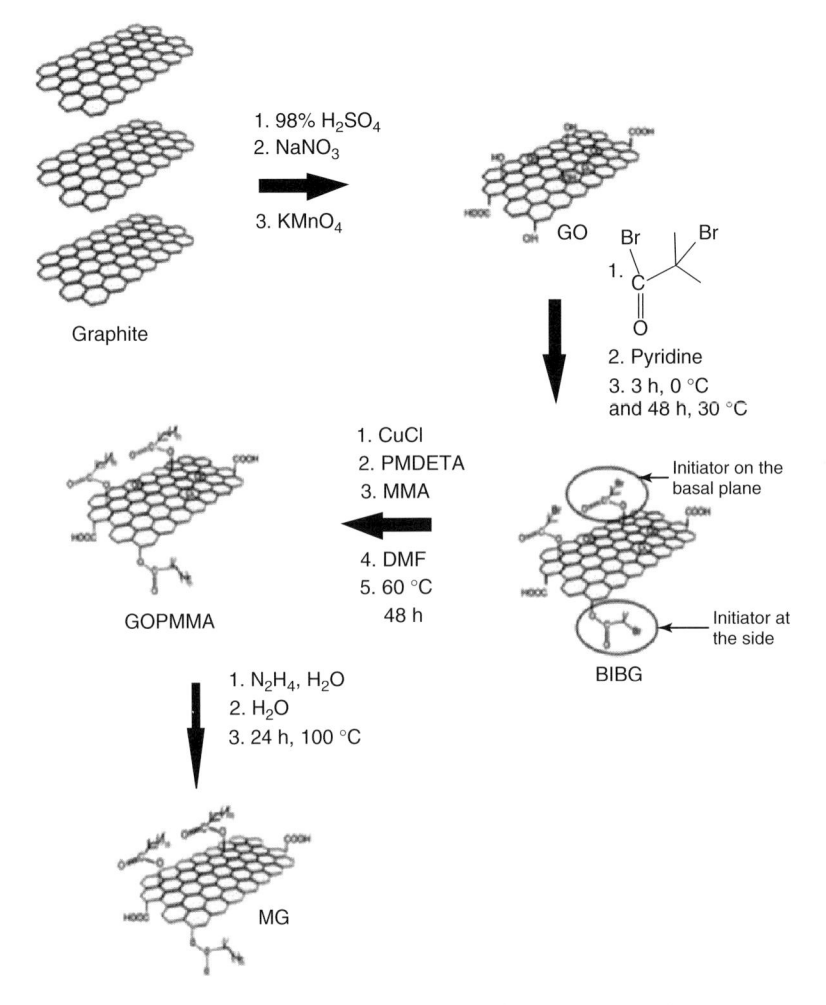

Figure 2.34 Synthesis reaction scheme for PMMA functionalized graphene. (Reproduced with permission from [102b], Elsevier (2010).)

hydrazine reducing agent was added and the PMMA-functionalized graphene (MG) was synthesized accordingly. Gonçalves *et al.* also reported that PMMA chains can be grafted from the GO surface via ATRP to yield GPMMA. The resulting GPMMA was soluble in chloroform. PMMA was then dispersed in the same medium to form polymer nanocomposite films of high flexibility [103].

The ring-opening polymerization in the presence of fillers is particularly suitable for fabricating PA-based composites. The graphene/PA-6 nanocomposites can be synthesized by *in situ* ring-opening polymerization of caprolactam in the presence of GO, and initiated by 6-aminocaproic acid at 250 °C. The PA-6 chains are grafted to the graphene sheets by condensation reaction between the carboxylic acid of GO with the amide groups of PA-6. During the polycondensation, GO is thermally reduced to graphene simultaneously [104].

The *in situ* emulsion polymerization of graphene/PS nanocomposites requires the use of a surfactant, that is SDS [105, 106]. This surfactant consists of a hydrophilic polar group and a hydrophobic chain. The graphene sheets with large surface area provide effective sites for anchoring the reaction species and PS particles. In the process, GOs were first dispersed in a flask containing water, followed by adding SDS and styrene monomer under sonication. Styrene monomer was absorbed onto the hydrophobic end of the surfactant and then onto the edges of the GOs having many oxygen functionalities. Subsequently, the $K_2S_2O_8$ (KPS) initiator was added to the mixture, and a subsequent refluxing at 80 °C for 5 h under a nitrogen atmosphere. KPS initiated the polymerization of styrene that resided in the adsorbed micelles. The PS microspheres were covalently bonded to the edges of GO sheets. Finally, hydrazine hydrate was added into the GO-PS dispersion (Figure 2.35). Indeed, TEM image reveals the presence of PS microspheres with

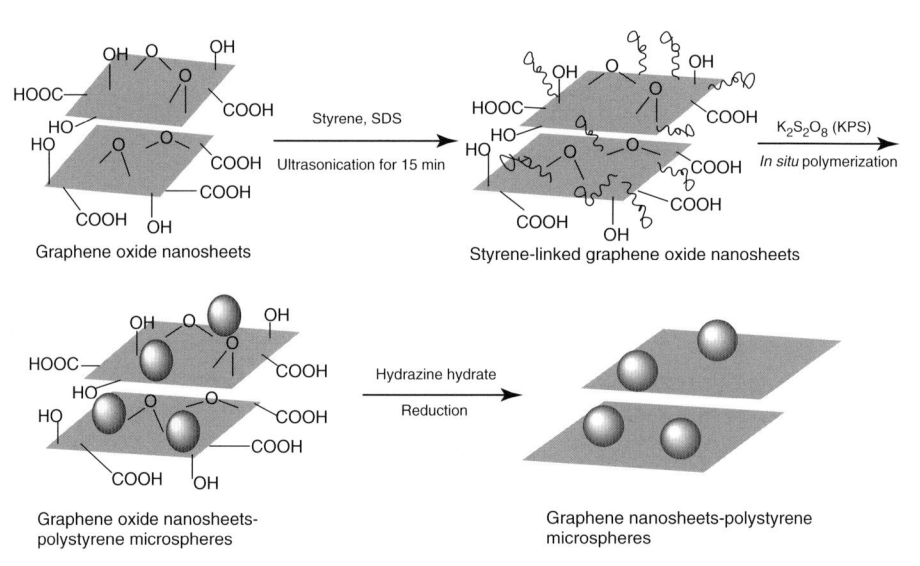

Figure 2.35 Schematic of the formation of PS-functionalized graphene sheets. (Reproduced with permission from Ref. [105], Elsevier (2010).)

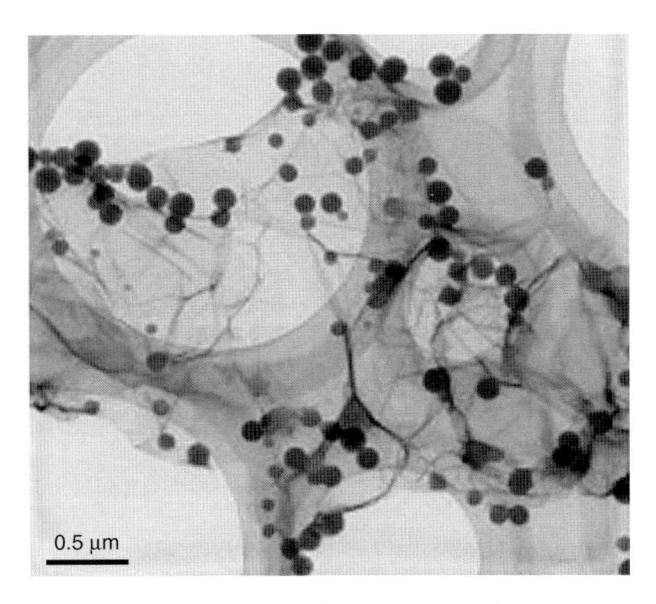

Figure 2.36 TEM micrograph of PS-graphene particles. (Reproduced with permission from Ref. [105], Elsevier (2010).)

diameters of ~90–150 nm attached to the surface of graphene (Figure 2.36). In another study, Patole *et al.* [106] also carried out *in situ* emulsion polymerization of graphene/PS nanocomposites using an 2,2′-azobisisobutyronitrile (AIBN) initiator instead. The PS-functionalized graphene shows good compatibility and enhanced interactions with the host PS to form conducting PS films with high flexibility (Figure 2.37). For nonpolar polyolefin, the graphene/PP nanocomposites can be prepared using Mg/Ti catalysts for *in situ* Ziegler-Natta polymerization of PP [107].

2.5.2
Nanocomposites with EG and GNP Fillers

EG displays a loose, porous, vermicular or worm-like feature (Figure 1.4a). Accordingly, monomers adsorb and penetrate into the pores readily to form polymer nanocomposites during the polymerization. The *in situ* polymerization of EG/polymer nanocomposites is well documented in the literature [108–110]. The polar feature and the presence of oxygen functionalities in the EGs promote their interactions with polar polymers. Therefore, this technique is quite suitable for dispersing EG fillers homogeneously in the polymer matrix. Figure 2.38 is an SEM micrograph showing the morphology of EG/poly(styrene-co-acrylonitrile) nanocomposite. The EG platelets, with thickness of a few hundreds of nanometres, are coated with the polymeric material. In the case of GNP fillers, a wide variety of polymers has been selected as the matrix materials of the GNP/polymer nanocomposites prepared by *in situ* polymerization [111–117].

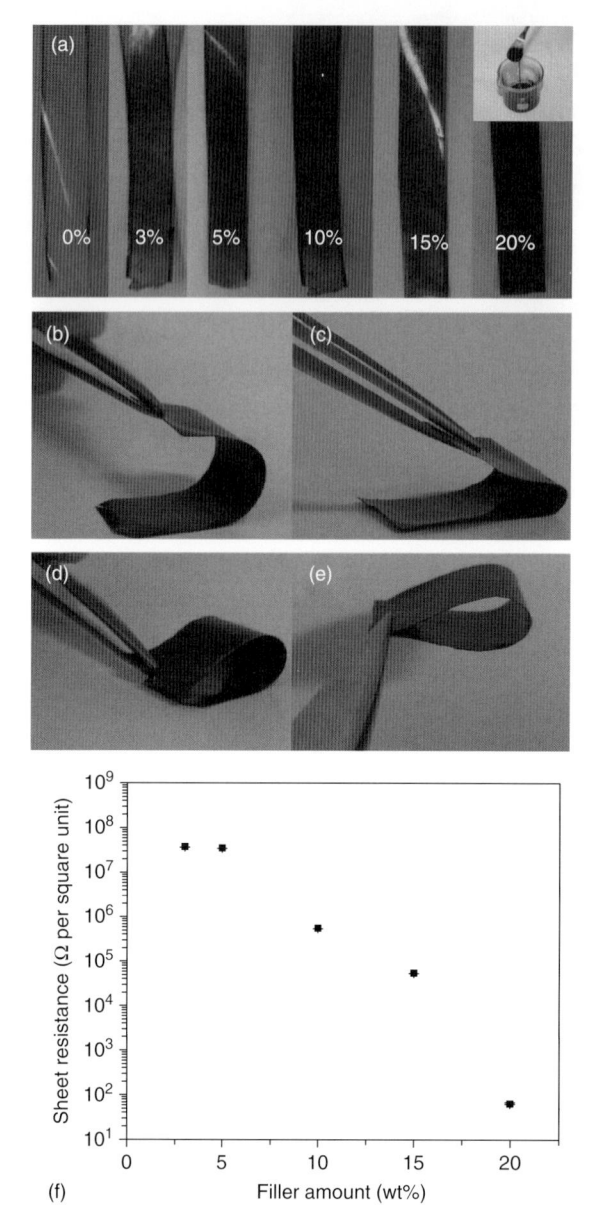

Figure 2.37 (a) Macrographs of conducting PS films containing different PS-functionalized graphene contents. The inset shows the slurry in THF used for fabricating PS films. (b–e) Photographs show the flexibility of the composite films. (f) Sheet resistance of the composite films as a function of the filler content. (Reproduced with permission from Ref. [106], Elsevier (2010).)

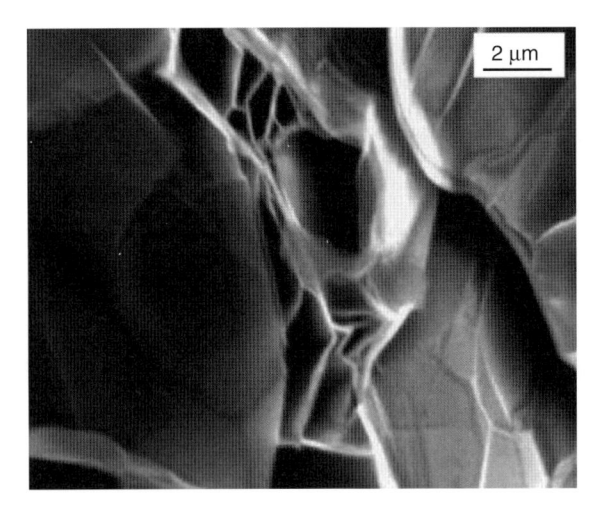

Figure 2.38 SEM micrograph of EG/poly(styrene-co-acrylonitrile) composite. (Reproduced with permission from Ref. [109], Elsevier (2004).)

2.5.3
Nanocomposites with CNT and CNF Fillers

CNTs can be functionalized covalently with polymers via "grafting to" and "grafting from" methods. In the former route, preformed polymers are attached to the nanotube surfaces through reactions between the functional groups of CNTs and the polymers [118, 119]. This involves the diffusion of polymer from solution to react with the nanotube surface. The limitation of this route is low grafting efficiency due to the slow diffusion of polymers to the nanotube surfaces. For the latter route, polymers are grown from CNTs typically using ATRP process [120–125]. Considering the polymer grafted onto the CNTs is the same or miscible with the matrix polymer, the resulting nanocomposites can exhibit strong interfacial bonding, enhanced elastic modulus and tensile strength [126, 127]. Hwang et al. [126] grafted MWNTs with MMA using emulsion polymerization technique. The PMMA-grafted MWNT was used as reinforcement for PMMA. Consequently, applied stress can be effectively transferred from the matrix to the nanotubes during mechanical testing. This leads to a significant improvement in the mechanical property of the PMMA-grafted MWNT/PMMA composites.

Jia et al. [128] were the first to conduct in situ polymerization of MWNT/PMMA nanocomposites using acid purified nanotubes, MMA monomer and a free radical initiator, AIBN. They reported that AIBN can open the π-bonds of MWNTs, thereby facilitating PMMA polymerization. Park et al. further confirmed the role of AIBN by opening π-bonds of nanotubes to produce radicals. The molecular weight of PMMA increased with increasing nanotube content [129]. Recently, Meng et al. [130] used BP to initiate polymerization of the MWNT/PMMA composites. In the process, carboxylated MWNT and BP were added into MMA monomer under sonication. The mixture was heated to 80–85 °C during which BP decomposed into two radicals.

As a result, the C=C bonds of the MMA monomers and the π-bonds of MWNTs opened up accordingly. This strategy has been adopted by Choi et al. to synthesize MWNT/PS nanocomposites. Sonication of vinyl monomers such as styrene in the presence of an initiator leads to radical initiation of polymerization [131]. More recently, Windle's group of Cambridge University (UK) [132] synthesized aligned MWNT/PS composites via radical polymerization of styrene monomers in the presence of nanotube-stacked layers and BP. They reported that fully densified composites can be prepared by varying the number of layers of aligned MWNT arrays or carpets. Stacks of nanotube carpets containing 1, 2, 3, 5, 10 or 150 individual layers were assembled between two aluminum cylinders coated with Kapton film. The stack was placed into a test tube before the addition of styrene and BP (Figure 2.39). The tube flushed with argon was sealed and heated at 90° for 24 h in an oil bath. The liquid styrene wetted the nanotubes and then impregnated the carpets. Figure 2.40a–d shows typical SEM images of cryo-fracture surfaces of the aligned composite. The aligned nanotubes are completely coated with the PS.

Ring-opening polymerization of caprolactam in the presence of carboxylated SWNT or amide-functionalized SWNT, and initiated by 6-aminocaproic acid at 250 °C has been employed by Haddon and coworkers [133, 134] to synthesize SWNT/PA-6 nanocomposites. Figure 2.41 shows the reaction scheme for the synthesis of SWNT/PA-6 nanocomposites using SWNT-COOH and SWNT-CONH$_2$. Higgins and Brittain [135] synthesized CNF/PC composites via *in situ* polymerization of cyclic oligomeric carbonates (Figure 2.42). Ring-opening polymerization of cyclic PC oligomers offers the advantages of low viscosity, lack of volatiles and better control over the molecular weight. In the process, the cyclic oligomers were mixed with CNFs followed by adding tetrabutylammonium tetraphenylborate catalyst. Ring-opening polymerization of oligomers in the presence of CNFs leads to the formation of CNF/PC nanocomposites. The nanofibers generally dispersed well in the polymer matrix.

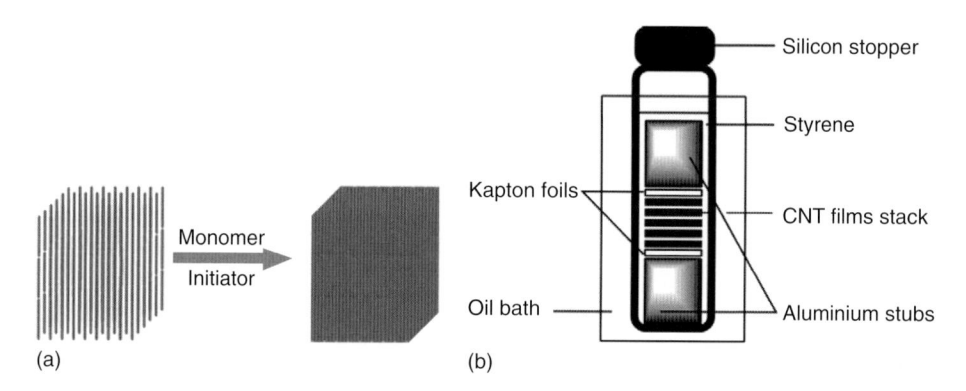

Figure 2.39 (a) Schematic representation of three MWNT arrays for forming aligned MWNT/PS composite. (b) The setup used to make the composite. (Reproduced with permission from Ref. [132], Elsevier (2011).)

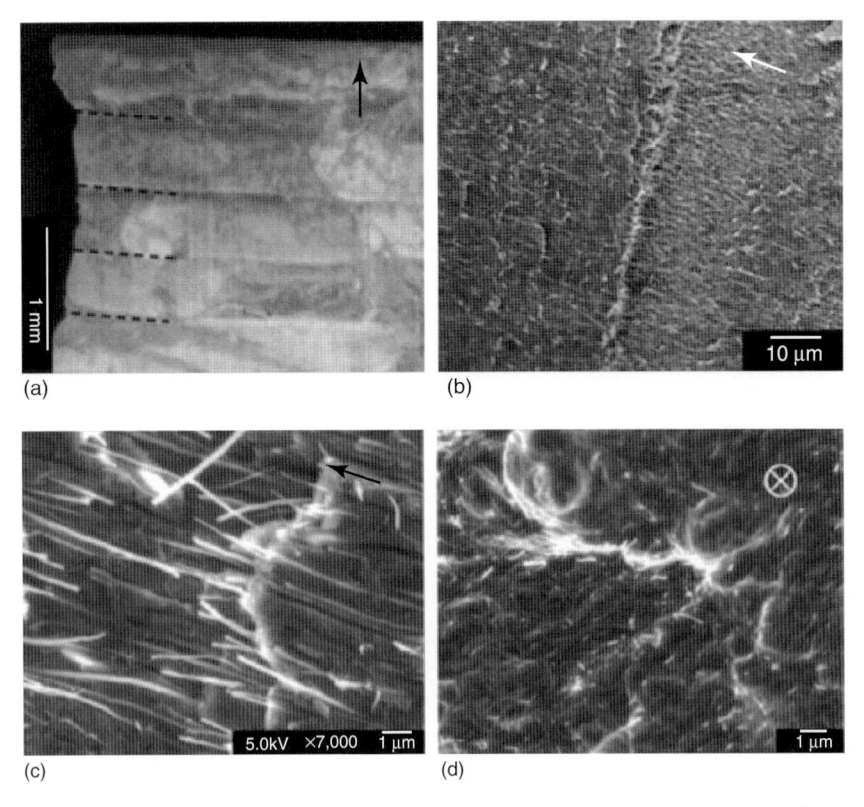

Figure 2.40 SEM micrographs of an MWNT/PS composite. (a) Side view containing five nanotube carpets. Dashed lines indicate carpet–carpet interface; (b) interface between the layers of two carpets; (c) higher magnification of the side view of the composite and (d) the top view of the composite showing the nanotube tips. The arrows indicate the alignment direction of nanotubes. (Reproduced with permission from Ref. [132], Elsevier (2011).)

In situ polymerization is also an effective technique for synthesizing CNT/polymer composites with the matrices made from high-performance polymers [136, 137]. Polyimide (PI) is widely used for many industrial applications such as electronic packaging and composite matrix materials because of its excellent thermal stability, flexibility, high glass transition temperature and good chemical resistance. PI can be synthesized from the reaction between a dianhydride and a diamine. Jiang *et al.* prepared MWNT/PI composite films by dispersing bare (chemical vapour deposition (CVD) synthesized and laser ablating tubes) nanotubes in N,N dimethylacetamide (DMAc) solvent under sonication [137]. The monomers, that is pyromellitic dianhydride and 4,4′-diaminodiphenylether were added to the nanotube suspension. By stirring the solution mixture for 3 h in a nitrogen atmosphere, polyamic-acid (PAA)-containing MWNT was obtained accordingly. The MWNT-PAA was subjected to solvent casting, drying and curing to form MWNT/PI nanocomposite films (Figure 2.43). SEM examinations

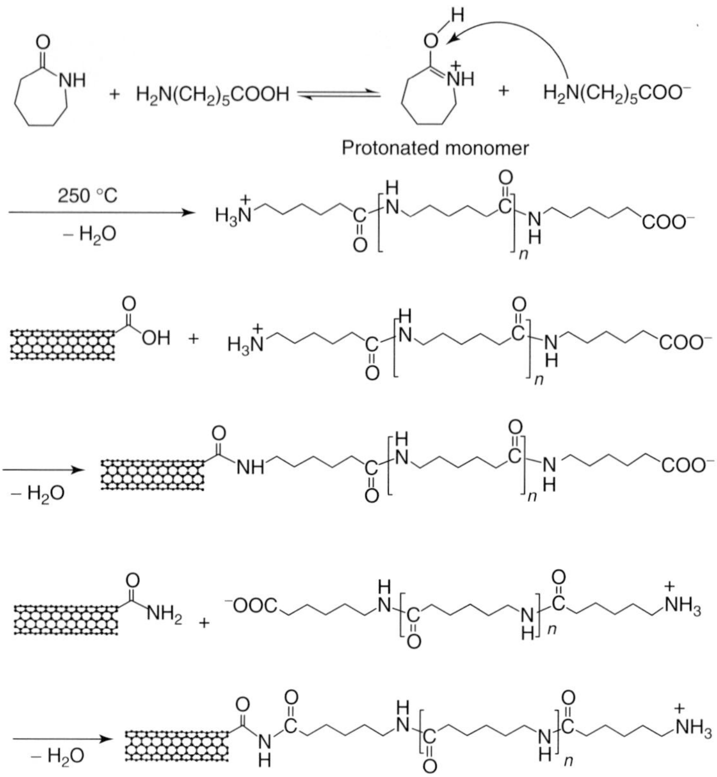

Figure 2.41 Synthesis of SWNT/PA-6 composites using SWNT-COOH and SWNT-CONH$_2$. (Reproduced with permission from Ref. [134], The American Chemical Society (2006).)

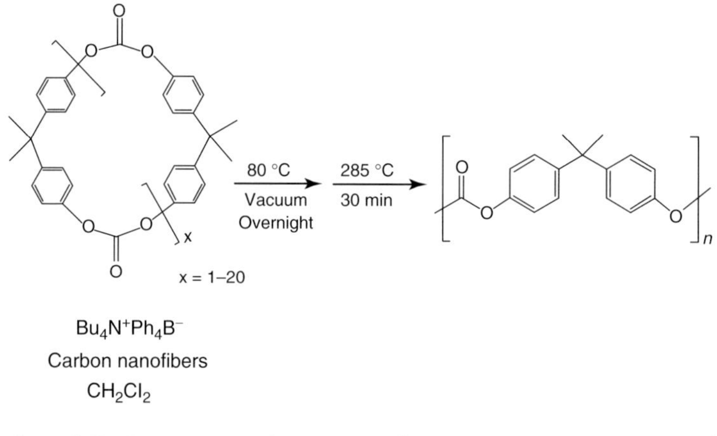

Figure 2.42 Ring-opening polymerization of CNF/PC nanocomposites. (Reproduced with permission from Ref. [135], Elsevier (2005).)

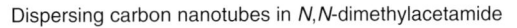

Figure 2.43 The strategy for the preparation of MWNT/polyimide nanocomposite film using bare nanotubes. (Reproduced with permission from Ref. [137], Elsevier (2005).)

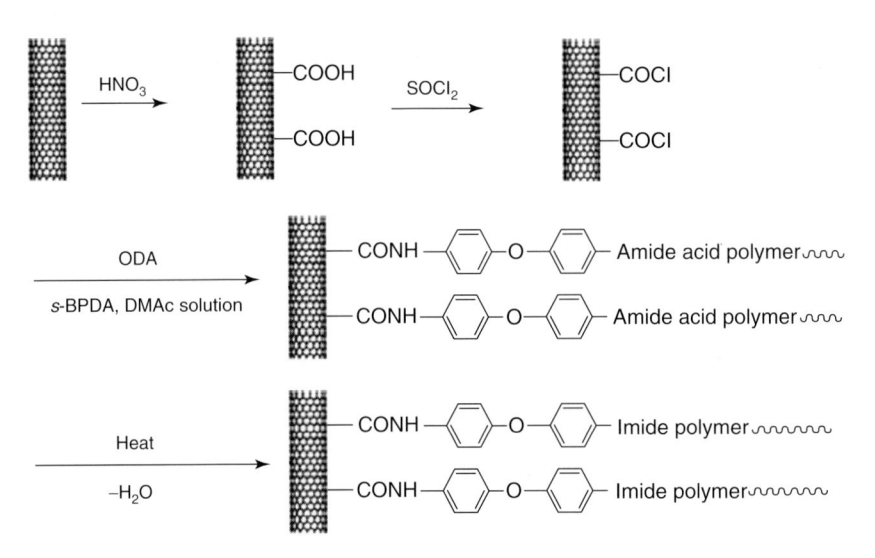

Figure 2.44 Schematic representation for the preparation of MWNT/polyimide nanocomposite film using functionalized nanotubes. (Reproduced with permission from Ref. [136], Wiley (2007).)

Table 2.1 Patent processes for fabricating polymer nanocomposites filled with 2D carbonaceous nanofillers.

Patent number and year	Type of nanocomposites	Fabrication processes	Inventor and assignee	Patent title
US 7935754 (2011)	TRG/PMMA	Solution mixing	Prud'Homme, R.K. Princeton University (USA)	Automotive body panel containing thermally exfoliated graphite oxide
US 7914844 (2011)	iGO/PS	Solution mixing	Stankovich, S., Nguyen, S. B., Ruoff, R. S. Northwestern University (USA)	Stable dispersions of polymer-coated graphitic nanoplatelets
US 7659350 (2010)	TRG/PMMA	*In situ* polymerization	Prud'Homme, R.K. Princeton University (USA)	Polymerization method for formation of thermally exfoliated graphite oxide containing polymer
US 7745528 (2010)	TRG/elastomer	Solution mixing; melt compounding	Prud'Homme, R.K. Princeton University (USA)	Functional graphene-rubber nanocomposites
US 7758783 (2010)	EG/polymer with matrix materials based on PE, epoxy and phenolic resin	Compression molding	Shi, J.J, Zhamu, A, Jang, B.Z. Nanotek Instruments, Inc., Dayton (USA)	Continuous production of exfoliated graphite composite compositions and flow field plates
US 7566410 (2009)	GNP/polymer with matrix materials based on PA-6, epoxy and phenolic resin	Solution coating; tow-preg; slurry molding; platelet/resin spraying	Song, L.L., Guo, J.S., Zhamu, A., Jang, B. Z. Nanotek Instruments, Inc., Dayton (USA)	Highly conductive nanoscaled graphene plate nanocomposite

2.5 In situ Polymerization

Table 2.2 Patent processes for producing polymer nanocomposites filled with carbon nanotubes and carbon nanofibers.

Patent number and year	Materials	Fabrication processes	Inventor and assignee	Patent title
US 7744844 (2010)	Fluorinated SWNT/polymer nanocomposites with matrix material mainly based on medium density polyethylene	Solution mixing	Barrera, E.V., Wilkins, R., Shofner, M., Pulikkathara, M.X., Vaidyanathan, R. Rice University, Houston (USA)	Functionalized carbon nanotube-polymer composites and interactions with radiation
US 7652084 (2010)	CNF/polyolefin nanocomposites	A two-step procedure of solution mixing and melt mixing	Chu, B., Hsiao, B.S. State University of New York, Albany (USA)	Nanocomposite fibers and films containing polyolefin and surface-modified carbon nanotubes
US 7361430 (2008)	SWNT/Nafion nanocomposite	Solution mixing	Gennett, B., Raffaelle, R.P., Landi, B.J., Heben, M.J. United States Department of Energy	Carbon nanotube-polymer composite actuators
US 7402264 (2008)	SWNT/polyimide	In situ polymerization	Ounaies, Z., Park, C., Harrison, J.S., Holloway, N.M., Draughon, G.K. National Aeronautics and Space Administration (USA)	Sensing/actuating materials made from carbon nanotube polymer composites and methods for making same
US 7285591 (2007)	SWNT/polymer nanocomposites with matrix materials based on PE and PS	Melt mixing via contacting SWNT dispersion with polymer melt	Winey, K.I., Haggenmueller, R., Du, F.M., Zhou, W. University of Pennsylvania, Philadelphia (USA)	Polymer-nanotube composites, fibers and processes
US 7153903 (2006)	SWNT/polymer nanocomposites with matrix material mainly based on PS	In situ miniemulsion polymerization	Barraza, H.J., Balzano, L., Pompeo, F., Rueda, O. L., O'Rear, E.A., Resasco, D.E. University of Oklahoma, Norman (USA)	Carbon nanotube-filled composites prepared by *in situ* polymerization
US 6762237 (2004)	CNT/polymer nanocomposite	Solution mixing	Glatkowski, P.J., Arthur, D.J. Eikos, Inc., Franklin, MA (USA)	Nanocomposite dielectrics

revealed that few laser ablating nanotube bundles can still be observed in the PI matrix. In this regard, Hu *et al.* [136] employed functionalized MWNTs to ensure better dispersion of the nanotubes in the PI matrix. The strategy for *in situ* polymerization of the MWNT/PI composites is shown in Figure 2.44. In the process, carboxylated nanotubes were reacted with thionyl chloride to form acylated MWNTs, followed by dispersing in DMAc under sonication in the presence of diamine (ODA), that is 4,4′-diaminodiphenylether. The suspension was then reacted with 3,3′,4,4′-biphenyltetracarboxylic dianhydride (s-BPDA) to yield MWNT-poly(amic acid). The acyl groups of the MWNTs participated in the reaction through the formation of amide bonds. The MWNT-poly(amic acid) was finally imidized at 350 °C for 1 h under vacuum to form MWNT/PI nanocomposite films. This process enabled uniform dispersion of MWNTs in the polymer matrix.

2.6
Patent Processes

The advancement in nanotechnology technologies has led to the development of innovated and advanced processing techniques for functionalizing carbonaceous nanofillers and fabricating polymer nanocomposites. These processes pave the way for manufacturing of high-performance functional products for industrial applications. Tables 2.1 and 2.2 list respectively the patent processes granted by the United States Patent and Trademark Office for fabricating iGO-, TRG-, EG- and GNP-filled nanocomposites, and CNT- and CNF-filled polymer composites.

References

1. Liang, J.J., Huang, Y., Zhang, L., Wang, Y., Ma, Y.F., Guo, T.Y., and Chen, Y.S. (2009) *Adv. Funct. Mater.*, **19**, 2297–2302.
2. Zhao, X., Zhang, Q.H., Chen, D.J., and Lu, P. (2010) *Macromolecules*, **43**, 2357–2363.
3. Yang, X., Li, L., Shang, S., and Tao, X. (2010) *Polymer*, **51**, 3431–3435.
4. Xu, X., Hong, W.J., Bai, H., Li, C., and Shi, G.Q. (2009) *Carbon*, **47**, 3538–3543.
5. Matsuo, Y., Tahara, K., and Sugie, Y. (1997) *Carbon*, **35**, 113–120.
6. Niyoki, S., Bekayarova, E., Itkis, M.E., McWilliams, J.L., Hamon, M.A., and Haddon, R.A. (2006) *J. Am. Chem. Soc.*, **128**, 7720–7721.
7. Cao, Y.W., Feng, J.C., and Wu, P.Y. (2010) *Carbon*, **48**, 1683–1685.
8. Yun, Y.S., Bae, Y.H., Kim, D.H., Lee, J.Y., Chin, I.J., and Jin, H.J. (2011) *Carbon*, **49**, 3553–3559.
9. Stankovich, S., Piner, R.D., Nguyen, S.B., and Ruoff, R.S. (2006) *Carbon*, **44**, 3342–3347.
10. Park, S., Dikin, D.A., Nguyen, S.B., and Ruoff, S. (2009) *J. Phys. Chem. C*, **113**, 15801–15804.
11. Lee, S.W., Dreyer, D.R., An, J., Velamakanni, A., Piner, R.D., Park, S., Zhu, Y.W., Kim, S.O., Bielawski, C.W., and Ruoff, R.S. (2010) *Macromol. Rapid Commun.*, **31**, 281–288.
12. McAllister, M.J., Li, J.L., Adamson, D.H., Schniepp, H.C., Abdala, A.A., Liu, J., Herrera-Alonso, M., Milius, D.L., Car, R., Prud'homme, R.K., and Aksay, I.A. (2007) *Chem. Mater.*, **19**, 4396–4404.

13. Mahmoud, W.E. (2011) *Eur. Polym. J.*, **47**, 1534–1540.
14. Ramanathan, T., Abdala, A.A., Stankovich, S., Dikin, D.A., Herrera-Alonso, M., Piner, R.D., Adamson, D.H., Schniepp, H.C., Chen, X., Ruoff, R.S., Nguyen, S.B., Aksay, I.A., Prud'Homme, R.K., and Brinson, L.C. (2008) *Nat. Nanotechnol.*, **3**, 27–331.
15. Choi, E.Y., Han, T.H., Hong, J.Y., Kim, J.E., Lee, S.H., Kim, H.W., and Kim, S.O. (2010) *J. Mater. Chem.*, **20**, 1907–1912.
16. Xu, Y., Bai, H., Lu, G.W., Li, C., and Shi, G.Q. (2008) *J. Am. Chem. Soc.*, **130**, 5856–5857.
17. An, X.H., Simmons, T., Shah, R., Wolfe, C., Lewis, K.M., Washington, M., Nayak, S.K., Talapatra, S., and Kar, S. (2010) *Nano Lett.*, **10**, 4295–4301.
18. Lomeda, J.R., Doyle, C.D., Kosynkin, D.V., Hwang, W.F., and Tour, J.M. (2008) *J. Am. Chem. Soc.*, **130**, 16201–16206.
19. Teng, C.C., Ma, C.C., Lu, C.H., Yang, S.Y., Lee, S.H., Hsiao, M.C., Yen, M.Y., Chiou, K.C., and Lee, T.M. (2011) *Carbon*, **49**, 5107–5116.
20. Liu, Y.T., Xie, X.M., and Ye, X.Y. (2011) *Carbon*, **49**, 3529–3537.
21. Li, J., Sham, M.L., Kim, J.K., and Marom, G. (2007) *Compos. Sci. Technol.*, **67**, 296–305.
22. Osbeck, S., Bradley, R.H., Liu, C., Idriss, H., and Ward, S. (2011) *Carbon*, **49**, 4322–4330.
23. Bradley, R.H. and Mathieson, I. (1997) *J. Colloid Interface Sci.*, **194**, 338–343.
24. Geng, Y., Li, J., Wang, S.J., and Kim, J.K. (2008) *J. Nanosci. Nanotechnol.*, **8**, 6238–6246.
25. Wang, W.P. and Pan, C.Y. (2004) *Eur. Polym. J.*, **40**, 543–548.
26. Lu, J., Do, I.W., Fukushima, H., Lee, I.S., and Drzal, L.T. (2010) *J. Nanomater.*, **2010** (Article ID 186486), 11.
27. Terrones, M., Botello-Mendez, A.R., Compos-Delgado, J., Lopez-Urias, F., Vega-Cantu, Y.I., Rodriguez-Macias, F.J., Elias, A.L., Munoz-Sandoval, E., Cano-marquez, A.G., Charlier, J.C., and Terrones, H. (2010) *Nano Today*, **5**, 351–372.
28. Stone, A.J. and Wales, D.J. (1986) *Chem. Phys. Lett.*, **128**, 501–503.
29. Hirsch, A. (2002) *Angew. Chem. Int. Ed.*, **41**, 1853–1859.
30. Hamon, M.A., Hui, H., Bhowmik, P., Itkis, H.M., and Haddon, R.C. (2002) *Appl. Phys. A*, **74**, 333–338.
31. Tasis, D., Tagmatarchis, N., Bianco, A., and Prato, M. (2006) *Chem. Rev.*, **106**, 1105–1136.
32. Gabriel, G., Sauthier, G., Fraxedas, J., Moreno-manas, M., Martinez, M.T., Miravittles, C., and Casabo, J. (2006) *Carbon*, **44**, 1891–1897.
33. Banerjee, S., Hemraj-Benny, T., and Wong, S.S. (2005) *Adv. Mater.*, **17**, 17–29.
34. Spitalsky, Z., Tasis, D., Papagelis, K., and Galiotis, C. (2010) *Prog. Polym. Sci.*, **35**, 357–401.
35. Georgakilas, V., Kordatos, K., Prato, M., Guldi, D.M., and Hirsch, A. (2002) *J. Am. Chem. Soc.*, **124**, 760–761.
36. Tagmatarchis, N. and Prato, M. (2004) *J. Mater. Chem.*, **14**, 437–439.
37. Coleman, K.S., Bailey, S.R., Fogden, S., and Green, M.L. (2003) *J. Am. Chem. Soc.*, **125**, 8722–8723.
38. Holzinger, M., Vostrowsky, O., Hirsch, A., Hennrich, F., Kappes, M., Weiss, R., and Jellen, F. (2001) *Angew. Chem. Int. Ed.*, **40**, 4002–4005.
39. Khabashesku, V.N., Margrave, J.L., and barrera, E.V. (2005) *Diamond Relat. Mater.*, **14**, 859–866.
40. Lee, Y.S. (2007) *J. Fluorine Chem.*, **128**, 392–403.
41. Zhang, M., Su, L., and Mao, L.Q. (2006) *Carbon*, **44**, 276–283.
42. Khare, B.N., Wilhite, P., and Meyyappan, M. (2004) *Nanotechnology*, **15**, 1650–1654.
43. Islam, M.F., Rojas, E., Bergey, D.M., Johnson, A.T., and Yodh, A.G. (2003) *Nano Lett.*, **3**, 269–273.
44. Richard, C., Balavoine, F., Schultz, P., Ebbesen, T.W., and Mioskowski, C. (2003) *Science*, **300**, 775–778.
45. Rastogi, R., Kaushal, R., Tripathi, S.K., Sharma, A.L., Kaur, I., and Bharadwaj, L.M. (2008) *J. Colloid Interface Sci.*, **328**, 421–428.
46. O'Connell, M.J., Boul, P., Ericson, L.M., Huffman, C., Wang, Y., Haroz,

E., Kuper, C., Tour, J., Ausman, K.D., and Smalley, R.E. (2001) *Chem. Phys. Lett.*, **342**, 265–271.
47. Curran, S.A., Ajayan, P.M., Blau, W.J., Carroll, D.L., Coleman, J.N., and Dalton, A.B. (1998) *Adv. Mater.*, **10**, 1091–1093.
48. Star, A., Liu, Y., Grant, K., Ridvan, L., Stoddart, J.F., Steuerman, D.W., Diehl, M.R., Boukai, A., and Heath, J.R. (2003) *Macromolecules*, **36**, 553–560.
49. Hwang, J., Jang, J., Hong, K., Kim, K.N., Han, J.H., Shin, K., and Park, C.E. (2011) *Carbon*, **49**, 106–110.
50. Giulianini, M., Waclawik, E.C., Bell, J.M., Crescenzi, M., Casttucci, P., Scarselli, M., and Motta, N. (2009) *Appl. Phys. Lett.*, **95**, 013304 (3 pp).
51. Vadukumpully, S., Paul, J., Mahanta, N., and Valiyaveettil, S. (2011) *Carbon*, **49**, 198–205.
52. Kim, H., Miura, Y., and Macosko, C.W. (2010) *Chem. Mater.*, **22**, 3441–3450.
53. Cai, D.Y., Yusoh, K., and Song, M. (2009) *Nanotechnology*, **20**, 085712 (5 pp).
54. Stankovich, S., Dikin, D.A., Dommett, G.H., Kohlhaas, K.M., Zimney, E.J., Stach, E.A., Piner, R.D., Nguyen, S.B., and Ruoff, R.S. (2006) *Nature*, **442**, 282–286.
55. Rafiee, M.A., Rafiee, J., Yu, Z.Z., and Koratkar, N. (2009) *Appl. Phys. Lett.*, **95**, 223103 (3 pp).
56. Liang, J.J., Wang, Y., Huang, Y., Ma, Y.F., Liu, Z.F., Cai, J.M., Zhang, C.D., Cao, H.J., and Chen, Y.S. (2009) *Carbon*, **47**, 922–925.
57. Kim, H.W., Kobayashi, S., Abdurrahim, M.A., Zhang, M.L., Khusainova, A., Hillmyer, M.A., Abdala, A.A., and Macosko, C.W. (2011) *Polymer*, **52**, 1837–1846.
58. Zheng, W.G. and Wong, S.C. (2003) *Compos. Sci. Technol.*, **63**, 225–235.
59. Ramanathan, T., Stankovich, S., Dikin, D.A., Liu, H., Shen, H., Nguyen, S.B., and Brinson, L.C. (2007) *J. Polym. Sci., Part B: Polym. Phys.*, **45**, 2097–2112.
60. Kumar, S., Sun, L.L., Caceres, S., Li, B., Perugini, A., Maguire, R.G., and Zhong, W.H. (2010) *Nanotechnology*, **21**, 105702 (9 pp).
61. Ramanujam, B.T., mahale, R.Y., and Radhakrishnan, S. (2010) *Compos. Sci. Technol.*, **70**, 2111–2116.
62. Li, Y.C., Li, R.K., and Tjong, S.C. (2010) *J. Mater. Res.*, **25**, 1645–1648.
63. Li, Y.C., Li, R.K., and Tjong, S.C. (2010) *Synth. Met.*, **160**, 1912–1919.
64. Yasmin, A., Luo, J.J., and Daniel, I.M. (2006) *Compos. Sci. Technol.*, **66**, 1182–1189.
65. Yu, A.P., Ramesh, P., Itkis, M.E., Bekyarova, E., and Haddon, R.C. (2007) *J. Phys. Chem. C*, **111**, 7565–7569.
66. Raza, M.A., Westwood, A., Brown, A., Hondow, N., and Stirling, C. (2011) *Carbon*, **49**, 4269–4279.
67. Qian, D., Dickey, E.C., Andrews, R., and Rantell, T. (2000) *Appl. Phys. Letters*, **76**, 2868–2870.
68. Sun, G.X., Chen, G.M., Liu, Z.P., and Chen, M. (2010) *Carbon*, **48**, 1434–1440.
69. Khan, M.U., Gomes, V.G., and Altarawneh, I.S. (2010) *Carbon*, **48**, 2925–2933.
70. Araujo, R., Paiva, M.C., Proenca, M.F., and Silva, C.J. (2007) *Compos. Sci. Technol.*, **67**, 806–810.
71. Ajayan, P.M., Stephan, O., Colliex, C., and Trauth, D. (1994) *Science*, **265**, 1212–1214.
72. Miyagawa, H. and Drzal, L.T. (2004) *Polymer*, **45**, 5163–5170.
73. Gong, X.Y., Liu, J., Baskaran, S., Voise, R.D., and Young, J.S. (2000) *Chem. Mater.*, **12**, 1049–1052.
74. Wang, D.H., Sihn, S.W., Roy, A.K., Baek, J.B., and Tan, L.S. (2010) *Eur. Polym. J.*, **46**, 1404–1416.
75. Baudot, C. and Tan, C.M. (2011) *Carbon*, **49**, 2362–2369.
76. Villmow, T., Kretzschmar, B., and Potschke, P. (2010) *Compos. Sci. Technol.*, **70**, 2045–2055.
77. Song, P.G., Cao, Z.T., Cai, Y.Z., Zhao, L.P., Fang, Z.P., and Fu, S.Y. (2011) *Polymer*, **52**, 4001–4010.
78. Kim, H.W. and Macosko, C.W. (2009) *Polymer*, **50**, 3797–3809.
79. Zhang, H.B., Zheng, W.G., Yan, Q., Wang, J.W., Lu, Z.H., Ji, C.Y., and Yu, Z.Z. (2010) *Polymer*, **51**, 1191–1196.
80. Kim, H.W. and Macosko, C.W. (2008) *Macromolecules*, **41**, 3317–3327.

81. Rafiq, A., Cai, D.Y., Jin, J., and Song, M. (2010) *Carbon*, **48**, 4309–4314.
82. Uhl, F.M., Yao, Q., and Wilke, C.A. (2005) *Polym. Adv. Technol.*, **16**, 533–540.
83. Chen, G.H. and Li, Y.C. (2007) *Polym. Eng. Sci.*, **47**, 882–888.
84. Kalaitzidou, K., Fukushima, H., and Drzal, L.T. (2007) *Composites Part A*, **38**, 1675–1682.
85. Chen, G.H., Chen, X.F., Wang, H.Q., and Wu, D.J. (2007) *J. Appl. Polym. Sci.*, **103**, 3470–3475.
86. Li, M.L. and Jeong, Y.G. (2011) *Composites Part A*, **42**, 560–566.
87. Chatterjee, S., Nuesch, F.A., and Chu, B.T. (2011) *Nanotechnology*, **22**, 275714 (8 pp).
88. Murariu, M., Dechief, A.L., Bonnaud, L., Paint, Y., Gallos, A., Fontaine, G., Bourbigot, S., and Dubois, P. (2010) *Polym. Degrad. Stab.*, **95**, 889–900.
89. Katbab, A.A., Hrymak, A.N., and Kasmadjian, K. (2008) *J. Appl. Polym. Sci.*, **107**, 3425–3433.
90. Andrews, R., Jacques, D., Minot, M., and Rantell, T. (2002) *Macromol. Mater. Eng.*, **287**, 395–403.
91. Chen, G.X., Li, Y., and Shimizu, H. (2007) *Carbon*, **45**, 2334–2340.
92. Zhang, W.D., Shen, L., Phang, I.Y., and Liu, T.X. (2004) *Macromolecules*, **37**, 256–259.
93. McNally, T., Potschke, P., Halley, P., Murphy, M., Martin, D., Bell, S.E., Brennan, G.P., Bein, D., Lemoine, P., and Quinn, J.P. (2005) *Polymer*, **46**, 8222–8232.
94. Lee, S.H., Cho, E.N., Jeon, S.H., and Youn, J.R. (2007) *Carbon*, **45**, 2810–2822.
95. Lee, G.W., Jagannathan, S., Chae, H.G., Minus, M.L., and Kumar, S. (2008) *Polymer*, **49**, 1831–1840.
96. Prashantha, K., Soulestin, J., Lacrampe, M.F., Krawczak, P., Dupin, G., and Claes, M. (2009) *Compos. Sci. Technol.*, **69**, 1756–1763.
97. Logakis, E., Pandis, C.H., Peoglos, V., Pissis, P., Pionteck, J., Potschke, P., Micusil, M., and Omastova, M. (2009) *Polymer*, **50**, 5103–5111.
98. Logakis, E., Pissis, P., Pospiech, D., Korwitz, A., Krause, B., Reuter, U., and Potschke, P. (2010) *Eur. Polym. J.*, **46**, 928–936.
99. Barick, A.K. and Tripathy, D.K. (2010) *Composites Part A*, **41**, 1471–1482.
100. Potts, J.R., Lee, S.H., Alam, T.M., An, J.H., Stoller, M.D., Piner, R.D., and Ruoff, R.S. (2011) *Carbon*, **49**, 2615–2623.
101. Luong, N., Hippi, U., Korhonen, J.T., Soininen, A.J., Ruokolainene, J., Johansson, L.S., Nam, J.D., Sinh, L.H., and Seppala, J. (2011) *Polymer*, **52**, 5237–5242.
102. (a) Edmondson, S., Osborne, V.L., and Huck, W.T. (2004) *Chem. Soc. Rev.*, **33**, 14–22; (b) Layek, R.K., Samanta, S., Chatterjee, D.P., and Nandi, A.K. (2010) *Polymer*, **51**, 5846–5856.
103. Gonçalves, G., Marques, P.A., Barros-Timmons, A., Bdkin, I., Singh, M.K., Emami, N., and Grácio, J. (2010) *J. Mater. Chem.*, **20**, 9927–9934.
104. Xu, Z. and Gao, C. (2010) *Macromolecules*, **43**, 6716–6723.
105. Hu, H.T., Wang, X.B., Wang, J.C., Wan, L., Liu, F.M., Zheng, H., Chen, R., and Xu, C.H. (2010) *Chem. Phys. Lett.*, **484**, 247–253.
106. Patrole, A.S., Patole, S.P., Kang, H., Yoo, J.B., Kim, T.H., and Ahn, J.H. (2010) *J. Colloid Interface Sci.*, **350**, 530–537.
107. Huang, Y.J., Qin, Y.W., Zhou, Y., Niu, H., Yu, Z.Z., and Dong, J.Y. (2010) *Chem. Mater.*, **22**, 4096–4102.
108. Chen, G.H., Wu, D.J., Weng, W.G., and Yan, W.L. (2001) *J. Appl. Polym. Sci.*, **82**, 2506–2513.
109. Zheng, G.H., Wu, J.S., Wang, W.P., and Pan, C.Y. (2004) *Carbon*, **24**, 2839–2847.
110. Li, L.W., Luo, Y.L., and Li, Z.Q. (2007) *Smart Mater. Struct.*, **16**, 1570–1574.
111. Chen, G.H., Weng, W.G., Wu, D.J., and Wu, C.L. (2003) *Eur. Polym. J.*, **39**, 2329–2335.
112. Weng, W.G., Chen, G.H., Wu, D.J., Chen, X.F., Lu, J.R., and Wang, P.P. (2004) *J. Polym. Sci., Part B: Polym. Phys.*, **42**, 2844–2856.
113. Chen, G.H., Wu, C.L., Weng, W.G., Wu, D.J., and Yan, W.L. (2003) *Polymer*, **44**, 1781–1784.

114. Du, X.S., Xiao, M., Meng, Y.Z., and Hay, A.S. (2004) *Polymer*, **45**, 6713–6718.
115. Lu, W., Lin, H., Wu, D.J., and Chen, G.H. (2006) *Polymer*, **47**, 440–444.
116. Srivastava, N.K. and Mehra, R.M. (2008) *J. Appl. Polym. Sci.*, **109**, 3991–3999.
117. Panwar, V. and Mehra, R.M. (2008) *Eur. Polym. J.*, **44**, 2367–2375.
118. Zhao, B., Hu, H., Yu, A., Perea, D., and Haddon, R.C. (2005) *J. Am. Chem. Soc.*, **127**, 8197–8203.
119. Li, H., Cheng, F., Duft, A.M., and Adronov, A. (2005) *J. Am. Chem. Soc.*, **127**, 14518–14524.
120. Qin, S., Qin, D., Ford, W.T., Resasco, D.E., and Herrera, J.E. (2004) *J. Am. Chem. Soc.*, **126**, 170–176.
121. Kong, H., Gao, C., and Yan, D. (2004) *J. Am. Chem. Soc.*, **126**, 412–413.
122. Shanmugharaj, A.M., Bae, J.H., Nayak, R.R., and Ryu, S.H. (2007) *J. Polym. Sci., Part A: Polym. Chem.*, **45**, 460–470.
123. Wu, H.X., Tong, R., Qiu, X.Q., Yang, H.F., Lin, Y.H., Cai, R.F., and Qian, S.X. (2007) *Carbon*, **45**, 152–159.
124. Matrab, T., Chancolon, J., L'hermite, M.M., Rouzaud, J.N., Deniau, G., Boudou, J.P., Chehimi, M.M., and Delamar, M. (2006) *Colloids Surf. Physicochem. Eng. Aspects*, **287**, 217–221.
125. Choi, W.S. and Ryu, S.H. (2011) *Colloids Surf., A Physicochem. Eng. Aspects*, **375**, 55–60.
126. Hwang, G.L., Shieh, Y.T., and Hwang, K.C. (2004) *Adv. Funct. Mater.*, **14**, 487–491.
127. Wang, M., Pramoda, K.P., and Goh, S.H. (2005) *Polymer*, **46**, 11510–11516.
128. Jia, Z., Wang, Z., Xu, C., Liang, J., Wei, B., Wu, D., and Zhu, S. (1999) *Mater. Sci. Eng. A*, **271**, 395–400.
129. Park, S.J., Cho, M.S., Lim, S.T., Choi, H.J., and Jhon, M.S. (2003) *Macromol. Rapid Commun. s*, **24**, 1070–1073.
130. Meng, Q.J., Zhang, X.X., Bai, S.H., and Wang, X.C. (2007) *Chin. J. Chem. Phys.*, **20**, 660–664.
131. Choi, J.J., Zhang, K., and Lim, J.Y. (2007) *J. Nanosci. Nanotechnol.*, **7**, 3400–3403.
132. Koziol, K.K., Boncel, S., Shaffer, M.S., and Windle, A.H. (2011) *Compos. Sci. Technol.*, **71**, 1606–1611.
133. Gao, J.B., Itkis, M.E., Yu, A.P., Bekyarova, E., Zhao, B., and Haddon, R.C. (2005) *J. Am. Chem. Soc.*, **127**, 3847–3854.
134. Gao, J.B., Zhao, B., Itkis, M.E., Bekyarova, E., Hu, H., Kranak, V., Yu, A.P., and Haddon, R.C. (2006) *J. Am. Chem. Soc.*, **128**, 7492–7496.
135. Higgins, B.A. and Brittain, W.J. (2005) *Eur. Polym. J.*, **41**, 889–893.
136. Hu, N.T., Zhou, H.W., Dang, G.D., Rao, X.H., Chen, C.H., and Zhang, W.J. (2007) *Polym. Int.*, **56**, 655–659.
137. Jiang, X.W., Bin, Y.Z., and Matsuo, M. (2005) *Polymer*, **46**, 7418–7424.

3
Thermal Properties of Polymer Nanocomposites

3.1
Crystallization

The crystalline morphology and structure as well as the degree of crystallinity of semicrystalline polymers affect their physical and mechanical properties significantly. The crystalline morphology and crystalline kinetics of semicrystalline polymers depend greatly on the processing conditions employed and the presence of carbonaceous nanofillers. During solidification from the melt, low loading levels of carbonaceous nanofillers act as heterogeneous nucleation sites for the polymer crystallites. This reduces nucleation activation energy, leading to an acceleration of the crystallization and a decrease of the spherulite size [1, 2]. In some cases, the crystal structure of the polymers can change to other polymorphs because of the carbonaceous nanofiller additions [3, 4]. For example, Manna and Nandi reported that multiwalled carbon nanotubes (MWNTs) can induce β-phase polyvinylidene fluoride (PVDF). The amount of this phase depends on the processing conditions and nanotube functionalization [3]. Grady et al. indicated that single-walled carbon nanotubes (SWNTs) promote the growth of β-form of polypropylene (PP) but decrease α-PP phase [4].

The process of crystallization can be monitored and controlled under a constant temperature condition, widely known as *isothermal crystallization*, or at a constant cooling rate, that is, *nonisothermal crystallization*. The crystallization of polymers and their nanocomposites is affected by the processing conditions including shear flow, temperature of the melt, filler addition, and kinetics of cooling. The crystallization kinetics of polymer nanocomposites is crucial for assessing their microstructural development during melt processing. The nonisothermal crystallization behavior of polymers and their composites is of practical interest because plastic products are manufactured under nonisothermal conditions. Differential scanning calorimetry (DSC) is one of the most popular techniques for measuring crystallization behavior of polymers. The DSC results are typically analyzed in terms of the degree of crystallinity, crystallization temperature, crystallization kinetics, and melting temperature. Semicrystalline polymers consist of ordered chain-folded lamellar crystallites dispersed within noncrystalline or amorphous regions; thus the determination of glass-transition temperature (T_g)

Polymer Composites with Carbonaceous Nanofillers: Properties and Applications, First Edition. Sie Chin Tjong.
© 2012 Wiley-VCH Verlag GmbH & Co. KGaA. Published 2012 by Wiley-VCH Verlag GmbH & Co. KGaA.

is considered of practical interest. Proper understanding of T_g is beneficial for the design and selection of polymers for specific applications. T_g is defined as the temperature at which amorphous polymers undergo a transition from a viscous, rubbery state to a rigid, glassy solid on cooling. The glass transition is not a phase transition as defined in thermodynamics but rather a kinetic effect [5].

3.2 Characterization Techniques for Crystallization

3.2.1 Dynamic Mechanical Analysis

Dynamic mechanical analysis (DMA) is a useful tool for characterizing the viscoelasticity of a polymer. It is commonly used to detect T_g and other small relaxation of polymers because of its high sensitivity. In DMA measurements, a dynamic (sinusoidal) stress is applied to a sample, and the strain (ε) of the material is measured or vice versa. The stress applied (σ) as a function of time (t) at a given frequency (ω) is given by the following equation

$$\sigma = \sigma_0 \sin(\omega t + \delta) \tag{3.1}$$

where σ_0 is the amplitude of the stress applied and δ the phase angle. For viscoelastic polymer, there exists a 90° phase lag of strain with respect to stress. Accordingly, the strain can be expressed as

$$\varepsilon = \varepsilon_0 \sin(\omega t) \tag{3.2}$$

Equation (3.1) can be rewritten in the following form

$$\sigma = (\sigma_0 \cos\delta) \sin(\omega t) + (\sigma_0 \sin\delta) \cos(\omega t) \tag{3.3}$$

The ($\sigma_0 \cos \delta$) represents the component of the stress that is in phase with the strain and ($\sigma_0 \sin \delta$) corresponds to the component of the stress that is out of phase with the strain. Through Hooke's law, the response of a sample is manifested in terms of the complex modulus, $E^* = E' + iE''$, and the loss factor, $\tan \delta = E''/E'$, as a function of temperature and frequency [6]. The storage modulus, E', is given by

$$E' = (\sigma_0 \cos \delta)/\varepsilon_0 \tag{3.4}$$

E' is related to the energy stored per cycle of sinusoidal deformation and characterizes the elastic behavior of the polymer. The loss modulus, E'', is defined as

$$E = (\sigma_0 \sin \delta)/\varepsilon_0 \tag{3.5}$$

E'' is a measure of the energy dissipated or lost as heat per cycle of sinusoidal deformation and characterizes viscous behavior of the polymer. By heating a polymer at a given frequency, the storage modulus decreases at the glass transition, while the loss modulus and the loss tangent reach apparent maximum. ASTM D4065 recommends the use of loss modulus peak as the T_g of the material [7]. However, the tan δ peak has been commonly used to describe the glass transition of semicrystalline polymers [8]. The specimens can be tested in different loading configurations in an analyzer including tension, compression, single cantilever, double cantilever, three-point bending and torsion. The selection of fixtures depends on the nature of materials to be tested. For example, tension fixture is preferred for testing thin films of low stiffness, while three-point bending configuration is often employed for composite materials. Two testing modes can be used for measuring viscoelastic behavior of polymers, that is, temperature sweep and frequency sweep. In the former mode, the frequency of stress or strain oscillation is kept constant while varying the specimen temperature. Semicrystalline polymer with both the amorphous and crystalline components shows separate transitions accordingly. Major peaks in tan(δ) and in E'' with respect to temperature can be attributed to the glass transition. Secondary relaxations associated with the temperature-dependent activation of chain motions from the side groups or branches of a polymer can also be detected [9]. The temperature sweep is a commonly used test mode for the DMA measurements. In the latter test mode, the specimen is measured at a fixed temperature but different frequencies.

3.2.2
Differential Scanning Calorimetry

DSC is widely used for measuring crystallization behavior of polymers and their nanocomposites. In DSC tests, the rate of heat flow into the specimen is proportional to its heat capacity. The T_g value can be estimated as the temperature at the mid-point of heat capacity change. It can be determined more accurately from the inflection point by intersecting tangent line of heat capacity curve with baselines extrapolated from onset temperature (T_{onset}) and end temperature (T_{end}) of the glass transition. For example, Sterzynski et al. [10] determined glass-transition temperatures of polyvinyl chloride (PVC) and solvent-cast MWNT/PVC films using DSC, DMA, and dielectric loss measurements. Figure 3.1a shows the DSC heating curve of the 0.01 wt% MWNT/PVC nanocomposite in which T_g is determined to be 69.4 °C. Figure 3.1b shows the loss peak spectra of this sample subjected to oscillation frequencies of 1 and 10 Hz. The T_g measured from loss tangent peak spectra is frequency dependent, having values of 67 and 73 °C under 1 and 10 Hz, respectively. The T_g measured at 1 Hz in DMA test agrees reasonably with that determined from DSC measurement.

The crystallization kinetics of a polymer under isothermal crystallization can be determined using DSC method. The following Avrami [11a] equation is widely used to describe isothermal crystallization kinetics

$$X_t = 1 - \exp\left(-kt^n\right) \tag{3.6}$$

where X_t is the relative crystallinity at time t, k the rate constant for the crystallization process, and n the Avrami exponent, depending on the nucleation type and the

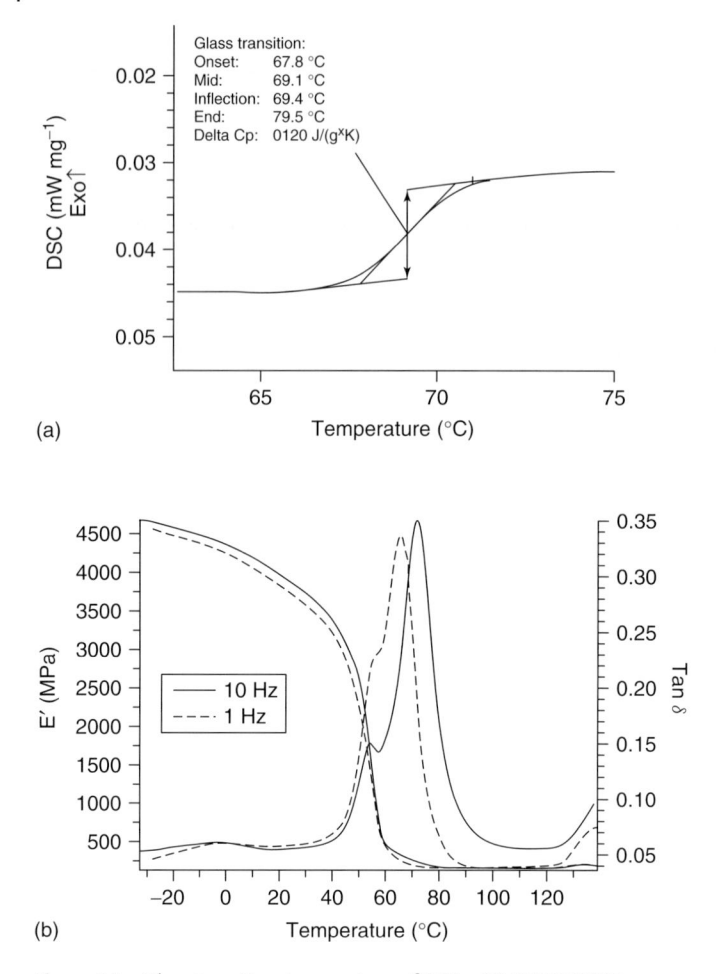

Figure 3.1 Glass-transition temperature of 0.01 wt% MWNT/PVC nanocomposite determined from (a) DSC and (b) DMA tests. (Source: Reproduced with permission from Ref. [10], Elsevier (2010).)

growth geometry of the crystals. For the three- and two-dimensional growths of the nuclei, n has values of 3 and 2, respectively. The double logarithm of the Avrami equation yields the following relation

$$\ln\left[-\ln(1 - X_t)\right] = \ln k + n \ln t \tag{3.7}$$

The k and n values can be determined from the intercepts and slopes, respectively, of the linear plots of $\ln\left[-\ln(1 - X_t)\right]$ versus $\ln t$.

In general, cooling rate is an important factor for crystallization under nonisothermal conditions. Ozawa [11b] modified the Avrami equation by incorporating cooling rate (β) factor

$$X_T = 1 - \exp(-K(T)/\beta^m) \tag{3.8}$$

where X_T is the relative crystallinity at temperature T, $K(T)$ the Ozawa crystallization rate constant, and m the Ozawa exponent. The double logarithm of the Ozawa equation leads to

$$\ln[-\ln(1-X_T)] = \ln K(T) - m \ln \beta \qquad (3.9)$$

Accordingly, the plot of $\ln[-\ln(1-X_T)]$ versus $\ln \beta$ at a fixed temperature produces a straight line. The parameters m and $K(T)$ can be determined from the slope and intercept of the line. However, the linearity is questionable if β varies over a large range. The deviation from straight lines can occur as a result of secondary crystallization. In this case, a combined Avrami and Ozawa equation has been used to describe nonisothermal crystallization kinetics of the MWNT/PCL (polycaprolactone) and GNP/PVDF (graphite nanoplatelet) nanocomposites [12, 13].

3.2.3
Nanocomposites with Graphene Nanofillers

Graphene oxide (GO) and thermally reduced graphene (TRG) additions can enhance melt crystallization temperature and crystallization kinetics of thermoplastics [14; Chapter 2, Ref. 8]. Very recently, Zhang and Qi investigated the crystallization behavior of solvent-cast TRG/PCL nanocomposites filled with 0.5 and 2 wt% TRG using DSC under isothermal and nonisothermal crystallization conditions [14]. For nonisothermal crystallization measurements, the specimen was first heated to 100 °C at 10 °C min^{-1}, held there for 3 min to eliminate previous thermal history, and then cooled to 0 °C at different cooling rates. For isothermal crystallization, the specimen was heated to 100 °C at 10 °C min^{-1} and then fast cooled to the selected crystallization temperature (T_c) at 60 °C min^{-1}. Figure 3.2 shows nonisothermal DSC cooling traces of pure PCL and its nanocomposites. The peak crystallization temperature (T_p) of PCL is $\sim$25.1 °C and increases to 33.6 and 33.8 °C by adding 0.5 and 2 wt% TRG, respectively. This demonstrates that TRGs acting as effective nucleating agents for PCL molecular chains on cooling from the melt. The effect of cooling rates on T_p values of neat PCL and its nanocomposites is shown in Figure 3.2b. For both pure PCL and nanocomposites, T_p tends to shift to lower temperatures as the cooling rate increases. This is because the specimens do not have sufficient time to crystallize at fast cooling rates. Figure 3.3a,b shows the plots of relative crystallinity versus crystallization time for pure PCL and 0.5 wt% TRG/PCL nanocomposite, respectively. These curves generally display sigmoidal features. Furthermore, the crystallization time increases with increasing T_c for pure PCL and nanocomposite specimens. The crystallization kinetics of these specimens can be well described by the Avrami equations (Eqs. (3.6) and (3.7)). In this case, the plots of $\ln[-\ln(1-X_t)]$ versus $\ln t$ for pure PCL and 0.5 wt% TRG/PCL nanocomposite at different T_cs should yield a series of straight lines (Figure 3.4a,b). The n and k values can be determined from the slope and intercept of best-fitting line, respectively. Table 3.1 lists the crystallization kinetic parameters for PCL and its nanocomposites. The *crystallization half-life time* ($t_{0.5}$) defined as

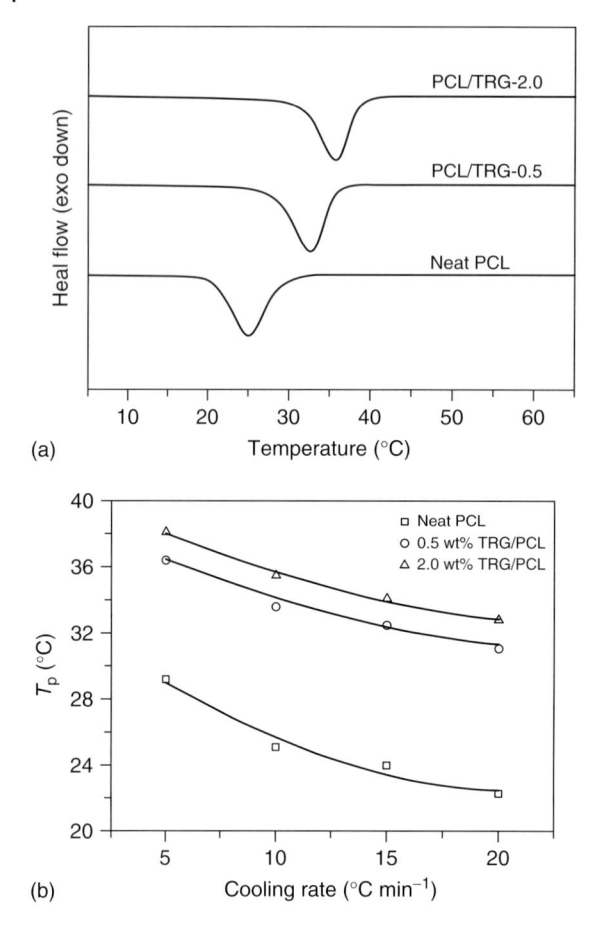

Figure 3.2 (a) Nonisothermal DSC cooling traces at 10 °C min^{-1} and (b) variation of T_p with cooling rate for neat PCL and its nanocomposites. (Source: Reproduced with permission from Ref. [14], The American Chemical Society (2011).)

the time require to achieve 50% of final crystallinity is also listed in this table. Mathematically, it can be expressed as

$$t_{0.5} = \left(\frac{\ln 2}{k}\right)^{1/n} \qquad (3.10)$$

The average values of n for pure PCL and its nanocomposites are 2.4 and 2.5, respectively. These correspond to heterogeneous nucleation followed by three-dimensional growth of PCL spherulites.

The thermomechanical properties of the graphene/polymers have received increasing attention recently [15–17; Chapter 2, Refs. 51, 100]. Figure 3.5a shows storage modulus versus temperature profiles for solution-mixed graphene/PVC nanocomposites [Chapter 2, Ref. 51]. Graphene layers with an average thickness of ~1.18 nm were prepared by ultrasonication of highly ordered pyrolytic graphite in

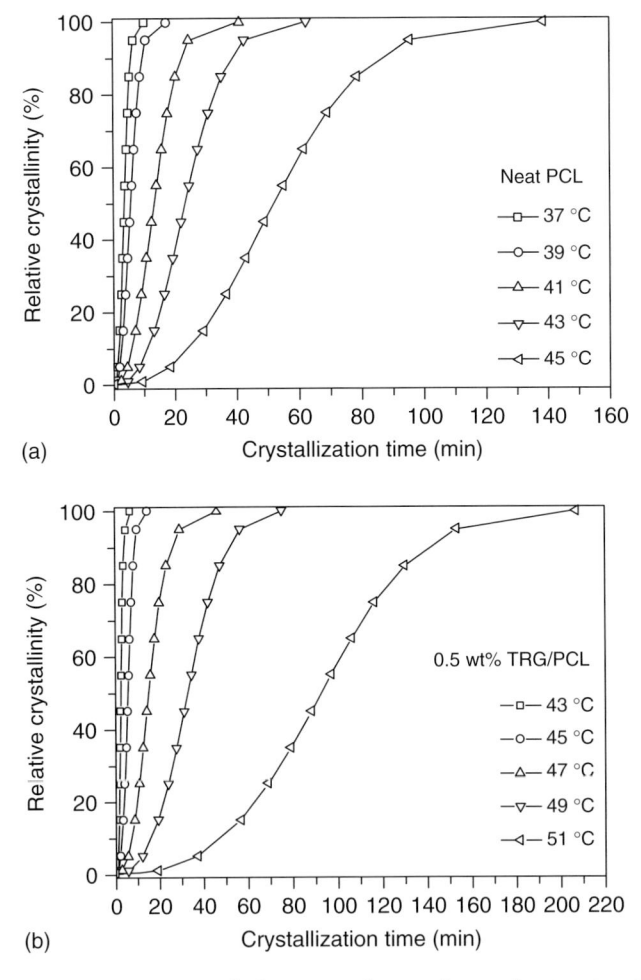

Figure 3.3 Variation of relative crystallinity with crystallization time for (a) neat PCL and (b) 0.5 wt% TRG/PCL nanocomposite at different T_cs. (Source: Reproduced with permission from Ref. [14], The American Chemical Society (2011).)

the presence of a cationic surfactant cetyltrimethylammonium bromide and acetic acid [15]. It can be seen that the storage modulus of PVC increases markedly with increasing graphene content. This implies that graphene nanofillers stiffen PVC effectively. The glass transition appears at ∼85 °C for pure PVC and shifts to 105 °C by adding 2 wt% graphene because of the restrictive motion of PVC segmental chains by graphene fillers (Figure 3.5b). Furthermore, the magnitude of tan δ peak of PVC reduces considerably at graphene loadings ≥1 wt%. The magnitude of loss tangent peak is related to damping capacity or vibrational energy dissipation. This demonstrates that the graphene addition improves the damping capacity of PVC markedly.

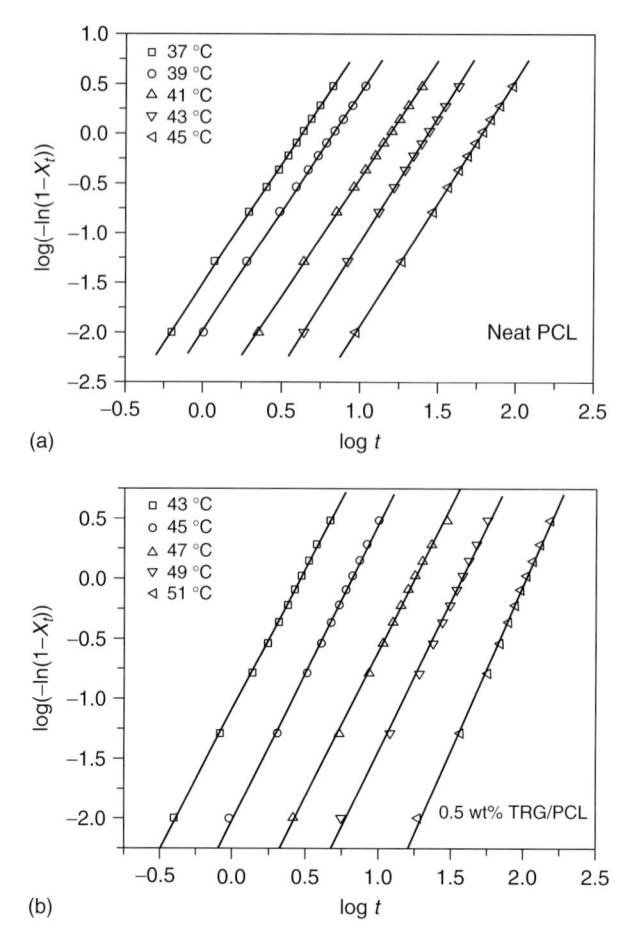

Figure 3.4 Avrami plots of (a) neat PCL and (b) 0.5 wt% TRG/PCL nanocomposite. (Source: Reproduced with permission from Ref. [14], The American Chemical Society (2011).)

Ruoff studied thermomechanical and static tensile behaviors of *in situ* polymerized GO/PMMA (poly(methyl methacrylate)) and rGO/PMMA (reduced graphene oxide) nanocomposites [Chapter 2, Ref. 100]. Figure 3.6a,b shows respective plots of storage modulus and loss modulus versus temperature for the GO/PMMA nanocomposites. The DMA results for both nanocomposite systems are listed in Table 3.2. The stiffening effect of GOs and rGO on PMMA is quite obvious, particularly for the composites filled with GOs. The T_g of PMMA can be increased to a large extent of 17.1 °C through the addition of only 0.05 wt% rGO. In another study, Martin-Gallego [17] also reported that the glass-transition temperature of epoxy can be increased significantly by adding functional graphene sheet (FGS). An increase of almost 40 °C in the T_g can be achieved by adding 1.5 wt% FGS (Figure 3.7).

Table 3.1 Isothermal crystallization kinetic parameters of pure PCL and TRG/PCL nanocomposites at various crystallization temperatures.

Specimen	T_c (°C)	n	k (min^{-n})	$t_{0.5}$ (min)
Neat PCL	37	2.39	3.14×10^{-2}	3.64
	39	2.39	1.05×10^{-2}	5.75
	41	2.38	1.50×10^{-3}	13.23
	43	2.52	2.45×10^{-4}	23.45
	45	2.48	3.84×10^{-5}	52.21
0.5 wt% TRG/PCL	43	2.34	8.02×10^{-2}	2.51
	45	2.47	9.51×10^{-3}	5.67
	47	2.42	9.40×10^{-4}	15.36
	49	2.5	1.14×10^{-4}	32.61
	51	2.76	2.65×10^{-6}	92.24
2.0 wt% TRG/PCL	43	2.49	1.33×10^{-1}	1.94
	45	2.45	2.56×10^{-3}	3.85
	47	2.47	2.31×10^{-3}	10.06
	49	2.29	5.06×10^{-4}	23.5
	51	2.55	1.28×10^{-5}	71.93

Source: Reproduced with permission from Ref. [14], The American Chemical Society (2011).

3.2.4
Nanocomposites with Carbon Nanotubes

The effects of carbon nanotube (CNT) additions on the crystallization behavior of thermoplastics including PCL, polyamide (PA), polyethylene (PE), polyethylene terephthalate (PET), PP, polystyrene (PS), PVC, and PVDF have been studied extensively [1–4, 10, 12, 18–22; Chapter 2, Ref. 68]. CNTs generally act as heterogeneous nucleation sites for macromolecular chains during crystallization. For semicrystalline polymers, the nucleating effect of nanotubes is more effective at very low filler contents, say 0.5–2 wt% [20]. At high filler contents, the nanotubes tend to agglomerate into large bundles, thereby hindering the mobility of polymers and retarding crystallization of the polymer chains in forming spherulites. Thus nanotubes can act either as effective nucleating agents to facilitate crystallization or as physical obstacles to retard crystallization, depending on their volume fractions in the nanocomposites. In some cases, CNT additions induce the phase transformation [3, 4, 23] and transcrystallinity in polymers [19].

A typical example for CNT-induced phase transition in PVDF homopolymer is given herein. PVDF exhibits different polymorph structures including α, β, γ, and δ forms [24]. Monoclinic α-phase with *TGTG* chain conformation is usually found during crystallization of PVDF from the melt. Polar β-phase has an orthorhombic cell with an all-trans conformation. The β polymorph is most attractive because of its piezoelectric, pyroelectric, and ferroelectric characteristics. These unique properties derive from the all-trans conformation of the polymer

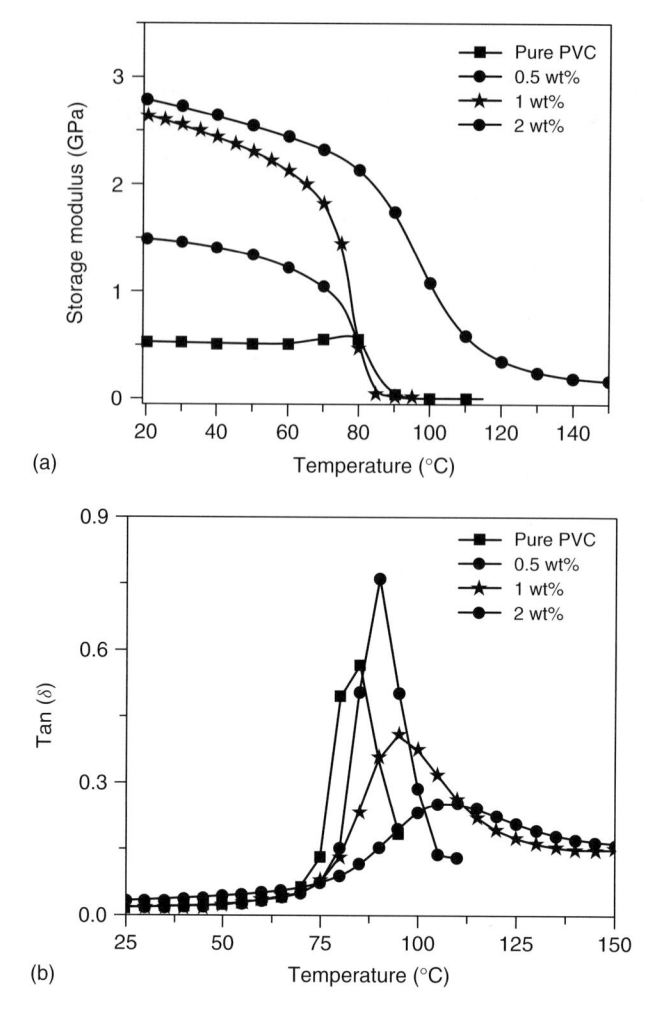

Figure 3.5 (a) Storage modulus and (b) loss tangent versus temperature at 1 Hz for solution-mixed graphene/PVC nanocomposites. (Source: Reproduced with permission from [Chapter 2, Ref. 51], Elsevier (2011).)

chains. To achieve piezoelectric properties, PVDF is mechanically stretched under the application of an electric field, giving rise to polar β-phase [25]. Polar γ-phase also exhibits an orthorhombic structure with a T_3GT_3G chain conformation [24]. It is produced during crystallization from the melt at higher isothermal crystallization temperatures [26].

Manna and Nandi [3] prepared MWNT/PVDF and ester-functionalized MWNT/PVDF nanocomposites by solvent casting. Solvent-cast FMWNT/PVDF (functionalized multiwalled carbon nanotube) nanocomposite films (designated as CP specimens) were melted in a hot stage microscope at 220 °C for 15 min followed by cooling. The heat-treated FMWNT/PVDF nanocomposite films

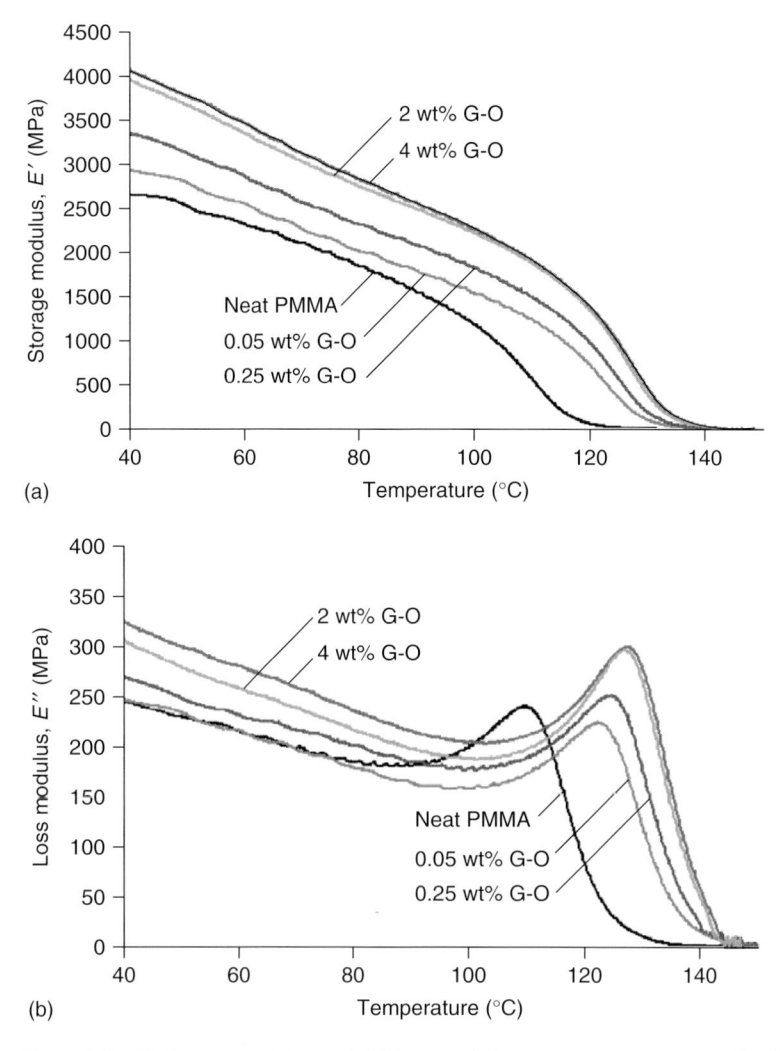

Figure 3.6 (a) Storage modulus and (b) loss modulus versus temperature profiles for *in situ* polymerized GO/PMMA nanocomposites. (Source: Reproduced with permission from [Chapter 2, Ref. 100], Elsevier (2011).)

were designated as melt-cooled samples (MCPs). Ester ($-COOC_2H_5$)-FMWNTs dispersed uniformly in the PVDF matrix due to the specific interactions between their C=O group with the CF_2 group of PVDF. Figure 3.8a,b shows respective X-ray diffraction (XRD) patterns of the CP and MCP specimens of different nanotube contents. Solvent-cast films show the presence of fully β-phase at FMWNT loadings ≥1 wt%. Below this concentration, a mixture of α and β polymorph is formed. Both α and β polymorphs also appear in the XRD traces of melt-cooled nanocomposite films. The β polymorph derives from its all-trans conformation matches closely with the zigzag carbon atoms of MWNT surface.

Table 3.2 DMA results for *in situ* polymerized GO/PMMA and rGO/PMMA nanocomposites. Storage modulus (E') values were taken at 30 °C.

rGO (wt%)	E' (GPa)	% change, E'	T_g (°C)	% change, T_g
0	2.97 ± 0.11	–	125.7 ± 0.84	–
0.025	3.01 ± 0.05	1.7	127.9 ± 0.15	2.0
0.05	3.02 ± 0.10	2.1	142.8 ± 0.78	13.9
0.25	3.18 ± 0.00	7.6	139.8 ± 0.20	11.5
0.5	3.16 ± 0.13	7.0	140.0 ± 0.87	11.6
1	3.73 ± 0.18	26.2	137.5 ± 0.53	9.6
2	3.88 ± 0.02	31.3	143.3 ± 0.10	14.3
4	3.64 ± 0.19	23.2	141.3 ± 0.78	12.7
8	3.75 ± 0.00	26.9	143.3 ± 0.80	14.3
GO (wt%)	E' (GPa)	% change, E'	T_g (°C)	% change, T_g
0.05	3.23 ± 0.19	9.3	132.1 ± 3.01	5.3
0.25	3.42 ± 0.09	15.6	139.4 ± 0.35	11.2
0.5	3.31 ± 0.01	14.3	138.9 ± 0.35	10.8
1	4.06 ± 0.12	37.4	139.5 ± 0.57	11.2
2	4.11 ± 0.11	39.1	141.5 ± 0.50	12.8
4	4.16 ± 0.05	40.6	141.1 ± 0.30	12.5
8	4.24 ± 0.27	43.4	143.8 ± 0.49	14.7

Source: Reproduced with permission from [Chapter 2, Ref. 100], Elsevier (2011).

Figure 3.9a–c shows storage modulus, loss modulus and loss tangent as a function of temperature of solvent-cast FMWNT/PVDF nanocomposites. The storage modulus increases nanotube content as expected. The loss modulus–temperature profiles exhibit two peaks with increase in temperature. The low-temperature peak at ∼ − 40 °C is attributed to the glass-transition temperature, while the high-temperature peak at ∼10 °C is associated with the relaxation of PVDF segments at the crystalline–amorphous interface. The increase of T_g due to the FMWNT addition results from the specific interaction between the nanotubes and PVDF segments. The loss tangent–temperature profiles also exhibit two peaks. It is noted that the peak temperatures determined from the loss modulus and loss tangent profiles are not the same. They differ in a value of ∼5 °C because of two different modes of the measurements. The former relates to the dissipation of energy as heat, while the latter corresponds to the reduction of vibration of the material, that is, damping.

Besides the phase transition, CNT additions can also induce transcrystallinity in polyolefins and syndiotactic polystyrene (sPS) [19, 27–29]. Transcrystallinity results from the formation of highly oriented, columnar crystals that grow epitaxially on the fiber surface with their *c*-axis aligned with the fiber axis. The growth direction of the lamellae with respect to the fiber depends on the degree of crystallographic

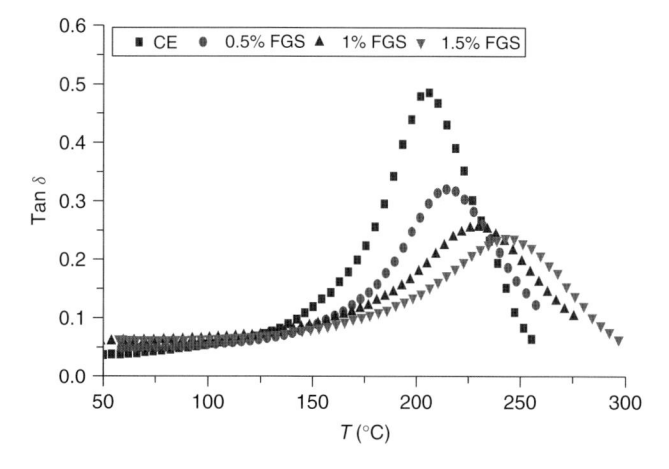

Figure 3.7 Loss tangent versus temperature profiles for UV cured epoxy (CE) filled with different FGS contents. (Source: Reproduced with permission from Ref. [17], Elsevier (2011).)

matching of fiber and matrix and the nucleation density on the fiber surface [30]. Transcrystallinity is commonly found in conventional carbon-fiber-reinforced polymer composites [31]. Transcrystalline layer is beneficial in enhancing effective load transfer across the fiber–matrix interface. For the CNT/polymer nanocomposites, a nanohybrid "shish-kebab" structure has been observed in nylon 66 during solution crystallization in which the MWNT and polymer lamellar crystals act as the "shish" and "kebab," respectively [32]. Transcrystallinity can also be induced in MWNT/PP, SWNT/PE, and MWNT/sPS nanocomposites on cooling from the melt [19, 27–29]. Generally, transcrystallinity can be formed by placing a single or few fibers between two polymer thin films followed by pressing in a hot stage optical microscope [19]. Figure 3.10a,b shows the transcrystalline structure of PP formed on the surface of a single MWNT and two MWNTs, respectively.

As aforementioned, CNTs generally act as effective nucleating agents for macromolecular chains during crystallization. Figure 3.11 shows nonisothermal DSC cooling curves of pure PP and MWNT/PP nanocomposites [33]. The onset crystallization temperature (T_{oc}), peak crystallization temperature and crystallization enthalpy (ΔH_c) and degree of crystallinity (X_c) of PP and its nanocomposites are listed in Table 3.3. X_c can be calculated from the following equation

$$X_c(\%) = \frac{\Delta H_c}{(1-p)\Delta H_m} \tag{3.11}$$

where ΔH_m is the heat of fusion for 100% crystalline isotropic PP ($\Delta H_m = 209\,\mathrm{J\,g^{-1}}$) and p the weight fraction of fillers in the composite. The results reveal that the MWNT additions lead to an increase in both T_{oc} and T_p values of PP. Furthermore, X_c of PP shows little changes by adding 0.1 wt% MWNT but decreases as the MWNT contents are increased to 0.5 and above. Although MWNTs act as

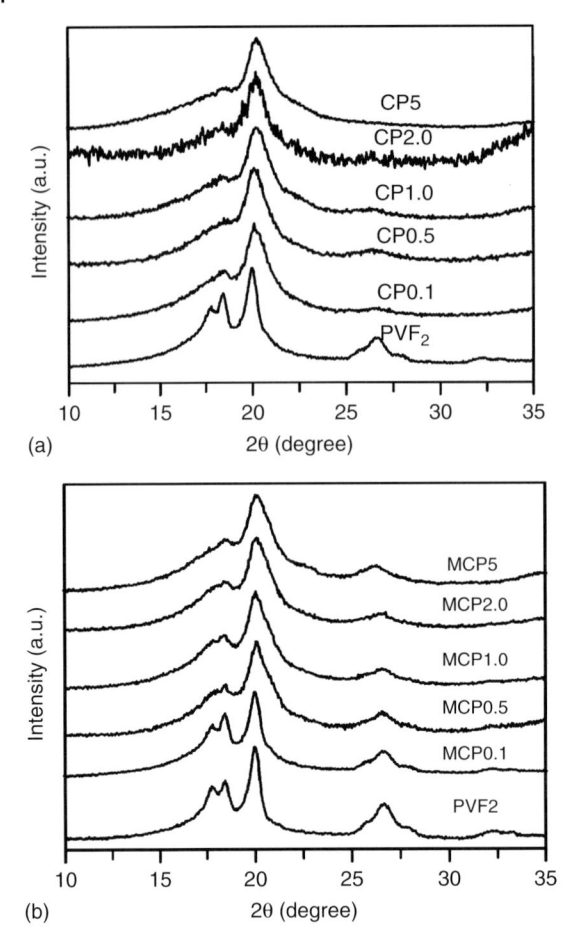

Figure 3.8 XRD traces of (a) solvent-cast FMWNT/PVDF and (b) melt-cooled FMWNT/PVDF nanocomposites containing 0.1, 0.5, 1.0, 2.0, and 5 wt% nanotube contents. (Source: Reproduced from Ref. [3] with permission from the American Chemical Society (2007).)

nucleating sites and promote crystallization, it appears that the crystals formed may have a lower degree of perfection than those in pure PP. In general, both increase and decrease in X_c of polymers can occur associated with the nanotube additions. The presence of nanomaterials in polymers can affect their crystallization behavior in two completive ways, that is, heterogeneous nucleation and mobility hindrance. The former leads to an increase while the latter causes a decrease of X_c. These two factors affect the formation of PP nuclei in the MWNT/PP nanocomposites, with the latter prevails at higher nanotube content. A reduction in X_c of nonpolar PE and other polymer with increasing nanotube contents has been reported in the literature [12; Chapter 2, Ref. 93].

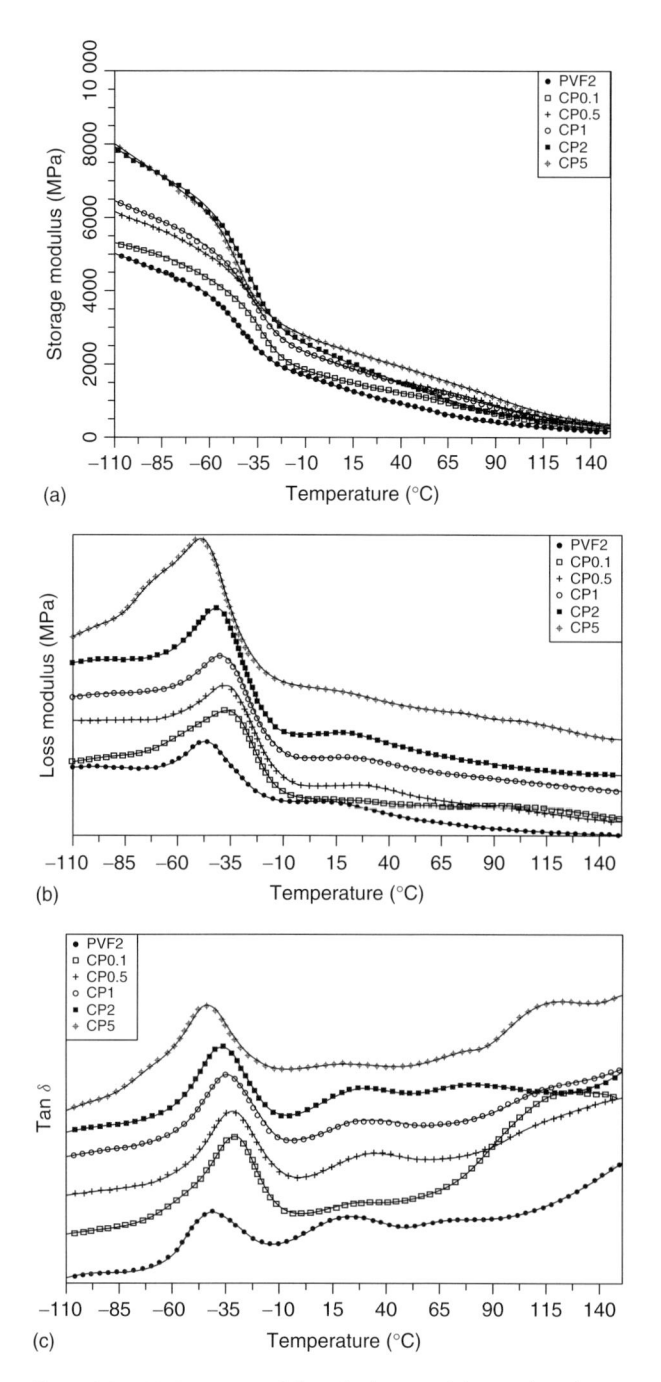

Figure 3.9 (a) Storage modulus, (b) loss modulus, and (c) loss tangent versus temperature profiles for solvent-cast FMWNT/PVDF nanocomposites. (Source: Reproduced from Ref. [3] with permission from the American Chemical Society (2007).)

Figure 3.10 Optical micrographs of transcrystalline PP crystals isothermally crystallized on (a) single and (b) two MWNT surfaces at 125 °C. (Source: Reproduced with permission from Ref. [20], Elsevier (2008).)

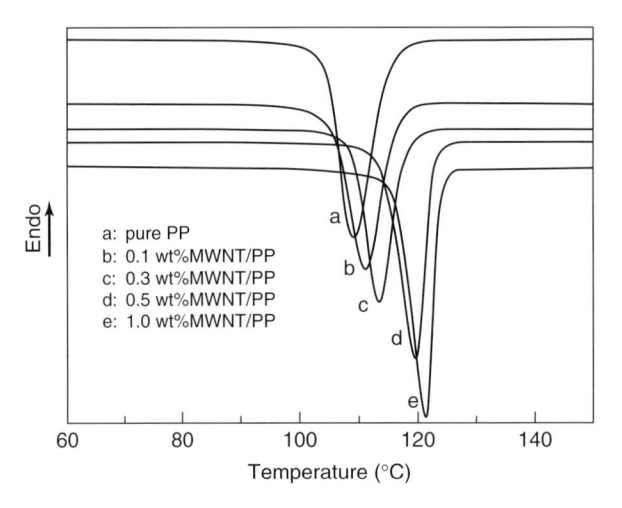

Figure 3.11 DSC cooling curves of pure PP and MWNT/PP nanocomposites at a cooling rate of 10 °C min^{-1} [33].

The inclusion of CNTs into polymers also enhances their crystallization kinetics markedly [29, 34–36]. For example, Chen and Wu [35] studied crystallization kinetics of solution-mixed MWNT/PCL nanocomposites containing 0.25–1 wt% MWNT under isothermal crystallization conditions. During isothermal measurements, the specimens were heated to 90 °C for 10 min to eliminate any thermal history and then rapidly cooled to the proposed crystallization temperatures (T_cs) in the range of 42–50 °C in steps of 2 °C. Figure 3.12 shows the plot of $\ln[-\ln(1 - X_t)]$ versus

Table 3.3 Nonisothermal crystallization parameters of pure PP and MWNT/PP nanocomposites [33].

Specimen	T_{oc} (°C)	T_p (°C)	$\Delta H_c (J g^{-1})$	X_c (%)
Pure PP	120.82	109.22	105.6	50.6
0.1 wt% MWNT/PP	121.10	111.27	102.9	49.3
0.3 wt% MWNT/PP	121.95	113.52	98.4	47.2
0.5 wt% MWNT/PP	125.82	119.58	99.3	47.8
1.0 wt% MWNT/PP	126.58	121.37	100.6	47.7

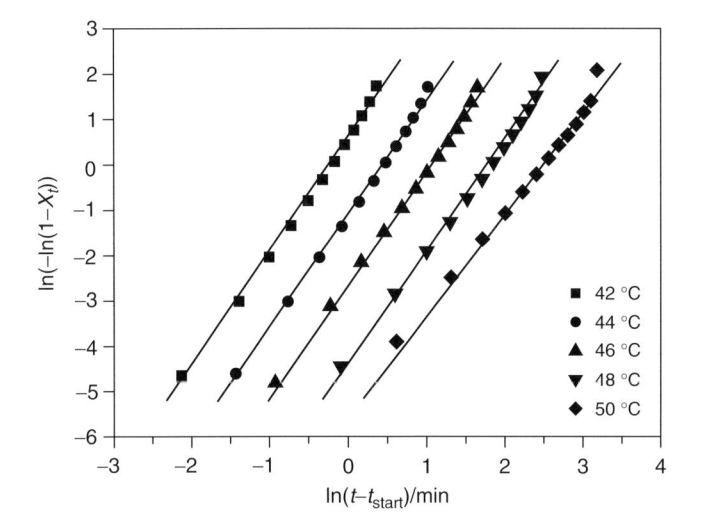

Figure 3.12 Avrami plot of 1 wt% MWNT/PCL nanocomposite. (Source: Reproduced with permission from Ref. [35], Elsevier (2007).)

$\ln t$ for the 1 wt% MWNT/PCL nanocomposite at various T_cs. The crystallization kinetic parameters for PCL and its nanocomposites are listed in Table 3.4. The half-times of crystallization ($t_{1/2}$) for these specimens are also shown in this table. It is evident that the Avrami equation describes the crystallization kinetics of the MWNT/PCL nanocomposites satisfactorily. The n values for pure PCL samples are in the range of 2.6–2.9, showing that the PCL polymer chains follow three-dimensional crystallization growth with heterogeneous athermal nucleation. The Avrami exponent of the MWNT/PCL nanocomposites lies within $2 < n < 3$ range, indicating heterogeneous nucleation followed by spherulitic crystalline growth. Moreover, the $t_{1/2}$ values of the MWCT/PCL nanocomposites are much smaller than those of PCL. These results imply that the isothermal crystallization rate of the MWNT/PCL nanocomposites at a given T_c is much faster than that of

Table 3.4 Isothermal crystallization kinetic parameters of pure PCL and MWNT/PCL nanocomposites at various crystallization temperatures.

		T_c (°C)				
		42	44	46	48	50
PCL	$t_{1/2}$ (min)	6.5	10.1	18.0	46.7	125.2
	n	2.62	2.72	2.76	2.86	2.58
	k	6.2E − 0.3	1.3E − 03	2.3E − 04	1.2E − 05	4.6E − 06
0.25 wt% MWNT/PCL	$t_{1/2}$ (min)	0.8	2.5	4.7	10.1	22.7
	n	2.57	2.52	2.63	2.55	2.51
	k	1.2E + 00	7.0E − 02	1.2E − 02	2.0E − 03	3.1E − 04
0.5 wt% MWNT/PCL	$t_{1/2}$ (min)	0.8	1.5	2.7	6.1	13.3
	n	2.51	2.50	2.54	2.52	2.44
	k	1.5E + 00	2.9E − 01	5.9E − 02	7.3E − 03	2.1E − 03
1 wt% MWNT/PCL	$t_{1/2}$ (min)	0.7	1.4	2.5	5.3	10.2
	n	2.45	2.44	2.49	2.42	2.39
	k	1.8E + 00	3.3E − 01	7.3E − 02	1.3E − 02	4.7E − 03

Source: Reproduced with permission from Ref. [35], Elsevier (2007).

PCL. Therefore, low loading levels of MWCNT act as effective nucleating agents and significantly speed up the crystallization rate of PCL. Figure 3.13a–d shows polarizing optical micrographs of PCL and its nanocomposites isothermally crystallized at 44 °C. The micrograph of pure PCL shows typical Maltese-cross spherulite morphology. The spherulite size of the MWNT/PCL nanocomposites decreases with MWNT additions. Similarly, Huang, and Valentini et al. reported that the Avrami equation can be used to analyze crystallization kinetics of the SWNT/PP and MWNT/sPS nanocomposites [29, 34].

3.3
Thermal Stability

The T_g of a polymer is affected by polymer chain flexibility and rigidity. The addition of rigid filler components usually results in higher T_g. Carbonaceous nanofiller additions lead to an increase in the glass-transition temperature of polymers. The shift in glass-transition temperature of polymers is often used for characterizing thermal stability of polymer nanocomposites. However, decomposition temperatures of thermogravimetric analysis (TGA), linear thermal expansion coefficient and heat deflection temperature (HDT) of polymers are more appropriate parameters for characterizing their thermal stabilities. These parameters are useful for designing and developing polymer nanocomposites with high thermal stability for industrial applications.

Figure 3.13 Polarizing optical micrographs of (a) PCL, (b) 0.25 wt% MWNT/PCL, (c) 0.5 wt% MWNT/PCL, and (d) 1 wt% MWNT/PCL nanocomposites isothermally crystallized at 44 °C. (Source: Reproduced with permission from Ref. [35], Elsevier (2007).)

3.3.1
Thermogravimetry Analysis

TGA is widely used in scientific and industrial sectors for detecting thermal degradation and stability of polymeric materials. The thermal degradation or weight loss measurement of a material is usually carried out in an inert atmosphere, such as nitrogen, helium, and argon, or in an oxidative environment, such as air and oxygen. The latter environment is used for assessing thermo-oxidative stability of the polymers [37]. The key parameters derived from TGA measurements, for example, the temperatures corresponding to a weight loss of 5 wt% ($T_{5\%}$) and a weight loss of 30 wt% ($T_{30\%}$) are often used for characterizing thermal stability of polymers exposed to an inert or oxidative atmosphere. More recently, Li and Jeong studied mechanical and thermal behaviors of melt-blended EG/PET (expanded graphite) nanocomposites [Chapter 2, Ref. 86]. Thermal stability of these nanocomposites was assessed using TGA under both nitrogen and oxygen environments. Figure 3.14a,b shows TGA thermograms of PET and its nanocomposites exposed to nitrogen and oxygen atmospheres, respectively. The $T_{5\%}$ and $T_{30\%}$ values of these specimens are listed in Table 3.5. These results indicate that the degradation temperatures of PET

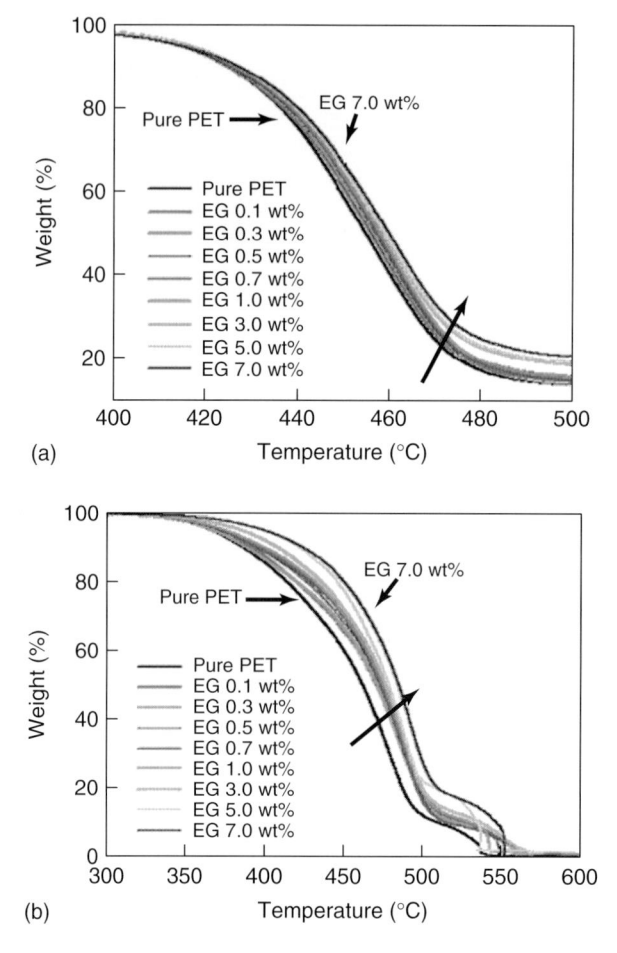

Figure 3.14 TGA curves for pure PET and EG/PET nanocomposites exposed to (a) nitrogen and (b) oxygen atmospheres. (Source: Reproduced with permission from [Chapter 2, Ref. 86], Elsevier (2011).)

and its nanocomposites exposed to oxygen are lower than those exposed in an inert nitrogen atmosphere. However, the increment of thermo-oxidative degradation temperatures due to the EG additions is higher than that under inert nitrogen gas exposure. For the 7 wt% EG/PET nanocomposite, its respective $T_{5\%}$ and $T_{30\%}$ values are 410.8 and 466.9 °C, being ~42 and 32 °C higher than those of pure PET.

Kim et al. [38] functionalized MWNTs with poly(L-lactide) (PLLA) to form PLLA-g-MWNTs and studied the effect of functionalized nanotube additions on electrical and thermal properties of PLLA-based composites. TGA measurements of pure PLLA, 2 wt% MWNT/PLLA and 2 wt% PLLA-g-MWNT/PLLA nanocomposites were carried out in an inert nitrogen atmosphere from 30 to 700 °C at heating rates of 5, 10, 20, 40 °C min^{-1}. The $T_{5\%}$ and $T_{50\%}$ values of these specimens under

Table 3.5 Thermal degradation temperatures for 5 and 30% weight loss for PET and EG/PET nanocomposites exposed to nitrogen and oxygen environments.

Sample	EG (wt%)	N_2		O_2	
		$T_{5\%}$ (°C)	$T_{30\%}$ (°C)	$T_{5\%}$ (°C)	$T_{30\%}$ (°C)
PET	0.0	414.0	443.9	368.7	435.1
EG/PET	0.1	414.3	444.8	371.0	444.2
	0.3	415.4	445.2	371.0	445.1
	0.5	415.4	445.4	373.8	448.6
	0.7	415.7	446.2	375.4	451.6
	1.0	415.9	446.7	377.1	452.1
	3.0	416.7	447.5	387.3	454.6
	5.0	416.9	447.7	410.0	461.2
	7.0	416.7	448.2	410.8	466.9

Source: Reproduced with permission from [Chapter 2, Ref. 86], Elsevier (2011).

Table 3.6 Thermal degradation temperatures for 5 and 50% weight loss for pure PLLA, 2 wt% MWNT/PLLA and 2 wt% PLLA-g-MWNT/PLLA nanocomposites exposed to an inert nitrogen atmosphere.

Heating rate (°C min^{-1})	PLLA		MWNT/PLLA		PLLA-g-MWNT/PLLA	
	$T_{5\%}$ (°C)	$T_{50\%}$ (°C)	$T_{5\%}$ (°C)	$T_{50\%}$ (°C)	$T_{5\%}$ (°C)	$T_{50\%}$ (°C)
5	263.4	317.8	284.1	328.1	290.6	342.2
10	280.5	336.7	309.2	347.9	305.5	355.3
20	295.7	353.8	328.4	364.9	313.2	368.2
40	319.3	378.1	337.3	375.5	329.9	388.3

Source: Reproduced with permission from Ref. [38], Elsevier (2007).

different heating rates are summarized in Table 3.6. Apparently, the $T_{5\%}$ and $T_{50\%}$ values of PLLA can be increased substantially by adding 2 wt% MWNT, especially for $T_{5\%}$. The $T_{5\%}$ and $T_{50\%}$ values can be further increased by incorporating 2 wt% PLLA-g-MWCNT into the PLLA matrix. The shift of degradation temperatures to higher values can be attributed to enhanced interfacial interaction between the PLLA matrix and PLLA-g-MWCNTs, resulting in the formation of protective layers or chars in the polymer matrix during the heating scan. Therefore, the thermal stability of the 2 wt% PLLA-g-MWCNT/PLLA nanocomposite is improved accordingly.

3.3.2
Linear Thermal Expansion

Polymers and their nanocomposites undergo dimensional changes when heated to higher temperatures. The coefficient of thermal expansion (CTE) defined as the change in unit length due to an increase or decrease in temperature can be used to describe thermal expansion behavior of a material, that is

$$\alpha = \frac{\Delta L}{L \Delta T} \quad (3.12)$$

where α is the CTE, ΔL the thermal expansion displacement, L the original length and ΔT the temperature change. The dimensional stability of polymers can be improved by incorporating CNTs and graphene with negative CTE values. Maniwa et al. [39] performed X-ray scattering studies of SWNT bundles in the temperature range of 300–950 K and determined thermal expansion coefficient of SWNT bundles to be $(-0.15 \pm 0.20) \times 10^{-5}$ (K^{-1}). More recently, Bao et al. [40] measured CTE of graphene to be -7×10^{-6} K^{-1} at $T = 300$ K. Despite these beneficial properties, limited information is available in the literature concerning thermal expansion behavior of polymers filled with CNT and graphene fillers [41–43; Chapter 2, Ref. 80].

Kuila et al. [42] functionalized GO with dodecyl amine and studied thermal as well as mechanical properties of solution-mixed GO/LLDPE (linear low-density polyethylene) nanocomposites. Figure 3.15a shows the dimensional change versus temperature of LLDPE and its nanocomposites. It is evident that the dimensional change of LLDPE can be reduced greatly by adding functionalized GO. The CTE of these specimens determined from 40 to 70 and from 80 to 110 °C and plotted as a function of filler content is shown in Figure 3.15b. The CTE of the nanocomposites decreases with increasing filler content. These results reveal that GO nanofiller additions are beneficial in reducing thermal expansion coefficient of LLDPE, thereby improving its dimensional stability markedly.

Wang et al. [43a] studied thermal expansion behavior of GO/epoxy nanocomposites (Figure 3.16). For the purpose of comparison, the thermal properties of 1 wt% graphite/epoxy and 1 wt% SWNT/epoxy composites were also investigated. The results are summarized in Table 3.7. Apparently, the addition of only 1 wt% GO to epoxy leads to a decrease of its CTE value from 8.2 ± 0.2 to $(7.2 \pm 0.6) \times 10^{-5}/°C$. The CTE of the composite can be further reduced to $(5.6 \pm 0.7) \times 10^{-5}/°C$ by adding 5 wt% GO, being 31.7% reduction compared with that of the epoxy resin. This is due to negative CTE of graphene. Similarly, a decrease in CTE is also observed by adding 1 wt% SWNT to epoxy resin as a result of the negative CTE of nanotubes. From Figure 3.16, the T_g can be estimated from the turning point of the curve where its slope changes significantly. The results are also listed in Table 3.7. It is evident that the addition of 1 wt% GO leads to a slight increase of glass-transition temperature of epoxy. Above T_g, the GO additions result in an increase of the α values, suggesting the occurrence of phonon mode vibration and Brownian motion of carbonaceous nanofillers in a cross-linking network of the epoxy resin [43b, 44].

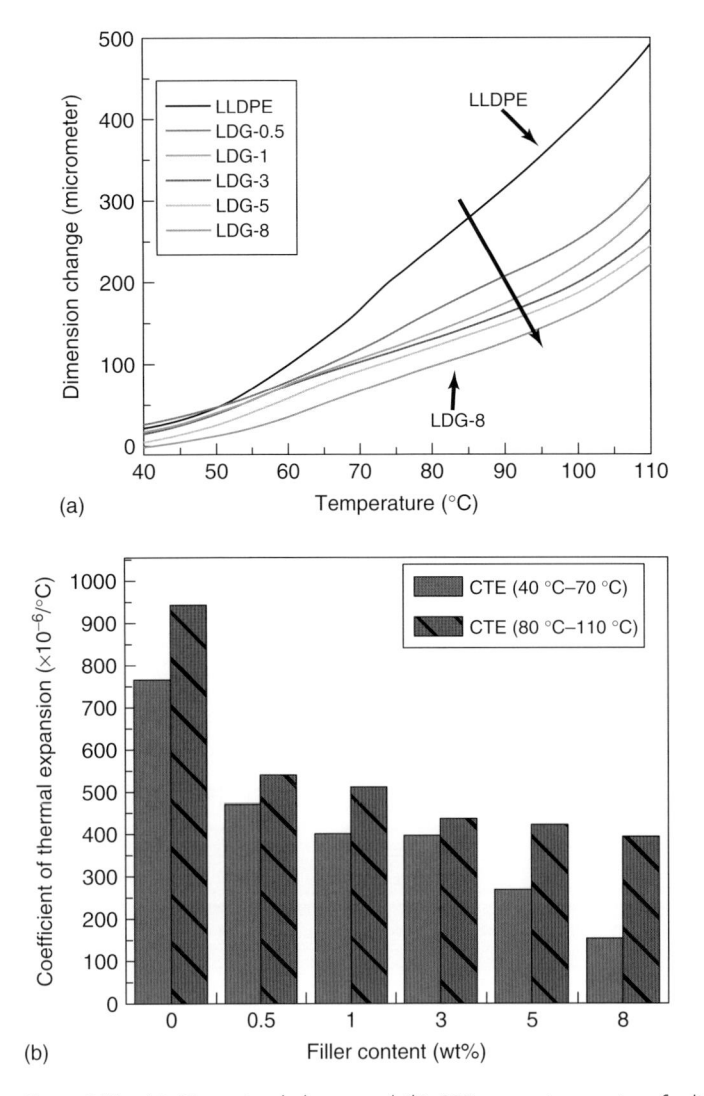

Figure 3.15 (a) Dimensional change and (b) CTE versus temperature for linear low-density polyethylene (LLDPE) and its nanocomposites containing dodecyl amine-modified graphene oxide (LDG) with concentrations of 0.5–8 wt%. (Source: Reproduced from Ref. [42] with permission from Elsevier (2012).)

3.3.3
Heat Deflection Temperature

HDT is an important indicator for heat resistance of materials deformed under an applied load. This property is mainly used for the product design and quality control of polymer products. The HDT measurement is described in ASTM D 648 [45]. During the measurement, a rectangular specimen is loaded under three-point

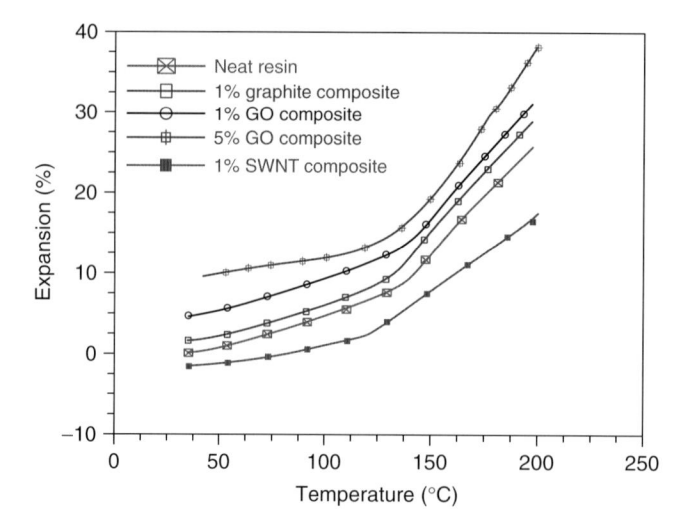

Figure 3.16 Thermal expansion behavior of epoxy composites filled with GO, SWNT, and graphite. (Source: Reproduced from Ref. [43] with permission from the American Chemical Society (2009).)

Table 3.7 Thermal expansion of epoxy and its composites.

	CTE below T_g ($\times 10^{-5}$/°C)	CTE above T_g ($\times 10^{-5}$/°C)	T_g(°C)
Epoxy resin	8.2 ± 0.2	28.2 ± 0.4	136.2 ± 1.8
1 wt% GO/epoxy	7.2 ± 0.6	30.4 ± 2.5	140.0 ± 2.1
5 wt% GO/epoxy	5.6 ± 0.7	31.1 ± 3.7	136.0 ± 1.9
1 wt% SWNT/epoxy	6.0 ± 0.6	28.1 ± 4.2	1315 ± 3.5
1 wt% graphite/epoxy	7.7 ± 0.1	28.9 ± 0.6	135.3 ± 0.8

Source: Reproduced with permission from [43a], The American Chemical Society (2009).

bending mode. The load on the specimen bar is adjusted to create a maximum fiber stress of 1.82 or 0.455 MPa. The temperature of medium is increased at 2 °C min^{-1} until the specimen deflects 0.25 mm from its initial room temperature deflection. This temperature is taken as the HDT at the specific fiber axis. The DMA equipment under three-point bending mode can be used to measure HDT of polymers. The deflection of the specimen is recorded by a sensor placed inside DMA. Very recently, Bao and Tjong [33] determined HDTs of PP and melt-compounded MWNT/PP nanocomposites using DMA techniques. The results showed that the HDT of PP increases from 66.83 to 76.76 and 77.22 °C by adding 0.1 and 1 wt% MWNT, respectively. Moreover, HDT of the PP–elastomer blends can also be increased by adding low loading levels of carbon nanofibers (CNFs) [46]. Similarly, Jia et al.

also indicated that MWNT additions are beneficial in enhancing HDTs of *in situ* polymerized MWNT/PMMA nanocomposites [Chapter 2, Ref. 128].

3.4 Thermal Conductivity

With the progressively decreasing size of electronic devices, thermal management is a key challenge for improving performance of the electronic products. Graphene and CNTs can serve as effective heat sinks in advanced electronic system because of their excellent thermal conductivity (Table 1.4). Thermal conductivities of an individual graphene sheet (4840–5300 W m^{-1} K^{-1}), SWNT (∼3500 W m^{-1} K^{-1}) and MWNT (>3000 W m^{-1} K^{-1}) are significantly higher than those of metals, for example, ∼147–370 W m^{-1} K^{-1} for copper [47]. Thus the additions of graphene sheet and CNT to polymers can remarkably improve their thermal conductivity. For example, the thermal conductivity of epoxy at room temperature improves by 125% by adding only 1 wt% pristine SWNT [48]. In this regard, graphene/polymer and CNT/polymer composites can meet increasing demands for high-performance thermal management materials used in heat sinks and electronic packaging. It is of practical interest to understand properly thermal conductivity of those polymer nanocomposites. Till present, the measured thermal conductivities of CNT/polymer and graphene/polymer composites are found to be much lower than the values estimated from intrinsic thermal conductivity of CNTs and graphene and from their volume fractions [43, 48–52]; Chapter 2, Ref. 19]. In other words, thermal conductivities of these polymer nanocomposites fail to reach their full potential as predicted by theoretical simulations.

3.4.1 Composites with CNTs

3.4.1.1 Thermal Interface Resistance

In general, there exists no thermal percolation concentration in the CNT/polymer nanocomposites beyond which thermal conductivity rises grammatically. However, CNT/polymer nanocomposites exhibit a percolation threshold in electrical conductivity as a result of the formation of a conducting path network. The large aspect ratio ($a = d/l = $ length/diameter) of CNTs enables the formation of electrically conducting network readily. The low thermal conductivity of CNT/polymer nanocomposites compared with CNTs arises mainly from the formation of interfacial thermal resistance between nanotubes and the polymer matrix [53–56]. Other factors such as dispersion and functionalization of nanotubes, length, and structural defects of nanotubes may affect thermal conductivity of CNT/polymer nanocomposites [57]. Shenogin *et al.* [54] studied the thermal properties of CNT/polymer composites using molecular dynamics (MD) simulations. They demonstrated that there is a weak coupling between phonon vibration in the CNTs and the polymer matrix. This weak coupling renders the interfacial contacts with a large thermal resistance. Those contacts act as effective scattering centers for phonons and barriers for heat flow. The lack of thermal percolation in CNT composites is largely due to the formation of a thin polymer film layer separating the CNTs, preventing a direct phonon transport across nanotubes. Even in the presence of a direct contact, the small contact area between two crossed CNTs limits the heat transfer across the junction. Thus the heat mainly transports through the polymer matrix, resulting in a high thermal interface resistance and suppression of percolation [58]. This eventually leads to a small ratio value between thermal conductivity of CNT (κ_f) to that of the matrix (κ_m), that is, κ_f/κ_m. In contrast, the ratio between electrical conductivity of CNT (σ_f) to that of the matrix (σ_m) is several orders of magnitude larger than the κ_f/κ_m. In this respect, the effective channel for the electric transport is along conducting network path of the CNT/polymer nanocomposites.

Several researchers have incorporated interfacial thermal resistance into their models for predicting thermal conductivity of CNT-based composites [53, 59–61]. For example, Nan et al. [53] used effective medium theory to predict thermal conductivity of the CNT-based composites. They reported that the presence of an interfacial thermal resistance (R_K) across the nanotube–matrix interface causes a marked decrease in thermal conductivity of the CNT-based nanocomposite. The R_K value is estimated in the order of 10^{-8} m^2 kW^{-1}. Considering a system of randomly oriented tubes in a host medium in which the tubes are covered with thin polymer films. The interfacial barrier layer limits heat transfer across the tubes, giving rise to thermal interfacial resistance in the system. The effective thermal conductivity of the composite (κ_e) can be expressed as follows:

$$\frac{\kappa_e}{\kappa_m} = 1 + \frac{\Phi a}{3} \frac{\kappa_f/\kappa_m}{a + \frac{2R_K \kappa_f}{d}} \tag{3.13}$$

where κ_m and κ_f are thermal conductivities of the composite matrix and CNT, respectively; Φ, d, and a are the volume fraction, diameter and aspect ratio of CNT, respectively; and R_K is the thermal interface resistance. For a perfect interface without any thermal interface resistance ($R_K = 0$) and $a > 1$, the thermal conductivity enhancement is further simplified as

$$\frac{\kappa_e}{\kappa_m} = 1 + \frac{\Phi}{3} \frac{\kappa_f}{\kappa_m} \tag{3.14}$$

Figure 3.17 shows the plots of thermal conductivity enhancement versus filler volume fraction for the composites filled with CNTs of different diameters using Eqs. (3.13) and (3.14). In the absence of R_K, the EMT model with a perfect interface overestimates thermal conductivity value (solid line). Thus the presence of thermal interface resistance degrades thermal conductivity of the CNT/polymer composites greatly. Further, this model with interfacial resistance agrees reasonably with the experimental data (■ symbol) for a nanotube suspension in oil.

Bryning et al. [49] employed Eq. (3.13) for predicting thermal conductivity of the SWNT/epoxy nanocomposites. The equation fits well with experimental data up to 10 vol% SWNT. Thus Eq. (3.13) generally gives reasonable predictions for thermal

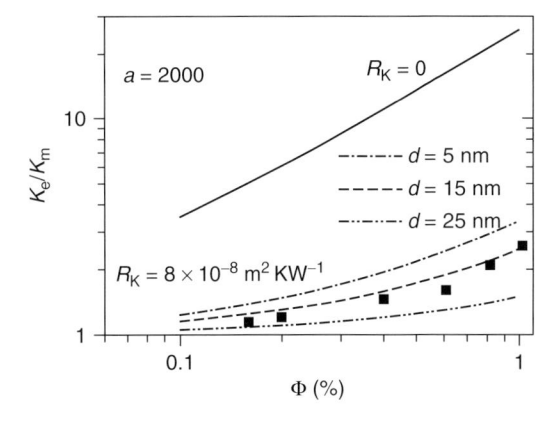

Figure 3.17 Thermal conductivity enhancement of the composites filled carbon nanotubes of different diameters for a perfect interface ($R_K = 0$) or under the presence of thermal interface resistance assuming $\kappa_f = 3000$ W m^{-1} K^{-1} and $\kappa_m = 0.4$ W m^{-1} K^{-1}. (Source: Reproduced from Ref. [53] with permission from American Institute of Physics (2004).)

conductivity of CNT–polymer composites at low filler volume fractions. Recently, Haggenmueller et al. [50] also reported a similar finding for thermal conductivity of SWNT/HDPE (high-density polyethylene) nanocomposites (Figure 3.18). The effective medium theory yields good fitting results for SWNT contents up to 6 vol%, yielding a R_K value of $(1.0 \pm 0.3) \times 10^{-8}$ m^2 kW^{-1}. At 20 vol% SWNT, the experimental value deviates significantly from theoretical prediction. Furthermore, the thermal conductivity of PE-based nanocomposites also depends on the degree of crystallinity of the polymer matrix. The κ values of HDPE and low-density polyethylene (LDPE) are determined to be 0.5 and 0.26 W m^{-1} K^{-1}, respectively. HDPE exhibits higher thermal conductivity because its crystallinity is larger than

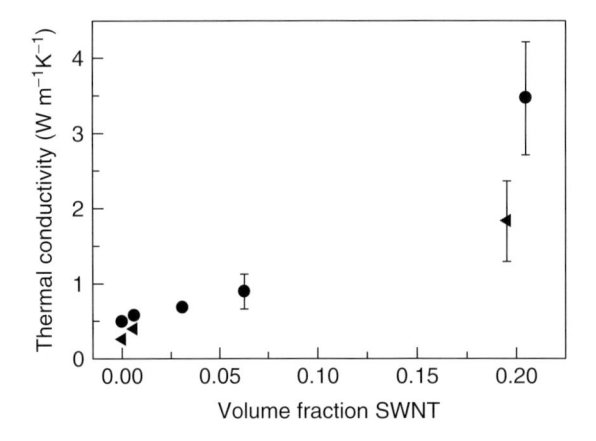

Figure 3.18 Thermal conductivity for SWNT/LDPE (tilted ◀) and SWNT/HDPE (●) composites at various nanotube loadings. (Source: Reproduced from Ref. [50] with permission from the American Chemical Society (2007).)

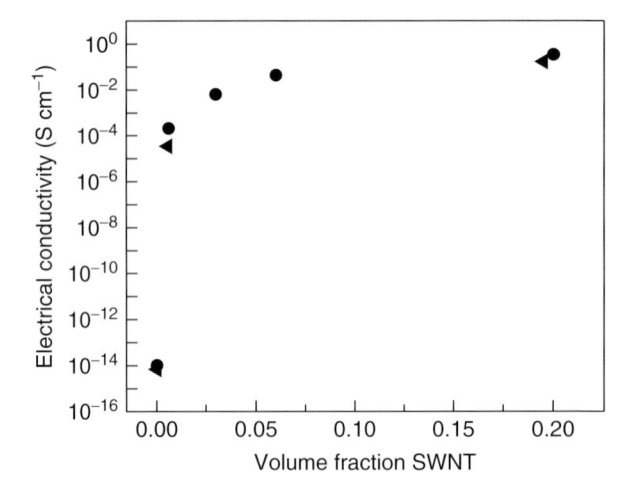

Figure 3.19 Electrical conductivity for SWNT/LDPE (tilted ◀) and SWNT/HDPE (●) composites at various nanotube loadings. (Source: Reproduced from Ref. [50] with permission from the American Chemical Society (2007).)

that of LDPE. In this regard, the SWNT/HDPE exhibit higher κ values than the SWNT/LDPE counterparts, particularly at a high filler loading of 20 vol%. Figure 3.19 displays electrical conductivity versus filler volume fraction plots for the SWNT/HDPE and SWNT/LDPE nanocomposites. Apparently, both nanocomposite systems display electrical percolation behavior but no thermal percolation in these systems. This is due to the presence of a large interfacial thermal resistance between the polymer matrix and SWNT. Assuming $\kappa_f = 3500$ W m^{-1} K^{-1} for SWNT, $\kappa_m = 0.5$ W m^{-1} K^{-1} for HDPE, $\sigma_f = 3 \times 10^4$ S cm^{-1} for SWNT [Chapter 1, Ref. 155] and $\sigma_m = 10^{-14}$ S cm^{-1} for HDPE, the κ_f/κ_m and σ_f/σ_m ratio values are determined to be 7×10^3 and 3×10^{18}, respectively. Apparently, the κ_f/κ_m value is about 14 orders of magnitude smaller than the σ_f/σ_m. The small value of κ_f/κ_m in the SWNT/HDPE nanocomposites results in the absence of thermal percolation [50].

3.4.1.2 Dispersion and Functionalization of Carbon Nanotubes

Apart from the interfacial resistance, poor dispersion of the CNTs in the polymer matrix is also a main factor affecting overall thermal properties of the polymer nanocomposites. Homogeneous dispersion of CNTs in the polymer matrix is rather difficult to achieve because of a high tendency of nanotubes for agglomeration. In general, good dispersion of CNTs in solvents and polymers can be achieved using ultrasonication, surfactant processing and nanotube functionalization. Proper selection of solvents and surfactants is critical for uniform dispersion of nanotubes in the nanocomposites prepared by solution mixing. Song and Young [62] dispersed MWNTs of different weight fractions in ethanol under sonication for 2 h. The nanotube dispersions were then mixed with the epoxy resin under sonication. For comparison, poorly dispersed MWNT/epoxy nanocomposites were

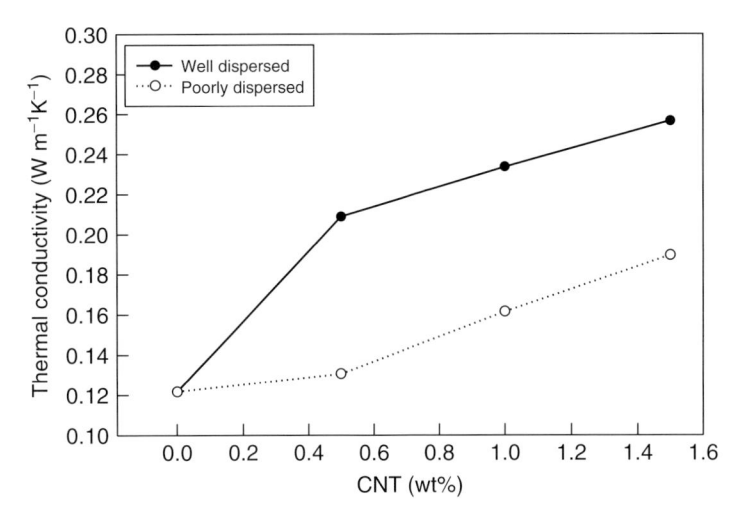

Figure 3.20 Thermal conductivity of MWNT/epoxy nanocomposites as a function of nanotube loading. (Source: Reproduced with permission from Ref. [62], Elsevier (2005).)

also prepared under sonication for 3 h without using the solvent. The nanotubes agglomerated into bundles in the polymer matrix as expected. Figure 3.20 shows that the MWNT/epoxy nanocomposites with well-dispersed nanotubes exhibiting higher thermal conductivity than poorly dispersed composites.

Covalent functionalization is beneficial in enhancing dispersion of the nanotubes in the polymer matrix. Covalent functionalization via acid treatment induces the formation of carboxylic and hydroxyl groups on the nanotube surfaces. Such functional groups introduce structural defects by disrupting extended π-conjugation network of CNTs, leading to the conversion of sp^2 carbon to sp^3 hybrid. Thus functional groups act as scattering centers for phonons, affecting heat flow considerably. In addition, functionalization also increases interfacial thermal resistance of the CNT–polymer nanocomposites. At present, there are no conclusive remarks for the effects of covalent functionalization on thermal conductivities of the CNT/polymer nanocomposites. Both improved [51, 63] and impaired [64–67] thermal conductivity performances due to covalent functionalization of nanotubes have been reported in the literature. Haddon and coworkers [51] purified SWNTs in nitric acid followed by high-shear mixing with epoxy. Purified SWNT/epoxy nanocomposites were found to exhibit higher thermal conductivity than those filled with pristine SWNTs. Very recently, Yang et al. [63] treated MWNTs in a mixed sulfuric/nitric acid 3/1 solution followed by grafting with triethylenetetramine (TETA). TETA functionalized nanotubes react covalently with epoxide molecules, thereby improving their dispersion in the polymer matrix. Figure 3.21 shows the effect of nanotube functionalization on thermal conductivity of the MWNT/epoxy nanocomposites. TETA-grafted MWNT/epoxy nanocomposites exhibit higher thermal conductivity than pristine MWNT/epoxy composites. Although TETA functional groups can act

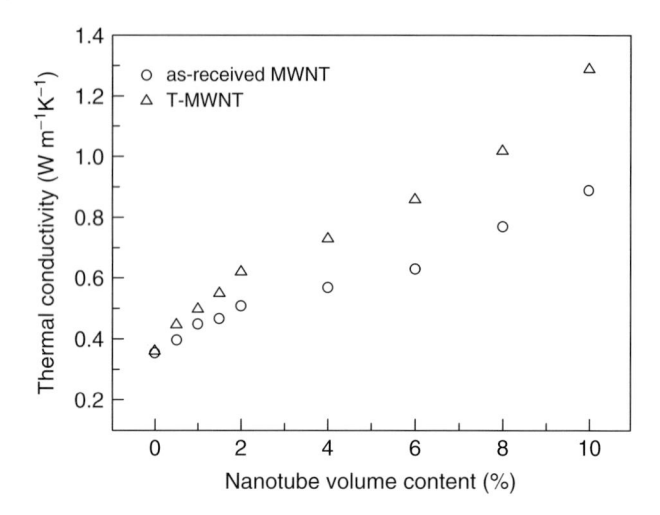

Figure 3.21 Thermal conductivity of MWNT/epoxy nanocomposites. (Source: Reproduced with permission from Ref. [63], Elsevier (2009).)

as phonon scattering centers, the improvement of the nanotube dispersion in the epoxy matrix overrides this adverse effect.

Gojny et al. [64] prepared CNT/epoxy composites filled with pristine double-walled carbon nanotube (DWNT), MWNT, SWNT, and amino-functionalized DWNT and SWNT. Those nanotubes were dispersed in the epoxy matrix using three-roll-milling process. Figure 3.22 shows thermal conductivity as a function of nanotube content for all CNT/epoxy nanocomposites. Pristine DWNT and MWNT additions produce the largest improvements in the thermal conductivity. However, the additions of 0.05 and 0.1 wt% SWNT (pristine) even reduce thermal conductivity of epoxy. For the SWNT content ≥ 0.3 wt%, thermal conductivity of the epoxy resin is gradually restored and followed by a small increase. The poor conductivity of pristine SWNT/epoxy composites derives from a huge surface area of the tubes, thereby producing many large phonon scattering points at the tube–matrix interface, yielding high thermal interface resistance. The MWNT of the largest diameter has the smallest interfacial areas, yielding the highest thermal conductivity in pristine MWNT/epoxy composites. From Figure 3.22, amino-functionalized DWNT and MWNT additions produce an adverse effect in thermal conductivity. A strong tube–matrix interaction enhances scattering of phonons at the interfaces, thereby affecting heat conduction.

In the case of amorphous thermoplastics, pristine CNT additions improve their thermal conductivity significantly [65, 67]. Bonnet et al. [65] reported that the introduction of 7.3 vol% SWNT to PMMA film increases its thermal conductivity from 0.23 to 0.32 $W\,m^{-1}\,K^{-1}$, being 55% enhancement over that of pure polymer. Similarly, Hong and Tai prepared pristine MWNT/PMMA and SWNT/PMMA, acid-functionalized MWNT/PMMA, and acid-functionalized SWNT/PMMA nanocomposites by means of solution compounding [67]. The

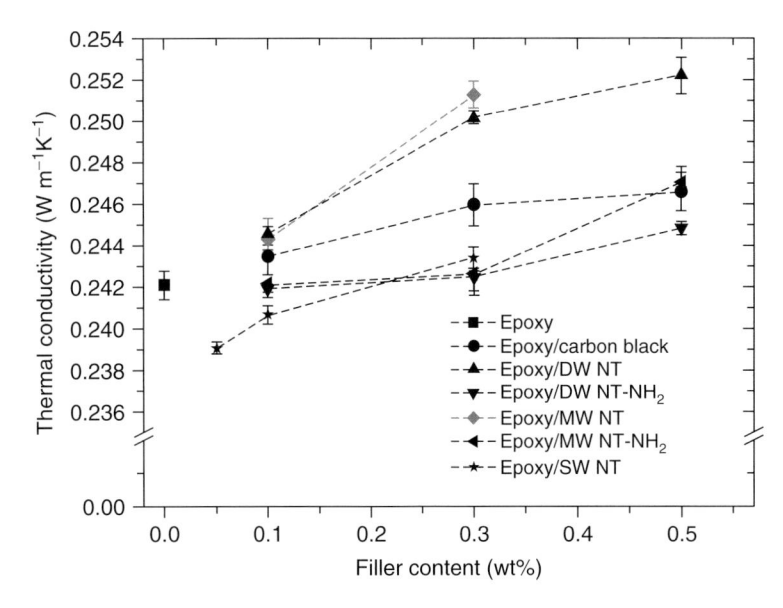

Figure 3.22 Thermal conductivity of epoxy composites as a function of filler content. (Source: Reproduced with permission from Ref. [64], Elsevier (2006).)

thermal conductivities of PMMA composites containing pristine nanotubes and functionalized nanotubes are shown in Figure 3.23a,b, respectively. From Figure 3.23a, the κ value of the 1.0 wt% SWNT/PMMA composite is 2.43 W m^{-1} K^{-1}, being a 10-fold increase over that of pure PMMA. Increasing SWCNT contents ≥ 2 wt% leads to a decrease in thermal conductivity. This derives from nonuniform dispersion and agglomeration of SWNTs in the PMMA matrix. The MWNT/PMMA composites exhibit even higher thermal conductivity. The thermal conductivity of the MWNT/PMMA composites increases with increasing nanotube content. At 4 wt% MWNT, the thermal conductivity reaches the highest value of 3.44 W m^{-1} K^{-1}. The high thermal conductivity in this nanocomposite results from better dispersion of MWCNTs in the PMMA matrix. The thermal conductivity of FMWNT/PMMA specimens is poorer than that of the PMMA composites with pristine MWNTs (Figure 3.23b). This is because carboxylic functional groups disrupt the conjugative network of the nanotubes and also serve as scattering centers for phonons.

3.4.2
Composites with GNP and Graphene Nanofillers

As mentioned before, pristine graphene is highly thermally conductive at room temperature. The thermal conductivity has been reported in the range of 4840–5300 W m^{-1} K^{-1} [Chapter 2, Refs. 130, 131]. The unusually high thermal conductivity of graphene has spurred much interest in its use as conducting

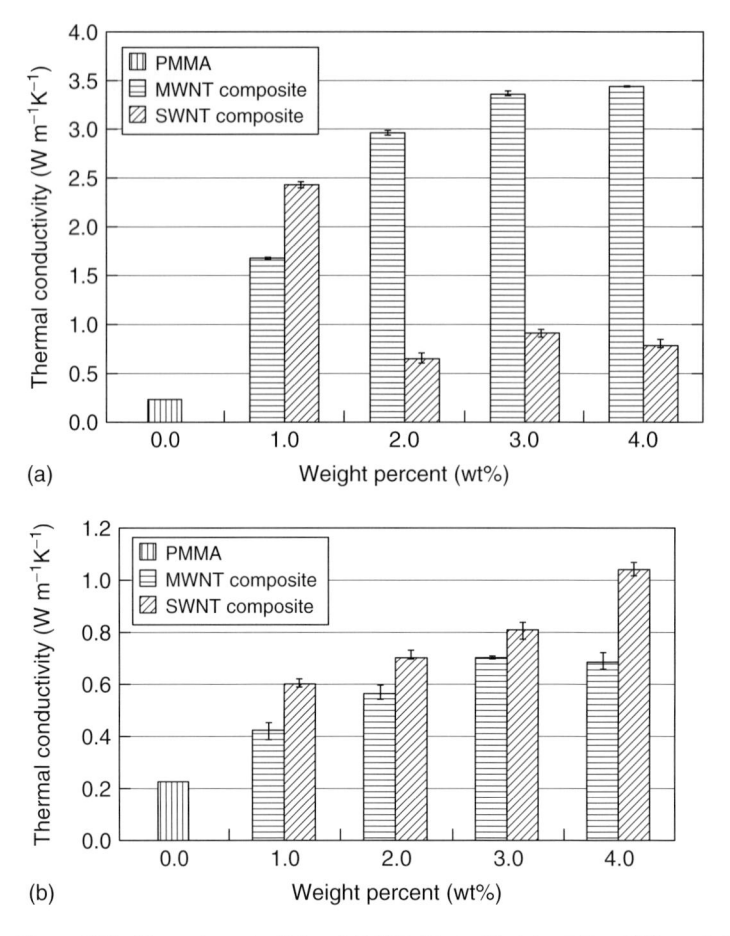

Figure 3.23 Thermal conductivity of PMMA filled with (a) pristine CNTs and (b) purified CNTs (2005). (Source: Reproduced with permission from Ref. [67], Elsevier (2008).)

nanofiller for polymers. However, measured thermal conductivity of GNP/polymer and graphene/polymer nanocomposites is far smaller than that of graphene [46, 68–73; Chapter 2, Refs. 19, 65]. For example, a κ value of 6.4 W m^{-1} K^{-1} can be achieved in the epoxy resin by adding GNP fillers of 25 vol% (Figure 3.24). Such an improvement in thermal conductivity requires the addition of large GNP contents. Similarly, Raza et al. [73] reported that thermal conductivity of silicone can be increased from 0.17 to 2.33 and 3.15 W m^{-1} K^{-1} by adding 20 and 25 wt% GNP, respectively. Therefore, more research efforts are needed in the future to achieve large thermal conductivity enhancements in the polymer nanocomposites at low filler loadings.

Figure 3.25 shows the thermal conducting behavior of the 1 wt% GO/epoxy and 5 wt% GO/epoxy nanocomposites [43]. It can be seen that the κ value of the 1 wt%

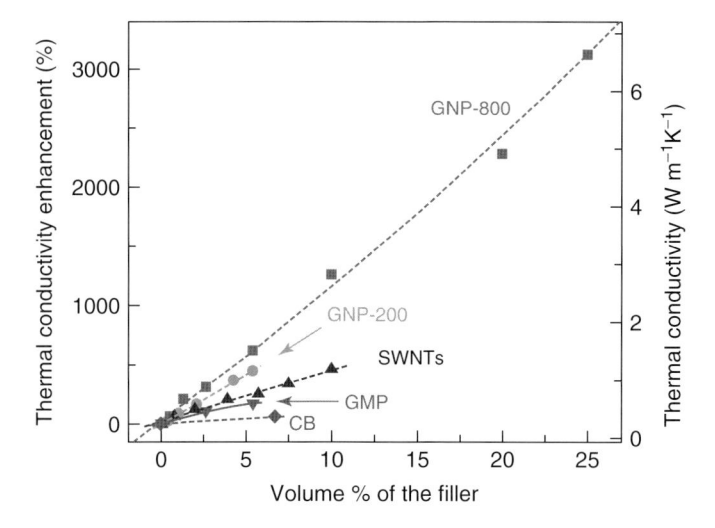

Figure 3.24 Thermal conductivity enhancement $[(\kappa - \kappa_0)/\kappa_0\ (\%)]$ where κ_0 is thermal conductivity of the polymer matrix of epoxy-based composites at 30 °C. GMPs, Graphitic microparticles; GNP-200: GNPs exfoliated at 200 °C; GNP-800: GNPs exfoliated at 800 °C. (Source: Reproduced from [Chapter 2, Ref. 65] with permission from the American Chemical Society (2007).)

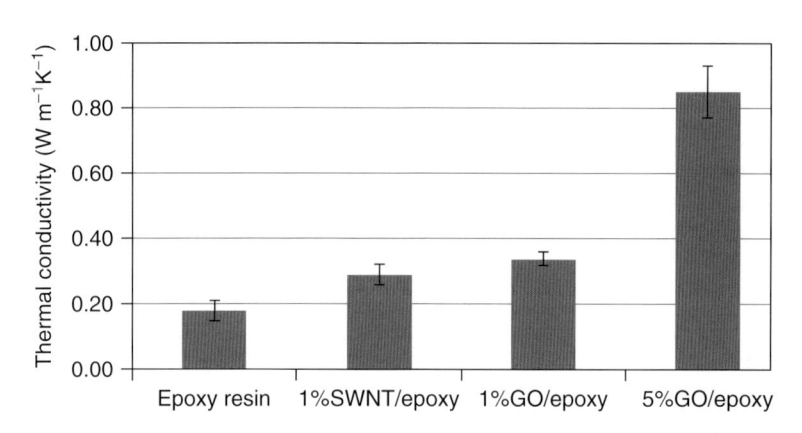

Figure 3.25 Thermal conductivity of epoxy filled with 1 wt% SWNT, 1 wt% GO, and 5 wt% GO. (Source: Reproduced from Ref. [43a] with permission from the American Chemical Society (2009).)

GO/epoxy nanocomposite is comparable to the 1 wt% SWNT/epoxy nanocomposite. The κ value reaches $\sim$0.83 W m^{-1} K^{-1} by adding 5 wt% GO, being fourfold increase over the neat epoxy with a κ value of $\sim$0.195 W m^{-1} K^{-1}. Since GO contains oxygenated groups, even higher thermal conductivity in the epoxy composites can be obtained by using rGO and thermally expanded graphene oxide (TEGO)

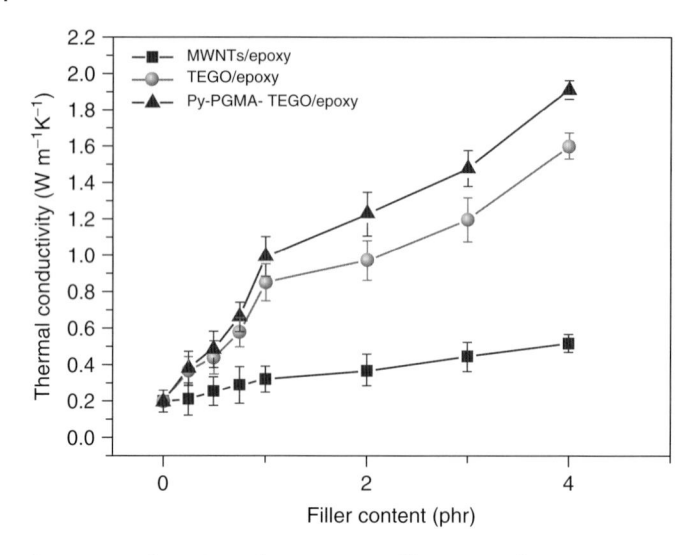

Figure 3.26 Thermal conductivity versus filler content for Py-PGMA-TEGO/epoxy, TEGO/epoxy, and MWNT/epoxy nanocomposites. (Source: Reproduced from [Chapter 2, Ref. 19] with permission from Elsevier (2011).)

fillers instead. Very recently, Ma and coworkers prepared graphene/epoxy composites through noncovalent functionalization of TEGO with Py-PGMA (poly(glycidyl methacrylate)) [Chapter 2, Ref. 19]. Figure 3.26 shows thermal conductivity versus filler content for Py-PGMA-TEGO/epoxy and TEGO/epoxy nanocomposites. For the purpose of comparison, the thermal conductivity plot of MWNT/epoxy composites is also shown in this figure. The Py-PGMA-TEGO/epoxy nanocomposites exhibit the highest thermal conductivity compared with the TEGO/epoxy and MWNT/epoxy composites. At 4 wt% TEGO, the κ value reaches 1.91 W m^{-1} K^{-1}, being 20 and 267% higher than the TEGO/epoxy and MWNT/epoxy nanocomposites, respectively. The Py-PGMA on TEGO surfaces plays an important role in preventing filler aggregation, facilitating their dispersion in the polymer matrix homogeneously.

To reduce interfacial thermal resistance, Haddon and coworkers [74] developed epoxy-based composites filled with GNP and SWNT hybrid fillers. A synergistic effect in thermal conductivity enhancement can be achieved in the composites with hybrid loadings of 10–20 wt% at a GNP/SWNT ratio of 1 : 3 (Figure 3.27a–c). The κ value of the hybrid composite with a filler loading of 20 wt% reaches 3.35 W m^{-1} K^{-1}. At a 10 wt% filler loading, the hybrid nanocomposite exhibits thermal conductivity enhancement of 775%, over more than a factor of 10 compared with that of the carbon black (CB) composite at the same loading. Haddon and coworkers attributed this to the bridging of GNPs by flexible SWNTs. This network provides additional channels for heat flows that bypass the polymer matrix (Figure 3.28a–c).

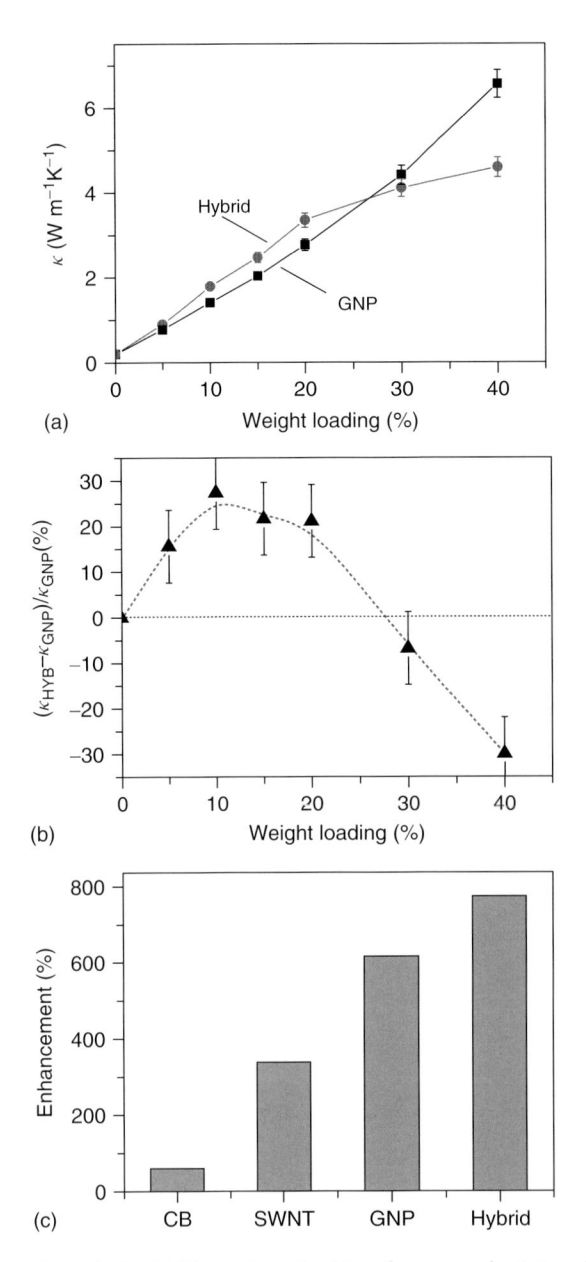

Figure 3.27 (a) Thermal conductivity of epoxy composites with GNP–SWNT hybrid filler (●; GNP:SWNT = 3 : 1) and GNP filler (■) as a function of the filler loading. (b) Synergistic effect $(\kappa_{HYB}-\kappa_{GNP})/\kappa_{GNP}$ (%) associated with the hybrid filler (HYB) as a function of the filler loading. (c) Thermal conductivity enhancement of epoxy composites for SWNT, GNP, and GNP–SWNT hybrid filler at 10 wt% loading in comparison with carbon black (CB). (Source: Reproduced from Ref. [74] with permission from Wiley-VCH (2008).)

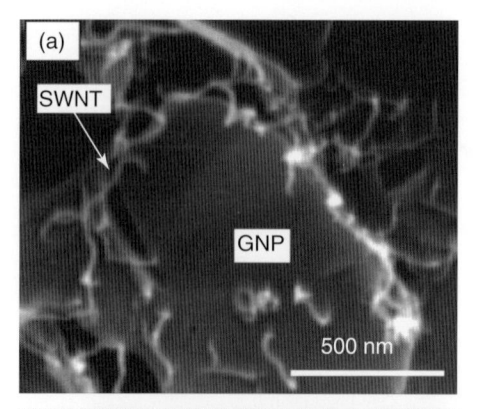

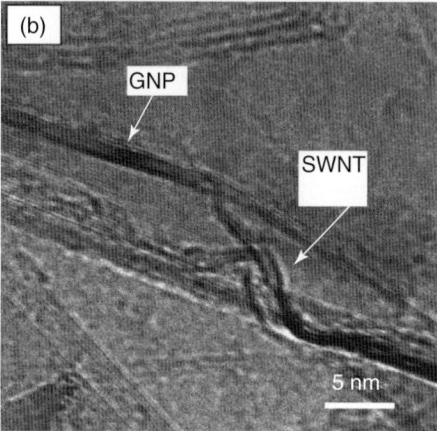

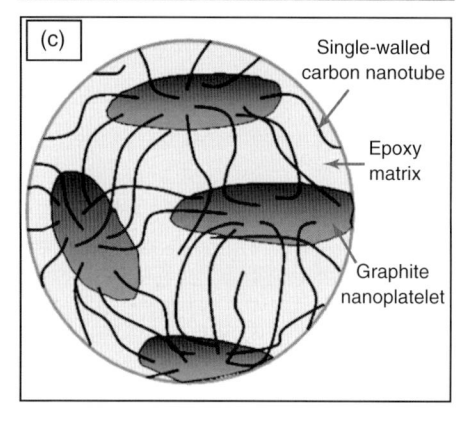

Figure 3.28 (a) Scanning electron microscopy (SEM) and (b) transmission electron microscopy (TEM) images of the cross section of GNP–SWNT hybrid filler/epoxy composite. SWNTs are bridging adjacent graphite nanoplatelets and SWNT ends are extended along the nanoplatelet surfaces. (c) Schematic representation of GNP–SWNT network in polymer matrix. (Source: Reproduced from Ref. [74] with permission from Wiley-VCH (2008).)

Nomenclature

a	Aspect ratio of filler
α	Coefficient of thermal expansion
β	Cooling rate
d	Diameter
δ	Phase lag between stress and strain
E'	Storage modulus
E''	Loss modulus
E^*	Complex modulus
ε	Strain
ΔH_c	Crystallization enthalpy
ΔH_m	Heat of fusion
$K(T)$	Ozawa crystallization rate constant
k	Crystallization rate constant
κ_e	Effective thermal conductivity of composite
κ_f	Thermal conductivity of carbon nanotube
κ_m	Thermal conductivity of polymer matrix
L	Original length of specimen
ΔL	Change in length due to thermal expansion
l	Length of carbon nanotube
m	Ozawa exponent
n	Avrami exponent
p	Weight fraction of filler
Φ	Volume fraction of filler
R_K	Interfacial thermal resistance
s	Critical conductivity exponent
σ	Stress
T	Temperature
ΔT	Temperature change
t	Time
$\tan \delta$	Loss tangent
ω	Angular frequency
X_c	Degree of crystallinity
X_T	Relative crystallinity

References

1. Bhattacharyya, A.R., Sreekumar, T.V., Liu, T., Kumar, S., Ericson, L.M., Hauge, R.H., and Smalley, R.E. (2003) *Polymer*, **44**, 2373–2377.
2. Anand, K.A., Agarwal, U.S., and Joseph, R. (2006) *Polymer*, **47**, 3976–3980.
3. Manna, S. and Nandi, A.K. (2007) *J. Phys. Chem. C*, **111**, 14670–14680.
4. Grady, B.P., Pompeo, F., Shambaugh, R.L., and Resasco, E. (2002) *J. Phys. Chem. B*, **106**, 5852–5858.
5. Rieger, J. (2001) *Polym. Test.*, **20**, 199–204.
6. Menard, K. (1999) *Dynamic Mechanical Analysis: A Practical Introduction*, CRC Press, Boca Raton, FA.
7. ASTM Standard (2004) D 4065-01. *Standard Practice for Plastics: Dynamic Mechanical Properties: Determination and Report of Procedure*, American Society for Testing and Materials, West Conshohocken, PA.
8. Rotter, G. and Ishida, H. (1992) *Macromolecules*, **25**, 2170–2176.

9. Mccrum, N.G., Buckley, C.P., and Bucknall, C.B. (1992) *Principles of Polymer Engineering*, Oxford University Press, Oxford.
10. Sterzynski, T., Tomaszewska, J., Piszczek, K., and Skorczewska, K. (2010) *Compos. Sci. Technol.*, **70**, 966–969.
11. (a) Avrami, M.J. (1939) *J. Chem. Phys.*, **7**, 1103–1112; (b) Ozawa, T. (1971) *Polymer*, **12**, 150–158.
12. Xu, G.Y., Du, L.C., Wang, H., Xia, R., Meng, X.C., and Zhu, Q.R. (2008) *Polym. Int.*, **57**, 1052–1066.
13. He, F., Fan, J.T., Lau, S.T., and Chan, L.W. (2011) *J. Appl. Polym. Sci.*, **119**, 1166–1175.
14. Zhang, J.B. and Qiu, Z.B. (2011) *Ind. Eng. Chem. Res.*, **50**, 13885–13891.
15. Vadukumpully, S., Paul, J., and Valiyaveettil, S. (2009) *Carbon*, **47**, 3288–3294.
16. Jang, J.Y., Kim, M.S., Jeong, H.M., and Shin, C.M. (2009) *Compos. Sci. Technol.*, **69**, 186–191.
17. Martin-Gallego, M., Verdejo, R., Lopez-Manchado, M.A., and Sangermano, M. (2011) *Polymer*, **52**, 4664–4669.
18. Zhang, Q.H., Lippits, D.R., and Rastogi, S. (2006) *Macromolecules*, **39**, 658–666.
19. Wang, B., Sun, G.P., He, X.F., and Liu, J.J. (2007) *Polym. Eng. Sci.*, **47**, 1610–1620.
20. Zhang, S.J., Minus, M.L., Zhu, L.B., Wong, C.P., and Kumar, S. (2008) *Polymer*, **49**, 1356–1364.
21. Wu, D.F., Sun, Y.R., and Zhang, M. (2009) *J. Polym. Sci., Part B: Polym. Phys.*, **47**, 608–618.
22. Antoniadis, G., Paraskevopoulos, K.M., Bikiaris, D., and Chrissafis, K. (2009) *J. Polym. Sci., Part B: Polym. Phys.*, **47**, 1452–1466.
23. He, L.H., Xu, Q., Hue, C.W., and Song, R. (2010) *Polym. Compos.*, **31**, 921–927.
24. Lovinger, J.L. (1983) *Science*, **220**, 1115–1121.
25. Newman, B.A. and Scheinbeim, J.I. (1983) *Macromolecules*, **16**, 60–68.
26. Morra, B.S. and Stein, R.S. (1982) *J. Polym. Sci., Part B: Polym. Phys.*, **20**, 2261–2275.
27. Haggenmueller, R., Fischer, J.E., and Winey, K.I. (2006) *Macromolecules*, **39**, 2964–2971.
28. Zhang, S.J., Lin, W., Zhu, L.B., Wong, C.P., and Bucknall, D.G. (2010) *Macromol. Chem. Phys.*, **211**, 1348–1354.
29. Huang, C.L. and Wang, C. (2011) *Eur. Polym. J.*, **47**, 2087–2096.
30. Quan, H., Li, Z.M., Yang, M.B., and Huang, R. (2005) *Compos. Sci. Technol.*, **65**, 999–1021.
31. Clark, R.L. Jr., Kinder, R.G., and Sauer, R.B. (1999) *Composites Part A*, **30**, 27–36.
32. Li, L.Y., Li, C.Y., Ni, C.Y., Rong, L.X., and Hsiao, B. (2007) *Polymer*, **48**, 3452–3460.
33. Bao, S.P. and Tjong, S.C. (2008) *Mater. Sci. Eng., A*, **485**, 508–516.
34. Valentini, L., Biagiotti, J., Lopez-Manchado, M.A., Santucci, S., and Kenny, J.M. (2004) *Polym. Eng. Sci.*, **44**, 303–311.
35. Chen, E.C. and Wu, T.M. (2007) *Polym. Degrad. Stab.*, **92**, 1009–1015.
36. Kim, J.Y., Choi, H.J., Kang, C.S., and Kim, S.H. (2010) *Polym. Compos.*, **31**, 858–869.
37. Chattopadhyay, D.K. and Webster, D.C. (2009) *Prog. Polym. Sci.*, **34**, 1068–1133.
38. Kim, H.S., Park, B.H., Yoon, J.S., and Jin, H.J. (2007) *Eur. Polym. J.*, **43**, 1729–1735.
39. Maniwa, Y., Fujiwara, R., Kira, H., Tou, H., Kataura, H., Suzuki, S., and Achiba, Y. (2001) *Phys. Rev. B*, **64**, 241402 (3 pp).
40. Bao, W.Z., Miao, F., Chen, Z., Zhang, H., Jang, W.Y., Dames, C., and Lau, C.N. (2009) *Nat. Nanotechnol.*, **4**, 562–566.
41. Xu, Y.S., Ray, G., and Abdel-Magid, B. (2006) *Composites Part A*, **37**, 114–121.
42. Kuila, T., Bose, S., Mishra, A.K., Khanra, P., Kim, N.H., and Lee, J.H. (2012) *Polym. Test.*, **31**, 31–38.
43. (a) Wang, S., Tambraparni, M., Qiu, J.J., Tipton, J., and Dean, D. (2009) *Macromolecules*, **42**, 5251–5255; (b) Wang, S., Liang, Z.Y., Gonnet, P.,

Liao, Y.H., Wang, B., and Zhang, C. (2007) *Adv. Funct. Mater.*, **17**, 87–92.

44. Wei, C.Y., Srivastava, D., and Cho, K. (2002) *Nano Lett.*, **2**, 647–650.
45. ASTM Standard (1972) D 648-72. *Standard Test Method for Deflection Temperature of Plastics Under Flexural Load*, American Society for Testing and Materials, West Conshohocken, PA.
46. Liao, C.Z. and Tjong, S.C. (2011) *Polym. Eng. Sci.*, **51**, 948–958.
47. Davis, J.R. (2001) *ASM Specialty Handbook: Copper and Copper Alloys*, ASM International.
48. Biercuk, M.J., Liaguno, M.G., Radosavljevic, M., Hyun, J.K., Johnson, A.T., and Fischer, J.E. (2002) *Appl. Phys. Lett.*, **80**, 2767–2769.
49. Bryning, M.B., Milkie, D.E., Islam, M.F., Kikkawa, J.M., and Yodh, A.G. (2005) *Appl. Phys. Lett.*, **87**, 161909 (3 pp).
50. Haggenmueller, R., Guthy, C., Lukes, J.R., Fischer, J.E., and Winey, K.I. (2007) *Macromolecules*, **40**, 2417–2421.
51. Yu, A., Itkis, M.E., Bekayarova, E., and Haddon, R.C. (2006) *Appl. Phys. Lett.*, **89**, 133102 (3 pp).
52. Jakubinek, M.B., White, M.A., Mu, M., and Winey, K.I. (2010) *Appl. Phys. Lett.*, **96**, 083105 (3 pp).
53. Nan, C.W., Liu, G., Lin, Y.H., and Li, M. (2004) *Appl. Phys. Lett.*, **85**, 3549–3551.
54. Shenogin, S., Xue, L., Ozisik, R., Keblinski, P., and Cahill, D.G. (2004) *J. Appl. Phys.*, **95**, 8136–8144.
55. Prasher, R. (2008) *Phys. Rev. B*, **77**, 075424 (11 pp).
56. Huxtable, S.T., Cahill, D.G., Shenogin, S., Xue, L.P., Ozisik, R., Barone, P., Usrey, M., Strano, M.S., Siddons, G., Shim, M., and Keblinski, M.P. (2003) *Nat. Mater.*, **2**, 731–734.
57. Han, Z.D. and Fina, A. (2011) *Prog. Polym. Sci.*, **36**, 914–944.
58. Shenogina, N., Shenogin, S., Xue, L., and Keblinski, P. (2005) *Appl. Phys. Lett.*, **87**, 133106 (3 pp).
59. Xu, Q.Z. (2006) *Nanotechnology*, **17**, 1655–1660.
60. Bagchi, A. and Nomura, S. (2006) *Compos. Sci. Technol.*, **66**, 1703–1712.
61. Hai, M., Duong, H.M., Papavassiliou, D.V., Mullen, K.J., and Maruyama, S. (2008) *Nanotechnology*, **19**, 065702 (8 pp).
62. Song, Y.S. and Youn, J.R. (2005) *Carbon*, **31**, 1378–1385.
63. Yang, K., Gu, M.Y., Guo, Y.P., Pan, X.F., and Mu, H.H. (2009) *Carbon*, **47**, 1723–1737.
64. Gojny, F.H., Wichmann, M.H., Fiedler, B., Kinloch, I.A., Bauhofer, W., Windle, A.H., and Schulte, K. (2006) *Polymer*, **47**, 2036–2045.
65. Bonnet, P., Sireude, D., Garnier, B., and Chauvet, O. (2007) *Appl. Phys. Lett.*, **91**, 201910 (3 pp).
66. Yang, S.Y., Ma, C.C., Tenng, C.C., Huang, Y.W., Liao, S.H., Huang, Y.L., Tien, H.W., Lee, T.M., and Chiou, K.C. (2010) *Carbon*, **48**, 592–603.
67. Hong, W.T. and Tai, T.H. (2008) *Diamond Relat. Mater.*, **17**, 1577–1581.
68. Ganguli, S., Roy, A.K., and Anderson, D.P. (2008) *Carbon*, **46**, 806–817.
69. Veca, L.M., meziani, M.J., Wang, W., Wang, X., Lu, F.S., Zhang, P.Y., Lin, Y., Fee, R., Connell, J.W., and Sun, Y.P. (2009) *Adv. Mater.*, **21**, 2088–2092.
70. Sun, X.B., Ramesh, P., Itkis, M.E., Bekyarova, E., and Haddon, R.C. (2010) *J. Phys. Condens. Matter*, **22**, 334216 (9 pp).
71. Debelak, B. and Lafdi, K. (2007) *Carbon*, **45**, 1727–1734.
72. Song, P., Cao, Z.H., Cai, Y.Z., Zhao, L.P., Fang, Z.P., and Fu, S.Y. (2011) *Polymer*, **52**, 4001–4010.
73. Raza, M.A., Westwood, A.V., Brown, A.P., and Stirling, C. (2012) *Compos. Sci. Technol.*, **72**, 467–475.
74. Yu, A.P., Ramesh, P., Sun, X.B., Bekyarova, E., Itkis, M.E., and Haddon, R.C. (2008) *Adv. Mater.*, **20**, 4740–4744.

4
Mechanical Properties of Polymer Nanocomposites

4.1
Background

Carbon fibers (CFs) possess outstanding tensile strength and modulus and superior electrical and thermal conductivity [1]. Accordingly, conventional polymer composites reinforced with CFs have been widely used as structural materials in aerospace, automobile, and transportation industries as well as artificial implants for orthopedics [2]. Because of their low density, high strength to weight ratio, goof thermal stability, and fatigue resistance, fiber-reinforced polymer composites can be substituted for metallic materials for some aerospace engineering applications. The incorporation of rigid CFs of high into polymers improves their mechanical stiffness and strength. Generally, high loadings CF loadings are needed to achieve desired mechanical strength and stiffness as well as fatigue resistance [3]. This leads to poor processability of the resulting polymer composites. Therefore, considerable research activities have been devoted in both academic and technological sectors for the development of reduced weight, high performance, and novel functional materials for structural engineering applications. Polymer composites reinforced with carbonaceous nanofillers offer significant advantages in the development of such advanced engineering materials. This is because improved mechanical properties and dimensional stability of polymer composites can be obtained by adding very low loading levels of carbonaceous nanofillers. The high aspect ratios, extraordinary high modulus, and strength of carbon nanotubes (CNTs) and graphenelike materials render them primary load-bearing component of the polymer nanocomposites. Furthermore, the additions of CNT, carbon nanofiber (CNF), and thermally reduced graphene (TRG) to polymers do not impair their tensile ductility and fracture toughness but rather improve them instead [4–6; Chapter 2, Ref. 81].

The reinforcing efficiency of the CNT/polymer and graphene/polymer nanocomposites depends greatly on the attainment of effective load transfer across the filler–polymer matrix interface. Several factors such as the aspect ratio, surface roughness, dispersion of fillers in the matrix, and processing conditions influence interfacial properties and hence effective reinforcement eventually. In the past decade, the studies of polymer nanocomposites reinforced with carbonaceous

Polymer Composites with Carbonaceous Nanofillers: Properties and Applications, First Edition. Sie Chin Tjong.
© 2012 Wiley-VCH Verlag GmbH & Co. KGaA. Published 2012 by Wiley-VCH Verlag GmbH & Co. KGaA.

nanofillers are mainly concentrated on the CNTs. However, the development and commercialization of the CNT/polymer composites are impeded by poor dispersion of CNTs in the polymer matrix and higher cost of the nanotubes. On the other hand, graphene-based materials with low production cost, outstanding mechanical and electrical properties, and ease of fabrication are ideal nanofillers for fabricating strong, tough, and conducting polymer nanocomposites.

As aforementioned, CNT and graphene exhibit Young's modulus of ~ 1 TPa and tensile strength of 130–150 GPa [Chapter 1, Refs. 120, 146]. However, TRGs exhibit distinct advantages over CNTs for reinforcing polymers [6]. The wrinkled surfaces of TRGs interlock very well with the surrounding polymeric material, thereby facilitating interfacial load transfer between graphene and the polymer matrix during mechanical testing. Load transfer from a relatively low modulus polymer matrix to a high modulus filler results in an increase of the composite modulus and strength. Two-dimensional (planar geometry) of graphene sheets with high specific surface area provide more contacts with the polymer matrix than one-dimensional CNTs. Therefore, TRGs of the TRG/epoxy composites can deflect microcracks more effectively than CNTs of the CNT/epoxy nanocomposites. This leads to the TRG/epoxy composites that have higher fracture toughness. Similarly, Ruoff and coworkers reported that TRGs are more effective than exfoliated graphites (EGs) and single-walled carbon nanotubes (SWNTs) for reinforcing poly(methyl methacrylate) (PMMA) because of their wrinkled features and strong chemical interactions between residual oxygenated groups of TRGs and polar PMMA matrix [Chapter 2, Ref. 14].

4.2
General Mechanical Behavior

For the CNT/polymer nanocomposites, control of the structure, alignment, and dispersion of nanotubes in the matrix are primary factors governing efficient load transfer at the polymer–nanotube interface. Furthermore, the compatibility of nanotubes with the polymer matrix is another issue that must be overcome for achieving homogeneous dispersion of nanotubes. Particularly, arc-grown multi-walled carbon nanotubes (MWNTs) are highly crystalline and defect free; thus their surfaces are inert and incompatible with polymers. The compatibility and interfacial bonding between CNTs and the polymer matrix can be improved through proper selection of the polymer matrix and chemical functionalization of nanotubes.

CNTs often disperse randomly as wavy tubes rather than straight ones in the polymer matrix of nanocomposites. The resulting composites could not fully take advantage of exceptionally high strength of individual nanotubes. Therefore, the reinforcing effect of nanotubes in nanocomposites has not reached to the full potential. On the basis of finite element simulations, Li and Chou reported that the nanotube waviness reduces the elastic modulus and tensile strength but enhances the ultimate strain of the composite, while the randomness of nanotube distribution reduces the elastic modulus and tensile strength of composites [7]. Apparently, the mechanical strength of the CNT/polymer nanocomposites can be markedly enhanced by reinforcing with vertically aligned nanotubes [8, 9].

The tensile yield stress of polymers is known to be strain rate and temperature dependent. The yield stress of polymers can be described by well-established Eyring theory taking yielding as an activation process for segmental molecular movement. Mathematically, Eyring equation takes the following form [10]

$$\frac{\sigma_y}{T} = \frac{\Delta H}{\upsilon T} + \frac{R}{\upsilon} \ln\left(\frac{2\dot{e}}{\dot{e}_0}\right) \quad (4.1)$$

where σ_y is the yielding stress, T the testing temperature in K, ΔH the activation energy for the polymer segments to jump under tension, R the gas constant, υ the activation volume, $\dot{e}$ the strain rate and $\dot{e}_0$ the preexponential factor. The plot of σ_y against $\ln(\dot{e})$ generates a straight line with a slope given by

$$\text{slope} = \frac{RT}{\upsilon} \quad (4.2)$$

The Eyring equation has been used successfully by Bao and Tjong [11, 12] to describe yielding behaviors of the MWNT/PP and CNF/PP nanocomposites tensile specimen deformed at different strain rates and temperatures. Apart from the tensile testing, dynamic mechanical analysis (DMA) can also be used to assess mechanical performance of the polymer nanocomposites. The storage modulus measured at room temperature can be roughly estimated to be identical to the Young's modulus determined from tensile testing.

For the nanocomposites with brittle matrix, flexural or bending testing is more appropriate for evaluating their strength and stiffness [13]. The flexural properties such as flexural strength and modulus can be determined using a composite beam specimen under a three-point bending mode according to the ASTM Standard D790-10 [14]. The flexural strength can be expressed as [15]

$$\sigma = \frac{3PS}{2WB^2} \quad (4.3)$$

where P is the maximum load at failure, S the specimen length between the two support points, W the width, and B the thickness of the specimen.

Polymer composites are currently being used as structural components for engineering applications. Therefore, their deformation and fracture characteristics are considered to be of technological importance. The fracture behavior of polymer composites can be determined traditionally using Charpy, Izod, and drop-weight impact tests. The procedures for conducting Izod impact test for polymers are described in detail by the ASTM Standard D256-10 [16]. These techniques determine absorbed energy in the impact specimens fractured with a falling pendulum from a known height. The impact energy value reveals the failure modes of notched polymer specimens in terms of brittle or ductile fracture under impact loading conditions, depending on the temperature and strain rate. However, impact tests do not provide property data for designing structural components. Impact energies are mainly used for making comparisons of the failure modes of two different materials subjected to similar impact conditions. Nevertheless, impact tests are often used for evaluating impact toughness of the CNT/polymer and graphene/polymer nanocomposites because of their simplicity and low equipment cost [4, 13].

4.3
Fracture Toughness

A proper understanding of the fracture behavior of polymer nanocomposites is essential to ensure long-term safety and reliability for their structural engineering applications. Fracture toughness, which reflects the fracture resistance of a material, is an inherent material parameter for designing structural components containing internal flaws. Accordingly, fracture mechanics approach is very useful to quantify the flaw tolerance of structural materials. For the materials that undergo elastic deformation or small scale yielding ahead of the crack tip, linear elastic fracture mechanics (LEFM) is more appropriate for characterizing their fracture toughness. According to the LEFM theory, the applied stress (σ) distribution in the vicinity of a crack tip is described by the stress intensity factor (K), defining as

$$K = \sigma(\pi a)^{1/2} \tag{4.4}$$

where a is the crack length. When K reaches a critical value, catastrophic crack growth occurs, leading to eventual fracture of a material. The critical stress intensity factor is termed as the *fracture toughness*, K_C. Under crack opening (mode I) test, the critical value of K is called K_{IC}, the plane-strain fracture toughness. To meet plane-strain fracture criterion, the thickness (B) of a specimen should satisfy the following relation

$$B = 2.5 \left(\frac{K_{IC}}{\sigma_y} \right) \tag{4.5}$$

where σ_y is the yield strength of a material.

The K_{IC} of plastic materials can be measured using single-edge-notched bending (SENB) and compact tension specimens. Under three-point bending mode, the stress intensity factor needed for the crack propagation in an SENB specimen is given by ASTM Standard D 5045-99 [17]

$$K_{IQ} = \frac{P}{B\sqrt{W}} f(a/W) \tag{4.6}$$

where $f(a/W)$ is a geometrical shape factor. K_{IQ} can be treated as K_{IC} when the plane-strain and linear elastic failure conditions are satisfied. The strain energy release rate, G, is the energy available for a unit increase in crack length. Fracture occurs when G reaches a critical value G_c. The critical stress intensity factor (K_{Ic}) is related to G_{Ic} by

$$K_{IC} = \sqrt{EG_{IC}} \tag{4.7}$$

where E is the Young's modulus. In general, K_c and G_c are used to characterize fracture toughness of brittle glassy polymers (e.g., PMMA and polystyrene (PS)), thermosets, and their composites. On the other hand, a large specimen thickness is needed for testing polymers with high toughness and low yield strength in order to maintain the plane-strain condition (Eq. (4.5)). This makes it very difficult to determine K_{IC} of tough polymers practically.

The toughness characterization of ductile semicrystalline polymers and their composites remains a challenge for materials scientists and mechanical engineers. For example, high-density polyethylene (HDPE) in the plane-stress failure experiences extensive ductile deformation that involves a localized necking process. In this case, the J-integral and essential work of fracture (EWF) concepts can be used for characterizing fracture toughness of ductile materials [18, 19]. The J-integral introduced by Rice is based on a path-independent contour integral around the crack tip under plastic deformation. The critical value of the J-integral (J_c) is accompanied by full crack tip blunting before crack growth. Two kinds of specimens are employed for the J-integral measurements, that is, multiple and single specimens. Multiple specimens are commonly used for the toughness measurements in which a set of identical specimens are loaded to various displacements to yield different crack extensions (Δa). A resistance curve ($J-R$) is then constructed by plotting J against Δa. According to ASTM Standard E813-81, the data of resistance curve ($J-R$) is best fitted with a linear line. J_c is defined as the crack initiation value and is determined at the point of intersection between the resistance curve and the blunting line with a slope of $2\sigma_y$, that is, $J = 2\sigma_y \Delta a$ [20a]. In the ASTM Standard E813-87, the resistance curve ($J-R$) is best fitted with a power law relation. J_c is determined from the intersection of power law regression curve with an offset line of 0.2 mm [20b].

Compared to tedious J-integral method, the EWF approach has gained popularity in recent years because of its simplicity and applicability to thinner specimens [21–24]. The EWF measurements can be performed using either double-edge-notched tension (DENT) or SENB specimens. The European Structural Integrity Society (ESIS) protocol recommends the use of DENT specimens under mode I quasistatic loading (low strain rate) tests [25]. The EWF approach involves the separation of the total energy consumed into EWF (W_e) and nonessential plastic work (W_p). W_e is the work needed to form new surfaces in inner process zone where the real fracture process occurs and is surface related. W_p is the energy dissipated by several deformation mechanisms in an outer plastic zone and is volume related (Figure 4.1). Thus the total work of fracture (W_f) can be expressed by the following equations

$$W_f = W_e + W_p \tag{4.8}$$

$$W_f = w_e LB + \beta w_p L^2 B \tag{4.9}$$

where L is the ligament length and β is a shape factor of the plastic zone. By dividing W_f with the net cross section LB, we have

$$w_f = \frac{W_f}{LB} = w_e + \beta w_p L \tag{4.10}$$

where w_f, w_e, and βw_p are the specific total, essential, and nonessential work of fracture, respectively. The W_f values of several DENT specimens with different

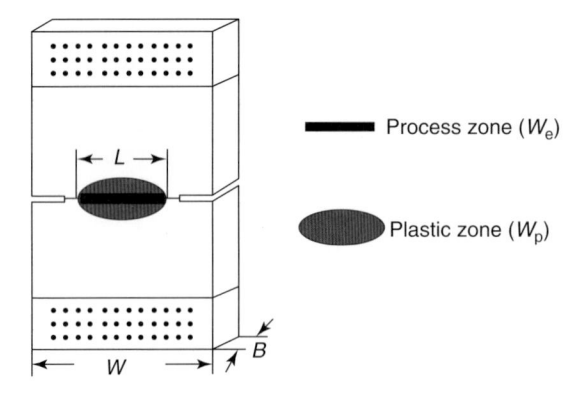

Figure 4.1 Geometry of DENT specimen for EWF tests.

ligament lengths can be determined from the areas under the load–displacement curves. Thus linear regression line can be obtained from the plot of w_f against L. The slope of this regression line gives the value for βw_p, reflecting the resistance against crack growth. The interpolation at $L = 0$ gives w_e, representing the fracture toughness of a material. The basic criteria for the validity of EWF approach are the ligament must be fully yielded before the onset of crack propagation and geometrical similarity of the load–displacement curves. The validity range of ligament L under plane-stress condition is given by Clutton [25]

$$\max(3B - 5B) \leq L \leq \min\left(W/3, 2r_p\right) \tag{4.11}$$

where $2r_p$ is the size of plastic zone that can be evaluated from the following equation

$$2r_p = \frac{1}{\pi} \frac{E w_e}{\sigma_y^2} \tag{4.12}$$

The EWF approach has been used successfully to characterize fracture toughness of polymer nanocomposites showing ductile failure [26–29].

4.4
Strengthening and Toughening Mechanisms

4.4.1
Interfacial Shear Stress

The stress transfer topic in conventional short-fiber-reinforced composites has been studied extensively by many researchers over the past two decades. The stress transfer from the matrix to the fiber reinforcement during mechanical loading takes place through an interfacial shear stress (IFSS; τ). For the short fiber composites, Kelly and Tyson [30] model can be used for estimating τ that depends greatly on the critical fiber length (l_c). Figure 4.2 displays the load in a fiber of different length

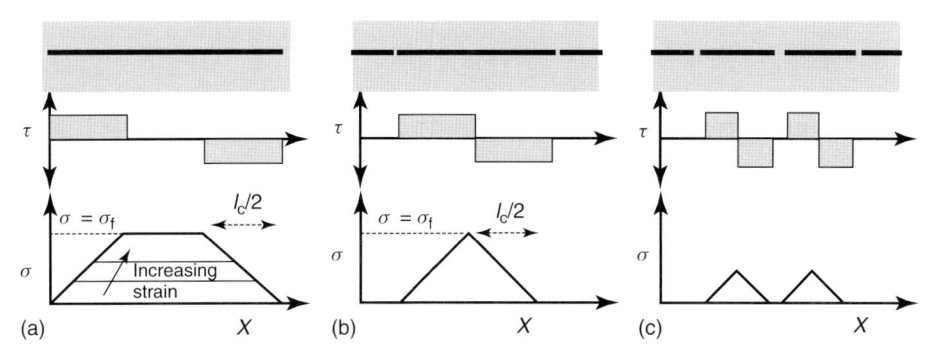

Figure 4.2 Effect of critical fiber length on the stress distribution of a discontinuous fiber. (a) $l > l_c$, (b) $l = l_c$, and (c) $l < l_c$. (Source: Reproduced with permission from [31], Elsevier (2010).)

(l) embedded in a resin subjected to longitudinal tensile loading with a constant IFSS [31]. In the case of $l > l_c$, the fiber stress is zero at fiber ends and increases linearly with fiber length because of increased strain levels in the matrix, reaching a maximum in fiber midpoint. However, interfacial stress is relatively high at both fiber ends. Thus fiber fracture prevails at this stage under tensile loading. For $l = l_c$, the fiber is just long enough to reach its fiber strength (σ_f). For $l < l_c$, the fiber cannot carry the maximum load because the fiber has insufficient length to reach its failure stress. However, the interfacial stress is sufficient to cause fiber/matrix debonding, resulting in fiber pullout. To achieve effective fiber reinforcement, the $l \gg l_c$ criterion must be met. By assuming a constant shear stress, the interface shear strength can be obtained according to the following equation

$$\tau = \frac{\sigma_f d}{2 l_c} \tag{4.13}$$

Accordingly, the critical aspect ratio (l_c/d) of the fiber plays an important role in efficient load transfer across the matrix–fiber interface. In practice, single fiber fragmentation test is used typically for evaluating τ. In this test, a single fiber is encapsulated in the central line of the matrix of a tensile specimen. On tensile loading, the fiber breaks into increasingly smaller fragments when the maximum fiber stress reaches its tensile strength.

The critical fiber length can be determined from the average fragment length $\bar{l}$ by

$$l_c = \frac{4}{3} \bar{l} \tag{4.14}$$

Compared to conventional fiber-reinforced polymer microcomposites, CNT/polymer and graphene/polymer nanocomposites offer large interfacial areas for stress transfer because of their high aspect ratios and surface areas. Furthermore, the interfacial interaction of the nanotubes with polymer matrix can result in a polymer interphase with morphology and properties different from the bulk. A typical example is the formation of transcrystalline polymer on the nanotube surface [Chapter 3, Ref. 20]. The polymer transcrystallinity is well recognized to enhance IFSS in the polymer microcomposites [32].

In general, stress transfer across the matrix–nanotube interface can be determined using experimental measurements and molecular dynamics (MD) simulations. It is rather difficult to conduct and device experiments for evaluating τ value of the CNT–polymer and graphene–polymer interfaces because of the nanoscale dimension of the fillers. For example, Tsuda et al. [33] determined IFSS of the MWNT/PEEK (poly(ether ether ketone)) composite using a special nanopullout testing system installed in a scanning electron microscopy (SEM). The τ value of the composite was measured to be 3.5–7 MPa. Wagner and coworkers [34a,b] also performed nanopullout experiments for measuring pullout strength between a single MWNT and a polymer matrix, that is, amorphous polyethylene (PE) butane using atomic force microsopy (AFM). A single MWNT was attached to an AFM tip before pushing it into a heated polymer film of ~300 nm thickness. They measured the force required to pull the tube out from the film on cooling. From the measured pullout force, the τ value of the composite was evaluated to be 47 MPa. Strong covalent bonding between the MWNT and PE-butene was responsible for such a high τ value. On the basis of MD simulations, Frankland et al. [35] reported that the interfacial shear strength of a CNT/polymer composite with weak nonbonded interactions can be increased by over an order of magnitude by introducing chemical bonding between the nanotube and polymer matrix through the formation of cross-links. In other words, chemical functionalization of nanotubes can enhance IFSS markedly because of the establishment of covalent bonding between the CNTs and the matrix. The τ value of nonbonded composite with amorphous polymer matrix increases from 2.7 to 30 MPa as a result of cross-linking. Therefore, strengthening the interface by covalent bonds can improve the mechanical properties of CNT/polymer composites greatly. A similar beneficial effect of covalent functionalization in enhancing τ value of the SWNT/PE composite was reported by Zheng et al. [36] using MD simulations.

4.4.2
Interfacial Interaction

Apart from nanopullout tests, Raman spectroscopy can provide useful information concerning stress transfer mechanism at the filler interface in the polymer nanocomposites. During mechanical loading, a large shift in a characteristic Raman peak allows the determination of local strain within CNTs and graphene in the polymer nanocomposites [37–42]. In this context, the application of tensile, compressive or shear stress to CNTs or graphene can induce local strain and change of interatomic distances, leading to changes in the C–C bond vibrational frequencies. The G'(or D*) band that derives from the second-order overtone of D-band at ~2700 cm^{-1} of CNTs is very sensitive to mechanical deformation. Thus this peak shift is often used to reveal interfacial stress transfer effect of the CNT/polymer composites. Schadler et al. reported that the D* peak of the MWNT/epoxy composites shifts markedly under compression but not in tension.

The D* peak was found to shift upwards by $7\,\text{cm}^{-1}$ by applying 1% compressive strain. A slightly positive shift occurred when the composites were under tension [38]. The large shift in compression compared to tension demonstrated that the MWNTs carry more strain in compression than in tension. Slippage of inner tubes of MWNTs took place during tension but not in compression, leading to efficient stress transfer during compression.

More recently, Koratkar and coworkers employed Raman spectroscopy to study stress transfer in the 0.1% TRG/polydimethyl-siloxane (PDMS) nanocomposite [42]. The Raman G-band ($\sim 1582\,\text{cm}^{-1}$) peak shift response of this nanocomposite under applied strains was examined. For comparison, the response of the 0.1% SWNT/PDMS nanocomposite was also recorded. They reported that Raman G-mode peak shift in the 0.1% TRG/PDMS nanocomposite is at least an order of magnitude higher than its SWNT counterpart under tensile and compression loadings. The G-peak of the 0.1% TRG/PDMS nanocomposite was found to shift by $\sim 2.4\,\text{cm}^{-1}$/composite strain %. In contrast, the peak shift was only $\sim 0.1\,\text{cm}^{-1}$/composite strain % in the 0.1% SWNT/PDMS counterpart. This implies that enhanced load transfer effectiveness for the TRG filler due to its rough/wrinkled surface features, facilitating mechanical interlocking at the TRG–polymer interfaces.

4.4.3
Micromechanical Modeling

The shear-lag model originally proposed by Cox [43a] gives a good estimation of the IFSS in the discontinuously fibers transferred from the matrix through the interface. Accordingly, this model has been used by some researchers to predict IFSS in the CNT/polymer composites [43b,c]. This model considers the stress distribution of a single fiber embedded in an elastic matrix with a perfectly bonded interface subjected to uniform displacement in the fiber direction. By further taking the fiber orientation into considerations, the Young's modulus of the discontinuously fiber composite is given by the following equation

$$E_c = \eta_l\, \eta_o\, E_f\, V_f + E_m\, (1 - V_f) \qquad (4.15)$$

where E_f is the fiber modulus, E_m the matrix modulus, V_f the fiber volume fraction, η_l the length efficiency factor that varies between 0 and 1, and η_o the orientation factor. The η_o value equals to unity for fully aligned fibers and 3/8 for random-two-dimensional orientation [44]. The strength of the composite (σ_c) based on the shear-lag model of Kelly and Tyson is given by Kelly and Tyson [30]

$$\sigma_c = \frac{l_c}{2l}\sigma_f V_f + \sigma_m (1 - V_f) \qquad (4.16)$$

where σ_f is the fiber strength and σ_m the matrix modulus. Equation (4.16) applies for a composite with fibers of an average length less than the critical fiber length. For $l \gg l_c$, this equation approaches the well-known rule of mixtures for the composite strength. In general, the composite strength can be expressed as follows [44]:

$$\sigma_c = \eta_l\,\eta_o\,\sigma_f\,V_f + \sigma_m\,(1 - V_f) \tag{4.17}$$

The Cox model generally gives a poor estimate of the strengthening effect of discontinuously reinforced composites due to the negligence of stress transfer effect at the fiber ends.

Alternatively, the elastic modulus of short-fiber-reinforced composites can be predicted from the Halpin-Tsai micromechanical model [45]. This model can also be used to describe stiffening effect of the CNT or graphene nanofillers of large aspect ratios in the polymer nanocomposites [6, 44, 45; Chapter 2, Refs. 1, 2, 67]. Mathematically, the longitudinal modulus (E_{CL}) and transverse modulus (E_{CT}) of discontinuously short-fiber-reinforced composite can be determined from the following relationships

$$\frac{E_{CL}}{E_m} = \frac{1 + \xi\,\eta_L\,V_f}{1 - \eta_L\,V_f} \tag{4.18}$$

$$\frac{E_{CT}}{E_m} = \frac{1 + 2\eta_T\,V_f}{1 - \eta_T\,V_f} \tag{4.19}$$

where E_m is the elastic modulus of matrix and V_f the volume fraction of the filler; η_L and η_T are given by the following expressions

$$\eta_L = \frac{(E_f/E_m) - 1}{(E_f/E_m) + \xi} \tag{4.20}$$

$$\eta_T = \frac{(E_f/E_m) - 1}{(E_f/E_m) + 2} \tag{4.21}$$

where E_f is the elastic modulus of filler; $\xi = 2(l/d)$ for the fiber- or tubelike filler in which l and d are its length and diameter and $\xi = 2\,l/3t$ for sheetlike filler in which t is its thickness. For randomly oriented short-fiber-reinforced composite, the elastic modulus of the composite is given by

$$E_{random} = \frac{3}{8}E_{CL} + \frac{5}{8}E_{CT} \tag{4.22}$$

4.4.4
Toughening Mechanism

For conventional short fiber composites, fiber additions affect fracture behavior of the polymers markedly. Several failure modes can occur in the polymer microcomposites subjected to mechanical loading including fiber breakage, crack bridging, crack deflection, fiber pullout, and matrix deformation/cracking, depending on the interfacial bonding strength and the matrix material selected. It is well recognized that crack bridging is an important failure mechanism contributing to the enhanced toughness of conventional composite materials. During mechanical loading, the propagating crack is bridged by fibers that are subjected to breakage and pullout. These fibers have not failed completely and still can carry the applied stress [46a,b]. Similarly, CNTs of large aspect ratios can also bridge propagating cracks through epoxy, polyamide (PA), and PS resin [47–49]. Figure 4.3 shows

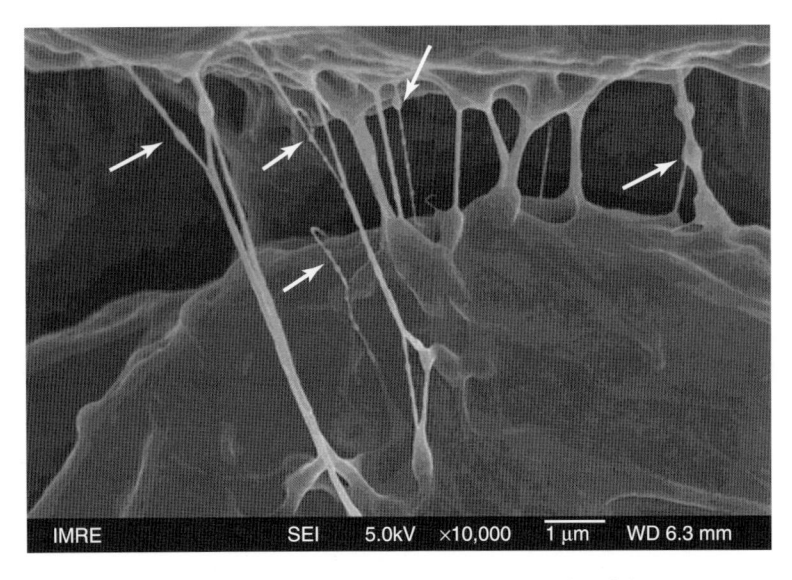

Figure 4.3 Scanning electron micrograph shows bridging of crack by MWNTs in melt-compounded 0.5 wt% MWNT/PA6 nanocomposite. (Source: Reproduced with permission from [49], The American Chemical Society).)

typical crack bridging in the 0.5 wt% MWNT/PA6 nanocomposite under tensile loading. Another failure event such as crack deflection is also responsible for the energy dissipation mechanisms, particularly for the composites with brittle polymer matrix (e.g., epoxy) [50]. In the absence of crack deflection, a crack can propagate unimpeded through the matrix catastrophically when it encounters the fibers. This results in no toughening because little energy is dissipated during the crack propagation. When the crack deflection occurs, the crack twists from its initial propagation direction and circumvents the fillers without penetrating them. This produces an increase in the total surface area, leading to higher energy absorption or enhanced fracture toughness. According to the literature, two-dimensional clay platelets and TRG sheets are effective to deflect propagating cracks in brittle epoxy matrix of the polymer nanocomposites [6, 51, 52].

4.5
Nanocomposites with Graphene Fillers

4.5.1
Thermoplastic Matrix

Graphene oxide (GO) prepared by chemical oxidation of graphite flakes contains hydroxyl and epoxide functional groups on their basal planes, as well as carbonyl and carboxyl groups at its sheet edges (Figure 1.5). These oxygenated groups render GO hydrophylic, facilitating its exfoliation in water and interaction with

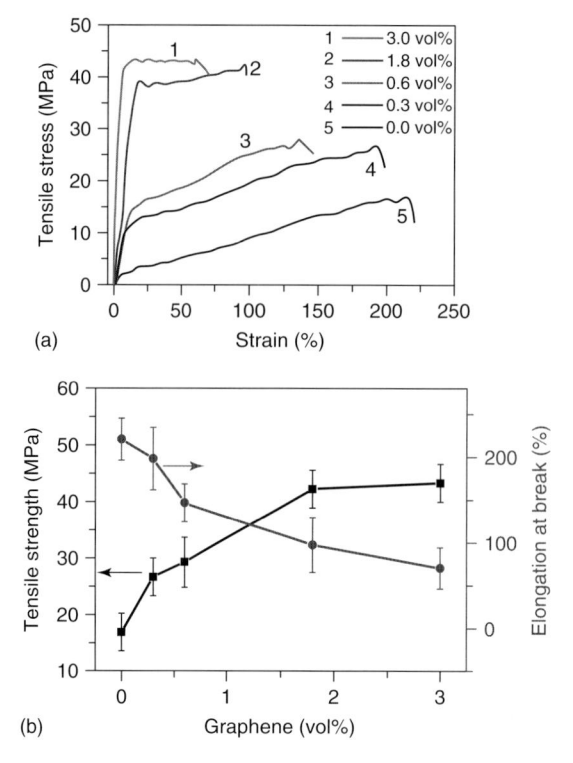

Figure 4.4 (a) Stress–strain curves, (b) tensile strength and elongation at break versus filler content. (Source: Reproduced with permission from [Chapter 2, Ref. 2], The American chemical Society (2010).)

water-soluble polymers such as poly(vinyl alcohol) (PVA) and poly(ethylene oxide) (PEO) [Chapter 2, Refs. 1–5]. This promotes the formation of a strong covalent interface for achieving efficient stress transfer. Therefore, homogeneous dispersion of GO in the PVA matrix and strong interfacial bonding lead to enhanced tensile strength and improved Young's modulus of PVA. Very recently, Zhao et al. [Chapter 2, Ref. 2] prepared solution-mixed rGO/PVA (reduced graphene oxide) nanocomposites reinforced with 0.5, 1, 3, and 5 wt% rGO, corresponding to 0.3, 0.6, 1.8, and 3 vol% rGO, respectively. Figure 4.4a shows typical stress–strain curves of rGO/PVA nanocomposites. The variations of tensile strength and elongation at break with rGO content of the PVA nanocomposites are shown in Figure 4.4b. It can be seen that the graphene additions improve the tensile strength of PVA at the expense of its ductility. The plot of Young's modulus as a function of rGO content for the rGO/PVA nanocomposites is depicted in Figure 4.5. There is a 10-fold increase in the modulus of PVA by adding 1.8 vol% rGO. For comparison, the Young's moduli of the nanocomposites that predicted from the Halpin-Tsai model using aligned fillers (Eq. (4.18)) and randomly oriented fillers (Eq. (4.22)) are also shown in this figure. The results reveal that the experimental data agree

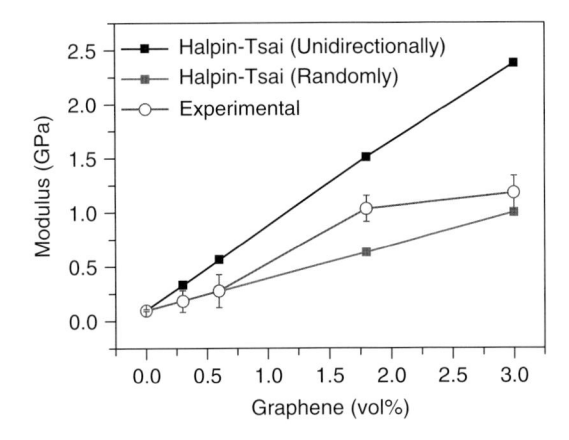

Figure 4.5 Young's modulus versus filler volume content of rGO/PVA nanocomposites. (Source: Reproduced with permission from [Chapter 2, Ref. 2], The American Chemical Society (2010).)

closely with those predicted from theoretical simulations for randomly dispersed rGO sheets, particularly for low filler contents.

The addition of low GO or rGO loadings to PVA also improves its tensile ductility since the oxygenated groups of GO and residual oxygenated groups of rGO can interact with water-soluble PVA. More recently, Wang et al. [53] reexamined tensile behavior of solution-mixed GO/PVA and rGO/PVA nanocomposites (Figure 4.6a,b). For the GO/PVA nanocomposites, the additions of 0.5, 1, and 1.5 wt% GO to PVA do not impair its tensile ductility but rather improve it. The tensile elongation increases from 165 to 222, 171, and 210% by adding 0.5, 1, and 1.5 wt% GO, respectively. For the rGO/PVA nanocomposites, only 0.5 wt% rGO is beneficial for improving tensile ductility of PVA.

As mentioned in Chapter 2, solution blending enables better dispersion of foliated graphene sheets in the PEO matrix than melt blending (Figure 2.27a, b). Figure 4.7a,b shows the stress–strain curves of solution-mixed and melt-blended TRG/PEO nanocomposites, respectively [Chapter 2, Ref. 13]. Solution-mixed TRG/PEO nanocomposites exhibit higher mechanical strength that melt-blended counterparts because of homogeneous dispersion and further exfoliation of the graphene sheets. The Young's modulus as a function of both nanocomposite systems is shown in Figure 4.8. The experimental elastic modulus results of solution-mixed FGS/PEO (functional graphene sheet) nanocomposites agree well with the longitudinal modulus predicted from the Halpin-Tsai equation. On the contrary, the Young's modulus data of melt-blended TRG/PEO nanocomposites agree reasonably with those predicted from the Halpin-Tsai equation for the composite with randomly oriented fillers.

Rafig et al. [6] prepared GO/PA-12 nanocomposites using melt-compounding process. GO was further exfoliated to graphene sheets in water under sonication before compounding. Figure 4.9a shows stress–strain curves of PA-12 and its nanocomposites. The variations of tensile strength, elongation at break, and Young's

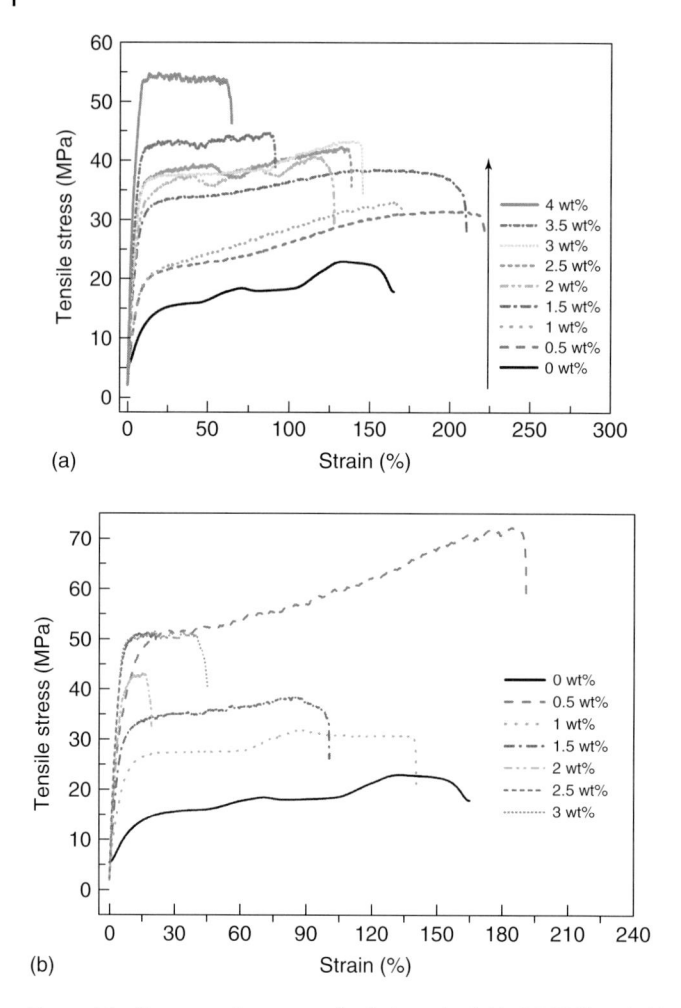

Figure 4.6 Stress–strain curves of solution-mixed (a) GO/PVA and (b) rGO/PVA nanocomposites. (Source: Reproduced with permission from [53], Wiley (2011).)

modulus with filler content of these nanocomposites are shown in Figure 4.9b,c. The tensile strength increases greatly with increasing filler content up to 1 wt% followed by a sharp decrease at 3 wt% GO. The elongation at break increases with filler content up to 0.6 wt%, thereafter decreasing markedly with increasing GO content. At 0.6 wt% GO, the tensile strength and elongation at break of PA-12 improve by ~35 and ~200%, respectively. The elastic modulus changes little with increasing filler content. The K_{IC} and drop-weight impact energy of the nanocomposites as a function of filler content are shown in Figure 4.10a,b, respectively. The fracture toughness of PA-12 increases from ~1.28 to ~2.2 MPa m$^{1/2}$, by adding 0.6 wt% GO being 72% enhancement in the toughness. Similarly, the addition of 0.6 wt% GO to PA-12 also improves its impact energy by 175%. The improvements in tensile

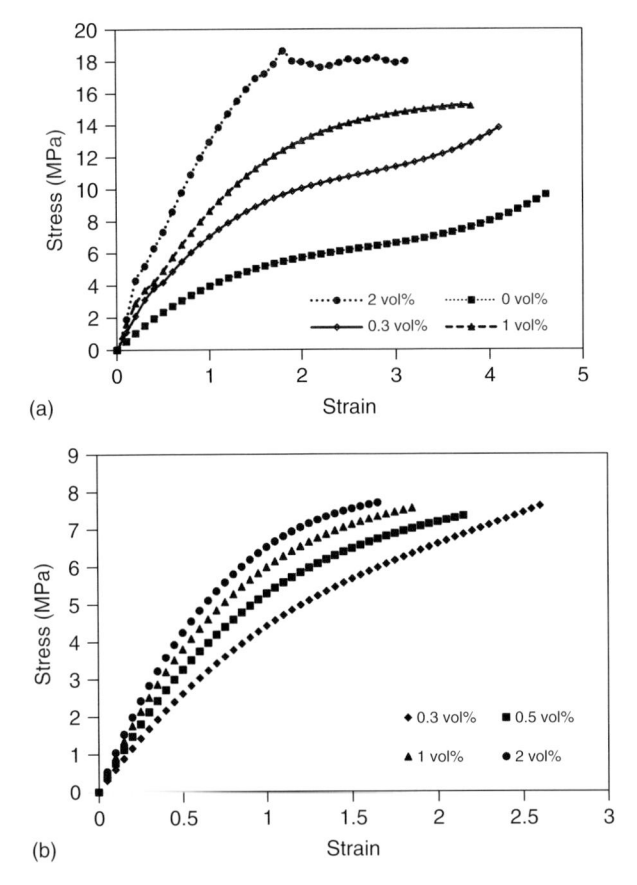

Figure 4.7 Stress–strain curves of (a) solution-mixed and (b) melt-blended TRG/PEO nanocomposites. (Source: Reproduced with permission from [Chapter 2, Ref. 13], Elsevier (2011).)

strength, elongation at break, fracture toughness, and impact energy of PA-12 are attributed to the establishment of hydrogen bonding between −NH group of PA-12 and oxygenated groups of the GOs and to the formation of γ-phase in PA-12 [6].

Potts et al. in situ polymerized GO/PMMA and rGO/PMMA nanocomposites and investigated their dynamic mechanical properties [Chapter 2, Ref. 100]. In addition, injection molding was used to fabricate specimens for static tensile measurements. They reported that both the GO and rGO additions increase elastic modulus and tensile strength of PMMA. However, GO sheets exhibit higher reinforcing effect than rGO because of the presence of oxygenated groups on the GO surface. This facilitates better interfacial adhesion between GO and PMMA when compared with rGO.

As aforementioned, TRGs exhibit wrinkled morphology, thereby favoring mechanical interlocking with the polymer chains and forming strong interfacial

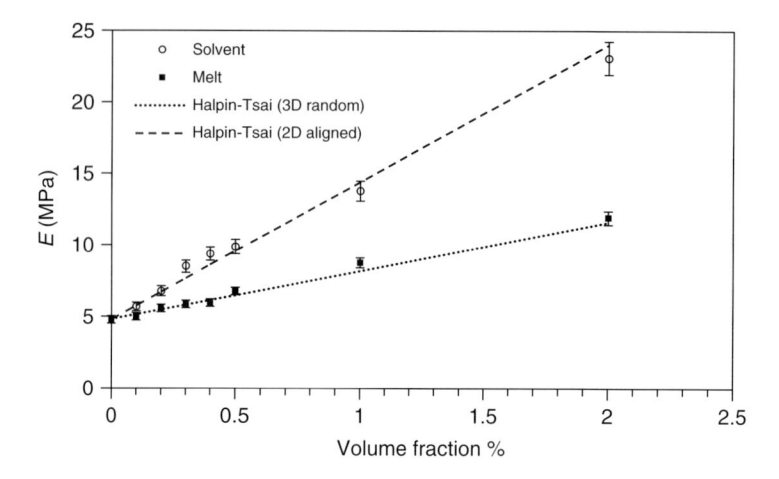

Figure 4.8 Young's modulus versus filler volume fraction for solution-mixed and melt-blended TRG/PEO nanocomposites. (Source: Reproduced with permission from [Chapter 2, Ref. 13], Elsevier (2011).)

interaction [Chapter 2, Ref. 14]. Figure 4.11 shows that the improvement in both thermal and mechanical properties of PMMA can be achieved by reinforcing with 1 wt% SWNT, 1 wt% EG, and 1 wt% TRG. However, TRG sheets are more effective for enhancing elastic modulus, ultimate strength and glass-transition temperature of PMMA when compared with the EG and SWNT fillers. All these parameters are determined from DMA measurements. The elastic modulus is taken from the storage modulus at room temperature. Strong interface between the filler and the matrix can be seen in the fracture surface of solvent-cast 1 wt% TRG/PMMA composite. In contrast, EG platelets exhibit poor bonding to the polymer matrix (Figure 4.12a,b). Furthermore, SWNTs show high tendency to form clusters, thus reducing their reinforcement in the PMMA matrix (Figure 4.12c).

For nonpolar polyolefins such as linear low-density (LLDPE), functionalization of polymer with cyano- (PE-CN), primary amino- (PE-NH$_2$), secondary amino- (PE-NHEt), and isocyanate-(PE-NCO) groups is needed for achieving better dispersion of TRGs in the polymer matrix [Chapter 2, Ref. 57]. Figure 4.13 shows the tensile modulus of solvent-blended LLDPE-, PE-CN-, PE-NH$_2$-, PE-NHEt-, and PE-NCO-based nanocomposites filled with 0.4 vol% TRG. The highest modulus of 9.7 MPa can be achieved by blending PE-NCO with TRG. No information was disclosed by the researchers for tensile strength of resulting nanocomposites. Alternatively, surface modification of GO can be employed to improve dispersion of graphene in nonpolar polyolefin matrix. Very recently, Kuila *et al.* [54] functionalized rGO with dodecyl amine (DA), terming as DA-G, and then blended with LLDPE via solution mixing to form DA-G/LLDPE nanocomposites. They reported that the tensile strength and modulus of LLDPE increases with increasing filler content at the expense of tensile ductility (Figure 4.14). The improvements in tensile strength derive from good interfacial bonding between the LLDPE matrix and the filler.

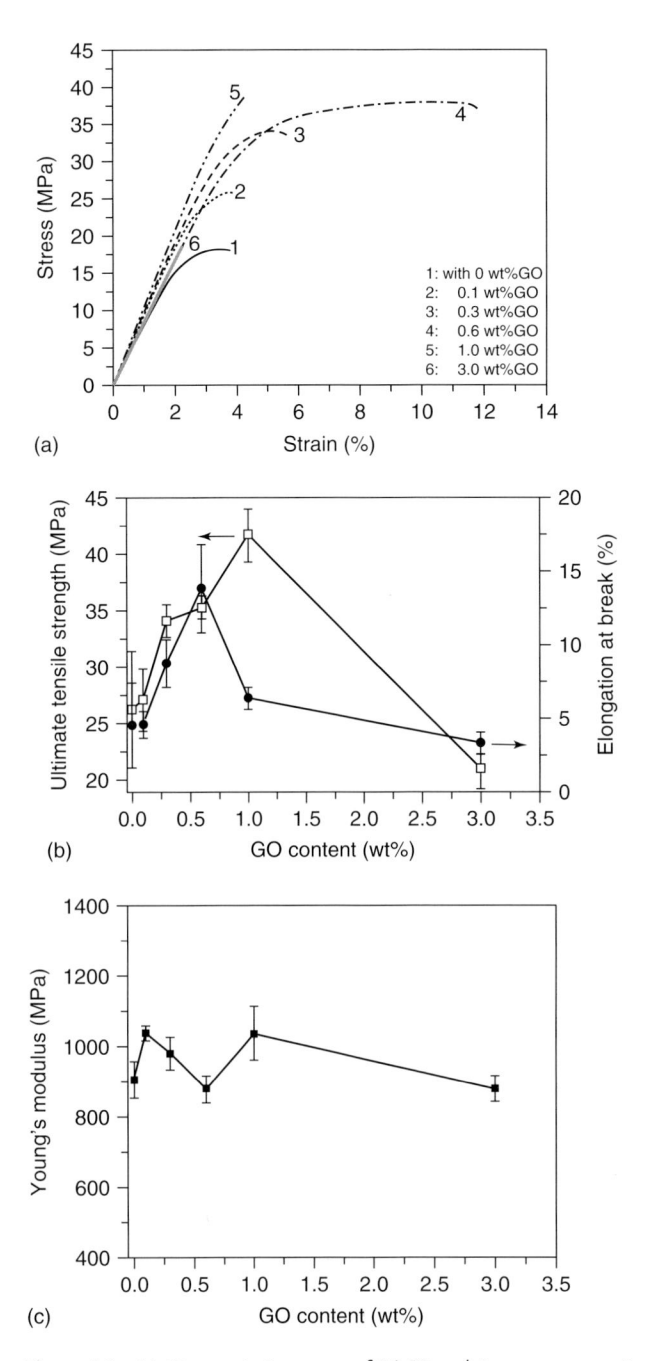

Figure 4.9 (a) Stress–strain curves of PA-12 and its nanocomposites. (b) Tensile strength and elongation at break, and (c) Young's modulus as a function of the filler loading. (Source: Reproduced with permission from [6], Elsevier (2010).)

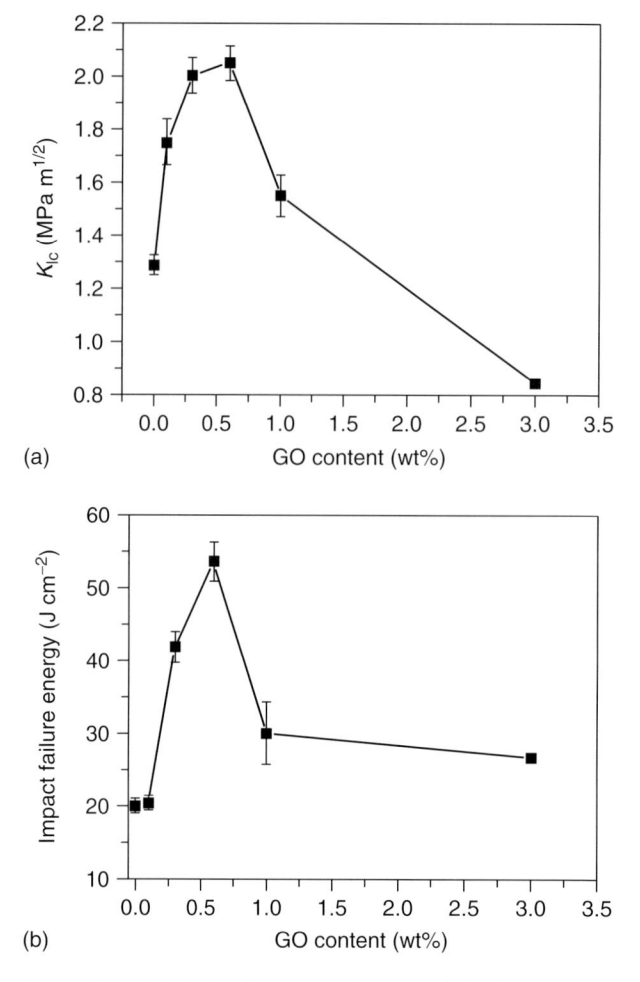

Figure 4.10 (a) Mode I fracture toughness and (b) drop-weight impact energy versus GO content for melt-compounded GO/PA-12 nanocomposites. (Source: Reproduced with permission from [6], Elsevier (2010).)

To achieve efficient stress transfer in nonpolar polypropylene (PP), Song et al. employed combined solution-mixing and melt-compounding processes to fabricate graphene/PP nanocomposites (Figure 2.26). Thus the fabrication processes consisted of forming PP-coated rGO initially via solution mixing followed by melt-blending PP-coated rGO with the PP polymer. Figure 4.15a shows the tensile stress–strain curves of graphene/PP nanocomposites, respectively. The elastic modulus as a function of filler content is shown in Figure 4.15b. In this figure, the filler content is expressed in volume percentage to meet the criterion of the Halpin-Tsai equation. The elastic modulus increases with filler content up to 0.42 vol% (i.e., 1 wt%), thereafter decreases as the graphene content increases. These modulus values show good agreement with the Halpin-Tsai model for the

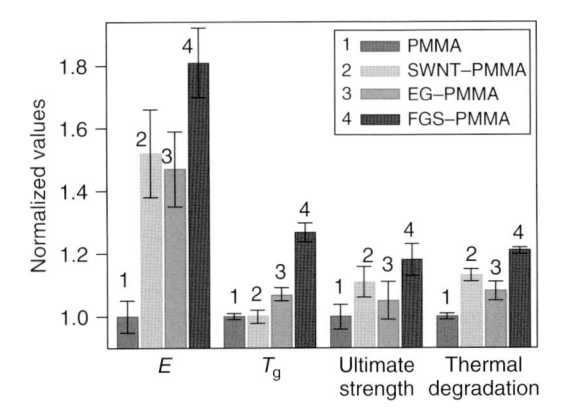

Figure 4.11 Summary of thermomechanical property improvements in solvent-cast PMMA nanocomposites by reinforcing with 1 wt% EG, SWNT or TRG. All property values are normalized to the values for pure PMMA and thus relative to unity on the scale above. Neat PMMA values are $E = 2.1$ GPa, $T_g = 105\,°C$, ultimate strength $= 70$ MPa, and thermal degradation temperature $= 285\,°C$. (Source: Reproduced with permission from [Chapter 2, Ref. 14], Nature Publishing Group (2008).)

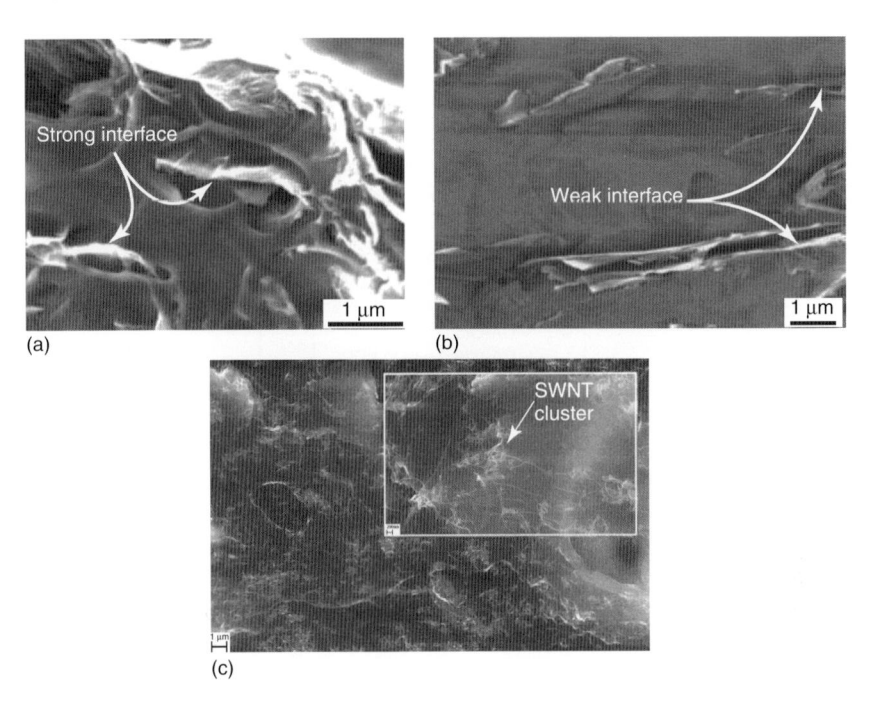

Figure 4.12 Scanning electron micrographs of (a) 1 wt% TRG/PMMA and (b) 1 wt% EG/PMMA nanocomposites reveal interfacial filler–matrix features. (c) Scanning electron micrograph of 1 wt% SWNT/PMMA nanocomposite with local clustering of the nanotubes. (Source: Reproduced with permission from [Chapter 2, Ref. 14], Nature Publishing Group (2008).)

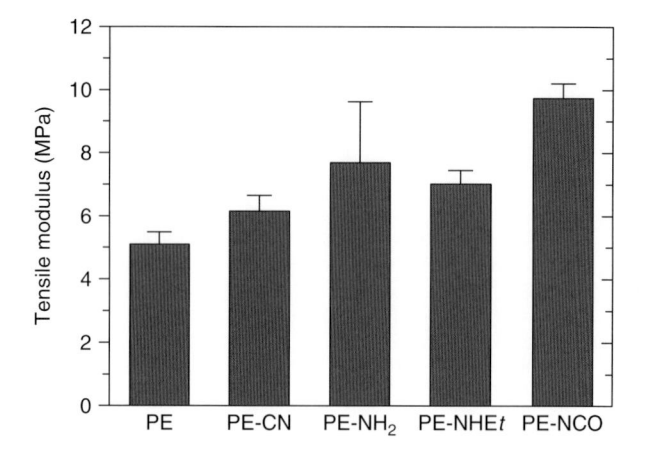

Figure 4.13 Tensile modulus of solvent-blended 0.4 vol% TRG in LLDPE and functionalized LLDPE. (Source: Reproduced with permission from [Chapter 2, Ref. 57], Elsevier (2011).)

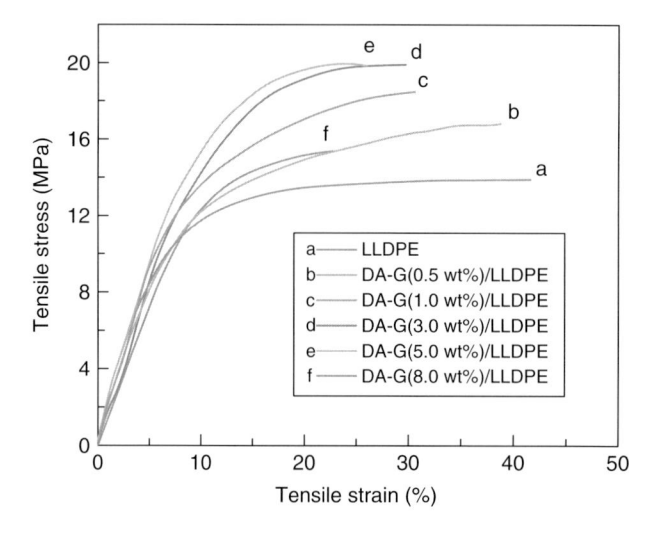

Figure 4.14 Stress–strain curves of LLDPE and its nanocomposites. (Source: Reproduced with permission from [54], Elsevier (2011).)

composite with aligned fillers. However, the experimental Young's modulus gradually approaches theoretical prediction for the randomly dispersed graphene sheets with further increasing graphene loadings. Table 4.1 lists the tensile properties of graphene/polymer nanocomposites. The addition of 0.1 wt% graphene to PP leads to improvements in tensile strength and Young's modulus by 38 and 23%, respectively. This demonstrates that tensile load can be efficiently transferred from the PP matrix to nanofillers through effective interfacial phase under tensile testing. Moreover, the elongation at break increases slightly indicating no deterioration in

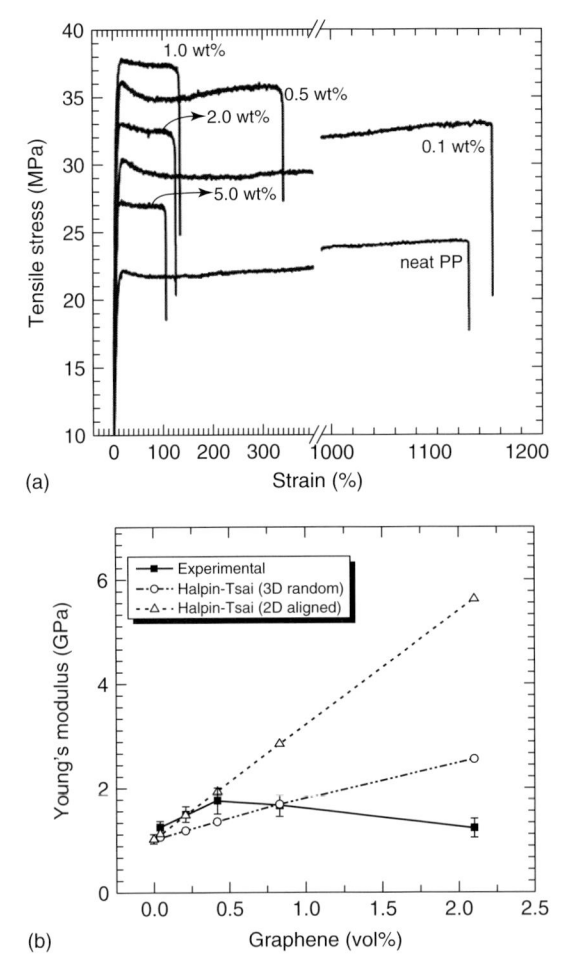

Figure 4.15 (a) Stress–strain curves and (b) Young's modulus versus filler content of graphene/PP nanocomposites prepared by combined solution-mixing and melt-blending processes. (Source: Reproduced with permission from [Chapter 2, Ref. 77], Elsevier (2011).)

the ductility of the nanocomposite at 0.1 wt% graphene loading. Maximum tensile strength and stiffness can be achieved by adding filler loading of 1 wt%.

4.5.2
Thermosetting Matrix

Epoxy resin (EP) with good chemical and corrosion resistance, high adhesion, and dimensional stability is an important class of thermosetting polymers finding widespread use as the matrix material of polymer composites for structural applications. However, EP is brittle because of its crosslinked molecular structure. It shows poor resistance to the crack growth under mode I loading and fatigue crack

Table 4.1 Mechanical properties of graphene/polymer nanocomposites prepared by different processes.

Polymer	Filler	Filler content (wt%)	Young' modulus (GPa)	Tensile strength (MPa)	% change, elastic modulus	% change, tensile strength	Processing method	Reference
PMMA	–	–	3.43 ± 0.06	58.1 ± 1.95	–	–	Injection molding	Chapter 2, Ref. 100
	rGO	0.25	3.75 ± 0.57	59.5 ± 4.25	9.5	2.4	Injection molding	Chapter 2, Ref. 100
	rGO	0.5	4.04 ± 0.22	59.0 ± 4.75	17.8	1.5	Injection molding	Chapter 2, Ref. 100
	rGO	1	4.28 ± 0.17	53.0 ± 5.55	25.0	−8.8	Injection molding	Chapter 2, Ref. 100
	rGO	2	4.65 ± 0.35	57.0 ± 4.55	35.8	−1.9	Injection molding	Chapter 2, Ref. 100
	GO	0.25	3.65 ± 0.06	62.1 ± 0.60	6.4	6.9	Injection molding	Chapter 2, Ref. 100
	GO	0.5	4.19 ± 0.03	66.4 ± 1.75	22.2	14.3	Injection molding	Chapter 2, Ref. 100
	GO	1	4.39 ± 0.18	70.9 ± 0.15	28.3	22.0	Injection molding	Chapter 2, Ref. 100
	GO	2	4.45 ± 0.00	66.8 ± 3.05	29.9	15.0	Injection molding	Chapter 2, Ref. 100
PI	–	–	1.8 ± 0.2	122 ± 6	–	–	In situ polymerization	Chapter 2, Ref. 101
	iGO	0.38	2.3 ± 0.4	131 ± 4	27.8	7.4	In situ polymerization	Chapter 2, Ref. 101
	iGO	0.75	2.4 ± 0.3	127 ± 5	33.3	4.1	In situ polymerization	Chapter 2, Ref. 101
PP	–	–	1.02 ± 0.10	24 ± 0.8	–	–	Melt compounding	Chapter 2, Ref. 77
	PP-grafted rGO	0.1	1.25 ± 0.12	33 ± 1.4	23	38	Solution and melt mixing	Chapter 2, Ref. 77
	PP-grafted rGO	0.5	1.50 ± 0.15	36 ± 1.5	47	50	Solution and melt mixing	Chapter 2, Ref. 77
	PP-grafted rGO	1	1.76 ± 0.25	37 ± 1.6	74	54	Solution and melt mixing	Chapter 2, Ref. 77
	PP-grafted rGO	2	1.66 ± 0.29	32 ± 1.0	63	33	Solution and melt mixing	Chapter 2, Ref. 77
PVDF	–	–	1.44	39.2	–	–	Solvent casting	Chapter 2, Ref. 102
	PMMA-grafted rGO	0.5	2.25	67.3	77	71	Solvent casting	Chapter 2, Ref. 102
	PMMA-grafted rGO	0.75	3.19	77.3	121	97	Solvent casting	Chapter 2, Ref. 102
	PMMA-grafted rGO	1	4.04	82.9	180	111.3	Solvent casting	Chapter 2, Ref. 102
	PMMA-grafted rGO	3	5.26	97.5	265	148.4	Solvent casting	Chapter 2, Ref. 102

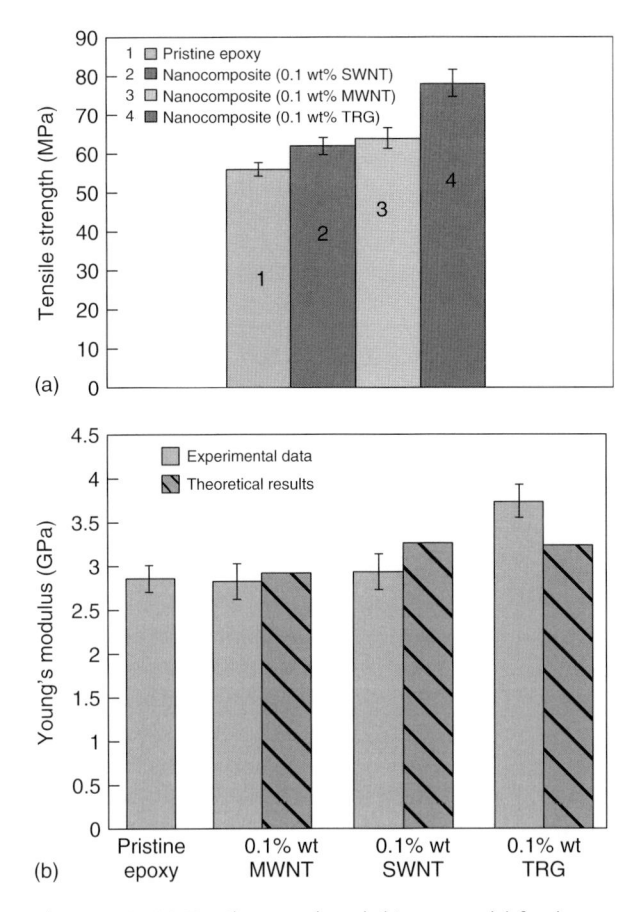

Figure 4.16 (a) Tensile strength and (b) Young's modulus of neat epoxy, 0.1 wt% MWNT/epoxy, 0.1 wt% SWNT/epoxy, and 0.1 wt% TRG/epoxy nanocomposite samples. Theoretical predictions using the Halpin-Tsai model for the nanocomposite samples with randomly oriented fillers are shown in (b). (Source: Reproduced with permission from [5a], The American Chemical Society (2009).)

propagation. Accordingly, carbonaceous nanomaterials with excellent mechanical strength and high aspect ratios are ideal filler material to deflect propagating cracks during static mode I and cyclic deformation tests. In general, structural materials subjected to cyclic deformation often fail at stresses much lower than their tensile strength. Koratkar and coworkers [5a,b] investigated the effect of TRG additions on tensile behavior, fracture toughness, and fatigue resistance of the EP. The measured mechanical properties of the 0.1 wt% TRG/epoxy nanocomposite were then compared with those of the 0.1 wt% MWNT/epoxy and 0.1 wt% SWNT/epoxy nanocomposites [5a]. They reported that the tensile strength and Young's modulus of the 0.1 wt% TRG/epoxy are far superior to those of pure EP. The 0.1 wt% TRG/epoxy nanocomposite also exhibited higher tensile strength and stiffness

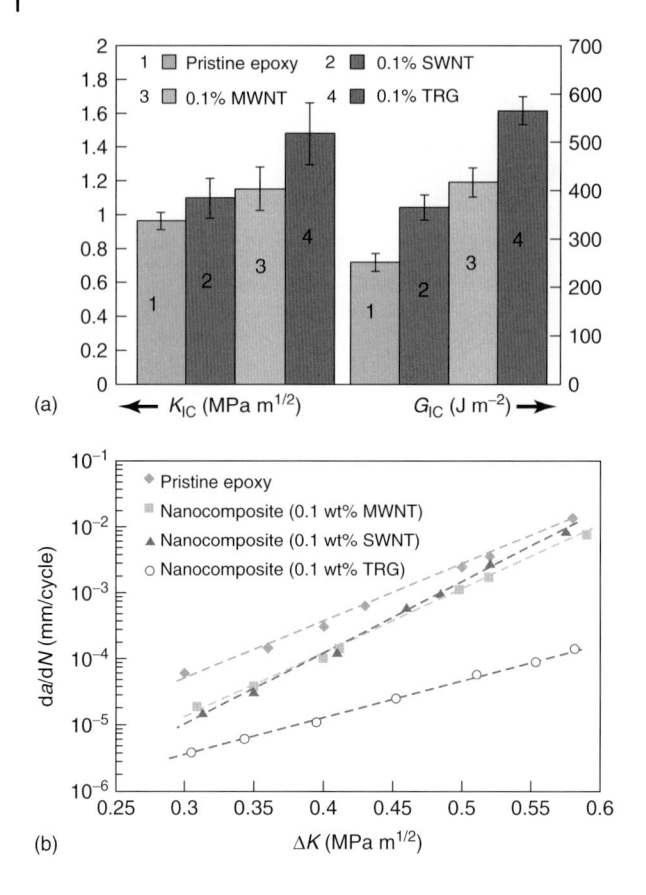

Figure 4.17 (a) Mode I fracture toughness (K_{IC}) and fracture energy (G_{IC}) and (b) fatigue crack growth rate versus stress intensity factor amplitude of neat epoxy, 0.1 wt% MWNT/epoxy, 0.1 wt% SWNT/epoxy, and 0.1 wt% TRG/epoxy nanocomposite samples. (Source: Reproduced with permission from [5a], The American Chemical Society (2009).)

than the 0.1 wt% MWNT/epoxy and 0.1 wt% SWNT/epoxy nanocomposites (Figure 4.16a,b). For the purpose of comparison, theoretical predictions for the Young's modulus of nanocomposite samples determined from the Halpin-Tsai model using randomly oriented fillers are shown in Figure 4.16b. The theoretical model under predicts experimental results of the 0.1 wt% TRG/epoxy nanocomposite by ∼13%. Furthermore, the 0.1 wt% TRG/epoxy nanocomposite also exhibits higher fracture toughness and critical energy release rate than those of pure epoxy, 0.1 wt% MWNT/epoxy, and 0.1 wt% SWNT/epoxy nanocomposites (Figure 4.17a). TRGs with wrinkled features enhance mechanical interlocking at the filler–matrix interface. This effect is responsible for reinforcing and stiffening effects of the epoxy. The two-dimensional geometry of TRGs can deflect propagating crack effectively, leading to enhanced fracture toughness and low fatigue crack growth rate of the 0.1 wt% TRG/epoxy nanocomposite. The fatigue resistance for a

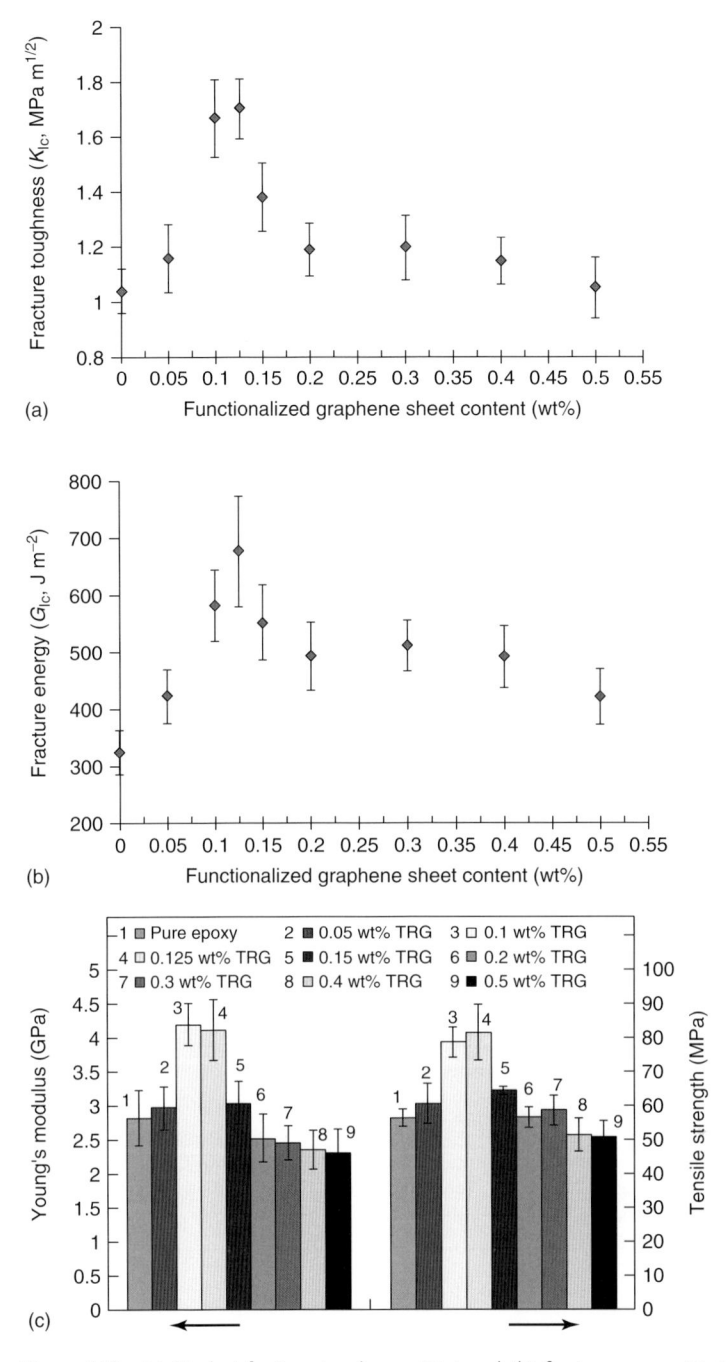

Figure 4.18 (a) Mode I fracture toughness (K_{IC}) and (b) fracture energy (G_{IC}) versus filler content of TRG/epoxy nanocomposites. (c) Tensile properties of neat epoxy and TRG/epoxy nanocomposites reinforced with various TRG contents. (Source: Reproduced with permission from [5b], Wiley-VCH (2010).)

composite with a crack length (*a*), which propagates under cyclic stresses can be determined from a plot of fatigue crack growth rate, d*a*/d*N*, versus applied stress intensity factor range, ΔK, in the slow crack growth regime. Figure 4.17b depicts clearly that the fatigue crack growth rate of the 0.1 wt% TRG/epoxy nanocomposite is considerably lower than that of pure epoxy and CNT-reinforced specimens. The effects of TRG loadings on the fracture toughness and tensile properties of epoxy are shown in Figure 4.18a–c. The K_{IC}, G_{IC}, and tensile strength of the TRG/epoxy nanocomposites are found to exhibit a peak maximum at 0.125 wt% TRG.

4.5.3
Elastomeric Matrix

The mechanical strength and stiffness of elastomeric materials are usually improved by incorporating carbon blacks (CBs). It is considered that GO with high

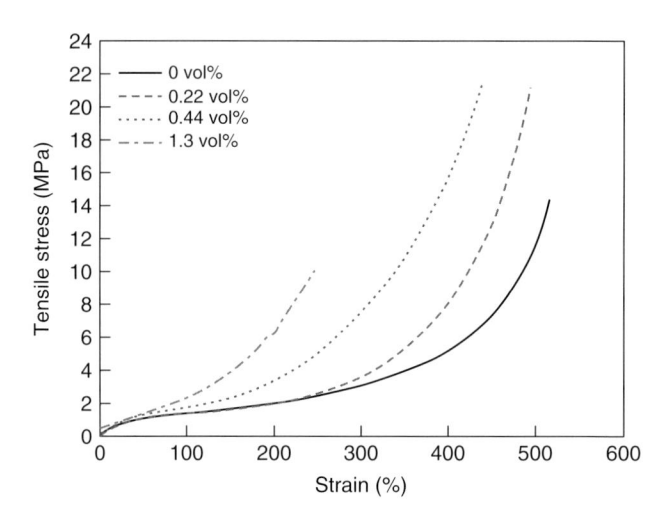

Figure 4.19 Stress–strain curves of HXNBR and solution-mixed GO/HXNBR nanocomposites. (Source: Reproduced with permission from [55], Elsevier (2011).)

Table 4.2 Tensile properties of GO/HXNBR nanocomposites.

GO content (vol%)	Modulus at 200% strain (MPa)	Tensile strength (MPa)	Elongation at break (%)
0	1.7 ± 0.2	14.8 ± 0.8	534 ± 13
0.22	2.2 ± 0.2	21.7 ± 1.8	485 ± 10
0.44	3.4 ± 0.1	22.4 ± 1.3	419 ± 18
1.3	6.5 ± 0.3	10.3 ± 0.5	248 ± 8

Source: Reproduced with permission from [55], Elsevier (2011).

aspect ratio and mechanical strength can further upgrade mechanical performance of rubber materials. Bai et al. [55] prepared GO/hydrogenated carboxylated nitrile-butadiene rubber (HXNBR) composites using solution-mixing process. Tensile measurements showed that the optimal addition of 0.44 vol% GO increases the tensile strength and modulus at 200% strain of HXNBR by more than 50 and 100% (Figure 4.19 and Table 4.2). This beneficial effect resulted from strong interaction between the carboxyl groups of HXNBR and the functional groups of GO.

4.6
Nanocomposites with EG and GNP Fillers

Expanded graphite or EG can be obtained by rapid heating of sulfuric-acid-based GIC precursor [Chapter 1, Ref. 18]. Rapid heating leads to a large expansion of the graphite flakes along their c-axis to yield wormlike EGs with a thickness of ~50–400 nm. Kim and Jeong [56] fabricated EG/polylactide (PLA) nanocomposites using melt-compounding process. The mechanical and electrical properties of such nanocomposites were investigated. Figure 4.20a,b shows stress–strain curves of melt-compounded EG/PLA and natural graphite (NG)/PLA composites, respectively. The variations of elastic modulus, tensile strength, and strain at break with filler content of these composites are shown in Figure 4.21a–c, respectively. Apparently, the modulus of the EG/PLA composites increases with increasing EG content, while the modulus of the NG/PLA composites increases slowly with NG content. Furthermore, the tensile strength of the EG/PLA composites also increases with

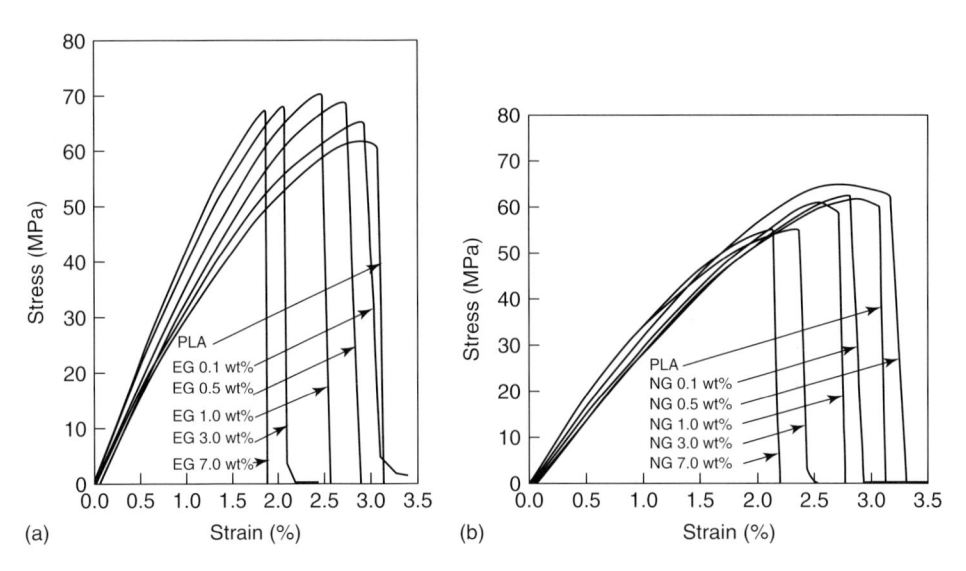

Figure 4.20 Stress–strain curves of melt-compounded (a) EG/PLA and (b) NG/PLA composites. (Source: Reproduced with permission from [56], Wiley (2010).)

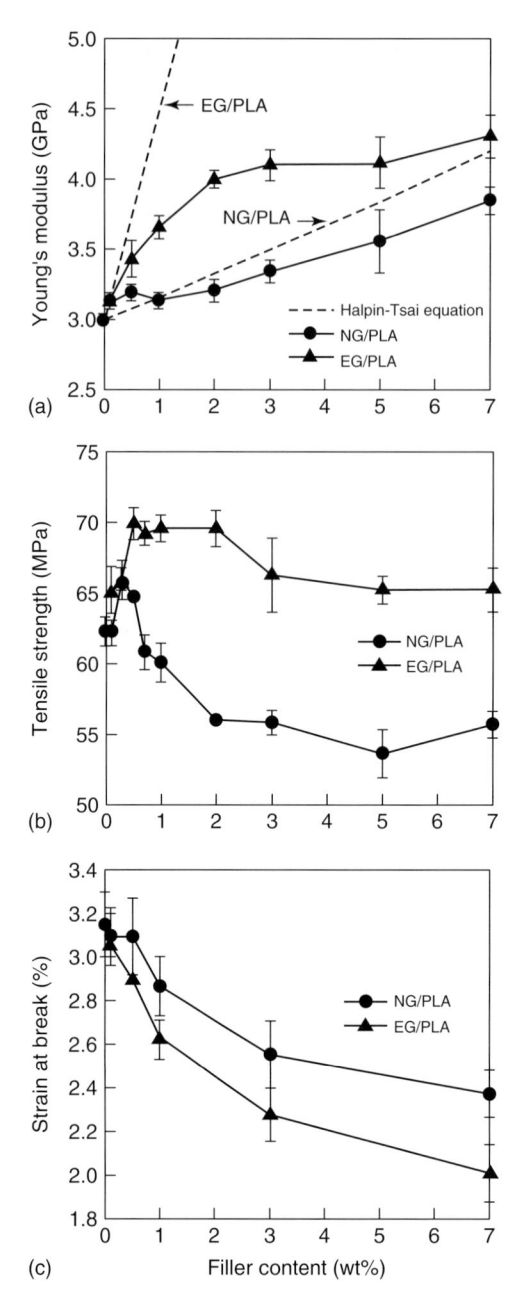

Figure 4.21 (a) Young's modulus, (b) tensile strength, and (c) strain at break versus filler content of EG/PLA and NG/PLA composites. The dashed lines in (a) denote the Young's moduli of both composites predicted by the Halpin-Tsai model. (Source: Reproduced with permission from [56], Wiley (2010).)

filler content. In contrast, the tensile strength of the NG/PLA composites shows an initial increase with NG content up to 0.0.5 wt% but decreases markedly as the NG content further increases. In other words, the tensile strength of NG/PLA composites is poorer than that of pure PLA at NG content ≥1 wt%, demonstrating that no stress transfer occurs across the matrix–EG interface. From Figure 4.21a, the experimental elastic modulus of the EG/PLA composites is far lower than that predicted from the Halpin-Tsai equation for randomly oriented fillers. Thus EGs are less effective than rGOs and TRGs to stiffen polymers as discussed above.

Very recently, Jana and Zhong [57] studied flexural behavior and fracture toughness of EG/epoxy nanocomposites. Figure 4.22 shows typical flexural stress–strain curves of such nanocomposites. The flexural properties and fracture toughness of EG/epoxy nanocomposites are summarized in Table 4.3. The results show that EG additions are beneficial in improving the flexural modulus, strength, and ductility as well as the fracture toughness of brittle epoxy. SEM examination of the fracture surface of pure epoxy reveals a relatively smooth surface feature. However, a much rougher fracture surface is observed by incorporating EG into the epoxy matrix. The increased surface roughness demonstrates that the path of the crack tip is tortuous, causing crack propagation more difficult. At 5 wt% EG loading, the flexural modulus, flexural strength, and fracture toughness of the EG/epoxy nanocomposite are increased by 13, 45.2 and 27.8%, respectively, over neat epoxy. At this filler loading, the K_{IC} value is 1.61 MPa m$^{1/2}$. From Figure 4.18a, the 0.1 wt% TRG/epoxy nanocomposite can also achieve this toughness value by adding only 0.1 wt% TRG. It is obvious that TRGs with much higher aspect ratios are more effective to deflect propagating crack in the epoxy matrix than the EGs.

Recently, much effort has been spent by the researchers to develop more effective fabrication processes such as master batch filling and combined solution mixing/melt compounding to improve dispersion of the EGs and graphite

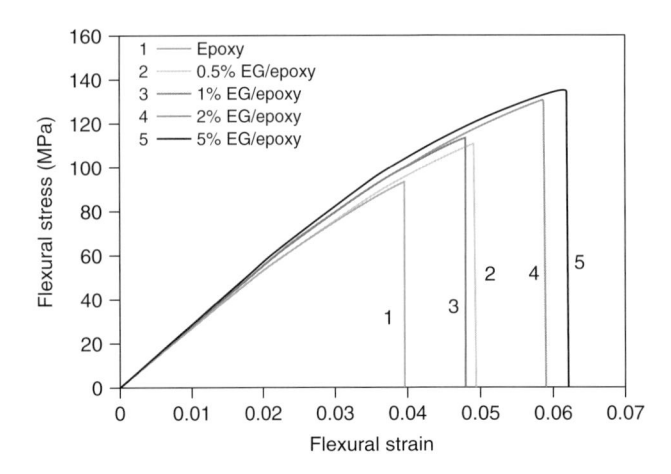

Figure 4.22 Flexural stress–strain curves of EG/epoxy nanocomposites. (Source: Reproduced with permission from [57], Elsevier (2009).)

nanoplatelets (GNPs) in nonpolar polyolefins. Li and Chen indicated that the tensile strength of EG/HDPE nanocomposites prepared from a master batch filling technique is considerably higher than that fabricated by direct melt blending [Chapter 2, Ref. 83]. Drzal and coworkers employed solution-mixing method to blend GNPs and PP powder initially in isopropyl alcohol under sonication before injection molding. Using this premixing method, GNPs were coated with a thin PP layer. Accordingly, the GNP/PP composites prepared via coating route exhibit high flexural modulus and strength (Figure 4.23a,b).

4.7
Nanocomposites with CNT and CNF Fillers

4.7.1
Thermoplastic Matrix

In general, the mechanical strength of the CNT/polymer nanocomposites depends largely on the attainment of efficient stress transfer from the polymer matrix to nanotubes, the dispersion, alignment, and aspect ratio of nanotubes. The stress transfer requires the formation of strong interfacial bonding and homogeneous dispersion of CNTs in the polymer matrix. Qian *et al.* conducted an earlier investigation on the tensile behavior of solvent-cast MWNT/PS composite film [Chapter 2, Ref. 67]. The film was prepared by mixing 1 wt% MWNT (pristine) with

Table 4.3 Mechanical properties of pure epoxy and EG/epoxy composites.

Properties	Epoxy	EG/epoxy composites			
		0.5 wt% EG	1 wt% EG	2 wt% EG	5 wt% EG
Flexural modulus (GPa)	2.48 ± 0.19	2.51 ± 0.15	2.66 ± 0.23	2.70 ± 0.03	2.80 ± 0.07
Increment		1.2%	7.3%	8.9%	13.0%
Flexural strength (MPa)	92.90 ± 17.20	110.68 ± 14.19	114.33 ± 9.71	127.55 ± 11.34	134.93 ± 7.05
Increment		19.1%	23.0%	37.3%	45.2%
Strain to failure (%)	0.049 ± 0.007	0.053 ± 0.009	0.052 ± 0.010	0.062 ± 0.006	0.063 ± 0.007
Increment		8.2%	6.1%	26.5%	28.6%
Fracture toughness (MPa m$^{1/2}$)	1.26 ± 0.03	1.49 ± 0.02	1.52 ± 0.07	1.51 ± 0.04	1.61 ± 0.04
Increment		18.3%	20.6%	19.8%	27.8%

Source: Reproduced with permission from [57], Elsevier (2009).

PS in toluene under ultrasonication and eventual solvent evaporation. Pristine MWNTs are known to be difficult to disperse in most solvents because of their high tendency for clustering. The use of an appropriate solvent, that is, toluene and ultrasonication during solution mixing enable better dispersion of nanotubes in the polymer matrix. The results showed that the addition of 1 wt% pristine MWNT to PS leads to 36–42% increase in Young's modulus and ~25% increase in tensile strength. By inducing thermal stresses in the composite film via electron-beam irradiation inside a transmission electron microscopy (TEM), MWNTs were mainly pullout from the propagating crack along the weak nanotube–matrix interface.

Covalent functionalization of nanotubes is an effective method to enhance interfacial matrix tube bonding. Apart from the covalent functionalization, grafting of polymers onto CNTs is an alternative route for enhancing polymer–nanotube interfacial adhesion. Very recently, Yuen et al. [59] purified MWNTs through chlorine oxidation (denoted as p-MWNTs) and then coated purified nanotubes with a PS layer via polymerization of styrene under microwave irradiation. The PS-coated nanotubes (m-MWNTs) were then blended with PS by means of melt

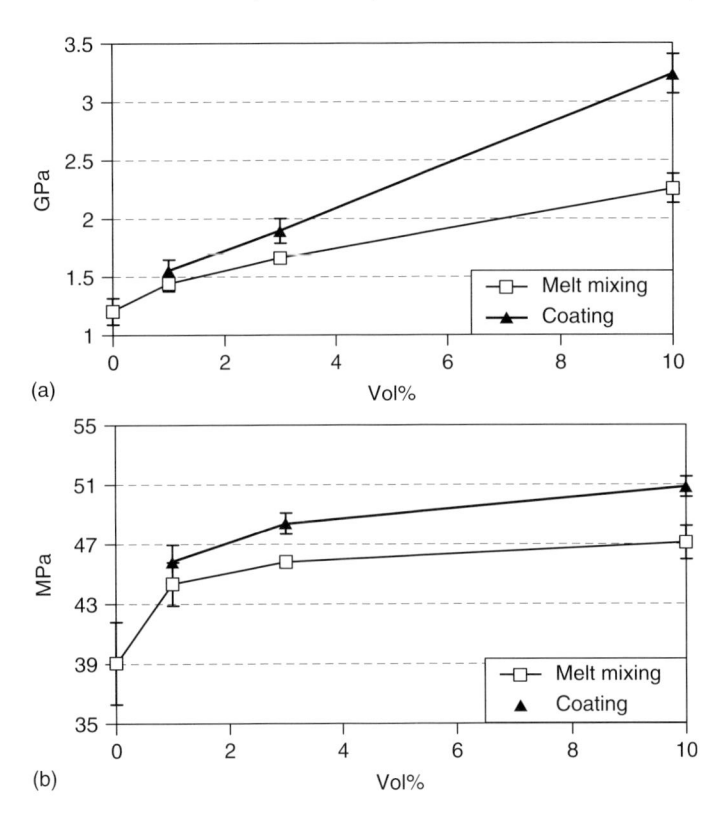

Figure 4.23 (a) Flexural modulus and (b) flexural strength as a function of GNP content of GNP/PP nanocomposites prepared by direct melt mixing and coating processes. (Source: Reproduced with permission from [58], Elsevier (2007).)

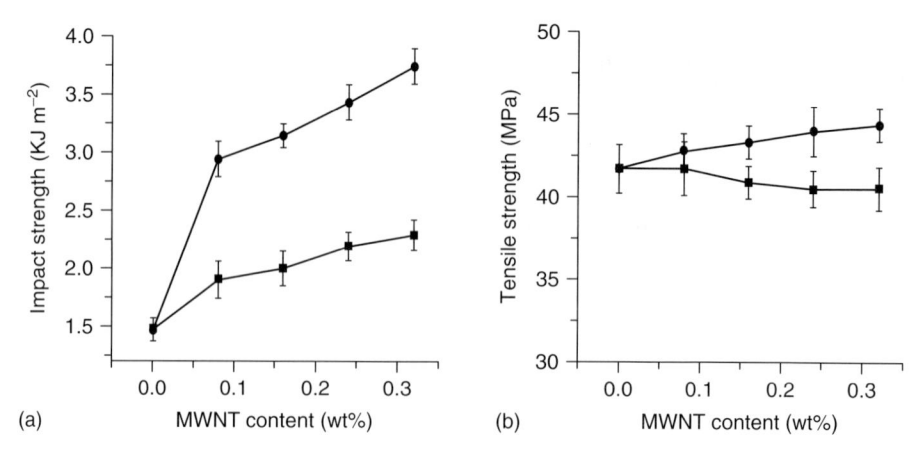

Figure 4.24 (a) Impact strength and (b) tensile strength versus MWNT content of p-MWNT/PS (■) and m-MWNT/PS (●) nanocomposites. (Source: Reproduced with permission from [59], Elsevier (2009).)

mixing and injection molding. Figure 4.24a,b shows the respective plots of impact strength and tensile strength versus nanotube content for the m-MWNT/PS and p-MWNT/PS nanocomposites. Apparently, the additions of low m-MWNT loadings to PS are beneficial in improving its impact strength and tensile strength due to the strong interfacial bonding. Microscopic examination of the fracture surface of the 0.32 wt% m-MWNT/PS nanocomposite reveals the presence of an intermediate PS layer between the PS matrix and m-MWNTs (Figure 4.25a,b). Owing to the PS coat layer and PS intermediate layer, the compatibility of nanotubes and PS matrix is greatly improved. On the other hand, the tensile strength of p-MWNT/PS nanocomposites decreases with increasing filler content, showing the absence of stress transfer mechanism due to the poor interfacial adhesion.

Jia et al. in situ polymerized MWNT/PMMA nanocomposites using both pristine and acid-purified nanotubes, MMA monomer, and 2,2′-azobisisobutyronitrile (AIBN) initiator [Chapter 2, Ref. 128]. The as-synthesized MWNTs were purified in hydrofluoric acid followed by refluxing in nitric acid for inducing carboxylic groups. For the composites filled with pristine nanotubes, the tensile strength decreases markedly with nanotube content as expected. However, acid-treated MWNT/PMMA nanocomposites show an increase in the tensile strength with increasing MWNT content up to 7 wt%. The tensile ductility of these nanocomposites remains nearly unchanged by adding acid-treated tubes (Table 4.4). The enhanced tensile strength derives from the interactions between functional groups of acid-purified nanotubes and the PMMA matrix. This manifests itself in a stronger interface and an efficient stress transfer. Acid-treated nanotubes also improve thermal stability of PMMA significantly as evidenced by a marked increase in the heat deflection temperature (HDT). HDT is an important indicator for heat resistance of materials under an applied load. Coleman and coworkers [60] also studied mechanical properties of in situ polymerized MWNT/PMMA nanocomposites using -OH-functionalized

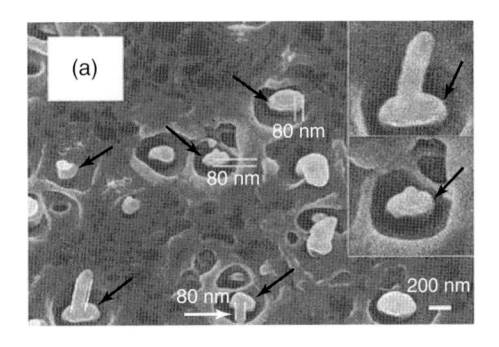

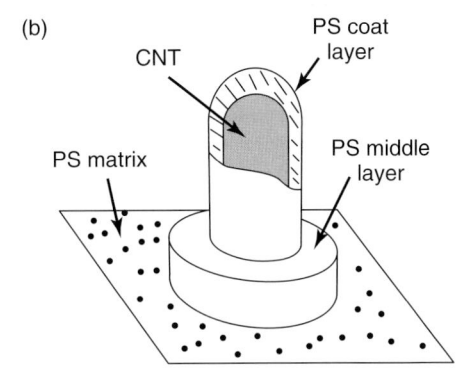

Figure 4.25 (a) SEM fractograph of 0.32 wt% m-MWNT/PS nanocomposite showing PS coat layer (black arrow) of ~80 nm thickness. (b) Schematic illustration of PS-coated nanotubes. (Source: Reproduced with permission from [59], Elsevier (2009).)

Table 4.4 Tensile and thermal properties of *in situ* polymerized MWNT/PMMA nanocomposites reinforced with pristine and acid-treated nanotubes.

MWNT content (wt%)	Tensile strength (MPa)	Toughness (kJ m^{-2})	Hardness HB (kg m^{-2})	Heat deflection temperature (K)
Pristine tube				
0	54.9	1.34	19.21	386–388
1.0	21.57	1.28	13.87	400–402
3.0	20.45	1.11	10.76	410–412
Acid-treated tube				
1.0	58.70	1.45	26.7	406–408
3.0	66.80	1.47	26.9	418–420
5.0	71.66	1.49	27.3	425–427
7.0	71.65	1.49	28.5	427–429
10.0	47.15	0.86	28.2	–

Source: Reproduced with permission from [Chapter 2, Ref. 128], Elsevier (1999).

MWNTs as reinforcing fillers. They reported that small MWNT loadings are very effective to increase tensile strength, Young's modulus, and strain-to-break of PMMA. Neat PMMA exhibits brittle failure with a strain-of-break of 0.57%. The addition of 0.075 vol% MWNT to PMMA increases its strain-of-break to 6.53%. The improvement in tensile ductility is attributed to the crack bridging effect of CNTs.

For polar PA resin and water-soluble PVA, functionalization of CNTs can also improve interfacial adhesion between the polymer matrix and nanotubes [49, 61, 62]. Liu et al. purified MWNTs in hydrochloric acid for dissolving catalyst particles followed by refluxing in nitric acid for forming carboxylic and hydroxyl groups. Purified MWNTs were then melt compounded with PA6 to produce

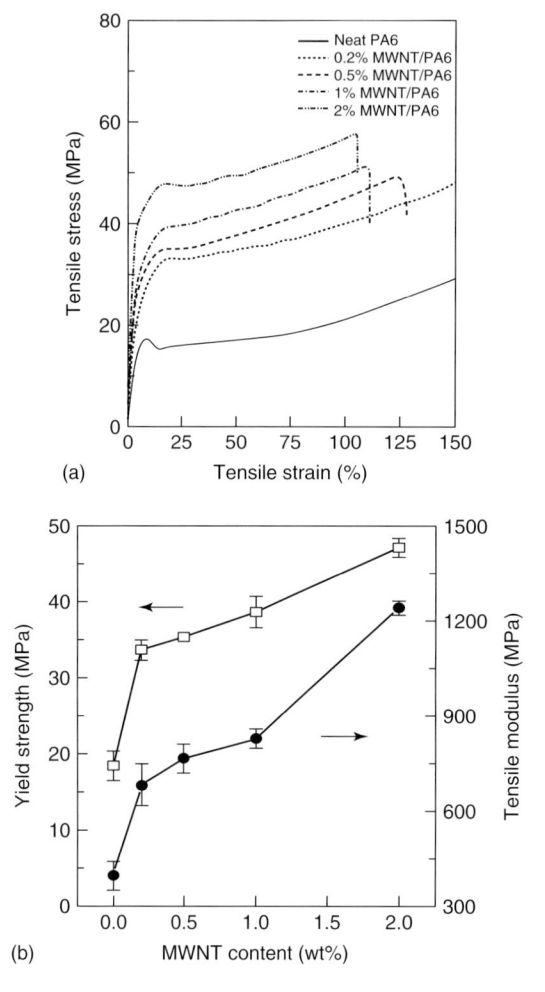

Figure 4.26 (a) Stress–strain curves and (b) yield strength and tensile modulus versus nanotube content of melt-compounded MWNT/PA6 nanocomposites. (Source: Reproduced with permission from [49], The American Chemical Society (2004).)

MWNT/PA6 nanocomposites. The tensile properties of such nanocomposites are depicted in Figure 4.26a,b. Both the yield strength and modulus of PA6 increase with increasing nanotube content up to 2 wt%. The yield strength of PA6 increases markedly from 18 to 33.7, 35.4, 40.3, and 47.2 MPa by adding 0.2, 0.5, 1, and 2 wt% MWNT, respectively. This result demonstrates effective load-carrying capacity of MWNTs. SEM fractograph of the 0.5 wt% MWNT/PA6 nanocomposite reveals that the nanotubes bridge propagating crack effectively during tensile testing (Figure 4.3)

Very recently, the tensile deformation behavior of melt-compounded MWNT/PP and CNF/PP nanocomposites as been studied by Bao and Tjong [12; Chapter 3, Ref. 33]. Figure 4.27a,b shows typical stress–strain curves of neat PP and 0.3 wt% MWNT/PP nanocomposite at 18 °C for different strain rates. The tensile properties of PP at low strain rates (e.g., 6.67×10^{-4} and $6.67 \times 10^{-3}\,\mathrm{s}^{-1}$) are characterized by well-known cold-drawing characteristic showing peak yield stress followed by stress softening. However, the yielding behavior is strain rate dependent. At higher strain rates, PP deforms with little or without cold drawing, leading to brittle failure. A similar strain rate dependent yielding behavior is observed for the 0.3 wt% MWNT/PP nanocomposite. Table 4.5 lists the tensile test results of PP and

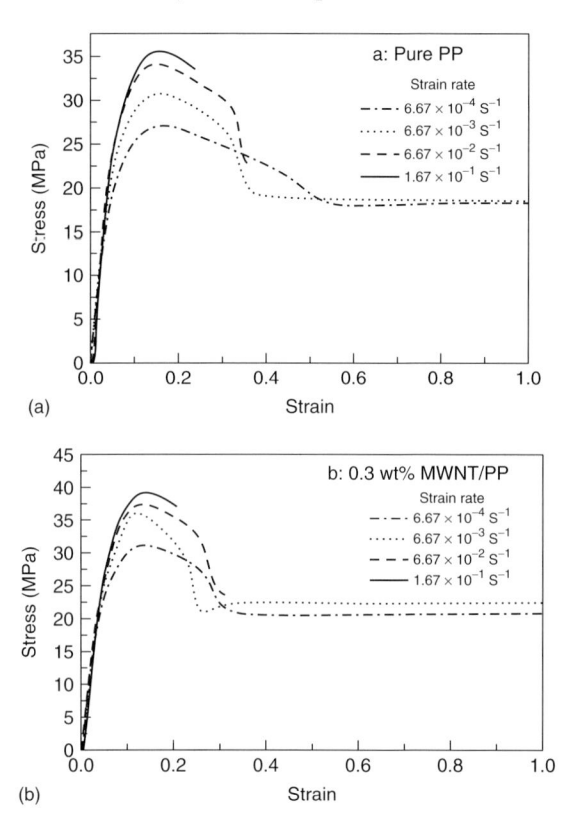

Figure 4.27 Representative stress–strain curves of (a) pure PP and (b) 0.3 wt% MWNT/PP nanocomposite at 18°C for various strain rates [Chapter 3, Ref. 33].

Table 4.5 Tensile properties of PP and its nanocomposites.

Specimens	Yield strength (MPa)	Young's modulus (MPa)	Izod impact strength (kJ m^{-2})	Elongation at break (%)	References
Pure PP	30.71 ± 0.18	1570 ± 24	4.01 ± 0.12	>500	Chapter 3, Ref. 33
0.1 wt% MWNT/PP	34.89 ± 0.39	1743 ± 65	–	>500	Chapter 3, Ref. 33
0.3 wt% MWNT/PP	35.98 ± 0.27	2107 ± 72	–	>500	Chapter 3, Ref. 33
0.5 wt% MWNT/PP	35.88 ± 0.39	2070 ± 66	–	>500	Chapter 3, Ref. 33
1 wt% MWNT/PP	35.65 ± 0.41	2056 ± 83	–	>500	Chapter 3, Ref. 33
0.1 wt% CNF/PP	34.42 ± 0.58	1732 ± 70	4.20 ± 0.11	>500	[12]
0.3 wt% CNF/PP	34.67 ± 0.50	1767 ± 73	4.52 ± 0.27	>500	[12]
0.5 wt% CNF/PP	35.17 ± 0.17	1882 ± 77	5.05 ± 0.21	>500	[12]
1 wt% CNF/PP	35.49 ± 0.69	1869 ± 56	5.12 ± 0.30	>500	[12]

its nanocomposites deformed at a strain rate of 6.67×10^{-3} s^{-1}. Apparently, the additions of very small amounts of MWNTs to PP improve its yield strength and Young's modulus considerably. Optimal yield strength and modulus of MWNT/PP nanocomposites can be attained at 0.3 wt% MWNT loading. Table 4.5 also reveals that CNF additions can produce comparable reinforcing effect at the same filler loading levels. The elastic modulus of the CNF/PP nanocomposites is slightly inferior to that of the MWNT/PP counterparts. This is because CNFs possess more crystalline defects and larger diameters. Furthermore, the CNF additions are also beneficial in enhancing impact strength of PP. It is noted that the yield stress of PP

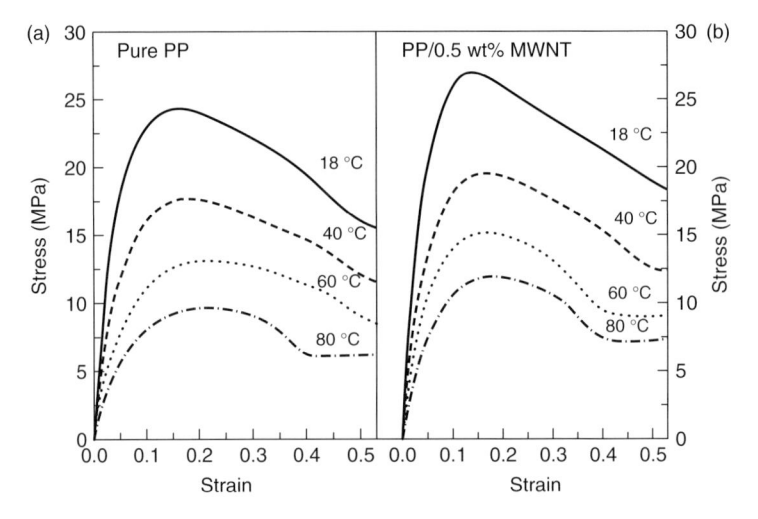

Figure 4.28 Stress–strain curves of pure PP and 0.5 wt% MWNT/PP nanocomposite at a strain rate of 6.67×10^{-5} s^{-1} for various temperatures [Chapter 3, Ref. 33].

and its nanocomposites also depends greatly on temperature. It decreases markedly as the temperature increases (Figure 4.28). Accordingly, the yielding behavior of PP and its nanocomposites can be well described by the Eyring equation (Eq. (4.1)). Figure 4.29a–c shows the representative Eyring plots of PP, 0.1 wt% MWNT/PP, and 1 wt% MWNT/PP specimens at 18 °C.

PP is a commodity plastic that finds widespread applications in industrial sectors owing to its versatile properties and low cost. However, neat PP is notch sensitive and experiences brittle fracture under impact loading, especially at low temperatures. Therefore, elastomer particles are incorporated into PP for improving its impact performance. The elastomer additions impair the stiffness and strength of PP. To restore mechanical strength of PP/elastomer blends, short glass fibers are added to generate rubber-toughened polymer microcomposites [63, 64]. Moreover, inorganic nanofillers such as clay nanoplatelets and silicon carbide nanoparticles have been used by Tjong's group to reinforce

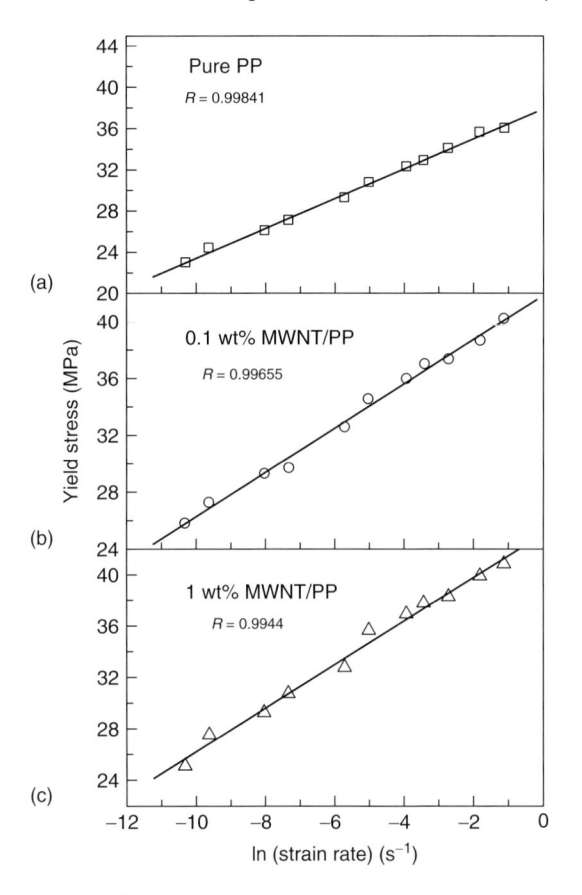

Figure 4.29 Strain rate dependent yield stress for (a) pure PP, (b) 0.1 wt% MWNT/PP, and (c) 1 wt% MWNT/PP samples at 18 °C. The straight lines are fitted by linear regression method and R is the coefficient of regression [Chapter 3, Ref. 33].

Table 4.6 Mechanical properties of PP/SEBS-g-MA 85/15 blend (designated as 15B) and its nanocomposites [Chapter 3, Ref. 46].

Specimen	E (MPa)	σ_y (MPa)		ε_b (%)		Impact strength (kJ m^{-2})
		10 mm min^{-1}	200 mm min^{-1}	10 mm min^{-1}	200 mm min^{-1}	
15B	1089 ± 7	21.92 ± 0.14	26.90 ± 0.10	No fracture	84 ± 5	2.55 ± 0.19
0.2 wt%CNF/15B	1123 ± 31	22.44 ± 0.12	27.53 ± 0.11	No fracture	93 ± 8	2.66 ± 0.16
0.5 wt%CNF/15B	1139 ± 2	22.47 ± 0.04	27.99 ± 0.12	No fracture	98 ± 10	3.59 ± 0.15
1 wt%CNF/15B	1150 ± 22	22.75 ± 0.11	28.16 ± 0.10	No fracture	106 ± 7	3.63 ± 0.28
2.5 wt%CNF/15B	1163 ± 9	23.13 ± 0.10	28.37 ± 0.14	No fracture	115 ± 11	3.66 ± 0.23

PP/elastomer blends [65, 66]. Apparently, CNFs of large aspect ratios and high mechanical strength can be used as efficient nanofillers to strengthen PP/maleated (styrene-ethylene-butadiene-styrene) (PP/SEBS-g-MA) (maleic anhydride) blends [Chapter 3, Ref. 46]. The effect of CNF additions on the mechanical properties of PP/SEBS-g-MA 85/15 (wt%) blend is listed in Table 4.6. It is evident that low CNF loadings not only can restore elastic modulus and yield strength of the PP/SEBS-g-MA 85/15 blend (designated as 15B) but can also enhance its elongation at break (ε_b) and impact strength. The load–displacement curves obtained from the EWF measurements using DENT specimens of the 15B and 2.5 wt% CNF/15B are shown in Figure 4.30a,b, respectively. The load–displacement curves of the 15B and 2.5 wt% CNF/15B specimens of various ligament lengths display a shape similarity. The resulting linear regression of w_f versus L for these specimens is shown in Figure 4.31. The w_e value can be extrapolated to zero ligament length from this plot. The fracture toughness parameters of PP/SEBS-g-MA 85/15 blend and its nanocomposites are given in Table 4.7. It is obvious that the w_e value of the PP/SEBS-g-MA blend increases markedly with increasing CNF content. In other words, CNFs act synergistically with the SEBS-g-MA elastomer to toughen hybrid composites.

4.7.2
Thermosetting Matrix

Schulte and coworkers [67, 68] studied systematically the effect of SWNT, purified double-walled carbon nanotube (DWNT), and MWNT additions on the tensile and fracture behaviors of the EP. The filler loading levels were relatively low, that is, 0.1–0.3 wt%. Purified DWNTs and MWNTs were further functionalized with amino groups by ball milling in ammonia. CB nanoparticles (30 nm) were also used as reinforcing fillers for comparison purposes. All the nanocomposites were fabricated using a three-roll mill (calendaring process) in which the edge rolls

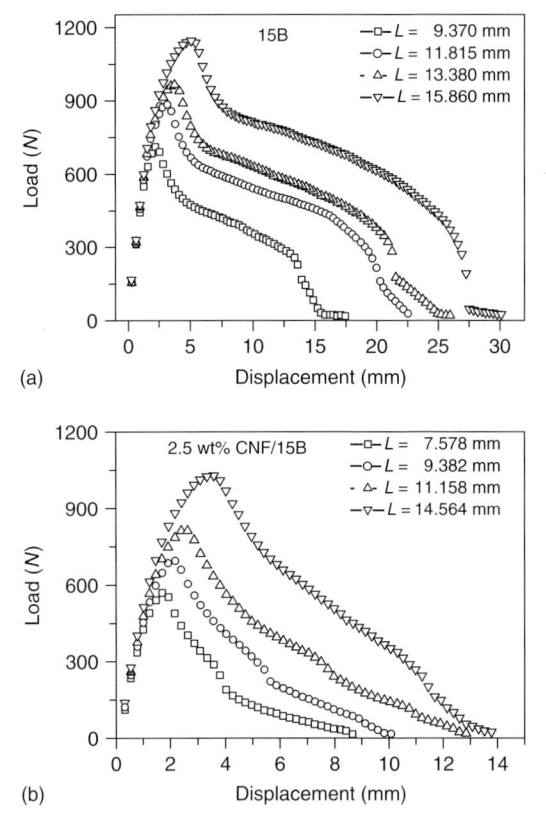

Figure 4.30 Load–displacement curves of (a) 15B and (b) 2.5 wt% CNF/15B specimens under EWF tests [Chapter 3, Ref. 46].

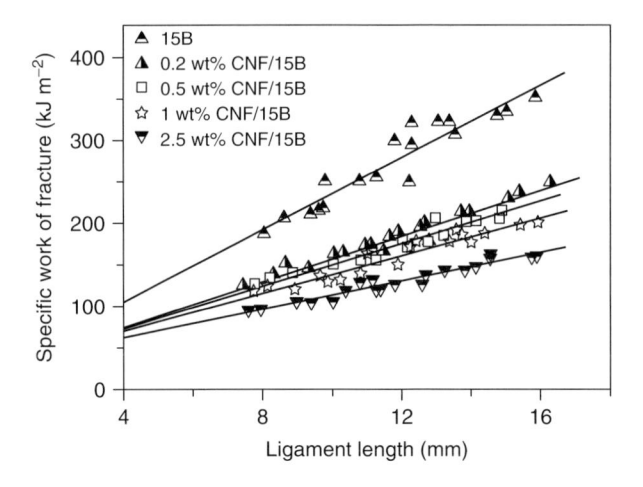

Figure 4.31 Specific work of fracture versus ligament length for PP/SEBS-g-MA 85/15 blend and its composites reinforced with different CNT contents [Chapter 3, Ref. 46].

Table 4.7 Tensile specific essential work of fracture (w_e) and specific plastic work (βw_p) of PP/SEBS-g-MA 85/15 blend and its nanocomposites [Chapter 3, Ref. 46].

Specimen	w_e (kJ m^{-2})	βw_p (MJ m^{-3})	Correlation coefficient
15B	17.47	21.83	0.945
0.2 wt%CNF/15B	20.73	13.67	0.970
0.5 wt%CNF/15B	23.00	12.74	0.975
1.wt%CNF/15B	25.68	11.28	0.965
2.5 wt%CNF/15B	27.78	8.57	0.969

rotated in the same direction while the centre roll rotated reversely [69]. A high shear stress can be generated by adjusting the gap distance between the rolls, thereby breaking up the nanotube agglomerates and producing better dispersion of nanotubes in the epoxy matrix [70]. The results of tensile and fracture toughness of epoxy and its nanocomposites are summarized in Table 4.8. Compared to MWNTs, SWNT additions result in larger enhancement in the tensile strength because of their largest aspect ratio. The addition of small amounts of MWNTs does not produce an increase in the tensile strength of epoxy but rather decreases slightly. The interactions between purified MWNTs and the epoxy matrix seem to be rather weak, leading to the pullouts of the nanotubes from the matrix during tensile testing [71]. On the other hand, the introduction of amino functional groups on the MWNTs facilitates the formation of covalent bonding with the epoxy. The amino functional group is highly reactive and reacts readily with EP, leading to the opening of epoxide ring to form a cross-link. However, Table 4.8 reveals that the incorporation of amino-functionalized MWNTs only produces a slight increase in the tensile strength. In contrast, the tensile strength of composite with amino-functionalized DWNT increases markedly. This behavior is mainly attributed to the structure of MWNTs consisting of concentric graphene shells. The stress is transferred only to its outer shell under tensile load. Because of the weak van der Waals forces between individual graphene shells of the MWNT, slipping occurs between the shells, leading to the so-called telescopic fracture. In other words, telescopic pullout is caused by the fracture of the outer MWNT shell due to a strong interfacial bonding and its associated pullout of the inner tubes. The outer shell of the tube still remains embedded in the matrix following pullout of inner tubes (Figure 4.32). The K_{IC} fracture toughness of epoxy-based nanocomposites measured from the compact tension specimens improves considerably by adding treated and untreated amine-functionalized nanotubes. The fracture toughness of the composite with 0.3 wt% DWNT–NH$_2$ ($K_{IC} = 0.92$ MPa m$^{1/2}$) is 42% higher than that of the neat epoxy ($K_{IC} = 0.65$ MPa m$^{1/2}$). The K_{IC} value can be increased to 0.93 MPa m$^{1/2}$ at 0.5 wt% DWNT–NH$_2$, being the highest fracture toughness achieved among all nanocomposites. In general, a strong interfacial bonding favors stress transfer effect but has a negative impact on the fracture toughness by restricting interfacial failure in discontinuously reinforced fiber/polymer composites [43b, 63, 64]. It

Table 4.8 Mechanical properties of pure epoxy and its nanocomposites.

Specimen	Filler content (wt%)	Young's modulus (MPa)	Ultimate tensile strength (MPa)	Fracture toughness K_{1c} (MPa m$^{1/2}$)
Epoxy	0.0	2599(±81)	63.80(±1.09)	0.65(±0.062)
CB/epoxy	0.1	2752(±144)	63.28(±0.85)	0.76(±0.030)
	0.3	2796(±34)	63.13(±0.59)	0.86(±0.063)
	0.5	2830(±60)	65.34(±0.82)	0.85(±0.034)
SWNT/epoxy	0.05	2681(±80)	65.84(±0.64)	0.72(±0.014)
	0.1	2691(±31)	66.34(±1.11)	0.80(±0.041)
	0.3	2812(±90)	67.28(±0.63)	0.73(±0.028)
DWNT/epoxy	0.1	2785(±23)	62.43(±1.08)	0.76(±0.043)
	0.3	2885(±88)	67.77(±0.40)	0.85(±0.031)
	0.5	2790(±29)	67.66(±0.50)	0.85(±0.064)
DWNT-NH$_2$/epoxy	0.1	2610(±104)	63.62(±0.68)	0.77(±0.024)
	0.3	2944(±50)	67.02(±0.19)	0.92(±0.017)
	0.5	2978(±24)	69.13(±0.61)	0.93(±0.030)
MWNT/epoxy	0.1	2780(±40)	62.97(±0.25)	0.79(±0.048)
	0.3	2765(±53)	63.17(±0.13)	0.80(±0.028)
	0.5	2609(±13)	61.52(±0.19)	–
MWNT-NH$_2$/epoxy	0.1	2884(±32)	64.67(±0.13)	0.81(±0.029)
	0.3	2819(±45)	63.64(±0.21)	0.85(±0.013)
	0.5	2820(±15)	64.27(±0.32)	0.84(±0.028)

Source: Reproduced from [68] with permission of Elsevier (2005).

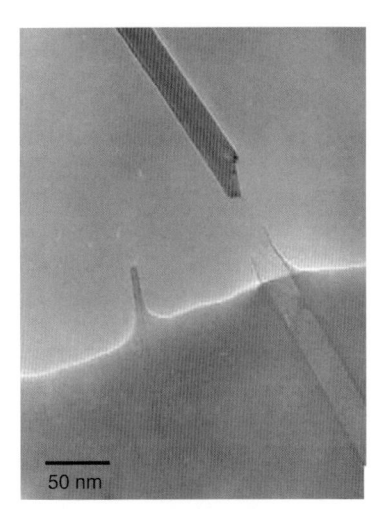

Figure 4.32 TEM image of amino-functionalized MWCNT/epoxy composites with telescopic pullout of nanotubes. (Source: Reproduced with permission from [71], Elsevier (2003).)

remains unclear why the composites with 0.3 and 0.5 wt% DWCNT–NH$_2$ exhibiting strong interfacial bonding still possess maximum fracture toughness. Nevertheless, the toughening mechanism of the epoxy nanocomposites with weak interfacial adhesion derives from the pullout of nanotubes from the matrix, dissipating energy during the crack propagation. However, telescopic pullout and crack bridging are responsible for toughening mechanisms in the epoxy nanocomposites with strong interfacial adhesion.

Apart from the amino functional groups, plasma oxygen oxidation of CNTs or CNFs can effectively enhance their interactions with the EP to form strong interfacial bonding [72–74]. For example, Kim et al. [72] first treated MWNTs in nitric acid and then reacted with either octadecylamine or plasma oxidation to improve interfacial bonding and dispersion of nanotubes in the epoxy matrix. Tensile results showed that the composites containing 1 wt% acid-, amine-, and

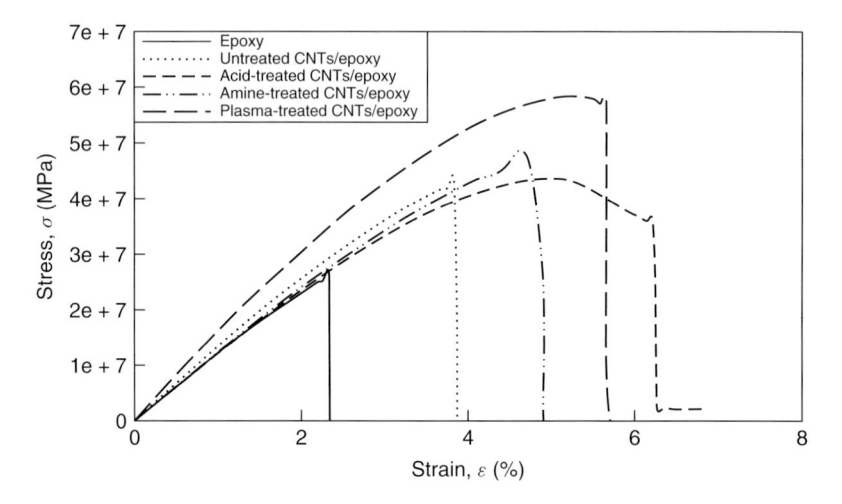

Figure 4.33 Stress–strain curves of epoxy, pristine MWNT/epoxy, and surface-modified MWNT/epoxy nanocomposites. (Source: Reproduced with permission from [72], Elsevier (2006).)

Table 4.9 Tensile properties of pure epoxy and its CNT/epoxy composites.

Sample	Young's modulus (GPa)	Tensile strength (MPa)	Elongation at break (%)
Epoxy	1.21	26	2.33
Untreated CNT/epoxy	1.38	42	3.83
Acid-treated CNT/epoxy	1.22	44	4.94
Amine-treated CNT/epoxy	1.23	47	4.72
Plasma-treated CNT/epoxy	1.61	58	5.22

Source: Reproduced from [72] with permission of Elsevier (2006).

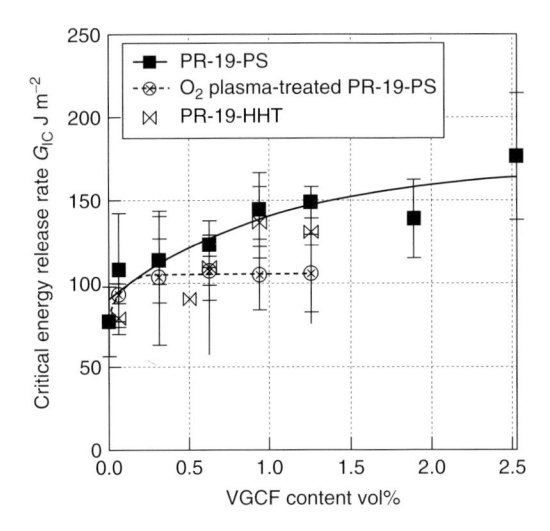

Figure 4.34 Critical energy release rate as a function of carbon nanofiber content for CNF/epoxy nanocomposites. (Source: Reproduced with permission from [73], Elsevier (2005).)

plasma-treated nanotubes exhibit higher stiffness, tensile strength, and strain to failure than the nanocomposite filled with 1 wt% pristine MWNT, particularly for the plasma-treated nanotubes (Figure 4.33 and Table 4.9). Miyagawa and Drzal demonstrated that strong interfacial bonding is detrimental to the fracture toughness of CNF/epoxy nanocomposites [73]. They employed Pyrograf-III PR-19-PS and PR-19-HHT with a diameter of 150 nm to reinforce EP. The PS-grade fiber is pyrolytically stripped at 600 °C to remove polyaromatic hydrocarbons from the fiber surface. The HHT-grade nanofiber is treated to temperatures up to 3000 °C to graphitize carbon on its surface [Chapter 1, Ref. 111]. Both nanofibers were subjected to oxygen plasma treatment. Figure 4.34 shows the variation of critical energy release rate versus nanofiber content for the epoxy composites filled reinforced with plasma-treated and plasma-untreated CNFs. For the composites with untreated PR-19-PS and PR-19-HHT, the G_{Ic} value increases considerably with fiber content. At 1.3 vol% filler content, the addition of untreated PR-19-PS to neat epoxy results in ~100% enhancement in the G_{Ic} value. However, the G_{Ic} value only increases by 38% using plasma-treated 1.3 vol% PR-19-PS. SEM fractograph reveals that fiber pullout is mainly responsible for the composite with untreated nanofibers (Figure 4.35a). As the nanofibers are stretched during the crack opening, slid with friction against the epoxy matrix, and finally pulled out, larger energy is dissipated due to the friction at the epoxy–fiber interfaces and to the plastic deformation of the epoxy matrix around nanofibers. In contrast, fiber fracture is observed in the SEM fractograph of the plasma-treated composite as a result of strong interfacial bonding (Figure 4.35b). Very little energy is dissipated without fiber pullout considering small volume contents of nanofiber used.

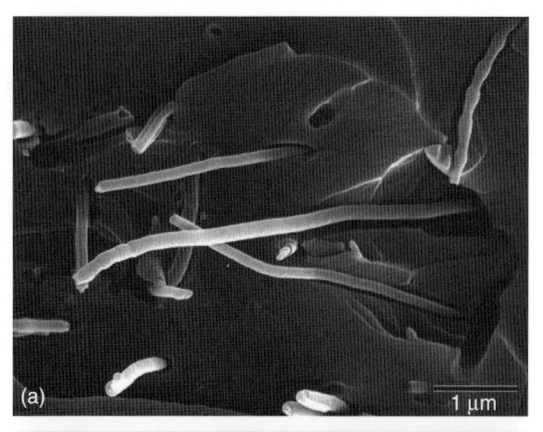

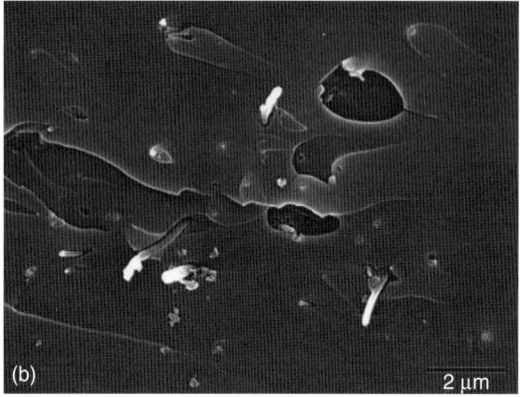

Figure 4.35 SEM fractographs of epoxy nanocomposites with (a) untreated 2.5 vol% PR-19-PS and (b) plasma-treated 1.3 vol% PR-19-PS. (Source: Reproduced with permission from [73], Elsevier (2005).)

4.8
Composites with Hybrid Fillers

The mechanical and physical properties of the polymer composites can be further improved by hybridizing carbonaceous nanomaterials with other inorganic nanofillers, and even with short carbon fibers (SCFs) and short glass fibers [75–79]. Filler hybridization offers a wide range of property improvements in the polymer composites that cannot be obtained with single filler alone. For example, Karger-Kocsis and coworkers [76] studied hybridized effect of SCFs (5–15 vol%) and CNFs (0.125–0.75 vol%) on the mechanical and fracture properties of EP. Figure 4.36a shows representative stress–strain curves of pure EP, 10 vol% SCF/EP, 0.75 vol% CNF/EP, and (10 vol% SCF + 0.75 vol% CNF)/EP specimens. The elastic modulus, tensile strength, and elongation at break of EP, SCF/EP microcomposites, and SCF + CNF/EP hybrids are depicted in Figure 4.36b–d, respectively. For the SCF/EP microcomposites, the Young's modulus increases

Figure 4.36 (a) Stress–strain curves of epoxy, 0.75 vol% CNF/EP, 10 vol% SCF/EP, and (10 vol% SCF + 0.75 vol%CNF)/EP specimens. (b) Elastic modulus, (c) tensile strength, and (d) elongation at break of epoxy and its composites. (Source: Reproduced with permission from [76], Elsevier (2010).).

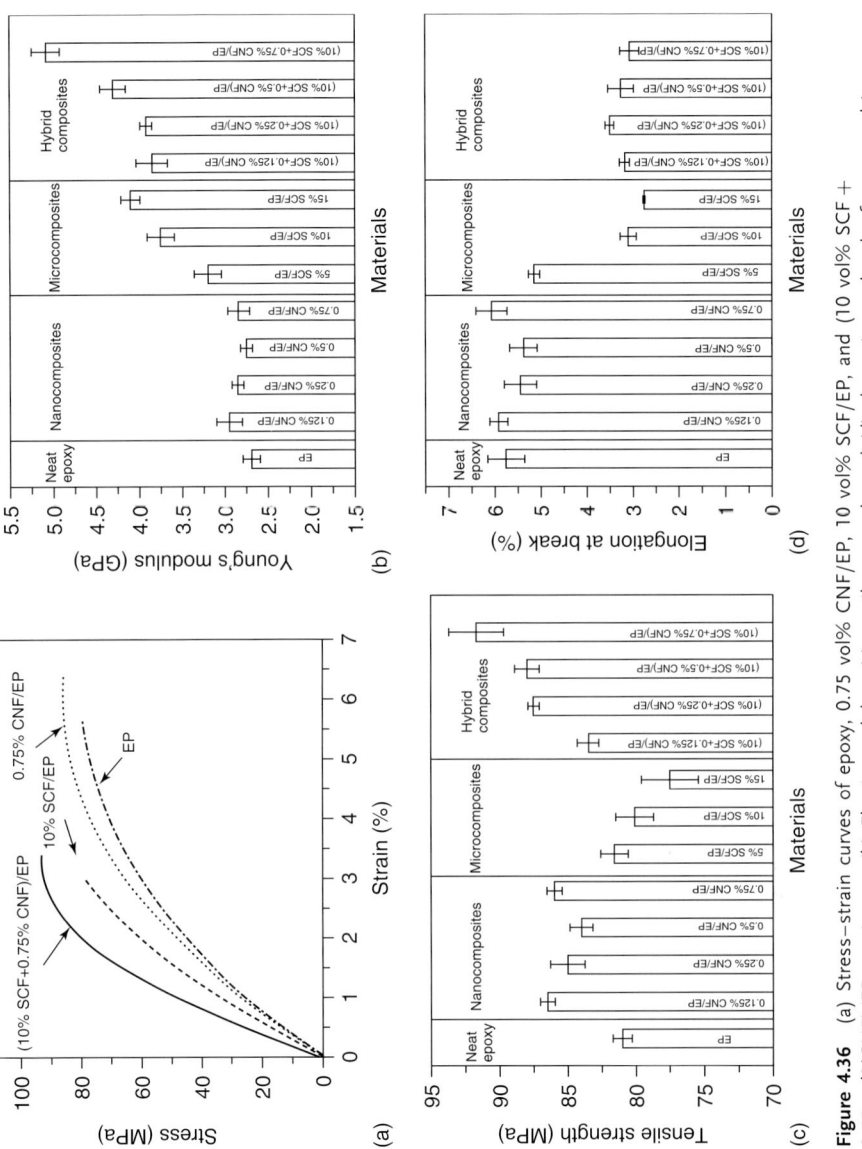

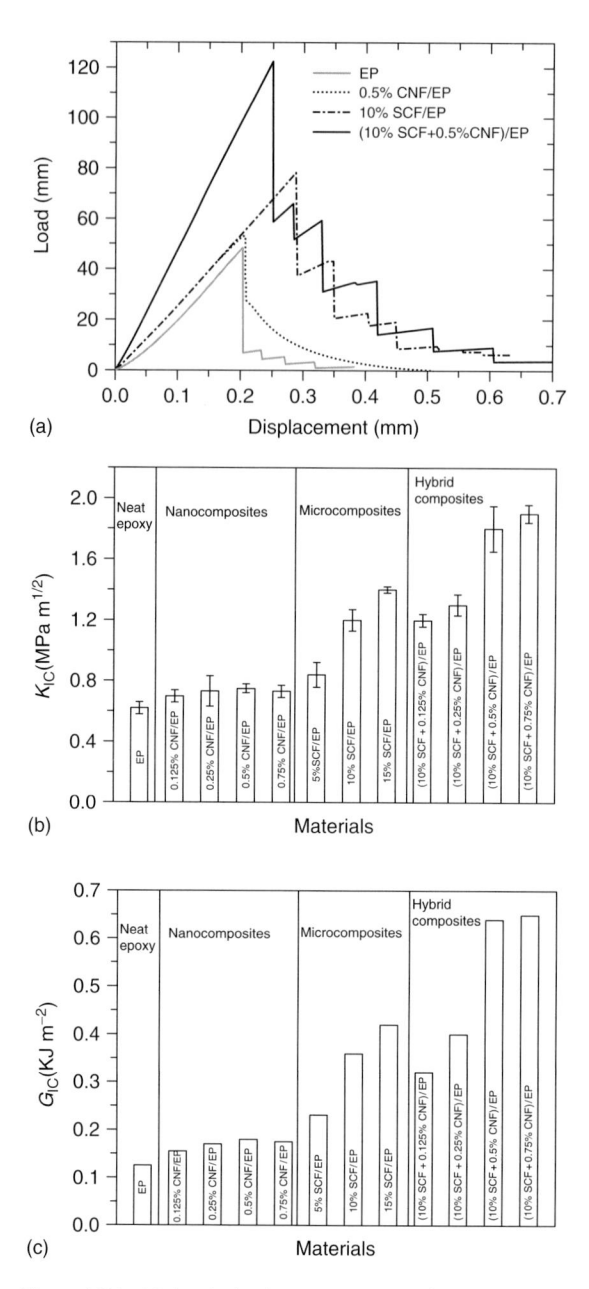

Figure 4.37 (a) Load–displacement curves of epoxy, 0.5 vol% CNF/EP, 10 vol% SCF/EP, and (10 vol%SCF + 0.5vol% CNF)/EP compact tension specimens. (b) Fracture toughness (K_{IC}) and (c) critical energy release rate (G_{IC}) for pure epoxy, CNF/EP, SCF/EP, and (SCF + CNF)/EP composites. (Source: Reproduced with permission from [76], Elsevier (2010).)

greatly, while the tensile strength and elongation at break decrease as the SCF content increases. The tensile ductility of the SCF/EP composites is considerably lower than that of epoxy, especially at high fiber loading of 15 vol%. On the other hand, the additions of very small amounts of CNF to EP lead to a moderate enhancement in tensile strength, while the tensile elongation remains nearly unchanged. The composites containing CNF and SCF hybrid fillers show remarkable improvements in the modulus and tensile strength. The incorporation of only 0.75 vol% CNF into 10 vol% SCF/EP composite increases its elastic modulus from 3.7 to 5.03 GPa (36% enhancement) and tensile strength from 80 to 91.4 MPa (14.25% improvement). Optimal enhancements in the modulus and strength can be achieved in the 10 vol% SCF/EP composite by adding CNF content ≥ 0.5 vol%, and these values are higher than those of respective SCF/EP and CNF/EP composites.

Figure 4.37a shows typical load–displacement curves of the epoxy and its composites under mode I fracture toughness tests using compact tension specimens. The K_{IC} and G_{IC} values of epoxy and its composites are shown in Figure 4.37b,c, respectively. For single composites, SCFs are more effective than CNFs to improve K_{IC} and G_{IC} of pure epoxy. The fracture toughness of epoxy is twice increased by adding 10 vol% SCF. Moreover, the fracture toughness of 10 vol% SCF/EP composite can be further enhanced by incorporating CNF ≥ 0.5 vol%. The distinctly different dimension scales of SCFs and CNFs are crucial for promoting synergistic effect on the fracture toughness. It can be concluded that the composites reinforced with hybrid fillers exhibit much higher stiffness, strength, and fracture toughness than the composites reinforced with either micro- or nanofillers.

Nomenclature

a	Crack length
B	Thickness of specimen
β	Shape factor of the outer plastic dissipation zone
d	Diameter of filler
da/dN	Fatigue crack growth rate
Δa	Crack extension
ΔH	Activation energy
E	Elastic modulus
E_{CL}	Longitudinal elastic modulus of composite
E_{CT}	Transverse elastic modulus of composite
E_f	Elastic modulus of fiber filler
E_f	Elastic modulus of fiber filler
E_M	Elastic modulus of the matrix of composite
$\dot{e}$	strain rate
G	Strain energy release rate
G_c	*Critical* strain energy release rate
J	*J*-integral
J_c	Critical *J*-value

K	Stress intensity factor
K_{IC}	Fracture toughness (mode I)
L	Ligament length
l	Length of filler
l_c	Critical fiber length
$\bar{l}$	Average fragment length of fiber
η_l	Fiber length efficiency factor
η_o	Fiber orientation factor
P	Applied load
R	Gas constant
r_p	Radius of plastic zone
S	Span length of specimen
σ	Applied stress
σ_c	Composite strength
σ_f	Fiber strength
σ_m	Matrix strength
σ_y	Yield stress
T	Temperature
τ	Interfacial shear stress
V_f	Volume fraction of filler
v	Activation volume
W	Width of specimen
W_e	Essential work of fracture
W_p	Plastic work of fracture
W_f	Total work of fracture
w_e	Specific essential work of fracture
w_f	Specific total work of fracture
w_p	Specific nonessential work of fracture

References

1. http://www.zoltek.com/carbonfiber/.
2. Godara, A., Raabe, D. and Green, S. (2007) *Acta Biomater.*, **3**, 209–220.
3. Evans, W.J., Isaac, D.H. and Saib, K.S. (1996) *Composites Part A*, **27**, 547–554.
4. Bortz, D.R., Merino, C. and Martin-Gullon, I. (2011) *Compos. Sci. Technol.*, **71**, 31–38.
5. (a) Rafiee, M.A., Rafiee, J., Wang, Z., Song, H., Yu, Z.Z. and Koratkar, N. (2009) *ACS Nano*, **3**, 3884–3890; (b) Rafiee, M.A., Rafiee, J., Srivastava, I., Wang, Z., Song, H., Yu, Z.Z. and Koratkar, N. (2010) *Small*, **6**, 179–183.
6. Rafig, R., Cai, D.Y., Jin, J. and Song, M. (2010) *Carbon*, **48**, 4309–4314.
7. Li, C.Y. and Chou, T.W. (2009) *Composites Part A*, **40**, 1580–1586.
8. Haggenmueller, R., Zhou, W., Fischer, J.E. and Winey, K.I. (2003) *J. Nanosci. Nanotechnol.*, **3**, 105–110.
9. Ogasawara, T., Moon, S.Y., Inoue, Y. and Shimamura, Y. (2011) *Compos. Sci. Technol.*, **71**, 1826–1833.
10. Ward, I.M. and Hadley, D.W. (1993) *An Introduction to the Mechanical Properties of Solid Polymers*, John Wiley & Sons, Ltd, New York.
11. Bao, S.P. and Tjong, S.C. (2008) *Mater. Sci. Eng., A*, **485**, 508–516.
12. Bao, S.P. and Tjong, S.C. (2009) *Polym. Compos.*, **30**, 1749–1760.
13. Yang, K., Gu, M.Y., Guo, Y.P., Pan, X.F. and Mu, G.H. (2009) *Carbon*, **47**, 1723–1737.
14. ASTM (2010) D790–10. *Standard Test Methods for Flexural Properties of Unreinforced and Reinforced Plastics and*

Electrical Insulating Materials, American Society for Testing and Materials, West Conshohocken.

15. Mallick, P.K. (1993) *Fiber-Reinforced Composites*, Marcel Dekker, New York.
16. ASTM (2010) D256-10. *Standard Test Methods for Determining the Izod Pendulum Impact Resistance of Plastics*, American Society for Testing and Materials, West Conshohocken.
17. ASTM Standard (1999) D 5045-99. *Standard Test Methods for Plane-Strain Fracture Toughness and Strain Energy Release Rate of Plastic Materials*, American Society for Testing and Materials, West Conshohocken.
18. Rice, J.R. (1968) *J. Appl. Mech.*, **35**, 379–381.
19. Broberg, K.B. (1968) *Int. J. Fract.*, **4**, 11–18.
20. (a) ASTM Standard (1981) E813-81. *Standard Test Method for J_{Ic}, a Measure of Fracture Toughness*, American Society for Testing and Materials, West Conshohocken; (b) ASTM Standard (1987) E813-87. *Standard Test Method for J_{Ic}, a Measure of Fracture Toughness*, American Society for Testing and Materials, West Conshohocken.
21. Martinez, A.B., Gamez-Perez, J., Sanchez-Soto, M., Velasco, J.J., Santana, O.O. and Li Maspoch, M. (2009) *Eng. Fail. Anal.*, **16**, 2604–2617.
22. Pegoretti, A., Catellani, L., Franchini, L., Mariani, P. and Penati, A. (2009) *Eng. Fract. Mech.*, **76**, 2788–2798.
23. Barany, T., Czigany, T. and Karger-Koccis, J. (2010) *Prog. Polym. Sci.*, **35**, 1257–1287.
24. Santana, O.O., Rodriguez, C., Belzunce, J., Gamez-Perez, J., Carrasco, F. and Li Maspoch, M. (2010) *Polym. Test.*, **29**, 984–990.
25. Clutton, E. (2001) in *Fracture Mechanics Testing Methods for Polymers, Adhesives and Composites* (eds D.R. Moore, A. Pavan, and J.G. Williams), Elsevier, New York.
26. Tjong, S.C. and Bao, S.P. (2007) *Compos. Sci. Technol.*, **67**, 314–323.
27. Satapahty, B.K., Weidisch, R., Potschke, P. and Janke, A. (2007) *Compos. Sci. Technol.*, **67**, 867–879.
28. Adhikari, A.R., Partida, E., Petty, T.W., Jones, R., Lozano, K. and Guerrero, C. (2009) *J. Nanomater.*, 6 (Article ID 101870).
29. Liao, C.Z. and Tjong, S.C. (2011) *Polym. Eng. Sci.*, **51**, 948–958.
30. Kelly, A. and Tyson, W.R. (1965) *J. Mech. Phys. Solids*, **13**, 329–350.
31. Duncan, R.K., Chen, X.Y., Bult, J.B., Brinson, L.C. and Schadler, L.S. (2010) *Compos. Sci. Technol.*, **70**, 599–605.
32. Wu, C.M., Chen, M. and Karger-Kocsis, J. (2001) *Polymer*, **42**, 129–135.
33. Tsuda, T., Ogasawara, T., Deng, F. and Takeda, N. (2011) *Compos. Sci. Technol.*, **71**, 1295–1300.
34. (a) Barber, A.H., Cohen, S.R. and Wagner, H.D. (2003) *Appl. Phys. Lett.*, **82**, 4140–4142; (b) Barber, A.H., Cohen, S.R., Kenig, S. and Wagner, H.D. (2004) *Compos. Sci. Technol.*, **64**, 2283–2289.
35. Frankland, S.J., Caglar, A., Brenner, D.W. and Griebel, M. (2002) *J. Phys. Chem. B*, **106**, 3046–3048.
36. Zheng, Q.B., Xia, D., Xue, Q.Z., Yan, K.Y., Gao, X. and Li, Q. (2009) *Appl. Surf. Sci.*, **255**, 3534–3543.
37. Zhao, Q. and Wagner, H.D. (2004) *Philos. Trans. R. Soc. London, A*, **362**, 2407–2424.
38. Schadler, L.S., Giannaris, S.C. and Ajayan, P.M. (1998) *Appl. Phys. Lett.*, **73**, 3842–3844.
39. Mu, M.F., Osswald, S., Gogotsi, Y. and Winey, k.Y. (2009) *Nanotechnology*, **20**, 335703 (7 pp).
40. Rahmat, M. and Hubert, P. (2011) *Compos. Sci. Technol.*, **72**, 72–84.
41. Gong, L., Kinloch, I.A., Young, R.J., Riaz, I., Jalil, R. and Novoselov, K.S. (2010) *Adv. Mater.*, **22**, 2694–2697.
42. Srivastava, I., Mehta, R.J., Yu, Z.Z., Schadler, L.S. and Koratkar, N. (2011) *Appl. Phys. Lett.*, **98**, 063102 (3 pp).
43. (a) Cox, H.L. (1952) *Br. J. Appl. Phys.*, **3**, 72–79; (b) Gao, X.L. and Li, K. (2005) *Int. J. Solids Struct.*, **42**, 1649–1667; (c) Haque, A. and Ramasetti, A. (2005) *Compos. Struct.*, **71**, 68–77.
44. Wang, W., Ciselli, P., Kuznetsov, E., Peijs, T. and Barber, A.H. (2008) *Philos. Trans. R. Soc. A*, **366**, 1613–1626.

45. Halpin, J.C. and Kardos, J.L. (1976) *Polym. Eng. Sci.*, **16**, 344–352.
46. (a) Lindhagen, J.E. and Berglund, L.A. (2000) *Compos. Sci. Technol.*, **60**, 885–893; (b) Lindhagen, J.E., Gamstedt, E.K. and Berglund, L.A. (2000) *Compos. Sci. Technol.*, **60**, 2883–2894.
47. Qian, D. and Dickey, E.C. (2001) *J. Microsc.*, **204**, 39–45.
48. Ajayan, P.M., Schadler, L.S., Giannaris, C. and Rubio, A. (2000) *Adv. Mater.*, **12**, 750–753.
49. Liu, T., Phang, I.Y., Shen, L., Chow, S.Y. and Zhang, W.D. (2004) *Macromolecules*, **37**, 7214–7222.
50. Savage, G. (1993) *Carbon–Carbon Composites*, Chapman & Hall, London, pp. 103–120.
51. Becker, O., Varley, R. and Simon, G. (2002) *Polymer*, **43**, 4365–4373.
52. Pluart, L.L., Duchet, J. and Sautereau, H. (2005) *Polymer*, **46**, 12267–12278.
53. Wang, J.C., Wang, X.B., Xu, C.H., Zhang, M. and Shang, X.P. (2010) *Polym. Int.*, **60**, 816–822.
54. Kuila, T., Bose, S., Hong, C.E., Uddin, M.E., Khanra, P., Kim, N.H. and Lee, J.H. (2011) *Carbon*, **49**, 1033–1051.
55. Bai, X., Wan, C.Y., Zhang, Y. and Zhai, Y.H. (2011) *Carbon*, **49**, 1608–1613.
56. Kim, I.H. and Jeong, Y.G. (2010) *J. Polym. Sci., Part B: Polym. Phys.*, **48**, 850–858.
57. Jana, S. and Zhong, W.H. (2009) *Mater. Sci. Eng. A*, **525**, 138–146.
58. Kalaitzidou, K., Fukushima, H. and Drzal, L.T. (2007) *Compos. Sci. Technol.*, **67**, 2045–2051.
59. Yuan, J.M., Fan, Z.F., Chen, X.H., Chen, X.H., Wu, Z.J. and He, L.P. (2009) *Polymer*, **50**, 3285–3291.
60. Blond, D., Barron, V., Ruether, M., Ryan, K.P., Nicolosi, V., Blau, W.J. and Coleman. J.N. (2006) *Adv. Funct. Mater.*, **16**, 1608–1614.
61. Paiva, M.C., Zhou, B., Fernando, K.A., Lin, Y., Kennedy, J.M. and Sun, Y.P. (2004) *Carbon*, **42**, 2849–2854.
62. Liu, L., Barber, A.H., Nuriel, S. and Wagner, H.D. (2005) *Adv. Funct. Mater.*, **15**, 975–980.
63. Tjong, S.C., Xu, S.A., Li, R.K. and Mai, Y.W. (2002) *Compos. Sci. Technol.*, **62**, 831–840.
64. Tjong, S.C., Xu, S.A. and Mai, Y.W. (2002) *J. Polym. Sci., Part B: Polym. Phys.*, **40**, 1881–1892.
65. Tjong, S.C. and Meng, Y.Z. (2003) *J. Polym. Sci., Part B: Polym. Phys.*, **41**, 2332–2341.
66. Liao, C.Z. and Tjong, S.C. (2010) *J. Nanomater.*, **2010**, (Article Number 327973), pp. 9.
67. Gojny, F.H., Wichmann, M.H., Kopke, U. and Schulte, K. (2005) *Compos. Sci. Technol.*, **64**, 2363–2371.
68. Gojny, F.H., Wichmann, M.H., Fiedler, B. and Schulte, K. (2005) *Compos. Sci. Technol.*, **65**, 2300–2313.
69. http://www.exakt.com.
70. Thostenson, E.T. and Chou, T.W. (2006) *Carbon*, **44**, 3022–3029.
71. Gojny, F.H., Nastalczyk, J., Roslaniec, Z. and Schulte, K. (2003) *Chem. Phys. Lett.*, **370**, 820–824.
72. Kim, J.A., Song, D.G., Kang, T.J. and Young, J.R. (2006) *Carbon*, **44**, 1898–1905.
73. Miyagawa, H. and Drzal, L.T. (2005) *Composites Part A*, **36**, 1440–1448.
74. Tang, L.C., Zhang, H., Han, J.H., Wu, X.P. and Zhang, Z. (2011) *Compos. Sci. Technol.*, **72**, 7–13.
75. Liu, L. and Grunlan, J.C. (2007) *Adv. Funct. Mater.*, **17**, 2343–2349.
76. Zhang, G., Karger-Kocsis, J. and Zou, J. (2010) *Carbon*, **48**, 4289–4300.
77. Grimmer, C.S. and Dharan, C.K. (2010) *Compos. Sci. Technol.*, **70**, 901–908.
78. Chandrasekaran, V.C., Advani, S.G. and Santare, M.H. (2010) *Carbon*, **48**, 3692–3699.
79. Fortunati, E., D'Angelo, F., Martino, S., Orlacchio, A., Kenny, J.M. and Armentano, L. (2011) *Carbon*, **49**, 2370–2379.

5
Electrical Properties of Polymer Nanocomposites

5.1
Background

The electrical conductivity of polymer nanocomposites reinforced with conducting carbonaceous nanofillers is considered of technological importance. Polymer nanocomposites show great promise for applications in electronic industry as functional materials for electromagnetic interference shielding, antistatic dissipation, embedded capacitor, cellular phone, sensor, and actuator. Polymer nanocomposites with excellent electrical conductivity are ideal candidates that can meet the requirements for those applications. High electrical conductivity can be achieved by proper selection of the polymer matrix materials and the composite processing conditions and the attainment of homogeneous dispersion of nanofillers in the polymer matrix. Polymers are insulators with very low electrical conductivity. The incorporation of low-volume fractions of carbonaceous nanofillers into insulating polymers can improve their electrical conductivity significantly. When the filler content reaches the percolation threshold, electrical conductivity of the polymer nanocomposites can increase by several orders of magnitude. Several factors such as the aspect ratio and type of fillers, functionalization and dispersion of fillers, as well as the processing conditions can affect the percolation threshold of conducting polymer nanocomposites markedly. Till present, the addition of reduced graphene oxide (rGO) and thermally reduced graphene (TRG) nanofillers to polymers can yield an ultrahigh low percolation threshold of ∼0.07–0.1 vol%. This can be attributed to the highly exfoliation of a layered graphite into thin graphene sheets of large aspect ratios. Low-cost graphenelike fillers are considered as a feasible substitute for expensive carbon nanotubes (CNTs) in forming functional polymer composites.

5.2
Percolation Concentration

5.2.1
Theoretical Modeling

Percolation is frequently used to describe the movement of fluid flow in a porous medium. Typical examples are the passage of water through soil and the movement of water through coffee beans and the filter. The theory concerns the transition from a nonconnected to connected state in a randomly disordered system consisting of two different components or phases. It explains a sudden increase in the value of one given physical property in the neighborhood of transition at a critical fraction of one component, referring to as the *percolation threshold* [1–4]. Percolation is particularly suitable for describing the flow of electrons in a network consisting of a random mixture of conducting and nonconducting elements [5]. The electrical charge and conductivity are typical "scalar" entities studied in the framework of percolation theory. Thus percolation can be extended to a two-phase composite-filler system for describing the mechanism of electrical conductivity near the insulating–conducting transition.

Percolation theory is typically used to statistically analyze the behavior of connected clusters in a random structure. The site percolation model is based on random connectivity of adjacent sites of the hard-core circles in an infinite lattice [6]. Kirkpatrick [5] employed this lattice model for predicting the conductance of a random resistor network. Let us consider a large, infinite periodic lattice, where each site is occupied randomly with probability P and the empty site with probability $(1 - P)$. The adjacent or nearest neighboring sites are occupied and connected to form clusters. The populated lattice can then be subdivided into distinct isolated clusters. At low concentrations of occupied sites, the lattice consists of few isolated clusters. At a critical probability P_c, percolation occurs in which an infinite cluster can span across the lattice, forming a connected path of occupied sites. Assuming the occupied sites are electrical conductors and the vacant sites insulating matrix, electrical current can flow between the nearest-neighbor conductor sites on contact. The electrical conductivity (σ_{dc}) is given by the following relation

$$\sigma \propto (P - P_c)^t \tag{5.1}$$

where P_c is the percolation probability of the lattice and t the critical exponent.

The percolation theory predicts the onset of a sharp transition of an infinite cluster in which long-range connectivity appears at a critical concentration. At the percolation threshold, Φ_c, a transition from insulating to conducting behavior occurs due to the formation of a conducting path network in the lattice. The conductivity rises rapidly as more percolation paths form until saturation is approached. The percolation concentration determined accordingly depends on the connectivity, lattice type (e.g., square, triangular, simple cubic), and lattice dimensionality. In continuum percolation theory, the Eq. (5.1) is rewritten as

$$\sigma \propto (\phi - \phi_c)^t \quad \text{for} \quad \Phi > \Phi_c \tag{5.2}$$

Numerical calculations for the hard-core circles give $\Phi_c \approx 15\%$ [6]. The percolation model based on the hard-core circles has its own limitation in predicting the percolation threshold of polymer composites with conducting nanofillers of large aspect ratios. The percolation threshold of polymer composites is sensitive to the shape and aspect ratio of filler particles. For example, TRG and rGO fillers with large aspect ratios facilitate the formation of conducting path network at very low percolation concentration of ~0.07–0.1 vol%. The percolation threshold of polymer nanocomposites is well below theoretical value of 15 vol%.

In general, the direct current (DC) conductivity and dielectric constant of a composite with conducting fillers observe the power law relation near the percolation threshold, that is, Eq. (5.2) and the following equations [7, 8]

$$\sigma_{dc} \propto (\phi_c - \phi)^{-s'} \quad \text{for} \quad \Phi < \Phi_c \tag{5.3}$$

$$\varepsilon'(f) \propto \left|\frac{\phi_c - \phi}{\phi_c}\right|^{-s} \quad \text{for} \quad \Phi > \Phi_c, \Phi < \Phi_c \tag{5.4}$$

where ε' is the permittivity and Φ the volume fraction of conductive filler. The corresponding filler content and percolation threshold in weight fraction are expressed in terms of p and p_c, respectively. The critical exponents s, s', and t are assumed to be universal. The currently accepted values of these exponents are $s' = t \approx 1.3$ for two-dimensional percolating system and $s' \approx 0.76$ and $t \approx 2.0$ for three-dimensional system. However, deviation of experimental conductivity exponents from the universal values is frequently reported in the literature [9, 10]. This possibly results from intimate contact between conducting particles, leading to electron tunneling from the fillers having concentrations above the percolation threshold [9, 11]. Experimentally, the filler volume fraction of a polymer composite can be determined from the filler weight fraction (p) according to the following relation [12]

$$\phi = \frac{p/\rho_f}{p/\rho_f + (1-p)/\rho_m} \tag{5.5}$$

where ρ_f and ρ_m are the density of filler and matrix of a composite, respectively.

Further, the polymer-layer-coated nanofiller surface produces the highest resistance region in the conducting pathway through the network. This polymer layer acts as a barrier to efficient charge transport between conducting nanofillers. In this regard, Sheng et al. [13a,b] proposed a model for conductivity based on fluctuation-induced tunneling through potential barriers of varying height due to local temperature fluctuations. They derived the following relationship for the conductivity

$$\sigma_{dc} \propto \exp\left[-\frac{T_1}{T+T_0}\right] \tag{5.6}$$

where T_1 is related to the energy required for an electron to overcome the insulator gap between conductive particles aggregate and T_0 is the temperature above which thermal fluctuations are significant. The DC conductivity due to tunneling conduction for a random distribution of the conducting particles can be further simplified to

$$\ln \sigma_{dc} \sim -p^{-1/3} \tag{5.7}$$

A linear relation between log σ_{dc} and $p^{-1/3}$ indicates that tunneling conduction may occur.

The responses of electrically conductive composites to an alternating current (AC) field near percolation are highly frequency dependent. The frequency-dependent conductivity and permittivity near the percolation threshold also obey a power-law-type relation

$$\sigma \propto f^u \tag{5.8}$$

$$\varepsilon \propto f^{-v} \tag{5.9}$$

where f is the frequency. The critical exponents u and v should satisfy the following relation [8, 14–16]

$$u + v = 1 \tag{5.10}$$

The frequency exponent u is related to the exponents s and t via the following equation

$$u = \frac{t}{s+t} \tag{5.11}$$

The validity of the percolation equations in polymer nanocomposites has been examined and reported in the literature [17, 18].

5.2.2
Excluded Volume

Balberg et al. [19, 20] then extended the percolation concept from the point-based lattice model to elongated rods or sticks of high aspect ratios in two- and three-dimensional systems using excluded volume concept. They simulated the percolation of sticks taking into account a distribution of lengths and different angle orientations. *Excluded volume* is defined as the volume around an object into which the center of another similarly shaped object is not allowed to penetrate. The concept is based on the idea that the percolation threshold is not linked to the true volume of the filler particles but rather to their excluded volume V_{ex}. This implies that the percolation volume fraction depends on the degree to which objects can be packed. Considering an individual object of volume V_o with its associated excluded volume, V_e, then the total excluded volume of a system, $\langle V_{ex} \rangle$, averaged over the orientation distribution is defined by the following relation

$$\langle V_{ex} \rangle = N_c \langle V_{ex} \rangle \approx \text{constant} \tag{5.12}$$

where N_c is the critical number density of objects at percolation. The total excluded volume is then linked to the percolation volume fraction by

$$\phi_c = 1 - \exp\left(-\frac{\langle V_{ex} \rangle V_o}{\langle V_e \rangle}\right) \tag{5.13}$$

Apparently, the percolation threshold is related with the aspect ratio of the inclusions or objects. Balberg [20] indicated that the excluded volume of the object is confined within two limits. Celzard et al. determined a lower limit of 1.4 for randomly oriented, infinitely thin cylinders and an upper bound of 2.8 for spheres with an aspect ratio of 1. Thus the critical volume fraction is given by

$$1 - \exp\left(-\frac{1.4V_o}{\langle V_e \rangle}\right) \leq \phi_c \leq 1 - \exp\left(-\frac{2.8V_o}{\langle V_e \rangle}\right) \tag{5.14}$$

On the basis of this double inequality, Liang and Tjong [21] determined the critical concentration of melt-compounded carbon nanofiber/polystyrene (CNF/PS) nanocomposites to be $0.276\% \leq \Phi_c \leq 5.8\%$.

For disk-shaped fillers, the percolation threshold is defined within the following bounds

$$1 - \exp\left(-\frac{1.8z}{\pi r}\right) \leq \phi_c \leq 1 - \exp\left(-\frac{2.8z}{\pi r}\right) \tag{5.15}$$

where r and z are the radius and thickness of the disk, respectively. As an example, a graphite nanosheet with $r = 6.05\,\mu m$ and thickness $z = 51.5\,nm$, the critical volume fraction of the GNP/unsaturated polyester (graphite nanoplatelet) composites is determined to be $0.487\% \leq \Phi_c \leq 0.765\%$ [22].

5.3
Electrical Conductivity and Permittivity

DC conductivity of conducting composites is commonly measured using the standard four-point probe method [Chapter 2, Ref. 54]. A constant current was passed through the two outer probes and an output voltage is measured across the inner probes with the voltmeter. In many practical applications, the current is AC and the applied voltage or electric field varies with time. The AC electrical conductivity and permittivity of percolating nanocomposites are often measured as a function of frequency using broadband impedance spectroscopy [23, 24]. The results can be represented in dielectric constant, conductivity, or electric modulus domain at broad frequency ranges. The dielectric sample is coated with conducting silver paste on its two parallel faces. An AC bridge is used to measure the conductance (G) and capacitance (C) of the sample as a function of frequency. The capacitance is directly related to the amount of electric charge stored on the sample surface under the influence of an applied electric field. Polarization is induced within the sample in response to an electric field. The amount of charge accumulated depends on the polarization of dielectric sample. During the measurements, a sinusoidal voltage signal with angular frequency $\omega(=2\pi f)$ is applied to the network system, where the sample under investigation is placed between the plates of a capacitor. The material can be regarded as an equivalent circuit of impedance $Z = (G + iC\omega)^{-1}$. The complex conductivity $\sigma^*(f) = \sigma'(f) - i\sigma''(f)$ and complex dielectric constant $\varepsilon^*(f) = \varepsilon'(f) - i\varepsilon''(f)$ of the sample can be determined from the impedance accordingly. The real part of the conductivity as a function of the angular frequency can be evaluated from the imaginary part of the dielectric constant $\varepsilon''(f)$ using the following relation

$$\sigma'(f) = \varepsilon_0 2\pi f \varepsilon''(f) \tag{5.16}$$

where ε_0 is the vacuum permittivity.

In general, the conductivity measured at $\omega \to 0 (f = 0\,\text{Hz})$ is often taken as the DC value. Thus the variation of conductivity with f is described by

$$\sigma'(f) = \sigma_{dc} + A f^u \tag{5.17}$$

where A is a temperature-dependent constant. Equation (5.17) is often referred to as *AC universal dynamic response* [18] or *AC universality law* [25] because many materials display such characteristic. Therefore, AC conductivity can be recognized as the combined effect of DC conductivity ($f = 0\,\text{Hz}$) caused by migrating charge carriers and frequency-induced dielectric dispersion. For the sample under investigation, $\varepsilon'(f)$ represents the real part of complex dielectric constant $\varepsilon^*(f)$ and usually regards as relative permittivity or dielectric constant, while $\varepsilon''(f)$ is associated with the loss factor. Therefore, loss tangent (tan δ) is defined as the ratio of $\varepsilon''(f)$ to $\varepsilon'(f)$. In other words, it is a measure of the ratio of the electric energy loss to energy stored under a cyclic electric field.

A typical example of the electrical properties of solution-mixed GNP/PVDF (polyvinylidene fluoride) nanocomposites determined using impedance spectroscopy is given in details herein. Figure 5.1a shows the frequency-dependent electrical conductivity curves of solution-mixed GNP/PVDF nanocomposites at room temperature [Chapter 2, Ref. 63]. For the composites with GNP contents ≤ 2 wt%, the electrical conductivity increases almost linearly with the increase of frequency, showing typical insulating behavior. At GNP content $= 2.5$ wt%, the conductivity displays a plateau in lower frequency region, commonly designating as the DC conductivity, and a frequency-dependent region beyond the onset of characteristic frequency (f_c). The characteristic frequency can be taken as the frequency where the conductivity reaches a value 10% higher than that of the plateau [26]. In other words, f_c is determined at which the AC conductivity reaches 110% of σ_{dc}, that is, $\sigma(f_c) = 1.1\,\sigma_{dc}$. The conductivity behavior of the 2.5 wt% GNP/PVDF nanocomposite can be fully described by Eq. (5.17). This leads to $\sigma_{dc} = 4.52 \times 10^{-6}\,\text{S m}^{-1}$, $f_c = 400\,\text{Hz}$, $u = 0.8$, and $A = 3.1 \times 10^{-9}$. At this stage, a conductive path network begins to form in the composite. Below the percolation threshold, σ_{dc} is extremely small and can be neglected.

The variation of ε' with the frequency of the GNP/PVDF nanocomposites at room temperature is shown in Figure 5.1b. For pure PVDF, the dielectric constant decreases slowly with increasing frequency. For the nanocomposites with high GNP loadings, the permittivity falls off rapidly at high-frequency regime. Moreover, the permittivity of the nanocomposites increases with increasing filler content. Above 2.5 wt% GNP, the permittivity shows a significant increase, particularly at low frequencies. The frequency dependence of the permittivity demonstrates the fact of strengthened interface polarization. The permittivity can reach up to ~ 3000 (at 1 kHz) by adding 4 wt% GNP. A large increase in the permittivity is attributed to the formation of many minicapacitors and to the interfacial polarization in the nanocomposites filled with GNPs. As recognized, heterogeneous composite material with a large difference in conductivity and permittivity between its constituent components can give rise to interfacial polarization, commonly known

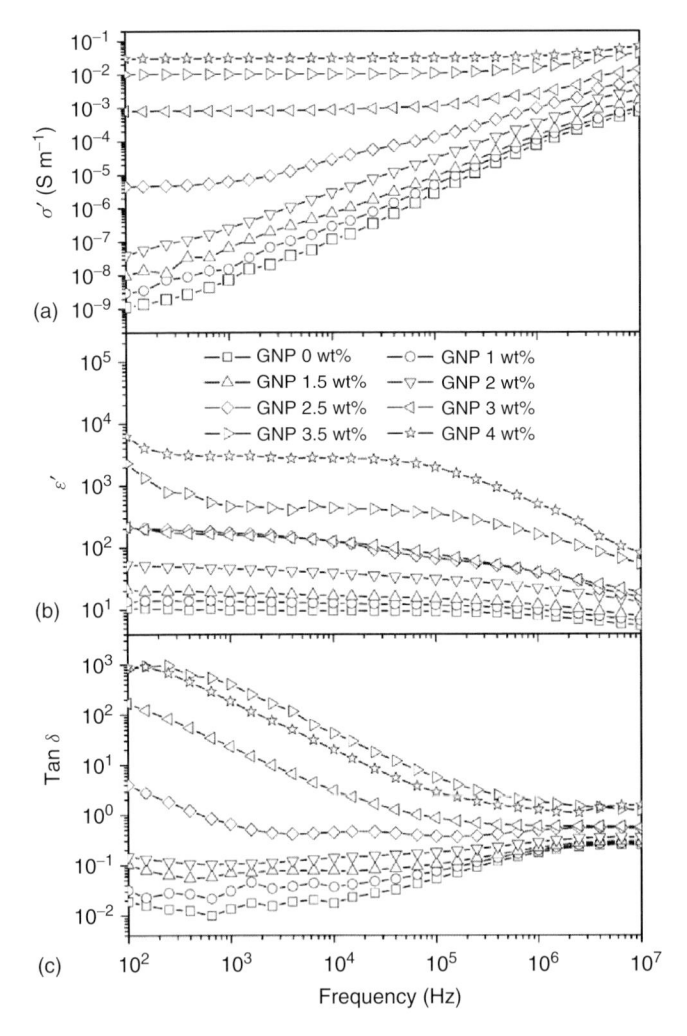

Figure 5.1 (a) Electrical conductivity, (b) dielectric constant, and (c) loss tangent as a function of frequency for GNP/PVDF nanocomposites [Chapter 2, Ref. 63].

as the Maxwell-Wagner-Sillars (MWS) polarization. The free carriers trapped at the interfaces between two media of different conductivity and permittivity form electric dipoles accordingly.

Figure 5.1c shows loss tangent versus frequency for the GNP/PVDF nanocomposites. The loss tangent is relatively low (i.e., <0.1) for the nanocomposites with GNP contents ≤2 wt%. Above this content, the dielectric loss increases markedly with increasing filler content, especially at low-frequency regime. The high loss tangent implies large energy consumption in the conducting path network during electron transport. In this case, dielectric loss results from the energy dissipation during the alignment and orientation of dipoles with the applied electric field. The

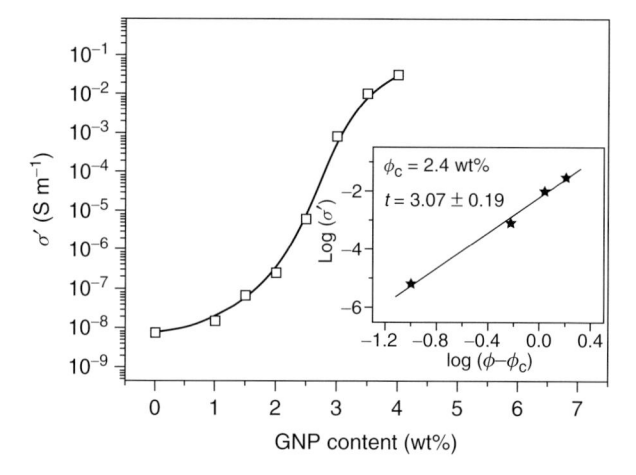

Figure 5.2 Electrical conductivity (1 kHz) as a function of GNP content for GNP/PVDF nanocomposites [Chapter 2, Ref. 63].

aligned and oriented dipoles not only can lead to the improvement of dielectric constant but also result in higher dielectric loss. As recognized, polymer nanocomposites with high dielectric constant and low dielectric loss show great potential for electronic applications. This poses a challenge for researchers to enhance the permittivity while maintaining dielectric loss at a relatively low level in the polymer nanocomposites filled with conducting nanofillers.

Figure 5.2 shows $\sigma'(f)$ as a function of GNP content for the GNP/PVDF nanocomposites. The $\varepsilon'(f)$ versus GNP content plot is shown in Figure 5.3. In these plots, the $\sigma'(f)$ and $\varepsilon'(f)$ values are taken from Figure 5.1a,b, respectively, at 1 kHz. These plots obviously show the percolation transition characteristics. Thus

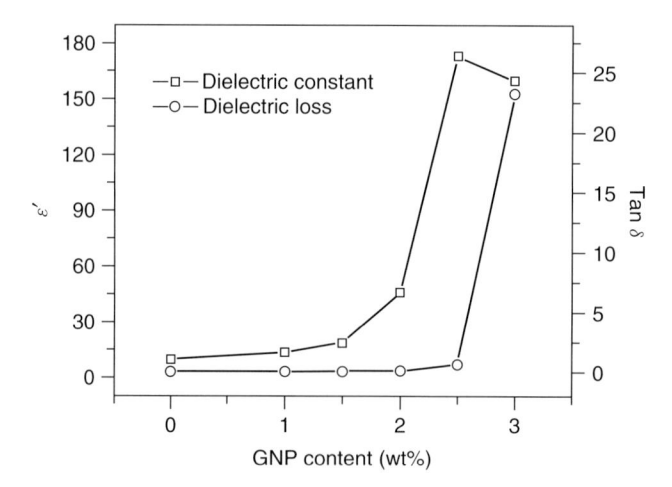

Figure 5.3 Permittivity and loss tangent at 1 kHz as a function of GNP content for GNP/PVDF nanocomposites [Chapter 2, Ref. 63].

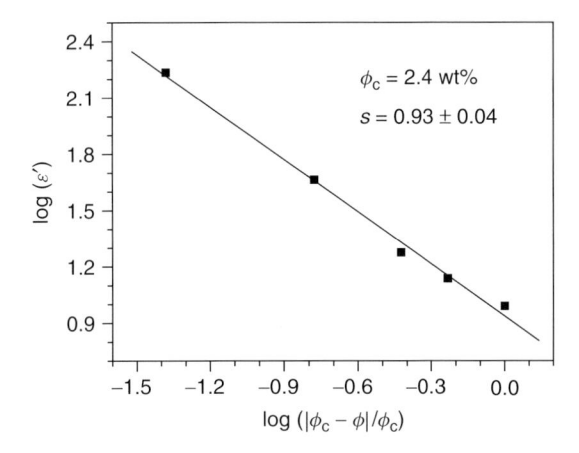

Figure 5.4 Best fitting of Eq. (5.4) for GNP/PVDF nanocomposites [Chapter 2, Ref. 63].

Eqs. (5.2) and (5.4) are used, respectively, to fit the conductivity and permittivity data presented in Figures 5.2 and 5.4. Best fitting of Eq. (5.2) yields a straight line in the plot of log σ' versus log($p - p_c$), producing $p_c = 2.4$ wt% (1.88 vol%) and $t = 3.07$ (inset Figure 5.2). Compared to pure PVDF, the conductivity increases by more than 7 orders of magnitude at GNP = 4 wt%. The critical exponent is somewhat larger than theoretical value ($t = 2$) for three-dimensional system. It can be seen from Figure 5.3 that the dielectric constant reaches an apparent maximum value at the percolation threshold. The 2.5 wt% GNP/PVDF nanocomposite exhibits a large dielectric constant of 173 and a low loss tangent of 0.65 at 1 kHz. Best fitting of Eq. (5.4) with the permittivity data produces $s = 0.93$ and $p_c = 2.4$ wt% for the PVDF/GNP nanocomposites (Figure 5.4). From the obtained t and s values for the PVDF/GNP system, the u value can be determined from Eq. (5.11) accordingly.

5.3.1 Graphene-Filled Polymer Composites

5.3.1.1 Thermoplastic Matrices

Ruoff and coworkers prepared graphene/PS nanocomposites via solution-mixing isocyanate-treated graphene oxide (iGO) and PS in dimethylformamide (DMF) followed by hydrazine reduction [Chapter 2, Ref. 54]. Such nanocomposites exhibit a low percolation threshold of 0.1 vol% rGO at room temperature (Figure 5.5). This percolation concentration is considerably lower than that reported for two-dimensional GNP fillers [Chapter 2, Refs. 24, 63]. This is due to the homogeneous dispersion and extremely large aspect ratio of graphene. At 0.15 vol% loading, the conductivity of the composites satisfies the antistatic criterion (10^{-6} S m^{-1}) for thin films. The value increases rapidly over a 0.4 vol% range. By increasing GO content to 1 vol%, a conductivity of ~0.1 S m^{-1} can be achieved, sufficient for electronic device applications. A low percolation threshold of 0.076 vol%

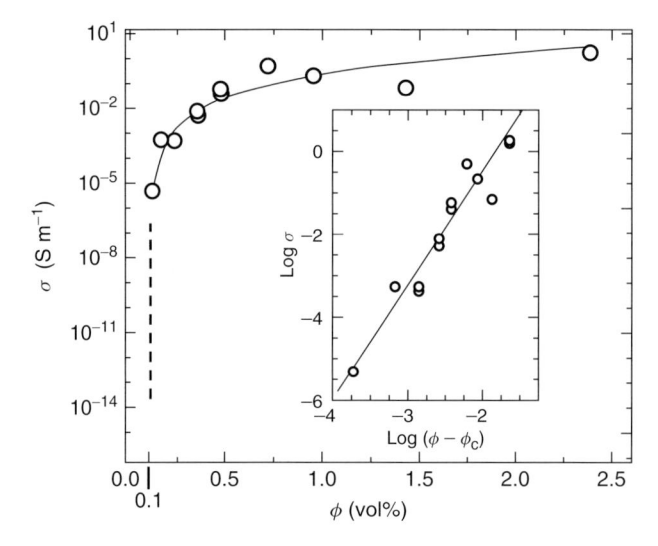

Figure 5.5 Electrical conductivity of the graphene/polystyrene composites as a function of the filler volume content. The conductivity is determined using a standard four-probe technique. Inset shows the plot of log σ_c versus log($\Phi - \Phi_c$) where Φ_c is the percolation threshold. Best fitting of Eq. (5.2) with the data yields $t = 2.74 \pm 0.20$ and $\Phi_c = 0.1$ vol%. (Source: Reproduced with permission from [Chapter 2, Ref. 54], Nature Publishing Group (2006).)

is also observed in the rGO/ultrahigh-molecular-weight polyethylene (UHMWPE) nanocomposites [27].

Since GOs are electrically insulating, chemical reduction is needed for restoring their electrical conductivity. Hydrazine reduction reagent is typically used for this purpose. The use of hydrazine must be with great care because of its toxicity and flammability. More recently, Li et al. [28] reported that simultaneous functionalization and reduction of GO can be realized by simple refluxing of GO with octadecylamine (ODA) without the use of any reducing reagents. The ODA-functionalized GO was then mixed with PS by solution blending, followed by drying and compression molding at 210 °C for 25 min. The thermal treatment under the compression molding conditions converts hydrophilic and insulating ODA-GO to hydrophobic and electrically conductive ODA-GO. The ODA functional groups render GO hydrophobic, thereby improving the compatibility and dispersion of GO in the PS matrix. Figure 5.6 shows the electrical conductivity versus filler content for the GO/PS and ODA-GO/PS nanocomposites. The ODA-GO/PS nanocomposites exhibit a sharp transition from insulating to conducting, showing percolation behavior. In contrast, there is a little improvement in the electrical conductivity of the GO/PS composites because of the poor dispersion of GO in the PS matrix.

As previously mentioned, TRG contains oxygen functionalities that can react with water-soluble polymers and polar thermoplastics during blending [29; Chapter 2, Ref. 79]. Zhang et al. prepared TRG/PET (polyethylene terephthalate) nanocomposites by melt compounding in a Brabender mixer and a subsequent compression

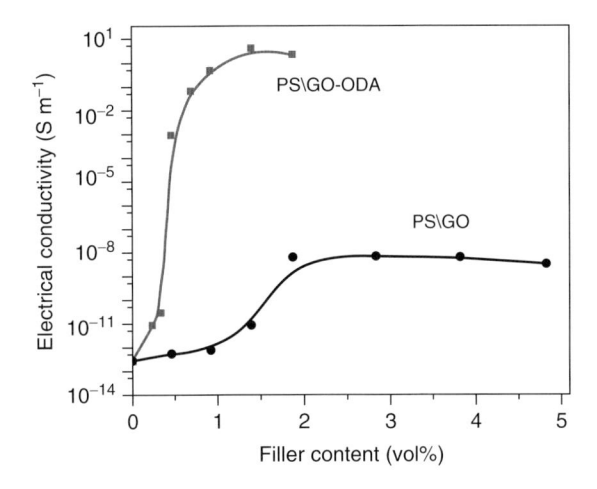

Figure 5.6 Electrical conductivity versus filler content of GO/PS and ODA-GO/PS nanocomposites. (Source: Reproduced with permission from [28], Elsevier (2011).)

molding. Figure 5.7 shows the electrical conductivity as a function of filler content for the TRG/PET nanocomposites. The nanocomposite system exhibits a low percolation threshold of 0.47 vol% and a conductivity exponent of 4.22. At 0.56 vol% TRG, the conductivity is 3.3×10^{-5} S m^{-1}, higher than the antistatic criterion of 10^{-6} S m^{-1}. By increasing TRG content to 3 vol%, the conductivity approaches 2.11 S m^{-1}. A large increase in electrical conductivity is associated with the formation of conductive path network. In contrast, graphite/PET microcomposites show

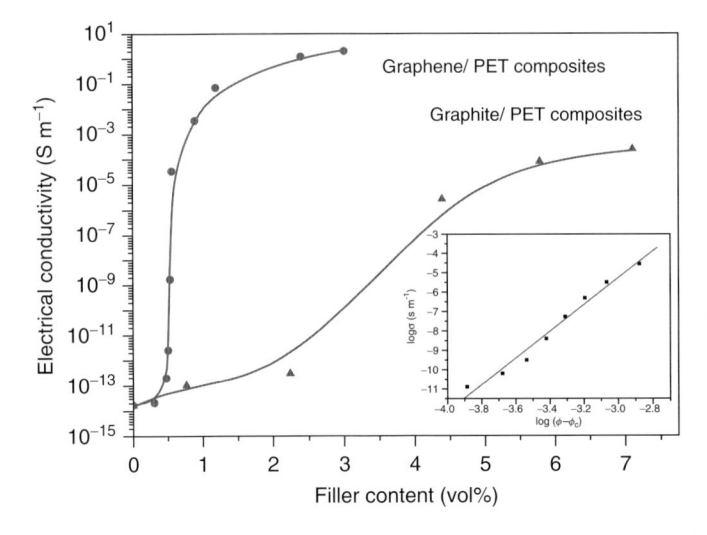

Figure 5.7 Electrical conductivity versus filler content of melt-compounded TRG/PET nanocomposites and graphite/PET microcomposites. (Source: Reproduced with permission from [Chapter 2, Ref. 79], Elsevier (2010).)

a broad transition percolation transition. The conductivity of the composite with 7.1 vol% graphite microparticles is 2.45×10^{-4} S m^{-1}.

More recently, Qi et al. [30] compared percolating behavior of the TRG/PS and MWNT/PS (multiwalled carbon nanotube) nanocomposites prepared by solution-mixing process (Figure 5.8). They reported that the TRG fillers are more effective to enhance the electrical conductivity of PS. Best fitting of Eq. (5.2) with experimental data of the TRG/PS nanocomposite yields $\Phi_c = 0.33$ vol% and $t = 2.58$, while for the MWNT/PS nanocomposites produces $\Phi_c = 0.50$ vol% and $t = 3.80$. The lower Φ_c value of the TRG/PS nanocomposite is caused by the homogeneous dispersion of TRG nanofillers in the PS matrix, facilitating the formation of a continuous network structure (Figure 5.9a). In contrast, some nanotubes agglomerate into clusters in the PS matrix (Figure 5.9b). The electrical conductivity of PS increases rapidly from $\sim \times 10^{-14}$ to ~ 0.15 S m^{-1} by adding 0.69 vol% TRG. The addition of 1.1 vol% TRG renders PS with a conductivity as

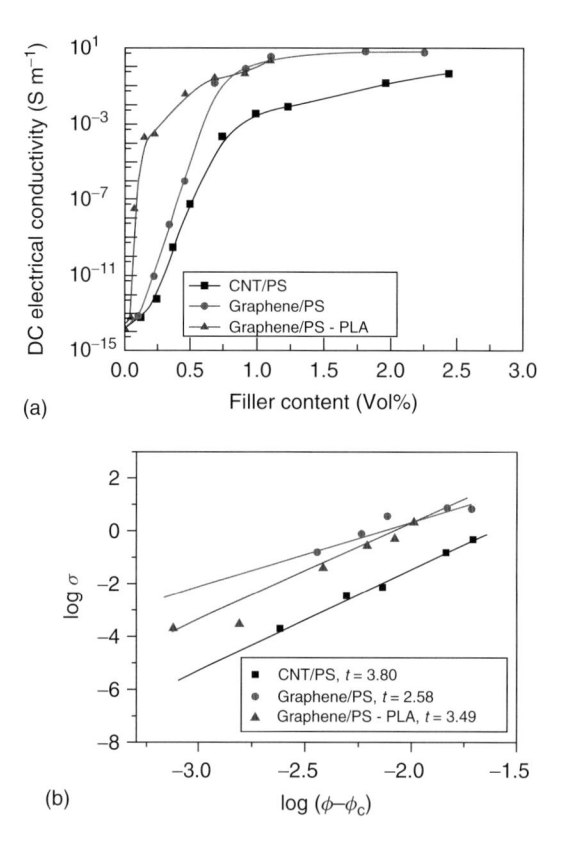

Figure 5.8 (a) Electrical conductivity versus filler content for TRG/PS, TRG/(PS-PLA), and MWNT/PS nanocomposites. (b) Double-logarithmic plot of electrical conductivity versus $(\Phi - \Phi_c)$ for these nanocomposites. (Source: Reproduced with permission from [30], The American Chemical Society (2011).)

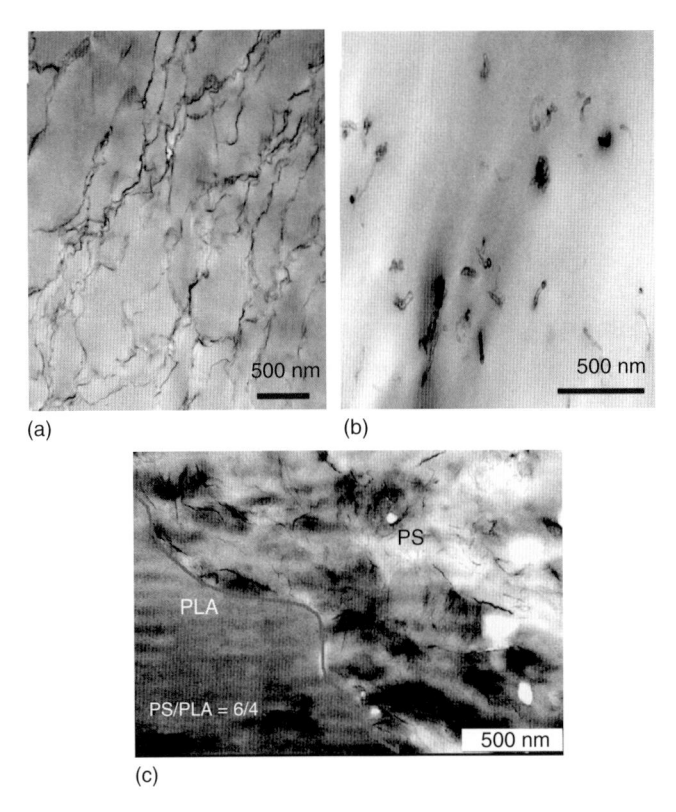

Figure 5.9 TEM images of (a) 0.69 vol% (~1.5 wt%) TRG/PS, (b) 0.69 vol% (~1.5 wt%) MWNT/PS, and (c) 0.46 vol% (~1.0 wt%) TRG/(PS-PLA) nanocomposites. (Source: Reproduced with permission from [30], The American Chemical Society (2011).)

high as ~3.49 S m^{-1}. For the 0.69 vol% MWNT/PS nanocomposite with the same filler content, its conductivity is only ~3×10^{-5} S m^{-1}, being nearly 5 orders of magnitude lower when compared to the TRG/PS nanocomposite. Qi et al. also demonstrated that the percolation threshold of the TRG/PS system can be further reduced by incorporating polylactic acid (PLA) into the polymer matrix. In this case, the matrix of the nanocomposites is made up of 60 wt% PS and 40 wt% PLA. The percolation threshold of the composites is determined to be 0.075 vol% TRG. Such a low percolation threshold derives from localized dispersion of TRG nanofillers in the PS phase (Figure 5.9c). The selective localization of TRG fillers facilitates the formation of a conductive path network at relatively low TRG content. Table 5.1 lists the percolation behavior of those graphene/thermoplastic nanocomposites discussed above.

Polycarbonate (PC) is a glassy thermoplastic showing good optical transparency, excellent dimensional and thermal stability, and good moldability. Therefore, PC finds a wide variety of applications in the manufacture of lenses, compact disks, bottles, and displays. Yoonessi and Gaier [31] fabricated TRG/PC nanocomposites

Table 5.1 Electrical conductivity of graphene/thermoplastic composites prepared by different processes.

Polymer	Filler	Percolation threshold (vol%)	Critical exponent (t)	Matrix conductivity (S m^{-1})	Maximum composite conductivity (S m^{-1})	Processing method	References
PS	rGO	0.1	2.74	–	1 at 2.5 vol% filler	Solution mixing	Chapter 2, Ref. 54
PP	rGO	0.07	1.26	–	7 × 10^{-2} at 0.6 vol% filler	Solution mixing	[27]
PS	TRG	0.33	2.58	10^{-14}	3.49 at 1.1 vol% filler	Solution mixing	[30]
PS/PLA	TRG	0.075	3.49	10^{-14}	3.49 at 1.1 vol% filler	Solution mixing	[30]
PET	TRG	0.47	4.22	10^{-14}	2.11 at 3 vol% filler	Melt compounding	Chapter 2, Ref. 79

using both solution blending and emulsion mixing, followed by compression molding at 287 °C. The emulsion mixing is based on the formation of graphene-coated PC microspheres as shown in Figure 5.10. This morphology provides an excellent conductive path for transporting electrons. PC microspheres can be synthesized from an oil-in-water (O/W) emulsion system, comprising an aqueous phase with poly(vinylpyrrolidone) (PVP) and a methylene chloride phase for dissolving PC [32]. TRGs were dispersed in a Triton X-1000 and added to the PC microemulsion under stirring. The TRGs dispersed in an aqueous phase with PVP can be selectively incorporated on the microsphere surfaces. These dried microspheres were compression molded to form TRG/PC nanocomposites. Both the DC and AC conductivity measurements were used to determine electrical properties of TRG/PC

Figure 5.10 Scanning electron micrograph of graphene nanosheets on the PC microspheres. (Source: Reproduced with permission from [31], The American Chemical Society (2010).)

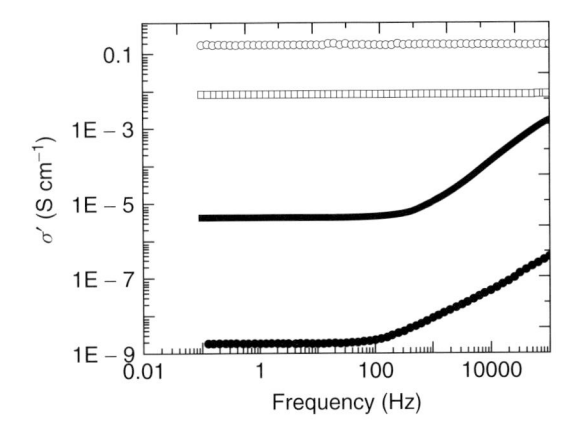

Figure 5.11 Frequency-dependent conductivities for solution-blended PC nanocomposites containing different TRG contents. ●: 0.41 vol% (0.75 wt%), ■: 0.55 vol% (1 wt%), □: 1.1 vol% (2 wt%), and ○: 2.2 vol% (4 wt%). (Source: Reproduced with permission from [31], The American Chemical Society (2010).)

nanocomposites. DC measurements showed that the percolation threshold values of solution-blended and emulsion-mixed nanocomposites are ∼0.27 and ∼0.14 vol%, respectively.

Figure 5.11 shows the real component of AC conductivity as a function of frequency for the TRG/PC nanocomposites prepared by solution blending. The filler concentration of the nanocomposite with 0.41 vol% TRG is above the percolation threshold. The 0.41 vol% TRG/PC and 0.55 vol% TRG/PC composites exhibit a frequency-independent plateau followed by a frequency-dependent conductivity at a critical frequency, f_c. The σ_{dc} values of these two composites are found to be 9.45×10^{-9} and 3.96×10^{-6} S cm^{-1}, respectively. A maximum conductivity of 0.226 S cm^{-1} can be achieved in the composite system by adding 2.2 vol% TRG. In contrast, microemulsion-mixed 2.2 vol% TRG/PC nanocomposite exhibits even a higher conductivity of 0.512 S cm^{-1} (Figure 5.12). At the electrical percolation threshold, the distance between TRG sheets dispersed in the nanocomposites is close enough for electron tunneling. Indeed, the linearity of the conductivity versus $\Phi^{-1/3}$ demonstrates that electron tunneling occurs between polymer-film-coated graphene nanosheets (Figure 5.13).

5.3.1.2 Thermosetting Matrices

Till present, very few information is available in the literature relating electrical behavior of the graphene/epoxy nanocomposites. Very recently, Liang et al. prepared rGO/epoxy nanocomposites by solvent casting and reported that these composites exhibit a low percolation threshold of 0.52 vol% and a conductivity exponent of 5.37 [Chapter 2, Ref. 56]. With the increasing role plays by graphenelike nanofillers in functional polymer composites, more research articles on the electrical properties of graphene/epoxy nanocomposites are expected to be reported in the literature in coming years.

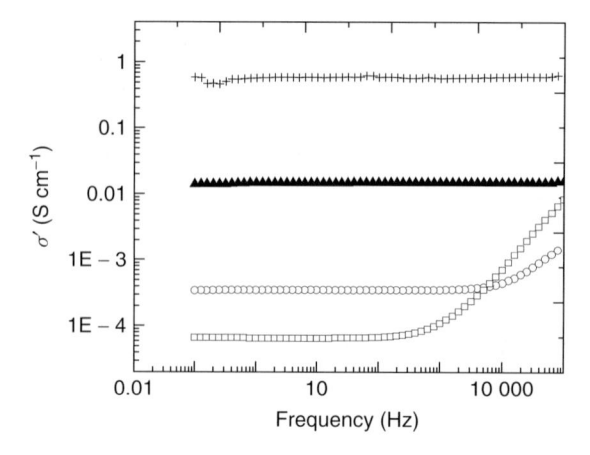

Figure 5.12 Frequency-dependent conductivities for emulsion-mixed PC nanocomposites containing different TRG contents. □: 0.27 vol% (0.5 wt%), ○: 0.55 vol% (1 wt%), ▲: 1.1 vol% (2 wt%), and +: 2.2 vol% (4 wt%). (Source: Reproduced with permission from [31], The American Chemical Society (2010).)

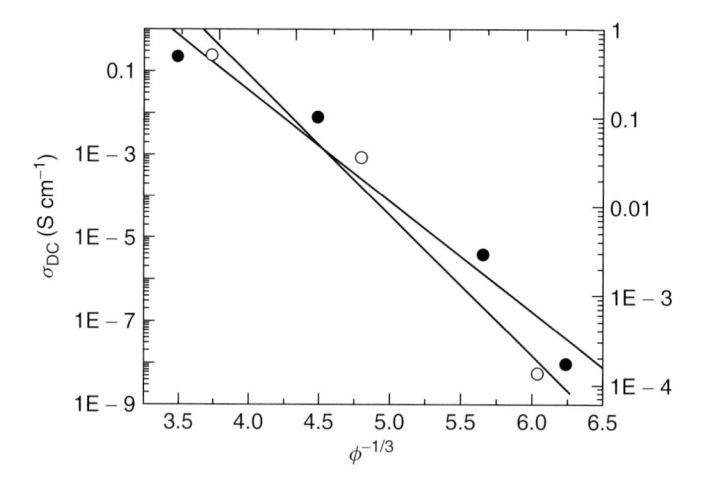

Figure 5.13 DC conductivity versus $\Phi^{-1/3}$ for emulsion-mixed nanocomposites (○) and solution-blended nanocomposites (●). (Source: Reproduced with permission from [31], The American Chemical Society (2010).)

5.3.1.3 Elastomeric Matrices

Macosko and coworkers studied the effect of processing methods on the filler dispersion, electrical and mechanical properties of the TRG/PU (polyurethane) nanocomposites [Chapter 2, Ref. 52]. Figure 5.14a–c depict transmission electron microscopy (TEM) images showing the dispersion of TRGs in the polymer matrix of the TRG/PU nanocomposites prepared from solution-mixing, *in situ*

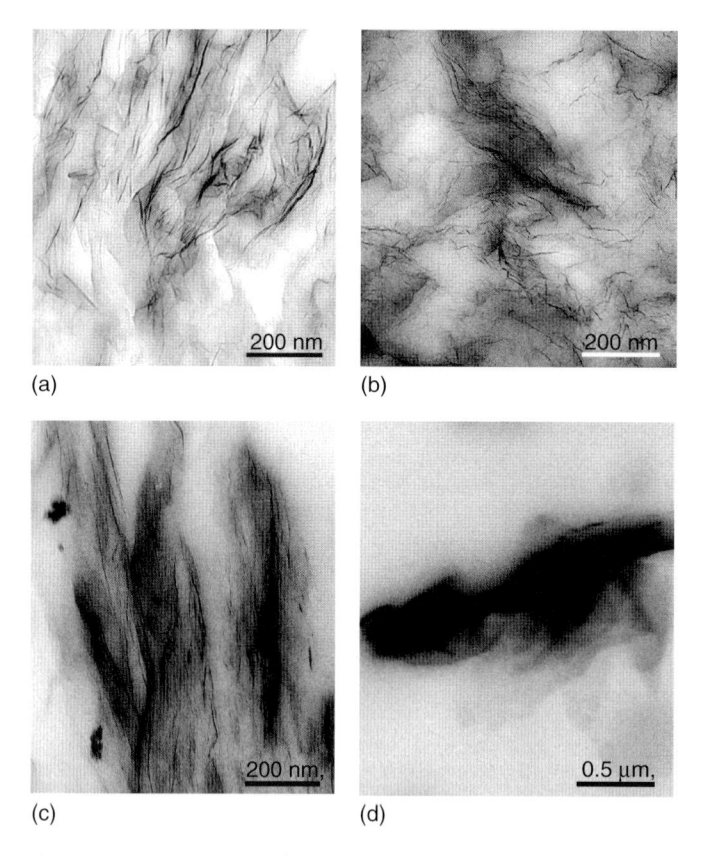

Figure 5.14 Transmission electron micrographs of 1.6 vol% TRG/PU nanocomposites prepared by (a) solution mixing, (b) *in situ* polymerization, and (c) melt compounding. (d) Transmission electron micrograph of 2.7 vol% graphite/PU microcomposite. (Source: Reproduced with permission from [Chapter 2, Ref. 52], The American Chemical Society (2010).)

polymerization and melt compounding processes, respectively. For the purpose of comparison, the microstructure of melt-compounded graphite/TPU (thermoplastic polyurethane) microcomposites is shown in Figure 5.14d. Unexfoliated graphite in a stacked morphology can be readily seen in this micrograph. However, both solution-mixing and *in situ* polymerization processes produce good dispersion of TRG sheets in the PU matrix. This is because oxygen functionalities of TRG can react with isocyanate (-NCO) groups of PU during the composite fabrication processes. On the contrary, melt-compounded TRG sheets are more oriented with some stacking. Figure 5.15 shows DC surface resistance as a function of filler content for melt-blended graphite/PU microcomposites and TRG/PU nanocomposites prepared from different processes. Both solution-mixed and *in situ* polymerized nanocomposite systems display a sharp decrease in surface resistance by adding 0.5 vol% TRG because of the well dispersion of TRG sheets in the PU matrix.

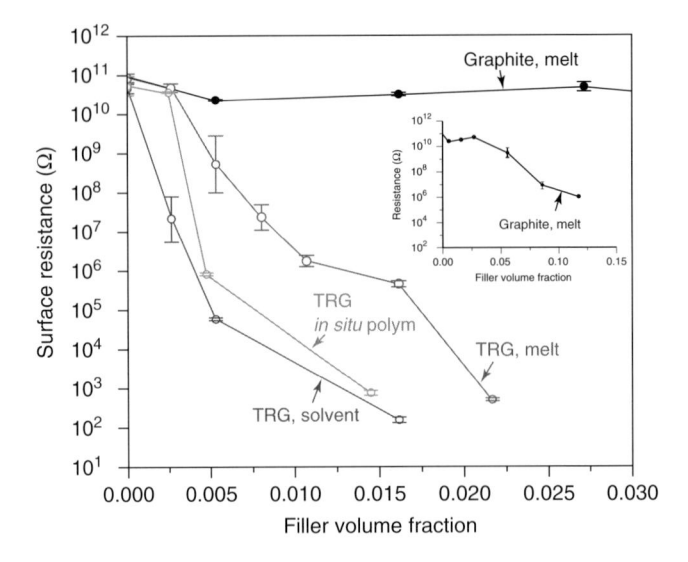

Figure 5.15 DC surface resistance of melt-blended graphite/TPU microcomposites and melt-blended, solution-mixed, and *in situ* polymerized TRG/PU nanocomposites. (Source: Reproduced with permission from [Chapter 2, Ref. 52], The American Chemical Society (2010).)

5.3.2
EG- and GNP-Filled Polymers

5.3.2.1 Thermoplastic Matrices

In general, the percolation threshold of thermoplastics with expanded graphite (EG) and GNP fillers is considerably higher than that of the composites containing graphene fillers. Furthermore, polymer nanocomposites prepared by both the solution-mixing and *in situ* polymerization techniques exhibit lower percolation threshold, compared with those fabricated by melt-compounding process. This is attributed to the wet chemical processes that enable better dispersion of nanofillers in the polymer matrix. The percolation behavior of solution-mixed GNP/PVDF nanocomposites is 1.88 vol% as discussed above. For the EG/PVDF nanocomposites, the percolation threshold of melt-compounded EG/PVDF nanocomposites is determined to be 6 vol% (Figure 5.16) [33]. This percolation concentration is three times higher than that of solution-mixed GNP/PVDF composites. Li and Jeong [34] reported that melt-compounded EG/PET nanocomposites also exhibit a high percolation threshold of ~5 wt% EG as shown in Figure 5.17. Figure 5.18 shows the electrical conductivity as a function of filler content of the EG/PP (polypropylene) nanocomposites fabricated by solution-mixing and melt-blending processes [35]. The solution-mixed EG/PP nanocomposites display a lower percolation threshold as expected. The PP molecular chains can penetrate into EG pores during solution mixing. The EG platelets are then coated with the polymeric materials, thereby

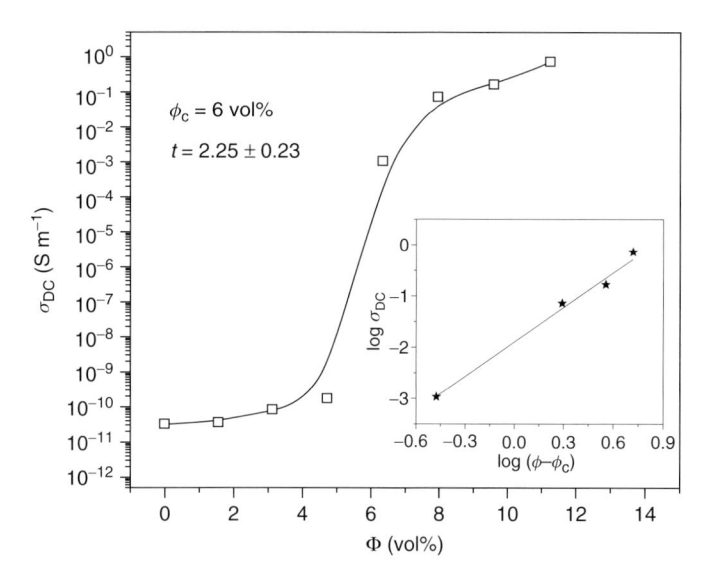

Figure 5.16 DC conductivity as a function of EG content of melt-compounded EG/PVDF nanocomposites [33].

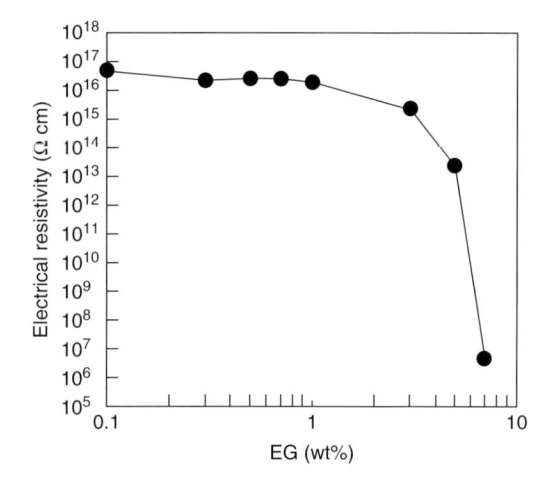

Figure 5.17 Electrical volume resistivity versus EG content of melt-compounded EG/PET nanocomposites. (Source: Reproduced with permission from [34], Elsevier (2011).)

retaining their original features. The EGs can make contact and form a conducting path network readily at lower concentrations.

Table 5.2 lists typical percolating behavior of the EG- and GNP-thermoplastic composites prepared by melt-compounding, solution-mixing, and *in situ* polymerization processes. GNPs with large aspect ratios facilitate the formation of a conductive path network at low percolation thresholds [36, 37]. The width (diameter) and thickness of GNPs generally depend on the thermal shock exposure

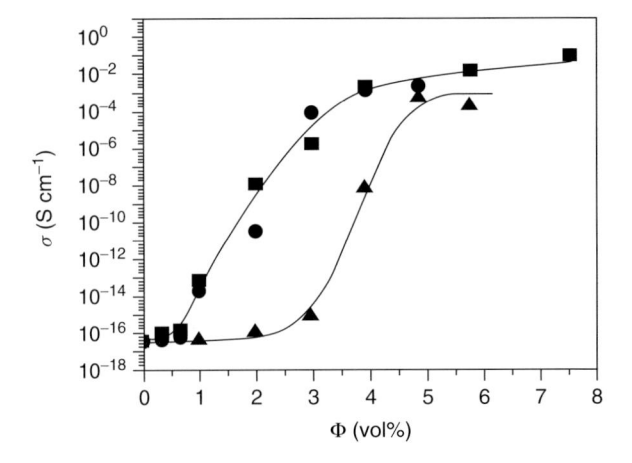

Figure 5.18 DC conductivity as a function of EG content of EG/PP nanocomposites prepared by solution-mixing (■) and melt-mixing (▲) processes. ● represents repeated experiment of solution-mixed nanocomposites. The conductivity versus filler content plot of solution-mixed nanocomposites shows good repeatability. (Source: Reproduced with permission from [35], Wiley (2003).)

Table 5.2 Electrical conductivity of EG- and GNP/thermoplastic composites prepared by different processes.

Polymer	Filler	Percolation threshold (vol%)	Critical exponent (t)	Matrix conductivity (S m^{-1})	Maximum composite conductivity (S m^{-1})	Processing method	References
PVDF	EG	6	2.25	4×10^{-11}	8×10^{-1} at 11.23 vol% filler	Melt compounding	Chapter 2, Ref. 63
PVDF	GNP	1.88	3.07	4×10^{-11}	3×10^{-2} at 4 vol% filler	Solution mixing	[33]
PVDF	GNP	1.01	1.97	–	1×10^{-2} at 3.12 vol% filler	Solution mixing	[36]
PS	GNP	1.10	3.07	6×10^{-11}	1.0 at 3.1 vol% filler	In situ polymerization	[37]

temperature employed for exfoliating the EG precursor [Chapter 2, Ref. 65]. It is noted that the rGO and TRG with only a few stacked graphene layers possess even higher aspect ratios than the GNP. Therefore, the percolation threshold of the graphene/polymer nanocomposites is very small, for example, in the range of 0.07–0.1 vol% (Table 5.1). Apparently, the more exfoliation of layered graphite

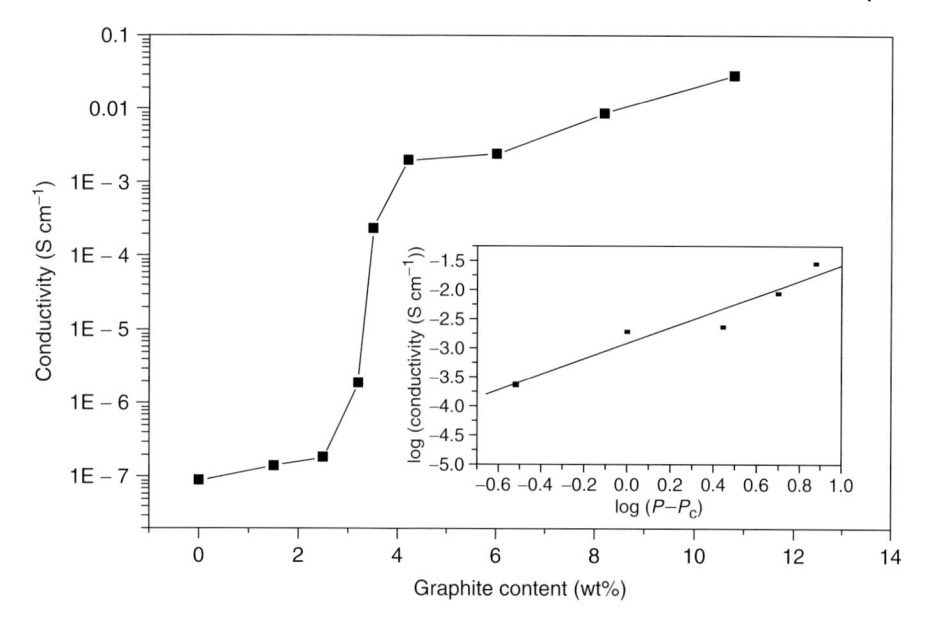

Figure 5.19 DC conductivity as a function of EG content of *in situ* polymerized EG/phenolic resin nanocomposites. (Source: Reproduced with permission from [38], Elsevier (2008).)

into independent graphene layers, the smaller percolation threshold of resulting nanocomposites is.

5.3.2.2 Thermosetting Matrices

Zhang et al. [38] studied the electrical behavior of the EG/phenolic resin nanocomposites prepared by *in situ* polymerization (Figure 5.19). Best fitting with the percolation conductivity equation gives $p_c = 3.2$ wt% and $t = 1.33 \pm 0.21$ (inset). The low value of $t = 1.33$ suggests that the charge carriers transport through a two-dimensional graphite surface. Recently, Li et al. used UV/ozone treatment to improve the filler–matrix interfacial bonding of the GNP/epoxy nanocomposites [Chapter 2, Ref. 21]. They reported that the electrical resistivity of treated nanocomposites decreases with increasing GNP content at a trend much faster than those containing untreated GNPs.

5.3.3
CNT- and CNF-Filled Polymer Composites

5.3.3.1 Thermoplastic Matrices

For the CNT/polymer nanocomposites, a wide range of values have been reported for the percolation threshold in the literature, depending on the types of nanofiller used, functionalization process, composite fabrication method, and the nature of polymer matrix [21, 39–51]. The dispersibility of electrically conductive nanofillers in an insulating polymer matrix affects the overall electrical conductivity of the polymer nanocomposites. The dispersion of CNTs in the polymer matrix by melt mixing is generally poorer than that obtained through solution mixing or *in situ* polymerization. Covalent functionalization of nanotubes is essential for facilitating dispersion of CNTs in the polymer matrix, thereby improving tensile properties of CNT/polymer nanocomposites. However, covalent functionalization is detrimental to the electrical conductivity of such nanocomposites. This is because covalent functionalization damages the $\pi-\pi$ conjugation network of the CNTs.

Figure 5.20a,b shows the electrical conductivity versus filler content for melt-compounded CNF/HDPE (high-density polyethylene) and MWNT/HDPE nanocomposites, respectively. The plateau regime can be readily seen in these figures and taken as the DC conductivity for both composite systems. For insulating samples, the value of the conductivity at the lowest measured frequency

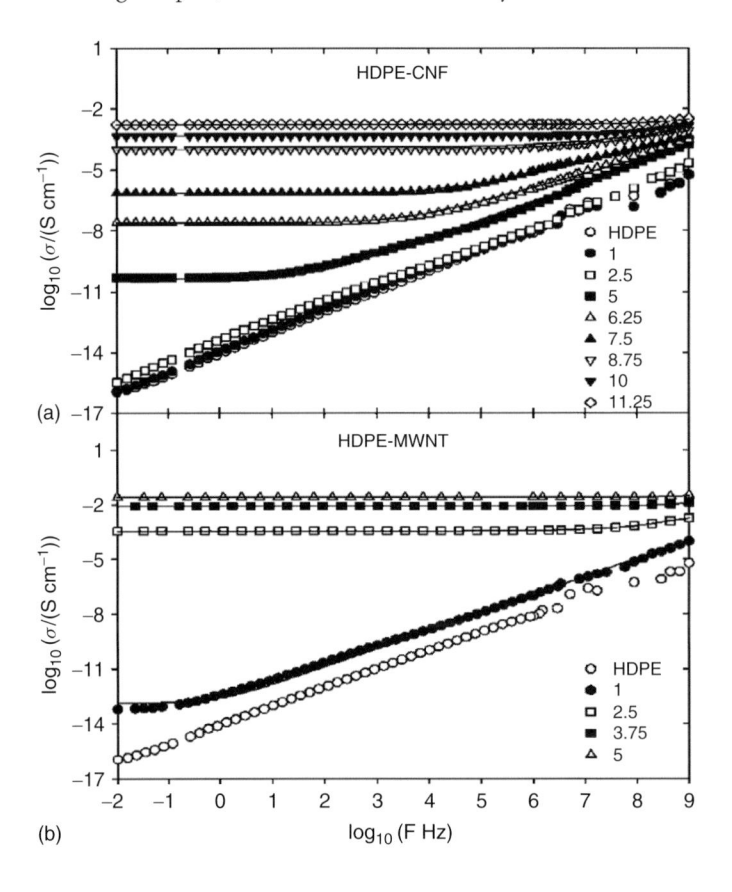

Figure 5.20 Broadband electrical conductivity as a function of frequency for melt-compounded (a) CNF/HDPE and (b) MWNT/HDPE nanocomposites. The filler volume content of both nanocomposites is given on the right-hand side of each figure. (Source: Reproduced with permission from [49], The American Chemical Society (2008).)

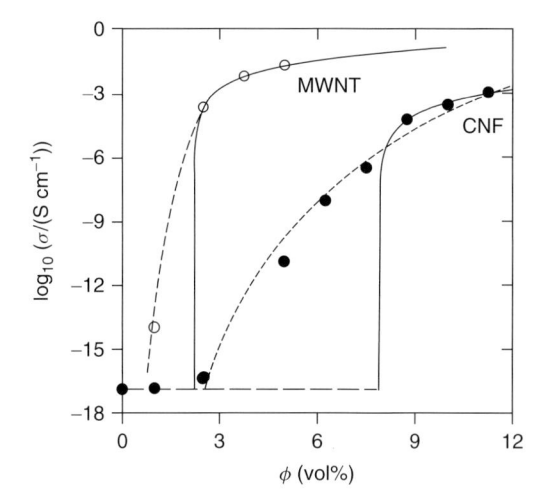

Figure 5.21 Semilogarithmic plot of the electrical conductivity versus filler content for melt-compounded CNF/HDPE (●) and MWNT/HDPE (○) nanocomposites. Solid curves are best fitting of Eq. (5.2) for experimental data. Dashed lines are the predictions according to tunneling theory. (Source: Reproduced with permission from [49], The American Chemical Society (2008).)

(10^{-2} Hz) is considered as σ_{dc} for the purpose of comparison. The plots of σ_{dc} versus filler content for CNF/HDPE and MWNT/HDPE nanocomposites are shown in Figure 5.21. The percolation threshold and conductivity exponent can be evaluated by fitting Eq. (5.2) with the experimental data, yielding $\Phi_c = 2.2$ vol% and $t = 1.8$ for the MWNT/HDPE and $\Phi_c = 7.8$ vol% and $t = 2$ for the CNF/HDPE

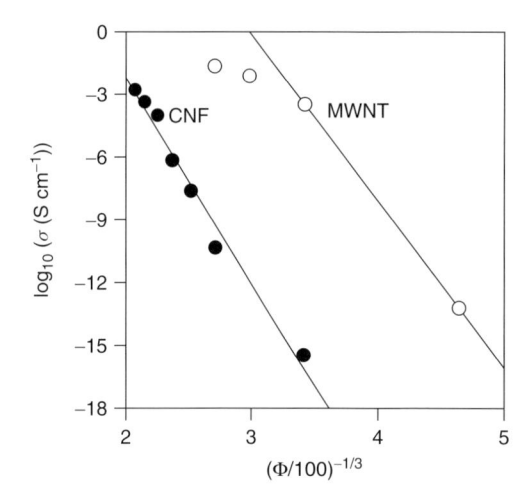

Figure 5.22 Logarithmic plot of conductivity versus $(\Phi/100)^{-1/3}$ for melt-compounded CNF/HDPE (●) and MWNT/HDPE (○) nanocomposites. (Source: Reproduced with permission from [49], The American Chemical Society (2008).)

nanocomposites, respectively. The higher percolation threshold of the CNF/HDPE nanocomposites is attributed to the less perfection in the crystalline structure of CNFs with lower aspect ratios. The transport behavior of electrons across the gaps between polymer-coated fillers can be well described by Eq. (5.7), that is, $\ln \sigma_{dc} \sim -\Phi^{-1/3}$, implying that tunneling conduction occurs in both composite systems (Figure 5.22).

In general, functionalized multiwalled carbon nanotube (FMWNT) and functionalized carbon nanofiber (FCNF) additions to polymers degrade their electrical conductivity markedly. Figure 5.23 shows σ_{cd} versus filler content for the MWNT/PVDF, carboxylated MWNT/PVDF, and CNF/PVDF nanocomposites prepared by solution mixing. The addition of FMWNT to PVDF decreases its DC conductivity. The corresponding log σ_{dc} versus $-\Phi^{-1/3}$ plots for these nanocomposite systems are shown in Figure 5.24. Recently, Dang et al. [47] modified MWNTs with 3,4,5 trifluorobromobenzene in order to improve their dispersion

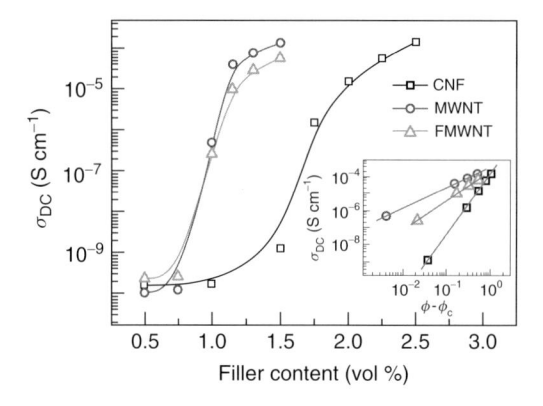

Figure 5.23 DC electrical conductivity versus filler volume content of PVDF nanocomposites filled with MWNT, functionalized MWNT (FMWNT), and CNF [43].

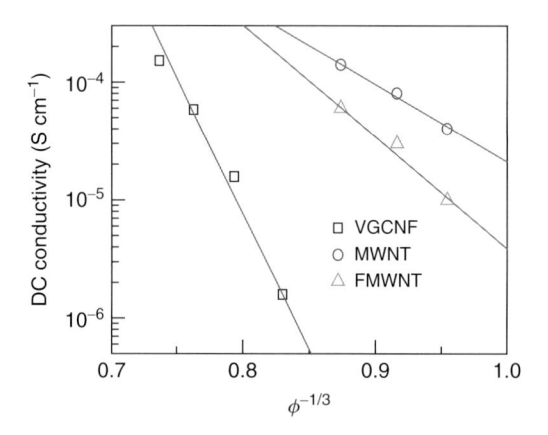

Figure 5.24 Semilog plot of σ_{dc} as a function of $\Phi^{-1/3}$ for PVDF nanocomposites filled with MWNT, functionalized MWNT (FMWNT), and CNF [44].

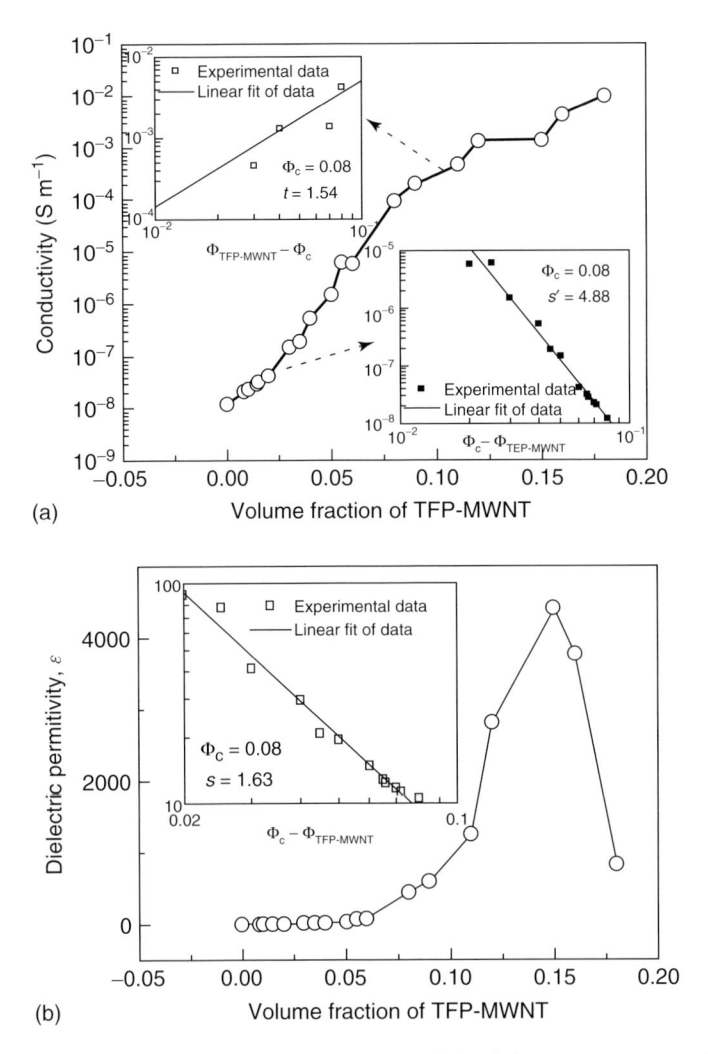

Figure 5.25 Variations of (a) conductivity and (b) dielectric permittivity with TFP-MWNT volume fraction for TFP-MWNT/PVDF nanocomposites. (Source: Reproduced with permission from [47], Wiley-VCH (2007).)

in PVDF. After treatment, trifluorophenyl (TFP)-functionalized MWNTs interact strongly with PVDF because of the fluoride groups formed on the nanotubes. The percolation threshold of solution-mixed FMWNT/PVDF nanocomposites is 8 vol%, and the maximum conductivity is $10^{-2}\,\text{S}\,\text{m}^{-1}$ at 17.5 vol% filler (Figure 5.25a). Despite the attainment of higher percolation threshold and lower conductivity for the FMWNT/PVDF system, the permittivity can reach as high as 5000 by adding 15 vol% FMWNT, that is, 500 times higher than that of pure PVDF (circa 10) (Figure 5.25b). The giant permittivity value originates from the trapping of charge carriers at the filler–matrix interfaces under an electric field.

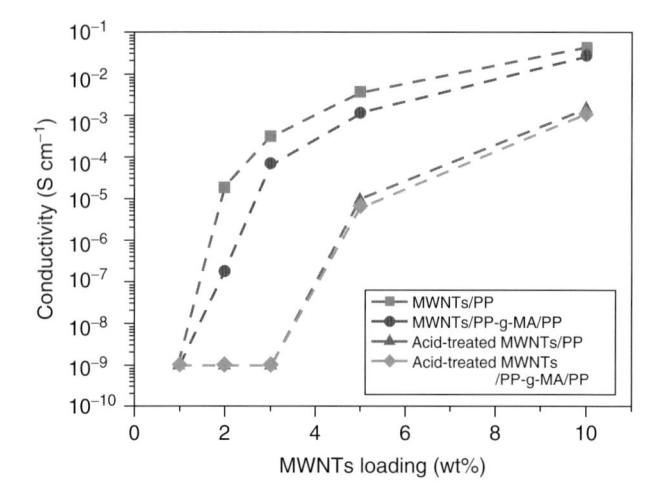

Figure 5.26 DC electrical conductivity versus filler content of pure PP and maleated PP composites filled with pristine MWNTs or acid-treated MWNTs. All nanocomposites were prepared by melt compounding. (Source: Reproduced with permission from [52], Elsevier (2010).)

Another example showing detrimental effects of covalently functionalized nanotube and compatibilizer additions on the electrical conductivity of PP is depicted in Figure 5.26 [52]. It can be seen that the MWNT/PP nanocomposites exhibit the highest electrical conductivity and the lowest percolation threshold. By incorporating PP-g-MA (maleic anhydride) into the PP matrix, the electrical conductivity is slightly reduced. The compatibilizer may hinder electron tunneling across the gaps between nanotubes being coated with an insulating layer of PP-g-MA. The electrical conductivity further decreases and the percolation threshold increases when acid-functionalized nanotubes are incorporated into PP.

Jeon et al. [53] studied the effect of matrix crystallinity on the electrical conductivity of semicrystalline PP and PP-ethylene random copolymers with 7.5, 14.7, and 20.8 mol% ethylene and filled with 0.15–1 wt% single-walled carbon nanotube (SWNT). These nanocomposites were prepared using solution mixing followed by precipitation in cold methanol. The degree of crystallinity of PP and PP copolymers determined from the differential scanning calorimetry (DSC) measurements is 70, 45, 23, and 10%. The nanotubes act as the nucleation sites for the polymer crystallites. Fast nucleation and growth of large PP crystallites on the surface of the nanotubes can prevent SWNT from clustering and curling. Long tube connectivity is found in the cryofractured surfaces of the PP/SWNT composites (not shown). Depletion of crystallites in the less crystalline matrices (<35% crystallinity) favors the formation of curly and poorly connected nanotubes. Figure 5.27a shows DC electrical conductivity versus SWNT content for PP and PP copolymer composites. Apparently, electrical conductivity decreases markedly by decreasing the crystallinity of PP matrix through the addition of ethylene. The conductivity decreases over 4 orders of magnitude for PP copolymers filled with 0.3 wt% SWNT. A reduction in the

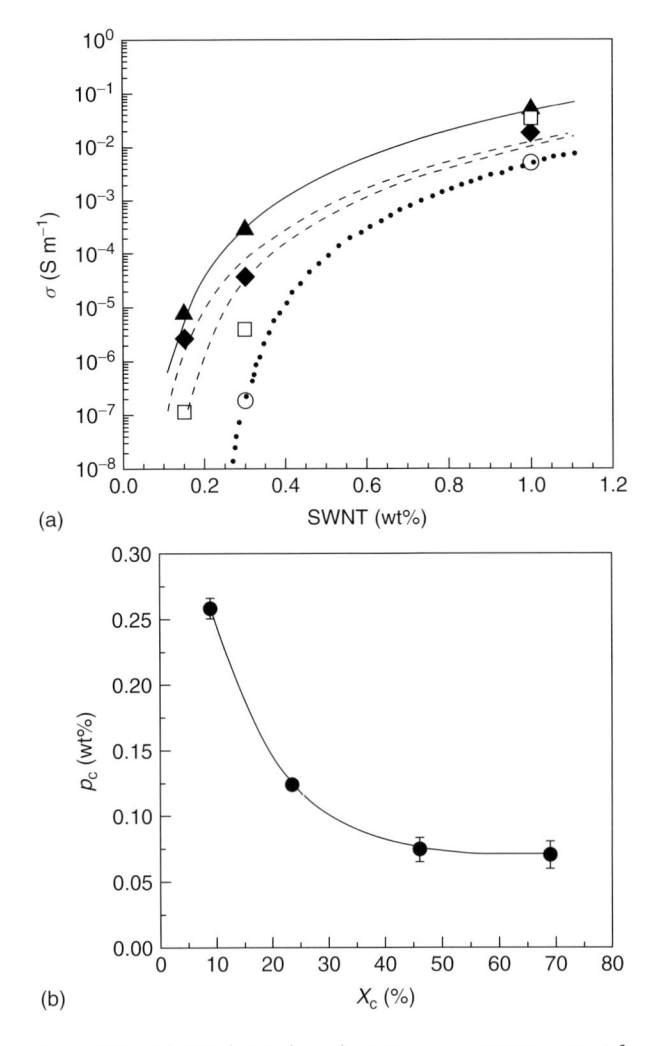

Figure 5.27 (a) DC electrical conductivity versus SWNT content for pure PP and PP copolymer nanocomposites. ▲: PP, ◆: PP-7.5 mol% ethylene, □: PP-14.7 mol% ethylene, and ○: PP-20.8 mol% ethylene. (b) Percolation threshold for electrical conductivity as a function of crystallinity. (Source: Reproduced with permission from [53], Wiley (2010).)

matrix crystallinity results in a gradual loss of SWNT connectivity and an increase in pc (Figure 5.27b). This leads to a decrease of electrical conductivity with increasing ethylene content in PP. Table 5.3 summarizes percolating characteristics of several CNT/thermoplastic and CNF/thermoplastic nanocomposites. Apparently, the percolation threshold of melt-compounded thermoplastics filled with pristine MWNTs is considerably higher than that of the composites prepared by solution mixing and *in situ* polymerization. Further, pristine nanotubes are preferred since functionalized nanotubes further degrade electrical conductivity of the nanocomposites.

Table 5.3 Electrical conductivity of CNT/thermoplastic composites prepared by different processes.

Polymer	Filler	Percolation threshold (vol%)	Critical exponent t	Matrix conductivity (S m^{-1})	Maximum composite conductivity (S m^{-1})	Processing method	References
PET	MWNT	0.10	2.9	–	2.83×10^{-1} at 2.5 vol% filler	In situ polymerization	Chapter 2, Ref. 98
PI	CNF	0.24	3.1	10^{-16}	3 at 3.5 vol% filler	In situ polymerization	[48]
PI	FCNF	0.68	3.2	10^{-16}	2×10^{-2} at 7.0 vol% filler	In situ polymerization	[48]
PS	MWNT	0.50	3.80	10^{-14}	10^{-1} at 2.0 vol% filler	Solution mixing	[30]
PVDF	MWNT	0.99	1.18	–	1.41×10^{-2} at 1.5 vol% filler	Solution mixing	[43]
PVDF	FMWNT	0.98	1.72	–	5.97×10^{-3} at 1.5 vol% filler	Solution mixing	[43]
PVDF	CNF	1.46	3.53	–	1.51×10^{-2} at 2.5 vol% filler	Solution mixing	[43]
PVDF	FMWNT	3.8	1.83	–	$\sim 8 \times 10^{-2}$ at 6.0 vol% filler	Solution mixing	[45]
PVDF	FMWNT	8.0	1.54	–	10^{-2} at 17.5 vol% filler	Solution mixing	[47]
HDPE	MWNT	2.2	1.8	$\sim 10^{-15}$	$\sim 10^{-1}$ at 5.0 vol% filler	Melt compounding	[49]
HDPE	CNF	7.8	2.0	$\sim 10^{-15}$	$\sim 10^{-1}$ at 11.25 vol% filler	Melt compounding	[49]
PA6	MWNT	1.7	8.4	–	1.34×10^{-2} at 14.1 vol% filler	Melt compounding of masterbatch	Chapter 2, Ref. 97

CNTs are always dispersed in wavy or coiled tube morphologies rather than long straight tubes in the matrix of polymer composites. The electrical properties of randomly oriented and coiled nanotubes are considerably lower than those of straight tubes [54]. Aligned CNTs can be synthesized from plasma-enhanced chemical vapor deposition (CVD) process [Chapter 1, Refs. 57–60]. For synthesizing aligned

CNT/thermoplastic nanocomposites, AC electric field is found to be effective to induce nanotube alignment during solution mixing or *in situ* polymerization. For example, Ma *et al.* [55] aligned pristine and oxidized MWNTs in the poly(methyl methacrylate) (PMMA) matrix using an AC electric field of $15\,\text{kV}\,\text{m}^{-1}$ at 500 Hz during *in situ* polymerization. More recently, Oliva-Aviles *et al.* [56] also employed an AC electric field to form aligned MWNT/polysulfone (PSF) nanocomposite film during solution mixing. PSF is an engineering thermoplastic with high thermal stability and low moisture absorption characteristics. Figure 5.28 shows optical micrographs of solution-mixed MWNT/PSF nanocomposite films fabricated with or without assistance of AC electric field (E_{AC}). For the composites prepared without the electric field, coiled MWNTs are randomly dispersed in the polymer matrix. For the nanocomposite films prepared with E_{AC}, most MWNTs align in the direction of the electric field in straight tube features. Figure 5.29 shows DC electrical conductivity as a function of MWNT content for the MWNT/PSF nanocomposites fabricated with and without E_{AC}. PSF is an insulator with a conductivity of

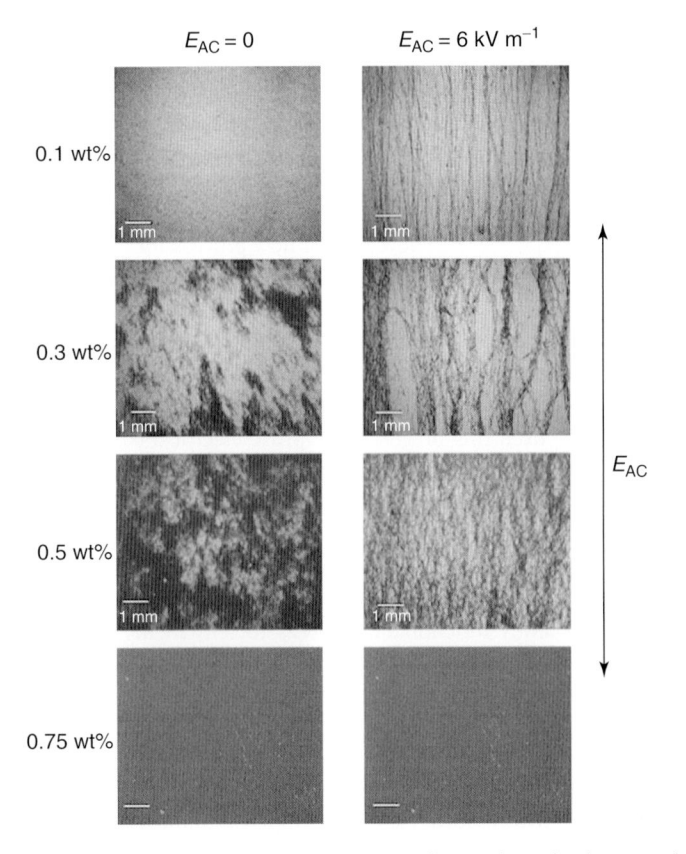

Figure 5.28 Optical images of MWNT/PSF films with randomly oriented and aligned MWNTs. The MWNT content of the composite films is given on the left-hand side. (Source: Reproduced with permission from [56], Elsevier (2011).)

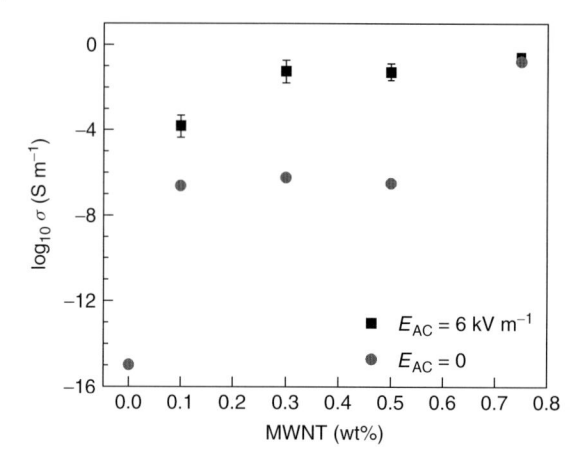

Figure 5.29 Electrical conductivity versus MWNT content of MWNT/PSF nanocomposite films fabricated with and without the application of AC electric field. (Source: Reproduced with permission from [56], Elsevier (2011).)

1.55×10^{-15} S m^{-1}. At 0.1 wt% MWNT, the conductivity of composite film with aligned MWNTs is 3 orders of magnitude higher than that of composite with randomly oriented nanotubes. By increasing MWNT content to 0.3 wt% above, the conductivity of aligned composite films is 5 orders of magnitude higher than that of the composites with randomly oriented nanotubes.

Very recently, Peng's group [57, 58] has developed a new process for fabricating aligned CNT/polymer nanocomposites. They first synthesized CNT arrays on silicon by means of CVD process (Figure 5.30). Uniform CNT sheets were then pulled out of the arrays and stabilized on the glass. Composite film was finally obtained by spin coating or casting polymer solutions onto the CNT sheets, followed by evaporation of solvent. The film can be readily pulled out of the substrate. Peng and Sun [58] used this process to improve electrical conductivity of aligned sulfonated PEEK/CNT (poly(ether ether ketone)) composite from 22 to 66.7 S cm^{-1} at room temperature by using longer CNTs.

5.3.3.2 Thermosetting Matrices

On the basis of the discussion in Section 5.3.3.1, it is clear that pristine CNTs are more effective than covalently functionalized CNTs for enhancing electrical conductivity of polymer nanocomposites. The exceptional case is the arc-grown CNTs containing a high level of amorphous carbon impurities. Figure 5.31 shows electrical conductivity as a function of filler content for epoxy-based composites filled with CVD-MWNTs, aligned-CVD-MWNTs, and arc-grown MWNTs [59]. The aligned-CVD-MWNTs exhibit a carpetlike morphology with a mean diameter of ~40 nm and a length of ~50 µm (Figure 5.32). The length of aligned nanotubes is kept similar to that of the CVD-MWNTs during synthesis for comparison purposes. From Figure 5.31, the arc-grown MWNT/epoxy composites show insulating behavior in the tested filler content regime because of their lower length (<1 µm)

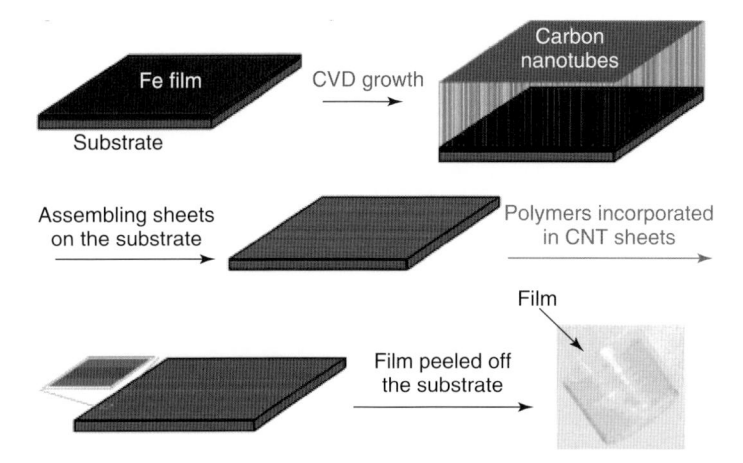

Figure 5.30 Schematic diagrams showing synthesis of aligned carbon nanotube/polymer composite film. (Source: Reproduced with permission from [57], The American Chemical Society (2008).)

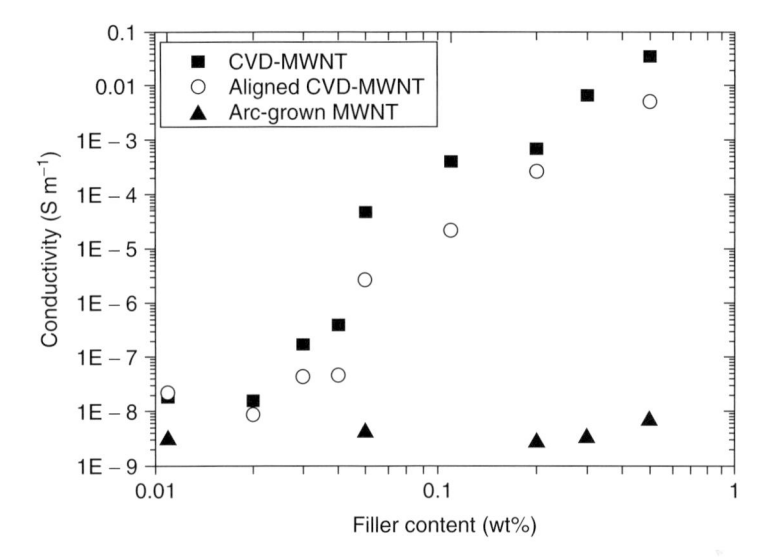

Figure 5.31 Percolation curves of epoxy composites filled with CVD-MWNT (■), aligned-CVD-MWNT (○), and arc-grown MWNT (▲). (Source: Reproduced with permission from [59], Elsevier (2010).)

and the presence of carbonaceous impurities. In contrast, both CVD-MWNT and aligned-CVD-MWNT nanocomposites display typical electrical percolation characteristics. The results for both composites are similar because the length of aligned-CVD-MWNTs is maintained almost the same as that of CVD-MWNTs. In general, polymer nanocomposites with aligned nanotubes are more conductive than those with entangled ones. The incorporation of aligned MWNTs into epoxy

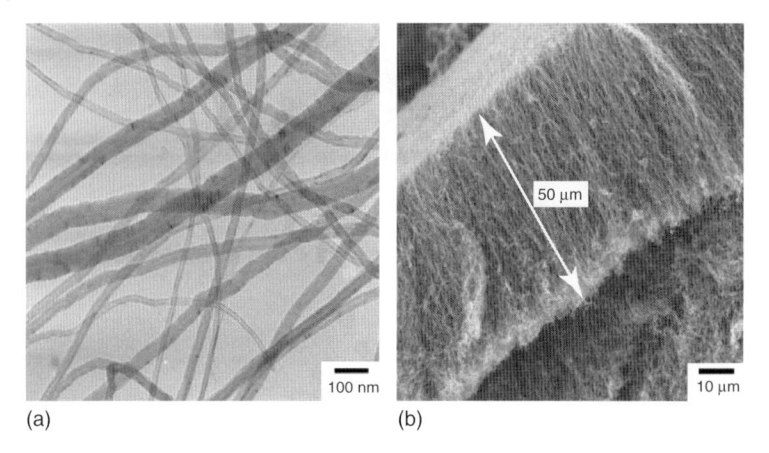

Figure 5.32 (a) TEM and (b) SEM images of aligned-CVD-MWNTs. (Source: Reproduced with permission from [59], Elsevier (2010).)

resin can lead to polymer nanocomposites that exhibit a very low percolation threshold of 0.0025 wt% [60].

The dispersion of CNTs in thermosetting resins also depends greatly on the nanotube functionalization. Acid oxidation treatment of CNTs degrades their electrical conductivity, and the degree of degradation depends on the oxidation time and temperature. Kim et al. [61] studied the effect of different oxidation conditions on the electrical conductivity of solvent-cast MWNT/epoxy nanocomposites. Table 5.4 lists the oxidation conditions for MWNTs in nitric acid and basic solution (H_2O_2/NH_4OH). These treatments result in the formation of carboxyl groups on the MWNT surfaces. The MWNTs oxidized under severe conditions are well purified, but their crystalline structure is partially damaged. The functionalized nanotubes are generally well dispersed in the epoxy matrix. Figure 5.33a,b shows the representative plots of DC conductivity versus nanotube content of A4-MWNT/epoxy and B0-MWNT/epoxy nanocomposites. The percolation threshold and critical exponent of functionalized MWNT/epoxy nanocomposites can be determined using the scaling law [$\sigma = \sigma_0(\Phi - \Phi_c)^t$]. The results are listed in Table 5.5. This table reveals

Table 5.4 Oxidation conditions for MWNTs.

MWNT designation	Solution	Concentration (%)	Treatment temperature (°C)	Treatment time (h)	pH
A0	HNO_3	28.5	Room temperature	4	3.5
A1	HNO_3	40.0	100	1	2.5
A2	HNO_3	40.0	100	2	2.5
A4	HNO_3	40.0	100	4	2.5
B0	H_2O_2/NH_4OH	28.5	Room temperature	4	10.5

Source: Reproduced with permission from [61], Elsevier (2005).

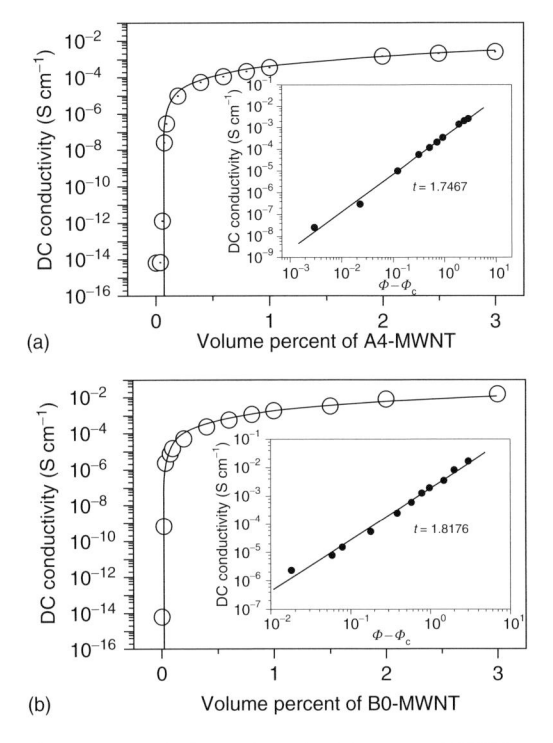

Figure 5.33 DC electrical conductivity versus MWNT content for (a) A4-MWNT/epoxy and (b) B0-MWNT/epoxy nanocomposites. (Source: Reproduced with permission from [61], Elsevier (2005).)

Table 5.5 Percolation parameters of FMWNT/epoxy nanocomposites.

Composites	σ_0 (S cm^{-1})	Φ_c (vol%)	t
A0-MWNT/epoxy	1.9×10^{-4}	0.017 ± 0.06	1.71 ± 0.03
A1-MWNT/epoxy	1.5×10^{-4}	0.024 ± 0.03	2.01 ± 0.04
A2-MWNT/epoxy	6.0×10^{-5}	0.028 ± 0.05	1.74 ± 0.06
A4-MWNT/epoxy	4.8×10^{-3}	0.077 ± 0.02	1.75 ± 0.05
B0-MWNT/epoxy	2.0×10^{-3}	0.021 ± 0.02	1.83 ± 0.06

Source: Reproduced with permission from [61], Elsevier (2005).

that the conductivity of solvent-cast MWNT/epoxy nanocomposites decreases as the oxidation time and temperature of nitric acid increase. The extent of structural damage in nanotubes increases with increasing immersion time and temperature of nitric acid. Moreover, oxidation MWNTs in a mixed H_2O_2/NH_4OH solution renders the epoxy nanocomposites with highest conductivity and low percolation threshold. It is noted that the homogeneous dispersion of MWNTs in the epoxy matrix enables the formation of a conductive path network at very low percolation threshold, despite the fact that the acid treatment induces structural defects in the nanotubes. The percolation threshold of FMWNT/epoxy nanocomposites is much lower than that of solvent-cast MWNT/thermoplastic composites.

As mentioned previously, noncovalent functionalization via the $\pi-\pi$ interaction can prevent the formation of structural defects in CNTs. Geng et al. [62] studied the effect of surfactant treatment on the electrical conductivity of MWNT/epoxy nanocomposites. Nonionic surfactant, that is, polyoxyethylene octyl phenyl ether (Triton X-100), was used to disperse MWNTs in acetone. At a critical micelle concentration (CMC), the nanotube surface was saturated with the surfactant. Above the CMC, micelles began to develop and adsorb on the nanotube surface. The nanotube suspension was then mixed with epoxy resin under sonication, followed by degassing in a vacuum oven (80 °C) and mold casting. Figure 5.34 shows electrical conductivity of pristine MWNT/epoxy and surfactant-treated MWNT/epoxy nanocomposites. For the purpose of comparison, the electrical conductivity behavior of silane-treated MWNT/epoxy composites is also shown in this figure. Both the pristine and surfactant-treated MWNT/epoxy nanocomposites display a percolation transition from insulating to conducting behavior. This implies that the adsorption of Triton surfactant on the nanotube surface does not disturb the charge-transfer process in the conducting path network. However, silane-treated MWNT/epoxy composites show insulating behavior as evidenced by their low electrical conductivity. The silane molecules with epoxy end groups are covalently bonded to the nanotube surfaces, which react with the amine hardener, leading to

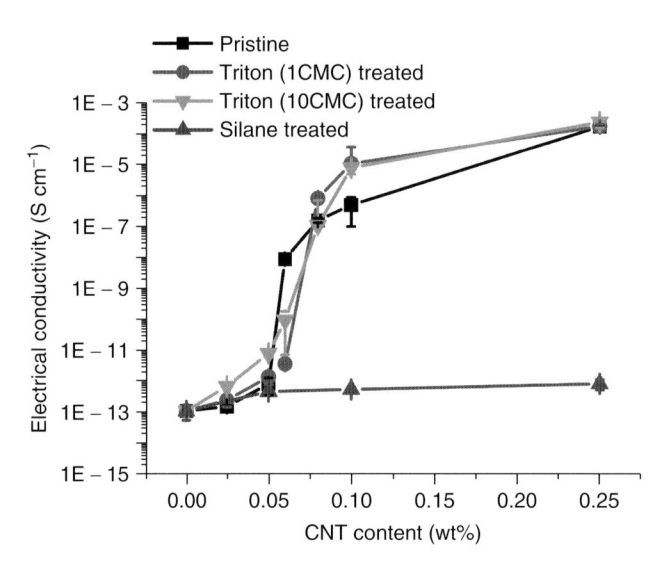

Figure 5.34 Electrical conductivity versus MWNT content for epoxy nanocomposites filled with pristine, Triton-treated, and silane-treated MWNTs. (Source: Reproduced with permission from [62], Elsevier (2008).)

the coating of MWNTs with epoxy. Such coated nanotubes limit the transport of electrons via tunneling effect, thereby producing very low electrical conductivity.

Considering the beneficial effect of aligned CNT addition for enhancing electrical conductivity of epoxy resin, some techniques have been developed to prepare aligned CNT/epoxy nanocomposites. A simple practice involves an initial synthesis of aligned CNT arrays/forests followed by infiltrating with the polymer resin [60, 63, 64]. Alternatively, Cheng et al. [65] prepared the epoxy/CNT nanocomposites using resin transfer molding (RTM) technique. In the process, CNT arrays were first grown on a silicon wafer. CNTs were then drawn from aligned arrays and joined end to end to form a continuous CNT sheet. Several CNT sheets were stacked together at different orientations to produce a preform followed by injecting epoxy resin into the RTM. Figure 5.35a shows schematic illustrations and scanning electron microscopy (SEM) images of the CNT sheet preforms with [0] and [0/90] orientations. The schematic diagrams of the RTM process for making CNT/epoxy composites are shown in Figure 5.35b. The specimens containing 2000 and 4000 CNT sheets were stacked in the same direction and were labeled as $[0]_{2000}$ and $[0]_{4000}$, respectively. The MWNT concentrations in the epoxy/$[0]_{2000}$, epoxy/$[0/90]_{2000}$, and epoxy/$[0]_{4000}$ composites are determined to be 9.94, 10.32, and 16.5 wt%,

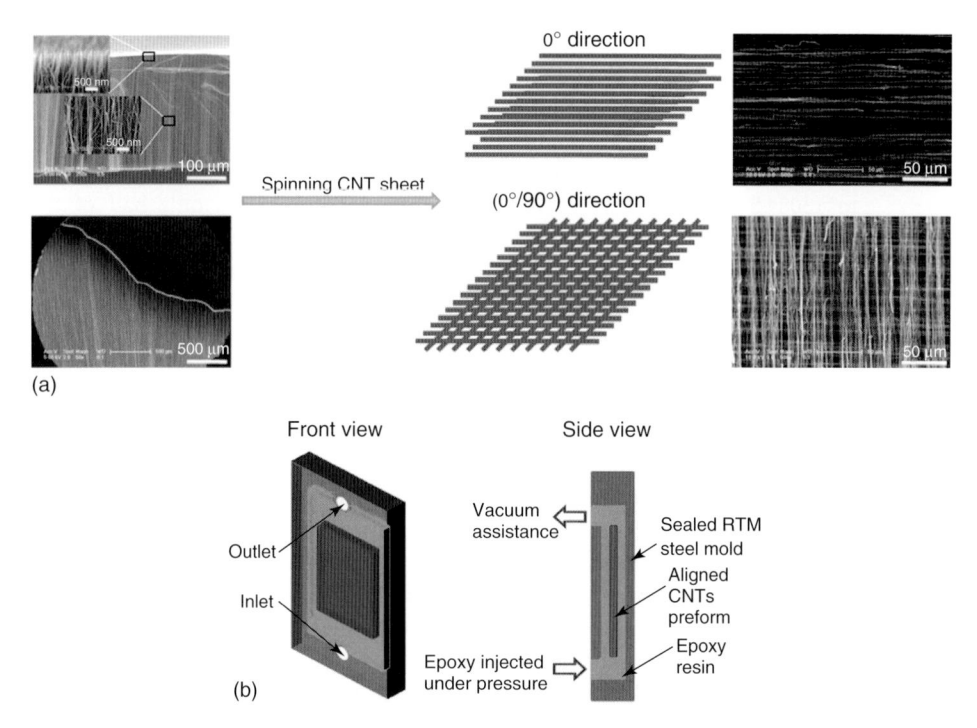

Figure 5.35 (a) Schematic diagrams and scanning electron micrographs of CNT performs with [0] and [0/90] alignment of CNT sheets. (b) Schematic illustrations of the RTM process for fabricating CNT/epoxy composites. (Source: Reproduced with permission from [65], Elsevier (2010).)

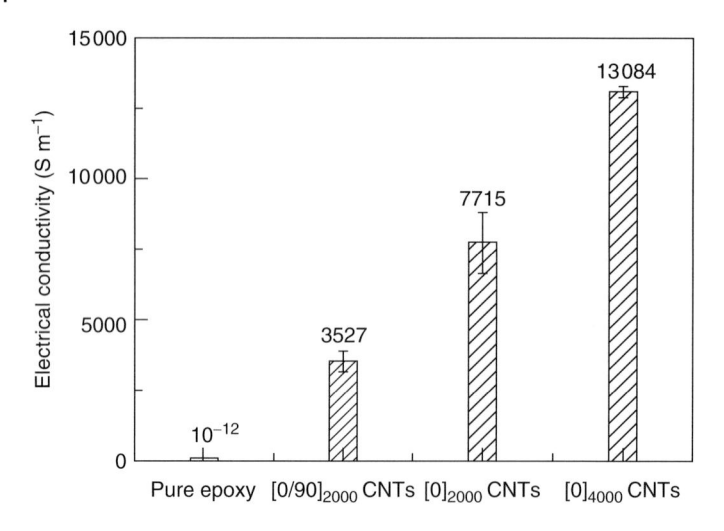

Figure 5.36 Electrical conductivities of pure epoxy and CNT/epoxy composites containing different CNT contents and alignments. (Source: Reproduced with permission from [65], Elsevier (2010).)

respectively. Figure 5.36 shows the electrical conductivities of pure epoxy and MWNT/epoxy composites with different CNT contents. The electrical conductivity of pure epoxy is 10^{-12} S m^{-1} and increases to 7715 S m^{-1} by incorporating $[0]_{2000}$. By increasing CNT layers to 4000, the electrical conductivity is significantly enhanced to 13084 S m^{-1}. For the MWNT/epoxy composite $[0/90]_{2000}$, only half of the MWNT sheets align along the measurement direction. The electrical conductivity is 3527 S m^{-1}, being about half of the electrical conductivity of the composite $[0]_{2000}$. It is obvious that the MWNT/epoxy composites with controllable alignment of nantubes exhibit extremely large electrical conductivity. Such nanocomposites find potential application as bipolar plates of fuel cells that require high electrical conductivity of larger than 100 S cm^{-1}.

5.3.3.3 Elastomeric Matrices

Elastomers are often reinforced with inorganic fillers and carbon blacks (CBs) in order to achieve substantial improvements in strength, stiffness, and electrical conductivity. Conductive elastomer composites with flexible mechanical properties find attractive application as materials for artificial muscles, pressure actuators, and sensors. The extent of property improvement depends on the functionalization and dispersion of nanotubes and fabrication techniques employed. Jang *et al.* [66] reported that covalent functionalization of MWNTs through acid treatment degrades the electrical conductivity of MWNT/PU nanocomposites. The degree of deterioration in electrical conductivity and percolation threshold of the MWNT/PU nanocomposites is directly related to the exposure time in both nitric acid and a mixture of nitric/sulfuric acids at 80 °C. Excessive acid treatment damages the

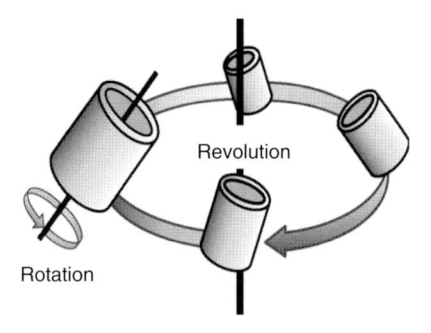

Figure 5.37 Schematic diagram of a rotation–evolution mixing facility. (Source: Reproduced with permission from [68], Elsevier (2011).)

crystalline structure of nanotubes, and the resulting composites show poor electrical behavior despite well dispersion of MWNTs in the polymer matrix.

Pedroni et al. [67] reported that solution-cast MWNT/styrene-butadiene-styrene composites exhibit a low percolation threshold of 0.50 wt%. More recently, Tsuchiya et al. [68] employed a rotation–revolution mixing technique without mechanical shearing to fabricate MWNT/styrene-butadiene rubber (SBR) composites (Figure 5.37). This solution-mixing technique was specially designed to tackle the difficulty of nanotube dispersion in highly viscoelastic rubber matrix. The mixer takes the advantage of a centrifugal force of about 400g, generating from a combination of rotation and revolution with opposite direction for blending rubber and pristine nanotubes. During the composite fabrication, pristine MWNTs and SWNTs from different sources and SBR were independently dispersed in tetrahydrofuran (THF). Sonicated nanotube suspensions were then mixed with the SBR/THF solution, followed by adding ZnO, sulfur, stearic acid, rubber aging indicator, and vulcanization accelerator. Mixing was conducted in a rotation–revolution mixer for 30 min. Figure 5.38a–d shows DC conductivity versus filler content for SBR composites filled with different MWNTs and SWNTs, respectively. The percolation threshold of these nanocomposites is <1 phr, especially for the composites with SWNTs (Super Growth) with the p_c value of 0.27 phr.

5.4
Current-Voltage Relationship

The applied electric field (E) has a large influence on the electrical conductivity of conducting polymer composites. At a small applied electric field, the current–voltage (I–V) response of the composites is usually linear. However, the response of a variety of percolating composite becomes nonlinear under the influence of a high electric field [69]. In this context, the I–V response of polymer composites can be described by the following equation [70]

$$I \propto V^n \tag{5.18}$$

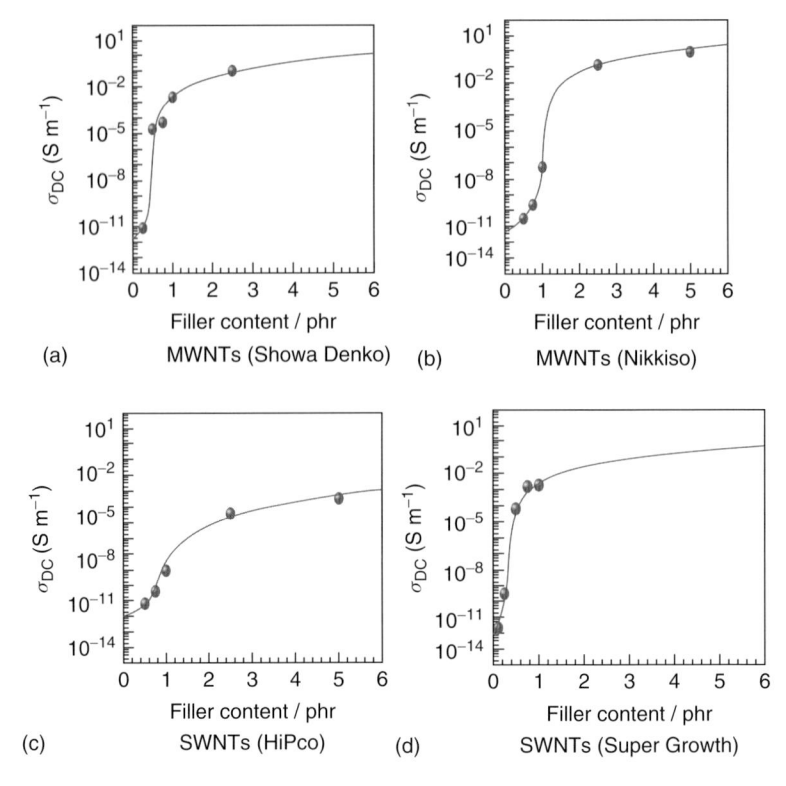

Figure 5.38 DC electrical conductivity versus filler content for SBR composites filled with (a) MWNTs (Showa Denko), (b) MWNTs (Nikkiso), (c) SWNTs (HiPco), and (d) SWNTs (Super Growth; National Institute of Advanced Materials Science and Technology, Japan). (Source: Reproduced with permission from [68], Elsevier (2011).)

where n is the nonlinearity exponent. For $n = 1$, the composites obey typical ohmic conduction behavior. However, the n values often deviate from unity for several polymer composite systems [71, 72]. The current–voltage relation of Eq. (5.18) can be expressed in terms of the current density ($J = I/A$ where A is the area) and electric field

$$J \propto E^\alpha \tag{5.19}$$

where α is the nonlinear coefficient.

Figure 5.39a shows the $I-V$ response for *in situ* polymerized GNP/PA6 nanocomposites [73]. The corresponding $J-E$ response is shown in Figure 5.39b. The current varies linearly with voltage at low-voltage regime. Above a crossover voltage (V_c) or current (I_c), a nonlinear behavior is observed. The nonlinearity is more pronounced for the nanocomposites, with their filler content approaching percolation threshold. Recently, Arlen et al. [48] studied the electrical and thermal behavior of *in situ* polymerized CNF/PI (polyimide) and FCNF/PI nanocomposites. Figure 5.40a shows the $I-V$ response of the 5 wt% (3.5 vol%) CNF/PI and 5 wt% FCNF/PI composites

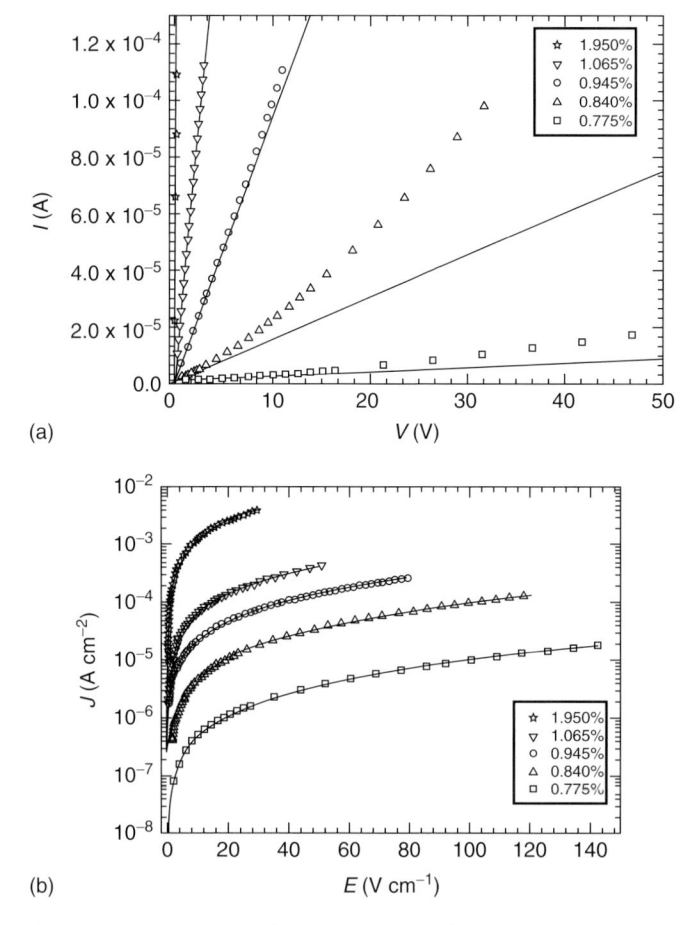

Figure 5.39 (a) I–V and (b) J–E responses of PA6/GNP nanocomposites filled with various GN contents. The straight line in (a) corresponds to linear portion of the curve. The GNP volume contents are given in insets. (Source: Reproduced with permission from [73], Wiley (2004).)

subjected to low applied voltages. The concentration of these composites is above the percolation threshold (Table 5.3). Both samples exhibit an ohmic response, that is, linear and symmetric I–V behavior from −0.2 to +0.2 V. The current value of the 5 wt% CNF/PI nanocomposite is 9 orders of magnitude higher than that of the 5 wt% FCNF/PI nanocomposite as expected. A nonlinear behavior can be observed in the I–V response of the 5 wt% FCNF/PI nanocomposite at high-voltage regime (Figure 5.40b). This implies that mechanism limiting transport is field dependent, such as hopping or tunneling across a narrow insulating gap [74].

Gefen et al. [74] studied nonlinear electrical transport in thin gold films. They proposed two phenomenological models to explain nonlinear current–voltage behavior, that is, nonlinear random resistor network (NLRRN) and dynamic random resistor network (DRRN). The NLRRN model describes a random resistor

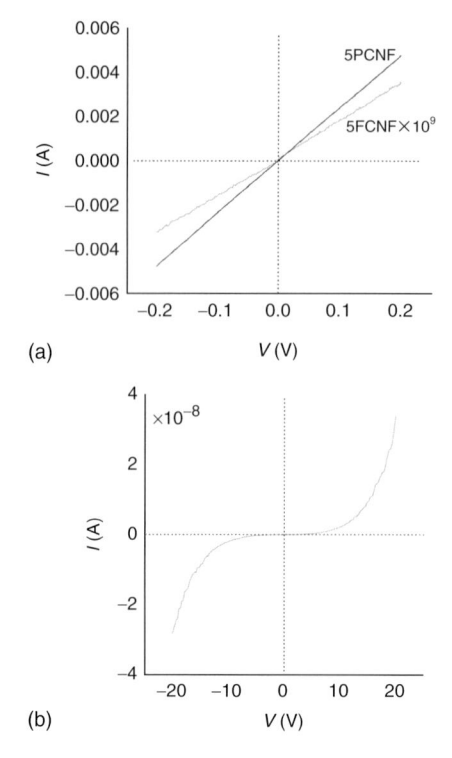

Figure 5.40 Current–voltage responses at room temperature for (a) 5 wt% pristine CNF/PI (denoting as 5PCNF) and 5 wt% FCNF/PI (denoting as 5FCNF) nanocomposites at low-voltage regime and (b) 5FCNF under high applied voltages. (Source: Reproduced with permission from [48], The American Chemical Society (2008).)

network, each element of which has a small nonlinear contribution that manifests itself when the field rises above a critical value. The DRRN model describes a network of lattice bonds with ohmic characteristics, but comprising a certain number of insulating channels. These originally insulating channels may become conducting under the influence of a sufficiently strong local field. This transition is due to the hopping or tunneling of charge carriers. Empirically, the electrical response can be described by the following expression [75, 76]

$$I = GV + G'V^b \tag{5.20}$$

where b is a constant with a value of 2 or 3; G and G' are the linear and high-order conductances, respectively. By using the current density expression, Eq. (5.20) becomes

$$J = \sigma E + \sigma' E^b \tag{5.21}$$

where σ' is the high-order conductivity. From Figure 5.39b, Chen et al. [73] reported that the relationship between J and E can be well fitted with $b = 2$.

For narrow insulating gaps between conducting nanofillers, very high field strength may develop, leading to electron tunneling across the fillers within the composite [12]. Considering various tunneling effects in a disordered system [77], internal field emission (IFE) mechanism may take place. This concept was first developed by Zener in the 1930s [78], and Shockley then extended it to semiconductors [79, 80]. Under IFE conditions, the associated current density j_{IFE} is given by the following expression

$$j_{IFE} = AE^n \exp(-B/E) \qquad (5.22)$$

where A, B, and n are constants; the value of n usually lies between 1 and 3. A is a function of tunneling frequency, that is, the number of attempts per second made by the carrier to cross the barrier, and B is a measure of the energy barrier between the insulating matrix and the filler material. It is generally known that the "Fowler-Nordheim" tunneling occurs at $n = 2$. In 1964, van Beek and van Pul reported that the nonohmic current–voltage relationship of CB/rubber composites can be explained by tunneling effects [81]. In this context, Eq. (5.22) fits reasonably well with the experimental data of those composites.

Taking the effect of IFE effect into consideration, the overall conductivity becomes $\sigma = \sigma_0 + \sigma_{IFE}(E)$, where σ_0 is the linear conductivity and electric field dependent. Therefore, the current density under the application of an external field can be expressed as

$$j(E) = \sigma_0 E + AE^n \exp(-B/E) \qquad (5.23)$$

More recently, He and Tjong [82, 83] demonstrated that electron tunneling in the CNF/HDPE nanocomposites under an applied electric field can be well described by the IFE mechanism. The filler contents of this composite system were fixed at 2.5, 2.6, 2.7, and 3.0 vol%, higher than the percolation threshold of 0.94 vol% [82]. Figure 5.41a–d shows the J–E responses for CNF/HDPE

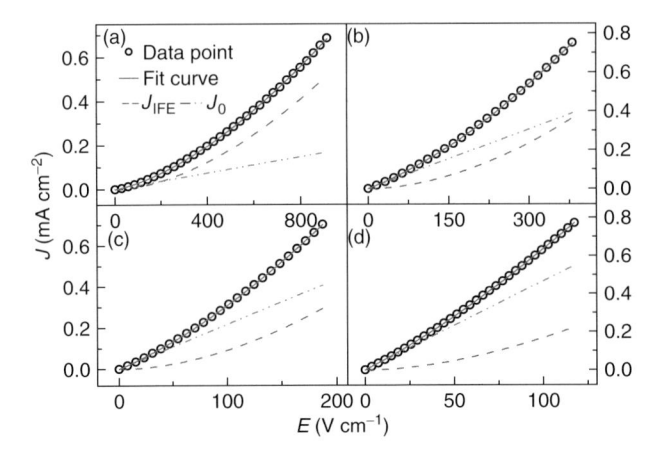

Figure 5.41 J–E responses of HDPE composites filled with (a) 2.5 vol%, (b) 2.6 vol%, (c) 2.7 vol%, and (d) 3 vol% CNF [83].

Table 5.6 Parameters characterizing $J-E$ characteristics of melt-compounded CNF/HDPE nanocomposites [83].

Nanocomposites	σ_0 (mS cm^{-1})	A	B (V cm^{-1})	n
2.5 vol% CNF/HDPE	1.83×10^{-4}	4.56×10^{-6}	22.61	1.71
2.6 vol% CNF/HDPE	1.02×10^{-3}	7.00×10^{-6}	6.33	1.83
2.7 vol% CNF/HDPE	2.17×10^{-3}	1.79×10^{-5}	3.09	1.86
3 vol% CNF/HDPE	4.66×10^{-3}	8.08×10^{-5}	9.12	1.68

nanocomposites. It is obvious that Eq. (5.23) fits well with numerical data of the HDPE/CNF nanocomposites. Therefore, the nonlinearity in the *j-responses* of these nanocomposites arises from the j_{IFE}. The fitted parameters are summarized in Table 5.6.

5.5
Positive Temperature Coefficient Effect

Apart from the applied electric field, temperature is another factor that affects the electrical conductivity of polymer nanocomposites [40, 84]. Conductive polymer composites usually exhibit a sharp increase in electrical resistivity at elevated temperatures close to the polymer melting point. This is often termed as the positive temperature coefficient (PTC) effect. On the contrary, a sudden decrease in resistivity with temperature is referred to as the negative temperature coefficient (NTC) effect. The PTC behavior is especially pronounced for the composites with semicrystalline polymer matrices. Because of the sharp increase in electrical resistivity, PTC composites find a wide range of industrial applications including switching devices, self-regulating heaters, current limiters, and fuses. For these applications, composite materials with high PTC intensity are needed because they must exhibit a fast response to the surrounding temperature fluctuations. On the other hand, conductive composite materials designed for electromagnetic interference shielding and electrostatic discharge protection should have a small PTC effect to ensure stable electrical performance.

The PTC effect is rather complicated and several models have been proposed for explaining it including the conducting path network, tunneling effect, electric field emission, and thermal expansion effect [85–89]. The latter is attributed to the difference of thermal expansion coefficients between the polymer matrix and conductive fillers. The increase in resistivity is due to the widening of the gaps between the filler clusters in the conducting path network and the change of the uniformity of the gaps. The NTC effect arises from the reaggregation or redistribution of conductive fillers above the polymer melting point. Several factors such as the type, size and content of fillers, dispersion of fillers, interactions between the polymer and fillers, and intrinsic nature of polymer can affect the PTC

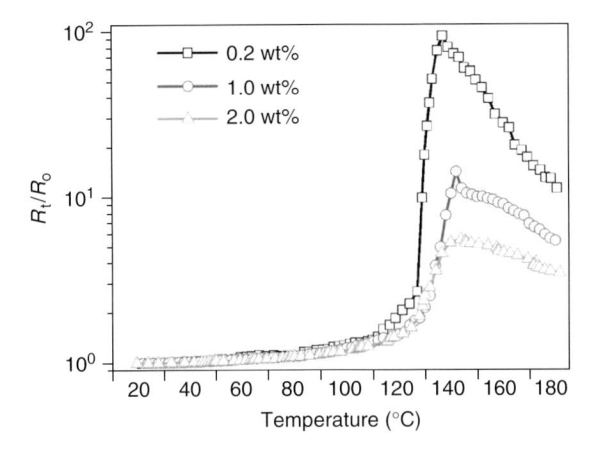

Figure 5.42 Temperature dependence of relative resistivity of MWNT/UHMWPE nanocomposites. (Source: Reproduced with permission from [92], Wiley (2008).)

effects [90, 91]. In general, the lack of electrical reproducibility and the NTC effect are the main limitations in using PTC materials.

Figure 5.42 shows the dependence of relative resistivity on temperature for solvent-cast MWNT/UHMWPE nanocomposites [92]. The PTC intensity is defined as log (R_t/R_o) where R_o and R_t are the electrical resistivity at room temperature and maximum resistivity, respectively, in the PTC region. The PTC intensity for the 0.2 wt% MWNT/UHMWPE, 1 wt% MWNT/UHMWPE, and 2 wt% MWNT/UHMWPE nanocomposites is 1.971, 1.021, and 0.743, respectively. The PTC intensity of nanocomposites decreases, while the melting temperature of polymer increases with increasing filler content. The NTC effect appears immediately next to the PTC in the resistivity–temperature curves. As mentioned above, the PTC effect depends on the nature of polymer and fillers employed. Jiang et al. [93] studied the PTC effect in solvent-cast 3 vol% CB (20 nm)/PVDF and 2 vol% MWNT/LDPE (low-density polyethylene) and 2.5 vol% MWNT/PVDF composites (Figures 5.43a,b and 5.44). Both the 3 vol% CB/PVDF and 2 vol% MWNT/LDPE composites exhibit obvious PTC and NTC effects. Further, the reproducibility of three heating cycles of these composite systems is quite poor, demonstrating low thermal stability. On the other hand, the 2.5 vol% MWNT/PVDF nanocomposite displays almost no NTC effect and shows high thermal stability after two heating cycles. The resistivity of this composite film only increases slightly in the third heating cycle. Compared to the 3 vol% CB/PVDF composite, the redistribution of MWNTs in molten PVDF is more difficult because of their large aspect ratio. In this case, more energy is needed for redistribution of the nanotubes. Therefore, the 2.5 vol% MWNT/PVDF nanocomposite exhibits no NTC effect. For the 2 vol% MWNT/LDPE nanocomposite, the density of LDPE (0.91–$0.94\,\mathrm{g\,cm^{-3}}$) is much lower than that of MWNT (1.50–$2.10\,\mathrm{g\,cm^{-3}}$). The density of PVDF ($1.70$–$1.80\,\mathrm{g\,cm^{-3}}$) is close to that of MWNT. Thus the gravity effect can lead to

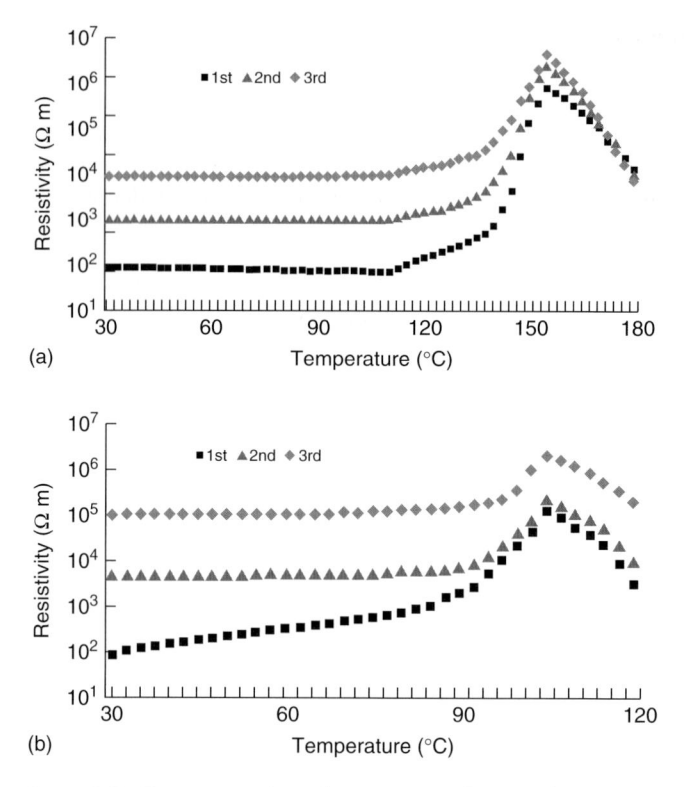

Figure 5.43 Temperature dependent resistivity of (a) 3 vol% CB/PVDF and (b) 2.0 vol% MWNT/LDPE composite films for three heating cycles. (Source: Reproduced with permission from [93], Wiley (2010).)

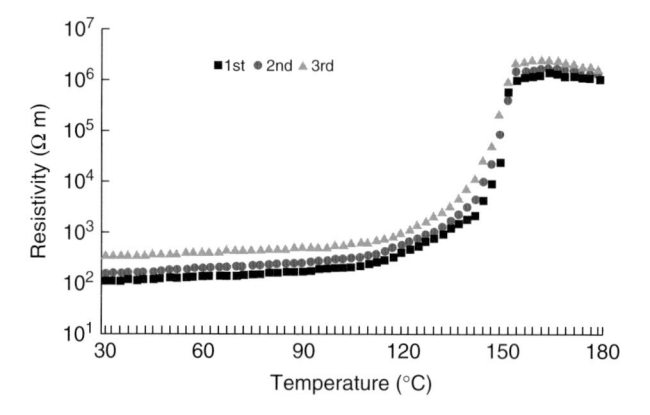

Figure 5.44 Temperature dependent resistivity of 2.5 vol% MWNT/PVDF composite film for three heating cycles. (Source: Reproduced with permission from [93], Wiley (2010).)

severe redistribution of MWNTs in molten LDPE, leading to the occurrence of NTC in the 2 vol% MWNT/LDPE composite.

It is interesting to note that PTC effect is not found in the FGS/PVDF (functional graphene sheet) nanocomposites [94]. Instead, the resistivity of such nanocomposites decreases from ambient to 170 °C. In other words, NTC effect prevails in these nanocomposites (Figure 5.45a). Ansari and Giannelis attributed to this the higher aspect ratio of FGS, which leads to contact resistance predominating over tunneling resistance. However, MWNT also exhibits very large aspect ratio,

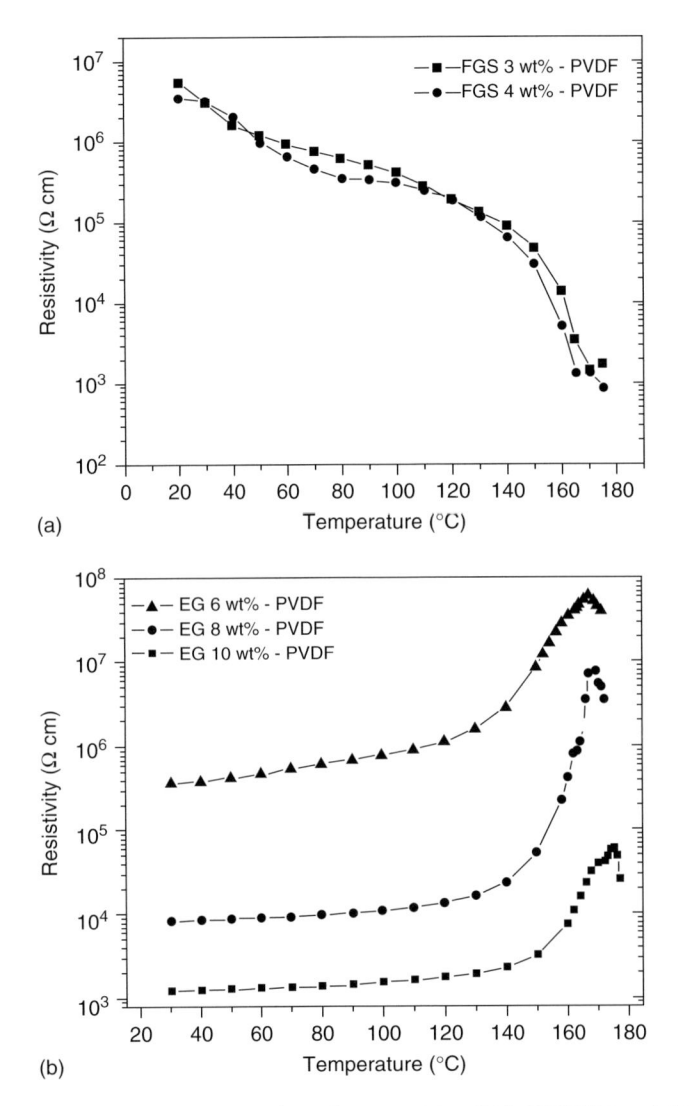

Figure 5.45 Temperature dependent resistivity of (a) FGS/PVDF and (b) EG/PVDF nanocomposites. (Source: Reproduced with permission from [94], Wiley (2009).)

and its incorporation into PVDF eliminates the NTC effect. It seems that large aspect ratio of FGS is not solely responsible for the NTC effect in the FGS/PVDF nanocomposites. By using EGs as nanofillers, both the PTC and NTC effects are apparent (Figure 5.45b). More study is needed to elucidate the mechanism responsible for the NTC effect in PVDF composites filled with two-dimensional EG and FGS nanofillers.

5.6
Hybrid Nanocomposites

Considering the high performance of CNTs, hybridization of CNTs with other nanofillers may generate novel polymer nanocomposites with tailored microstructures and electrical properties [95–103]. From the literature, silver nanoparticles have been used by several researchers to form conducting polymer nanocomposites [10, 97, 103]. Silver exhibits good electrical and thermal conductivities, and its nanoparticles find a variety of applications in catalysis, antimicrobials, conductive inks, and electronic devices [104–106]. Figure 5.46 shows DC electrical conductivity versus silver nanoparticle content for melt-compounded Ag/PP nanocomposites [97]. The electrical conductivity of the nanocomposites increases gradually with increasing filler content. No abrupt increase in electrical conductivity can be seen. The conductivity only reaches 4×10^{-8} S m^{-1} by adding 50 wt%Ag. Further, there is no improvement in electrical conductivity of the nanocomposites subjected to annealing treatment at 90 °C for 80 h. For the crystalline polymer composites, annealing treatment usually produces a more perfect crystalline structure of the polymer matrix. However, the conductivity rises sharply to 3×10^{-4} S m^{-1} by adding 2 wt% MWNT to the 10 wt% Ag/PP nanocomposite. For comparison, it requires the addition of 3 wt% MWNT to PP for achieving an insulating/conducting transition

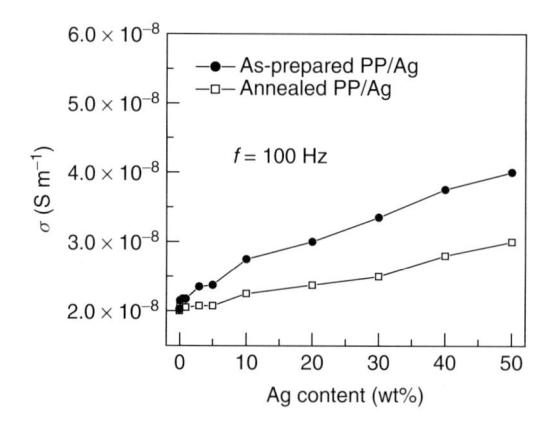

Figure 5.46 Electrical conductivity versus filler content for as-prepared and annealed PP/Ag nanocomposites [97].

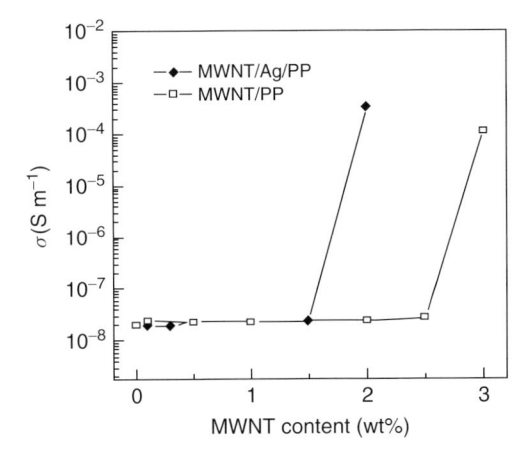

Figure 5.47 Variation of electrical conductivity with MWNT content for binary MWNT/PP and ternary MWNT/Ag/PP nanocomposites. The Ag content is fixed at 10 wt% for MWNT/Ag/PP nanocomposites [97].

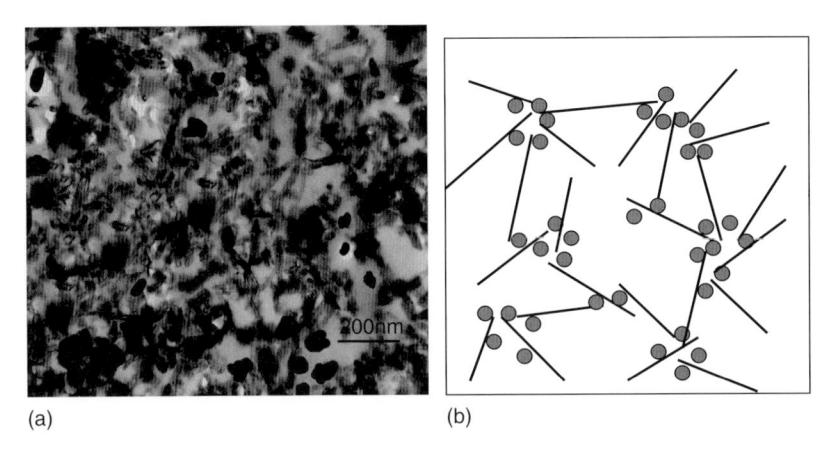

Figure 5.48 (a) Transmission electron micrograph of 2.0 wt% MWNT/10 wt% Ag/PP composite and (b) schematic illustration of MWNT/Ag/PP composite. The gray circles denote Ag nanoparticles and solid lines represent MWNTs [97].

in binary MWNT/PP system (Figure 5.47). The beneficial effect of MWNT addition can be attributed to the MWNTs of large aspect ratios and can bridge isolated Ag nanoparticles and clusters to form a conductive network (Figure 5.48a,b). A similar bridging effect of SWNTs for improving electrical conductivity of solvent-cast SWNT/PCL (polycaprolactone) nanocomposites was also reported by Fortunati et al. [103] very recently.

Finally, the influence of nanofiller hybridization on the PTC effect of conducting polymer composites is discussed herein. Lee et al. [95] studied the effect of CNT additions on the PTC behavior of CB (24 nm)/HDPE composites prepared by combined solution and melt-mixing process. Figure 5.49a shows electrical

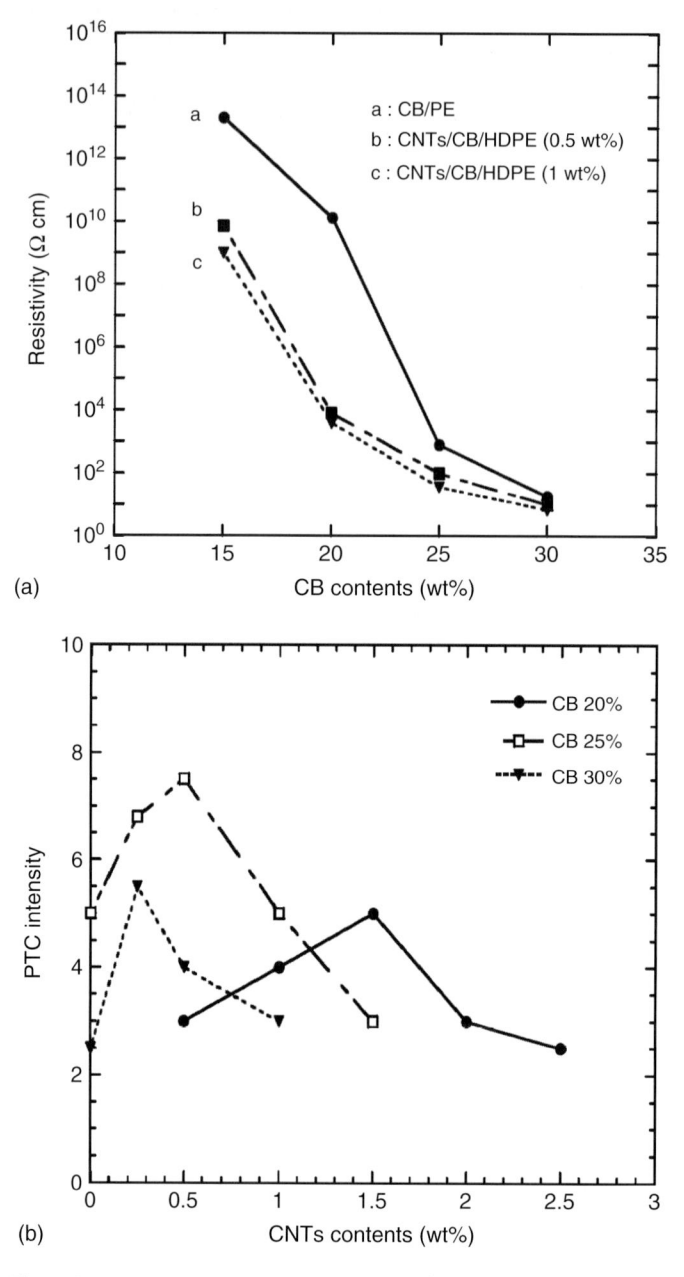

Figure 5.49 (a) Resistivity versus CB content for CB/HDPE composites with and without MWNTs at room temperature and (b) PTC intensity versus CNT content for hybrid nanocomposites with various CB and MWNT contents. (Source: Reproduced with permission from [95], Elsevier (2006).)

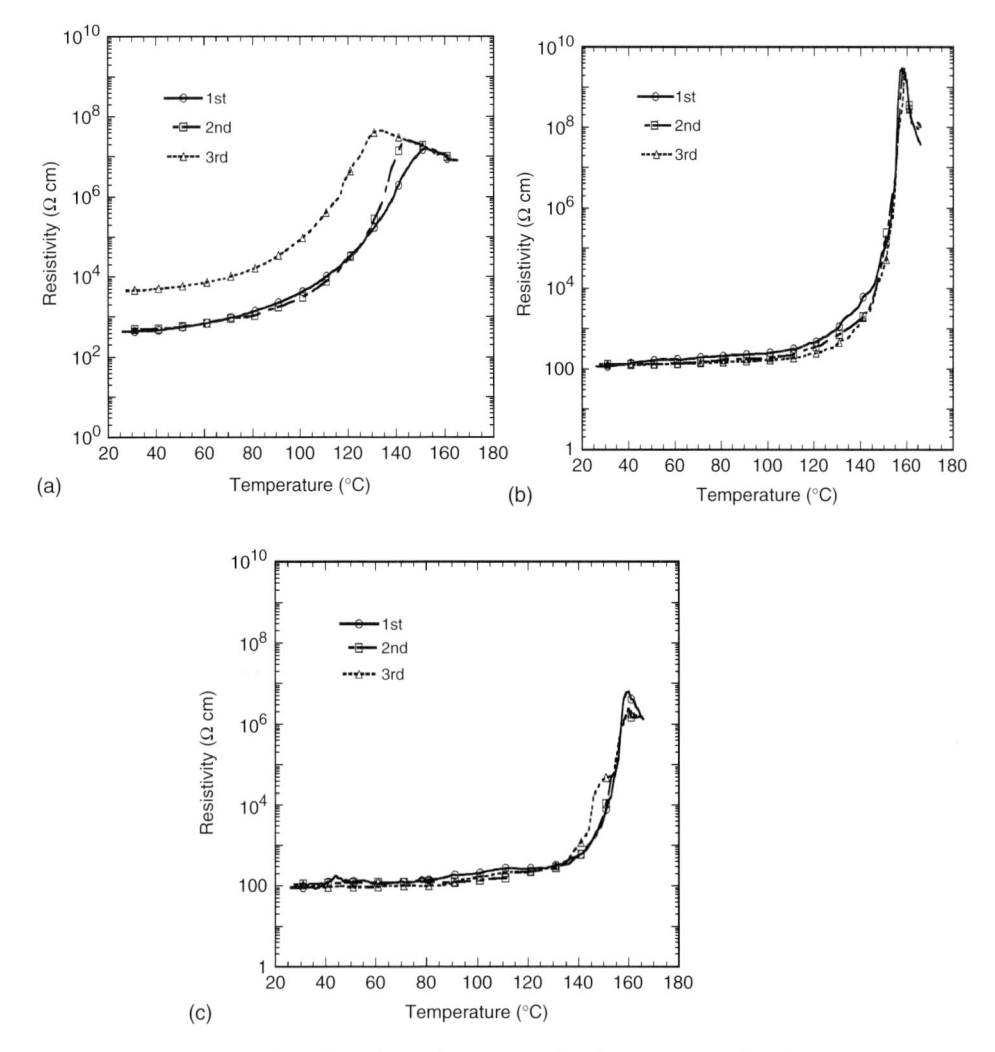

Figure 5.50 Temperature dependent electrical resistivity after three heating cycles of (a) 25 wt% CB/HDPE, (b) 25 wt%CB/0.5 wt% MWNT/HDPE, and (c) 25 wt% CB/1 wt% MWNT/HDPE nanocomposites. (Source: Reproduced with permission from [95], Elsevier (2006).)

resistivity versus CB content for the CB/HDPE, CB/0.5 wt%MWNT/HDPE, and CB/1 wt%MWNT/HDPE nanocomposites at room temperature. A large reduction in electrical resistivity can be seen in binary CB/HDPE composites by adding CB content ≥ 25 wt%. With a small addition of MWNTs, the resistivity of CB/HDPE composites decreases markedly, especially in the range of 15–20 wt% CB. The PTC intensities of hybrid nanocomposites with various MWNT and CB contents are shown in Figure 5.49b. It can be seen that the hybrid composite containing 25 wt% CB and 0.5 wt% MWNT exhibits the highest PTC intensity. Therefore, particular attention is paid to the 25 wt% CB/HDPE composite specimens with and without MWNTs. Figure 5.50a–c shows the temperature dependence of electrical resistivity for the 25 wt%CB/HDPE, 25 wt%CB/0.5 wt%MWNT/HDPE, and 25 wt%CB/1 wt%MWNT/HDPE under three cyclic heating runs. The 25 wt% CB/HDPE composite shows poor reproducibility in the third heating cycle. In contrast, the other two hybrid nanocomposites display good repeatability for three heating cycles as a result of the nanotube additions.

Tjong and coworkers [101] demonstrated that the PTC transition temperature in PP composites can be regulated in a broad temperature range by adding 1 wt% MWNT and various organoclay contents. Montmorillonite (MMT) clay silicate is widely used to reinforce polymers because of its availability at relatively low cost. Hydrophilic MMT generally shows a high tendency for agglomeration in the polymer matrix of nanocomposites. The clay surface can be converted from hydrophilic to organophilic via cation exchange of its Na^+ ions with organic modifiers, forming the so-called organoclay. Figure 5.51 shows the plot of normalized resistivity versus temperature of the 1 wt%MWNT/xMMT/PP nanocomposites

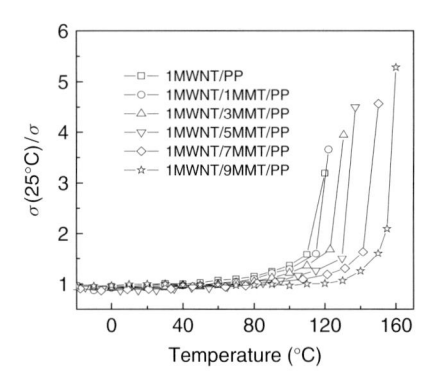

Figure 5.51 Variation of normalized resistivity (reciprocal of conductivity) with temperature of 1 wt% MWNT/x wt%MMT/PP nanocomposites [101].

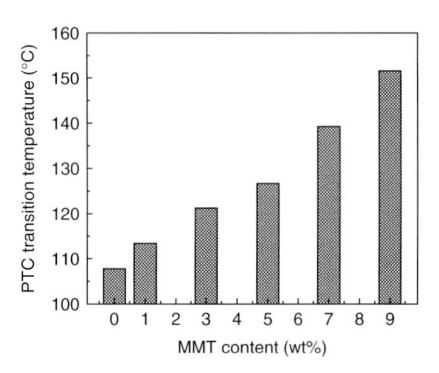

Figure 5.52 PTC temperature versus MMT content of 1 wt%MWNT/x wt%MMT/PP composites [101].

containing various MMT contents ($x =$ MMT weight percentage). Both binary and ternary nanocomposites exhibit PTC effect with different transition temperatures. Figure 5.52 summarizes the PTC temperatures of the 1 wt%MWNT/xMMT/PP nanocomposites containing various MMT contents. It is quite obvious that the PTC temperature of these nanocomposites increases with increasing MMT loadings. Therefore, MMT clay can be used to tune PTC temperatures of conducting MWNT/PP nanocomposites in a broad temperature range.

References

1. Stauffer, D. and Aharony, A. (1992) *Introduction to Percolation Theory*, 2nd edn, Taylor & Francis, London.
2. Sahimi, M. (1994) *Applications of Percolation Theory*, Taylor & Francis, London.
3. Essam, J.W. (1980) *Rep. Prog. Phys.*, **43**, 833–912.
4. Broadbent, S.R. and Hammersley, J.M. (1957) *Proc. Camb. Philos. Soc.*, **53**, 629–641.
5. Kirkpatrick, S. (1973) *Rev. Mod. Phys.*, **45**, 574–588.
6. Scher, H. and Zallen, R. (1970) *J. Chem. Phys.*, **53**, 3759–3761.
7. Webman, I., Jortner, J., and Cohen, M.H. (1977) *Phys. Rev. B*, **16**, 2593–2596.
8. Nan, C.W. (1993) *Prog. Mater. Sci.*, **37**, 1–116.
9. Mamunya, Y.P., Davydenko, V.V., Pissis, P., and Lebedev, E.V. (2002) *Eur. Polym. J.*, **38**, 1887–1897.
10. Gonon, P. and Boudefel, A. (2006) *J. Appl. Phys.*, **99**, 024308 (8 pp).
11. Balberg, I. (1987) *Phys. Rev. Lett.*, **59**, 1305–1308.
12. Mallick, P.K. (1993) *Fibre-Reinforced Composites*, Marcel Dekker, Inc., New York.
13. (a) Sheng, P., Sichel, E.K., and Gittleman, J.L. (1978) *Phys. Rev. Lett.*, **40**, 1197–1200; (b) Sheng, P. (1980) *Phys. Rev. B*, **21**, 2180–2195.
14. Song, Y., Noh, T.W., Lee, S.I., and Gaines, J.R. (1986) *Phys. Rev. B*, **33**, 904–909.
15. Bergman, D.J. and Imry, Y (1977) *Phys. Rev. Lett.*, **39**, 1222–1225.
16. Clerc, J.P., Giraud, G., Laugier, J.M., and Luck, J.M. (1990) *Adv. Phys.*, **39**, 191–309.
17. Panda, M., Srinivas, V., and Thakur, A.K. (2008) *Appl. Phys. Lett.*, **92**, 132905 (3 pp).
18. Barrau, S., Demont, P., Peigney, A., Laurent, C., and Lacabanne, C. (2003) *Macromolecules*, **36**, 5187–5194.
19. Balberg, I., Anderson, C.H., Alexander, S., and Wagner, N. (1984) *Phys. Rev. B*, **30**, 3933–3943.
20. Balberg, I. (1985) *Phys. Rev. B*, **31**, 4053–4055.
21. Liang, G.D. and Tjong, S.C. (2008) *IEEE Trans. Dielectrics Electr. Insul.*, **15**, 214–220.
22. Lu, W., Lin, H.F., Wu, D.J., and Chen, G.H. (2006) *Polymer*, **47**, 4440–4444.
23. Kremer, F. and Schönhals, A. (2003) *Broadband Dielectric Spectroscopy*, Springer-Verlag, Berlin.
24. Bello, A., Laredo, E., Marvel, J.R., Grimau, M., Arnal, M.L., and Muller, A.J. (2011) *Macromolecules*, **44**, 2819–2828.
25. Psarras, G.C. (2006) *Composites Part A*, **37**, 1545–1553.
26. Kilbride, B.E., Coleman, J.N., Fraysse, J., Fournet, P., Cadek, M., Drury, A., Hutzler, S., Roth, S., and Blau, W.J. (2002) *J. Appl. Phys.*, **92**, 4024–4030.
27. Pang, H., Chen, T., Zhang, G.M., Zeng, B.Q., and Li, Z.M. (2010) *Mater. Lett.*, **64**, 2226–2229.
28. Li, W.J., Tang, X.Z., Zhang, H.B., Jiang, Z.G., Yu, Z.Z., Du, X.S., and Mai, Y.W. (2011) *Carbon*, **49**, 4724–4730.
29. Mahmoud, W. (2011) *Eur. Polym. J.*, **47**, 1534–1540.
30. Qi, X.Y., Yan, D., Jiang, Z.G., Cao, Y.K., Yu, Z.Z., Yavari, F., and Koratkar, N. (2011) *ACS Appl. Mater. Interfaces*, **3**, 3130–3133.

31. Yoonessi, M. and Gaier, J.R. (2010) *ACS Nano*, **4**, 7211–7220.
32. Jung, R., Park, W.I., Kwon, S.M., Kim, H.S., and Jin, H.J. (2008) *Polymer*, **49**, 2071–2076.
33. Li, Y.C., Li, R.K., and Tjong, S.C. (2010) *J. Nanomater.* (Art. No.: 261748), 10 pages.
34. Li, M. and Jeong, Y.G. (2011) *Composites Part A*, **42**, 560–566.
35. Shen, J.W., Chen, X.M., and Huang, W.Y. (2003) *J. Appl. Polym. Sci.*, **88**, 1864–1869.
36. He, F., Lau, S., Chan, H.L., and Fan, J. (2009) *Adv. Mater.*, **21**, 710–715.
37. Srivastava, N.K. and Mehra, R.M. (2008) *J. Appl. Polym. Sci.*, **109**, 3991–3999.
38. Zhang, X.L., Shen, L., Xia, X., Wang, H.T., and Du, Q.G. (2008) *Mater. Chem. Phys.*, **111**, 368–374.
39. Spitalsky, Z., Tasis, D., Papagelis, K., and Galiotis, C. (2010) *Prog. Polym. Sci.*, **35**, 357–401.
40. Liang, G.D. and Tjong, S.C. (2006) *Mater. Chem. Phys.*, **100**, 132–137.
41. Tjong, S.C., Liang, G.D., and Bao, S.P. (2007) *Scr. Mater.*, **57**, 461–464.
42. He, L.X. and Tjong, S.C. (2010) *Eur. Phys. J. E*, **32**, 249–254.
43. He, L.X. and Tjong, S.C. (2010) *Curr. Nanosci.*, **6**, 520–524.
44. He, L.X. and Tjong, S.C. (2011) *J. Nanosci. Nanotechnol.*, **12**, 10668–10672.
45. Li, Q., Xue, Q.Z., Hao, L.Z., Gao, X.L., and Zheng, Q.B. (2008) *Compos. Sci. Technol.*, **68**, 2290–2296.
46. Wang, L. and Dang, Z.M. (2005) *Appl. Phys. Lett.*, **87**, 042903 (3 pp).
47. Dang, Z.M., Wang, L., Yin, Y., Zhang, Q., and Lei, Q.Q. (2007) *Adv. Mater.*, **19**, 852–857.
48. Arlen, M.J., Wang, D., Jacobs, J.D., Justice, R., Trionfi, A., Hsu, J.W., Schaffer, D., Tan, L.S., and Vaia, R.A. (2008) *Macromolecules*, **41**, 8053–8062.
49. Linares, A., Canalda, J.C., Cagiao, M.E., Garcia-Gutierrez, M.C., Nogoles, A., Martin-Gullon, I., Vera, J., and Ezquerra, T.A. (2008) *Macromolecules*, **41**, 7090–7097.
50. Bauhofer, W. and Kovacs, J.Z. (2009) *Compos. Sci. Technol.*, **69**, 1486–1498.
51. Liu, Q.M., Tu, J.C., Wang, X., Yu, W.X., Zheng, W., and Zhao, Z.D. (2012) *Carbon*, **50**, 339–341.
52. Pan, Y.Z., Li, L., Chan, S.W., and Zhao, J.H. (2010) *Composites Part A*, **41**, 419–426.
53. Jeon, K., Warnock, S., Ruiz-Orta, C., Kismarahardja, A., Brooks, J., and Alamo, R.G. (2010) *J. Polym. Sci., Part B: Polym. Phys.*, **48**, 2084–2096.
54. Li, C., Thostenson, E.T., and Chou, T.W. (2008) *Compos. Sci. Technol.*, **68**, 1445–1452.
55. Ma, P.C., Zhang, W., Zhu, Y., Ji, L., Zhang, R., Koratkar, N., and Liang, J. (2008) *Carbon*, **46**, 706–710.
56. Oliva-Aviles, A.I., Aviles, F., and Sosa, V. (2011) *Carbon*, **49**, 2989–2997.
57. Peng, H. (2008) *J. Am. Chem. Soc.*, **130**, 42–43.
58. Peng, H. and Sun, X. (2009) *Chem. Phys. Lett.*, **47**, 103–105.
59. Sumfleth, J., Prehn, K., Wichmann, M.H., Wedekind, S., and Schulte, K. (2010) *Compos. Sci. Technol.*, **70**, 173–180.
60. Sandler, J.K., Kirk, J.E., Kinloch, I.A., Schaffer, M.S., and Windle, A.H. (2003) *Polymer*, **44**, 5893–5899.
61. Kim, Y.J., Shin, T.S., Choi, H.D., Kwon, J.H., Chung, Y.C., and Yoon, H.G. (2005) *Carbon*, **43**, 23–30.
62. Geng, Y., Liu, M.Y., Li, J., Shi, X.M., and Kim, J.K. (2008) *Composites Part A*, **39**, 1876–1883.
63. Wardle, B.L., Saito, D.S., Garcia, E.J., Hart, A.J., de Villoria, R.G., and Verploegen, E.A. (2008) *Adv. Mater.*, **20**, 2707–2714.
64. Boncel, S., Koziol, K.K., Walczak, K.Z., Windle, A.H., and Schaffer, M.S. (2011) *Mater. Lett.*, **65**, 2299–2303.
65. Cheng, Q.F., Wang, J.P., Wen, J.J., Liu, C.H., Jiang, K.L., Li, Q.Q., and Fan, S.S. (2010) *Carbon*, **48**, 260–266.
66. Jang, P.G., Suh, K.S., Park, M., Kim, J.K., Kim, W.N., and Yoon, H.G. (2007) *J. Appl. Polym. Sci.*, **106**, 110–116.
67. Pedroni, L.G., Soto-Oviedo, M.A., Rosolen, J.M., Felisberti, M.I., and Nogueira, A.F. (2009) *J. Appl. Polym. Sci.*, **112**, 3241–3248.
68. Tsuchiya, K., Sakai, A., Nagaoka, T., Uchida, K., Furukawa, T., and

Yajima, H. (2011) *Compos. Sci. Technol.*, **71**, 1098–1104.
69. Gupta, A.K. and Sen, A.K. (1998) *Phys. Rev. B*, **57**, 3375–3388.
70. Sodolski, H., Zieliski, R., Slupkowski, T., and Jachym, B. (1975) *Phys. Status Solidi A*, **32**, 603–609.
71. Narkis, M., Ram, A., and Flashner, F. (1978) *Polym. Eng. Sci.*, **18**, 649–652.
72. Lin, H., Lu, W., and Chen, G.H. (2008) *Physica B*, **400**, 229–236.
73. Chen, G., Weng, W.G., Wu, D.J., and Wu, C.L. (2004) *J. Polym. Sci., Part B: Polym. Phys.*, **42**, 155–167.
74. Gefen, Y., Shih, W.H., Laibowitz, R.B., and Viggiano, J.M. (1986) *Phys. Rev. Lett.*, **57**, 3097–3100.
75. Gefen, Y., Shih, W.H., Laibowitz, R.B., and Viggiano, J.M. (1987) *Phys. Rev. Lett.*, **58**, 2727–2727.
76. Chakrabarty, R.K., Bardhan, K.K., and Basu, A. (1991) *Phys. Rev. B*, **44**, 6773–6779.
77. Chynoweth, A.G. (1960) *Progress in Semiconductors*, John Wiley & Sons, Inc., New York.
78. Zener, C. (1934) *Proc. R. Soc. A*, **145**, 523–529.
79. McAfee, K.B., Ryder, E.J., Shockley, W., and Sparks, M. (1951) *Phys. Rev.*, **83**, 650–651.
80. Chynoweth, A.G., Feldmann, W.L., Lee, C.A., Logan, R.A., Pearson, G.L., and Aigrain, P. (1960) *Phys. Rev.*, **118**, 425–434.
81. van Beek, L.K. and van Pul, B.I. (1964) *Carbon*, **2**, 121–126.
82. He, L.X. and Tjong, S.C. (2010) *Synth. Met.*, **160**, 2085–2088.
83. He, L.X. and Tjong, S.C. (2011) *Synth. Met.*, **161**, 540–543.
84. He, L.X. and Tjong, S.C. (2011) *J. Nanosci. Nanotechnol.*, **11**, 3916–3921.
85. Meyer, J. (1973) *Polym. Eng. Sci.*, **13**, 462–486.
86. Meyer, J. (1974) *Polym. Eng. Sci.*, **14**, 706–716.
87. Ohe, K. and Naito, Y.A. (1971) *Jpn. J. Appl. Phys.*, **10**, 99–108.
88. Klason, C. and Kubat, J. (1975) *J. Appl. Polym. Sci.*, **19**, 831–845.
89. Allack, H.K., Brinkman, A.W., and Woods, J. (1993) *J. Mater. Sci.*, **28**, 117–120.
90. Luo, Y.L., Wang, G.C., Zhang, B.Y., and Zhang, Z.P. (1998) *Eur. Polym. J.*, **34**, 1221 – 1227.
91. Tang, H., Liu, Z.Y., Piao, J.H., Chen, X.F., Lou, Y.X., and Li, S.H. (1994) *J. Appl. Polym. Sci.*, **51**, 1159 – 1164.
92. Gao, J.F., Yan, D.X., Huang, H.D., Dai, K., and Li, Z.M. (2009) *J. Appl. Polym. Sci.*, **114**, 1002–1010.
93. Jiang, S.L., Yu, Y., Xie, J.J., Wang, L.P., Zeng, Y.K., Fu, M., and Li, T. (2010) *J. Appl. Polym. Sci.*, **116**, 838–842.
94. Ansari, S. and Giannelis, E.P. (2009) *J. Polym. Sci., Part B: Polym. Phys.*, **47**, 888–897.
95. Lee, J.H., Kim, S.K., and Kim, N.H. (2006) *Scr. Mater.*, **55**, 1119–1122.
96. Liang, G.D. and Tjong, S.C. (2006) *Adv. Eng. Mater.*, **9**, 1014–1017.
97. Liang, G.D., Bao, S.P., and Tjong, S.C. (2007) *Mater. Sci. Eng. B*, **142**, 55–61.
98. Liu, L. and Grunlan, J.C. (2007) *Adv. Funct. Mater.*, **17**, 2343–2348.
99. Li, J., Wong, P.S., and Kim, J.K. (2008) *Mater. Sci. Eng., A*, **483–484**, 660–663.
100. Ma, P.C., Liu, M.Y., Zhang, H., Wang, S.Q., Wang, R., Wang, K., Wong, Y.K., Tang, B.Z., Hong, S.H., Paik, K.W., and Kim, J.K. (2009) *ACS Appl. Mater. Interfaces*, **1**, 1090–1096.
101. Bao, S.P., Liang, G.D., and Tjong, S.C. (2009) *IEEE Trans. Nanotechnol.*, **8**, 729–736.
102. Dang, Z.M., Yao, S.H., Yuan, J.K., and Bai, J.B. (2010) *J. Phys. Chem. C*, **114**, 13204–13209.
103. Fortunati, E., Angelo, F.D., Martino, S., Orlacchio, A., Kenny, J.M., and Armentano, I. (2011) *Carbon*, **49**, 2370–2379.
104. Yang, J.X., Hasell, T., Wang, W.X., and Howdle, S.M. (2008) *Eur. Polym. J.*, **44**, 13331–11336.
105. Balogh, L., Swanson, D.R., Tomalia, D.A., Hagnauer, G.L., and McManus, A.T. (2001) *Nano Lett.*, **1**, 18–21.
106. Natsuki, J. and Abe, T. (2011) *J. Colloid Interface Sci.*, **359**, 19–23.

6
Carbonaceous Nanomaterials and Polymer Nanocomposites for Fuel Cell Applications

6.1
Overview

Global warming, energy shortage, and environmental pollution are the dilemmas faced by many countries today because of tremendous increase in world population. Energy insecurity and rising cost of conventional fossil energy sources pose a threat to the industrial development and economic stability of most countries. The development of green, renewable electricity generation to replace fossil energy is driven by the demand to reduce pollutant and carbon dioxide emissions [1]. The use of fuel cell for energy production as a substitute for fossil energy is considered to be an effective route to reduce greenhouse gas emissions. A fuel cell converts chemical energy into electrical energy by direct electrochemical reaction of a fuel and an oxidant with a high efficiency and low carbon dioxide emission.

On the basis of the type of electrolyte (or mobile ions), operating temperature, and fuel/oxidant used, fuel cells can be classified into proton exchange membrane, phosphoric acid, direct methanol, alkaline, molten carbonate, and solid oxide [2]. Proton exchange or polymer exchange membrane fuel cell (PEMFC) uses hydrogen as the fuel for generating useful clean energy with water as the only by-product. It exhibits several advantages including low operating temperature, short response time, convenient fuel supply, and long lifetime. PEMFCs generally operate at low temperatures ranging from 60 to 120 °C, and hence fast start-up time can be readily achieved. In this regard, PEMFCs show practical applications as the power sources for consumer electronics and transportation vehicles. Electronic devices are being designed to have more complex structures, multifunctional performance, and smaller sizes, as the demand for increased speed accelerates. These devices need microfuel cells to power them [3]. PEMFCs are recommended by the US Department of Energy (DOE) as the replacements for internal combustion engines for transportation vehicles [4]. However, the commercialization of PEMFCs is significantly hindered from the high cost of bipolar plates and platinum electrocatalysts. In this regard, carbon nanomaterials offer significant promise as solutions to long-standing technological issues in the development of fuel cells. The use of synthetic, but organic in nature, material such as carbon nanotube (CNT) for clean and sustainable energy development is considered of technological importance.

Polymer Composites with Carbonaceous Nanofillers: Properties and Applications, First Edition. Sie Chin Tjong.
© 2012 Wiley-VCH Verlag GmbH & Co. KGaA. Published 2012 by Wiley-VCH Verlag GmbH & Co. KGaA.

CNT and graphene-based materials can strengthen and enhance electrical conductivity of polymers at very low loading levels. By using less materials overall, the consumption of our natural resources can be reduced. Accordingly, carbonaceous nanomaterials can be incorporated into polymer composites to form bipolar plates of high performances. Moreover, carbonaceous nanomaterials of superior electrical conductivity, high mechanical strength, and large surface area are ideal supports for platinum catalysts needed for the oxygen reduction reaction (ORR).

6.2
Polymer Exchange Membrane Fuel Cell

PEMFC is made up of proton exchange membrane, catalyst layer, gas diffusion layer (GDL), anode, cathode, and bipolar plates (Figure 6.1) [5]. The membrane, anode and cathode catalyst layers, and two GDLs are commonly termed as membrane electrode assembly (MEA). The main function of the GDL is to distribute the reactants from the gas flow channels uniformly along the surface of catalyst layer. Furthermore, GDL must ensure proper transport of product water, electron, and heat of reaction. At the anode, hydrogen gas is oxidized into proton and electron through the following reaction:

$$H_2 \longrightarrow 2H^+ + 2e \tag{6.1}$$

The reaction requires platinum catalysts to facilitate dissociation of hydrogen molecules. Proton migrates through the membrane to the cathode, and electron flows along an external circuit thereby creating current. The membrane is formed from a solid polymer material having good proton-conducting but poor electron-conducting characteristics. Nafion® (Du Pont) composed of a hydrophobic

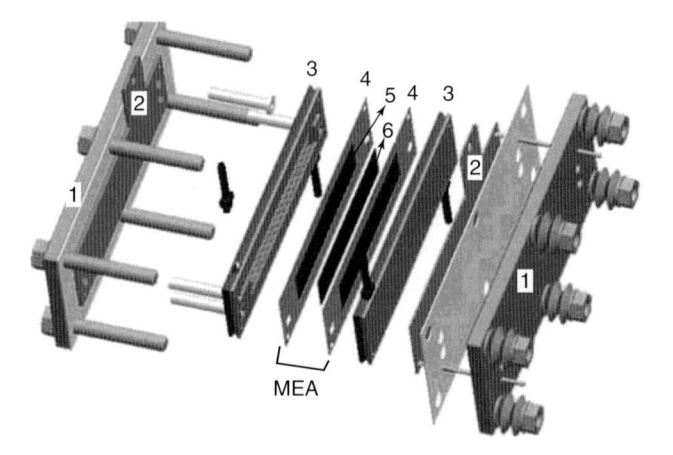

Figure 6.1 PEM fuel cell configuration: (1) End plate (EP), (2) current collector (CC), (3) bipolar plate (BP), (4) gas diffusion layer (GDL), (5) electrode, (6) proton electrolyte membrane (PEM). (Source: Reproduced from Ref. [5] with permission from Intech (2010).)

fluoroethylene backbone and side chains terminated with hydrophilic sulfonic acid (SO_3H) groups is a commonly used membrane. This is attributed to its reasonable proton conductivity and acceptable chemical stability. At the cathode, proton and electron recombine and react with oxygen to yield water. The ORR takes place as

$$O_2 + 4H^+ + 4e \longrightarrow 2H_2O \tag{6.2}$$

The ORR reaction is also catalyzed by platinum metal particles supported on carbon blacks (CBs).

Bipolar plates generally require high mechanical strength to support fuel cell stack assembly and withstand clamp forces of stacking, and to resist vibration during portable electronic and transportation applications. The plates make up 60–80% of the fuel cell stack weight and 29–45% of the total cost of the fuel cell [6, 7]. In this regard, materials engineers and chemists are continuously searched for low-cost bipolar plate materials with light weight, small volume, low hydrogen permeability, good mechanical strength, and high electrical and thermal conductivity. Apparently, carbonaceous nanomaterials play an important role in the development of advanced bipolar plates, platinum catalysts, and proton exchange membrane with desired chemical, physical, and mechanical properties [8].

6.3
Conventional Bipolar Plates

The preceding discussion has established that bipolar plates are key components affecting the general performance of fuel cells. Thus the design and material selection of bipolar plates are crucial in achieving long-term durability and stability of fuel cells. Accordingly, bipolar plates of a fuel cell stack should possess the following properties:

a) to distribute the fuel and oxidant within the cell,
b) to facilitate water management within the cell to avoid flooding,
c) to separate individual cells in the stack,
d) to carry current away from the cell, and
e) to facilitate the heat management [9].

To meet these performance requirements, bipolar plates must exhibit several properties including: (i) high electrical conductivity; (ii) good mechanical properties; (iii) thermal stability at fuel cell operating temperature; (iv) chemical stability in the presence of fuel, oxidant, water, and acidic conditions; (v) high thermal conductivity; (vi) low permeability to fuel and oxidant; (vii) low cost; (viii) low weight; and (ix) good processability [6, 8]. In order to perform the functions of the bipolar plates listed earlier, the composite plates must also satisfy DOE and industrial target requirements (Table 6.1) [4, 10, 11].

Graphite with excellent electrical and thermal conductivity, low density, and high corrosion resistance is an ideal conducting material for making bipolar plates. Graphite can be either natural or synthetic. At present, China is the main producer

Table 6.1 Target values of bipolar plates in PEMFCs [4, 10].

Governmental department/ company	Electrical conductivity (S cm^{-1})	Crush strength (kPa)	Flexural strength (MPa)	Tensile strength (MPa)	Impact strength (J m^{-1})	Thermal conductivity (W(m K)$^{-1}$)	Corrosion resistance (μA cm^{-2})	H$_2$ permeation flux (cm^3(cm^2 s)$^{-1}$)
DOE	100	–	>25	–	–	–	<16	<2 × 10^{-6}
Plug power	–	>4200	>59	>41	>40.5	>10	–	–

of natural graphite, and supplies 40% of the world's output. Natural flake graphite has higher electrical conductivity than synthetic graphite because it possesses higher degree of crystallinity [12]. The main drawbacks of using graphite for making bipolar plates are its inherent brittleness, high cost for machining gas flow channels, and high gas permeability. Consequently, graphite bipolar plates must have a thickness of several millimeters to meet the mechanical property and gas tightness requirements. This makes graphite bipolar plates heavy and bulky.

Metallic alloys are well recognized to exhibit superior workability and high durability for shock and vibration than graphite [13]. They also possess excellent mechanical strength and ductility and high electrical and thermal conductivity. Stainless steels, aluminum- and titanium-based alloys are typical metallic materials used for fabricating bipolar plates. Among them, the properties of stainless steel bipolar plates are widely studied because of their low cost and ease of manufacture. Both etching and stamping techniques have been employed for forming flow channels onto metallic plates. The stability of metal bipolar plates depends on the nature of metals, the electrode potential, and relative humidity. As recognized, metallic alloys are susceptible to corrosion when exposed to acidic and humid environments of fuel cells. For stainless steel bipolar plates, corrosion generally occurs at the anode where the materials are operated in their active state. So stainless steels dissolve easily in severe environments without forming a passive oxide layer. The release of metal ions such as Fe^{3+} and Cr^{3+} during corrosion may further poison polymer electrolyte membrane and lower its conductivity or contaminate electrode catalysts [14]. The metal surfaces must be protected by coatings to resist corrosion. The coatings increase electrical resistance at the interface, resulting in a degradation of the stack performance. At the cathode, the electrode potential is high enough to yield a passive oxide layer. This passive layer protects the metal surface from corrosion but increases the contact resistance between the plates and MEA assembly. Accordingly, the overall power output of PEMFCs drops to a certain extent because of the voltage loss between the plates and MEA assembly.

To replace bipolar plates with the sole graphite, polymer–graphite composite plates consisting of polymer and graphite or other carbon-based fillers are fabricated [15, 16]. The advantages of employing polymer composites for bipolar plate applications are mainly due to their low cost, high flexibility, good processability, high gas tightness, and tailored mechanical properties. Thermoplastic polymers

Table 6.2 Properties of polymer composite bipolar plates [7].

Manufacturer	Polymer	% Graphite + fibers	Conductivity (S cm^{-1})		Mechanical strength	
			In-plane	Through-plane	Tensile (MPa)	Flexural (MPa)
GE	PVDF	74	119	–	–	36.2
GE	PVDF	64 + 16 CF	109	–	–	42.7
LANL	Vinyl ester	68	60	–	23.4	29.6
Premix	Vinyl ester	68	85	–	24.1	28.2
BMC	Vinyl ester	69	30	–	26.2	37.9
Commercial	–	–	105	–	19.3	20.7
BMC 940	Vinyl ester	–	100	50	30.3	40
Plug Power	Vinyl ester	68	55	20	26.2	40
DuPont	–	–	–	25–33	25.1	53.1
SGL	–	–	100	20	–	40.0
H$_2$ Economy	–	–	67	–	–	29.4
Virginia Tech	PET	65 + 7 glass fiber	230	18–25	36.5	53
Virginia Tech	PPS	70 + 6 carbon fiber	271	19	57.5	95.8

such as polyvinylidene fluoride (PVDF), polypropylene (PP), polyethylene (PE), polyethylene terephthalate (PET), poly(phenylene sulfide) (PPS), liquid crystalline polymer (LCP), and thermosetting resins including phenolics, epoxies, and vinyl esters (VEs) have been used to fabricate composite bipolar plates [6]. Insulating polymers with poor electrical conductivity generally require very high graphite content for making bipolar plates. The electrical conductivity of polymer–graphite bipolar plates generally increases with increasing graphite particle size [17].

For a massive large-scale production, polymer composite bipolar plates can be fabricated by means of injection molding and compression molding. Injection molding is extensively used for mass production of polymer composites. Its disadvantage for making composite bipolar plates is the high viscosity of molten polymers at high filler loadings, leading to excessive mold wear and poor processability. Therefore, compression molding is more effective to produce polymer composite bipolar plates commercially [18a,b]. The physical and mechanical properties of conventional polymer composite bipolar plates fabricated by commercial companies and research institutions are listed in Table 6.2. Apparently, the electrical conductivity and flexural strength of most polymer composites are inferior to the DOE and Plug Power target values as listed in Table 6.1.

Conventional polymer composite plates require high graphite flake/particle loadings to meet the minimum requirement for electrical conductivity. Such loading level is significantly higher than the percolation threshold needed to form a conductive path network in the polymer matrix. It also exceeds the critical pigment volume concentrations (CPVCs) (50–70 v/o) need to formulate "electrically conductive" plastics and coatings [19]. As a result, very small amounts of the polymer matrix material can bind graphite particles and fill the interstices between them at such extremely high filler loadings. Thus highly graphite-filled composite plate materials are porous, extremely brittle (with tensile elongation of less than 0.4%), and possess high gas permeability [18a,b]. It deems necessary to develop novel composite materials with low filler loadings having reduced weight, compact size, and excellent electrical and thermal conductivity. CNTs, expanded graphites (EGs), and graphite nanoplatelets (GNPs) are excellent carbon nanomaterials for improving electrical, mechanical, and thermal properties of polymer composite bipolar plates [20].

6.4
Polymer Nanocomposite Bipolar Plates

6.4.1
CNT-Filled Composite Plates

The electrical conductivity of percolating polymer/CNT nanocomposites is in the order of $\sim 10^{-4} - 10^{-1}$ S cm^{-1}. Such values satisfy the conductivity requirement for electronic device applications but cannot meet the DOE target for bipolar plates as listed in Table 6.1. To increase the conductivity of the polymer/CNT nanocomposites to a high value of 100 S cm^{-1}, the nanotube content must be significantly increased. This can be achieved by reinforcing polymers with aligned CNT laminates as mentioned in Chapter 5 (Figure 5.37). However, such aligned CNT/polymer composite laminates are still in the earlier stage of development. Till present, polymer composite bipolar plates with electrical conductivity ≥ 100 S cm^{-1} are fabricated by adding low CNT loading levels to conventional graphite/polymer composites as shown in Figure 6.2a–c [21].

Ma and coworkers studied the effects of functionalized multiwalled carbon nanotube (FMWNT) additions on the electrical and mechanical performance of polymer/graphite 30/70 and polymer/graphite 20/80 composites systematically. Both thermoplastic PP and thermoset VE were, respectively, used as the polymer matrix materials for these composites [22a–f]. VE is a cured thermosetting resin with a network structure having high resistance to moisture and chemical attacks. Pristine multiwalled carbon nanotubes (MWNTs) disperse poorly in the polymer matrix of composites, thus functionalization of nanotubes is needed to achieve homogeneous dispersion. Oxidation of CNTs in strong acids produces carboxylic groups at their surface defect sites. Such acid-treated MWNTs can disperse in poly(oxypropylene)-diamines (POP) of different molecular weights (Figure 6.3) [22a]. This is because carboxylic groups of MWNTs attract organic species bearing $-NH_2$ groups of POP-diamines through electrostatic interactions, thereby forming a zwitterion. The composite bipolar plates with thermosetting matrix can then be prepared by mixing FMWNTs, VE, t-butyl perbenzoate (TBPB) initiator, styrene monomer, thickening agent (MgO), release agent (zinc stearate) and graphite powder in a kneader by means of bulk-molding compound (BMC) process (Table 6.3). The BMC was thickened for 36 h followed by hot pressing.

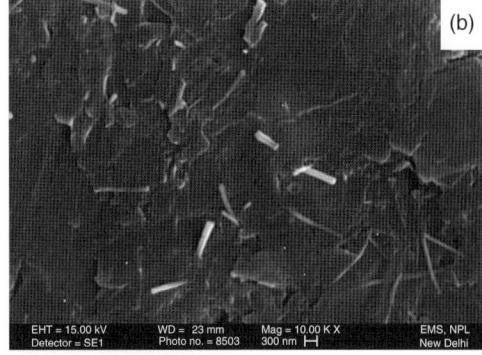

Figure 6.2 Scanning electron microscopy (SEM) micrographs of (a) phenolic resin/graphite 35/65 (vol%), (b) phenolic resin/graphite/MWNT 35/64/1 (vol%), and (c) phenolic resin/graphite/MWNT 35/63/2 (vol%) composites. (Source: Reproduced from Ref. [21] with permission from Elsevier (2010).)

As mentioned previously, carboxylic functional groups of MWNTs can be converted to carbonyl chloride (−COCl) groups by reacting with thionyl chloride. Acylchloride can further react with amide-based molecules via amidation or esterification. Ma and coworkers [22e] used this strategy to functionalize MWCNTs followed by either reacting acylchloride with POA-MA, that is, poly(oxyalkylene)-amines (POA) bearing maleic anhydride (MA) (Figure 6.4), or with POA-DGEBA, that is, POA bearing diglycidyl ether bisphenol A (DGEBA) via amidation reaction (Figure 6.5). Figure 6.6a–f shows transmission electron microscopy (TEM) micrographs of pristine MWNTs, MWNTs/POA400-DGEBA, and MWNTs/POA2000-DGEBA. Apparently, the walls of pristine MWNTs are relatively smooth and clean (Figure 6.6a,b). An organic coating on the exterior wall of MWNTs/POA400-DGEBA can be readily seen at high magnification, leading to a remarkable increase in the diameter of MWNTs (Figure 6.6d). For MWNTs/POA2000-DGEBA, the nanotubes tend to coil each other owing to longer polymer chains grafted to the MWNT surfaces (Figure 6.6e) [22e].

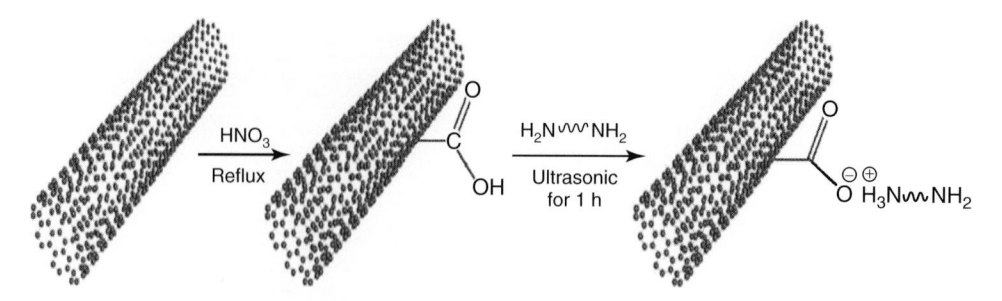

Figure 6.3 Synthesis scheme for MWNTs-POP. (Source: Reproduced from Ref. [22a] with permission from Elsevier.)

Table 6.3 Formulation of BMC process.

Components	Composition	
	Resin composition	BMC composition (wt%)
Vinyl ester[a] (wt%)	75	30
Low profile agent (wt%)	8	–
Styrene monomer (wt%)	17	–
TBPB (phr)	1.8	–
Zinc stearate (phr)	3.5	–
Magnesium oxide (phr)	1.8	–
MWNTs (phr)	0–2	–
Graphite powder (wt%)	–	70
Total		100

[a]Chemical structure of phenolic-novolac, epoxy-based, vinyl ester is as follows:

$$R: -CH_2-CH(OH)-CH_2-O-C(=O)-C(CH_3)=CH_2$$

Source: Reproduced from Ref. [22a] with permission from Elsevier.

6.4.2
Graphene-Sheet-Filled Composite Plates

Graphene and chemically modified graphene sheets exhibit large aspect ratio, high electrical conductivity, and superior mechanical properties comparable with or even better than CNTs. Furthermore, graphene-based materials can be easily obtained via simple chemical treatment of raw graphite. The electrical conductivity of free standing reduced graphene oxide (rGO) papers prepared by vacuum filtration

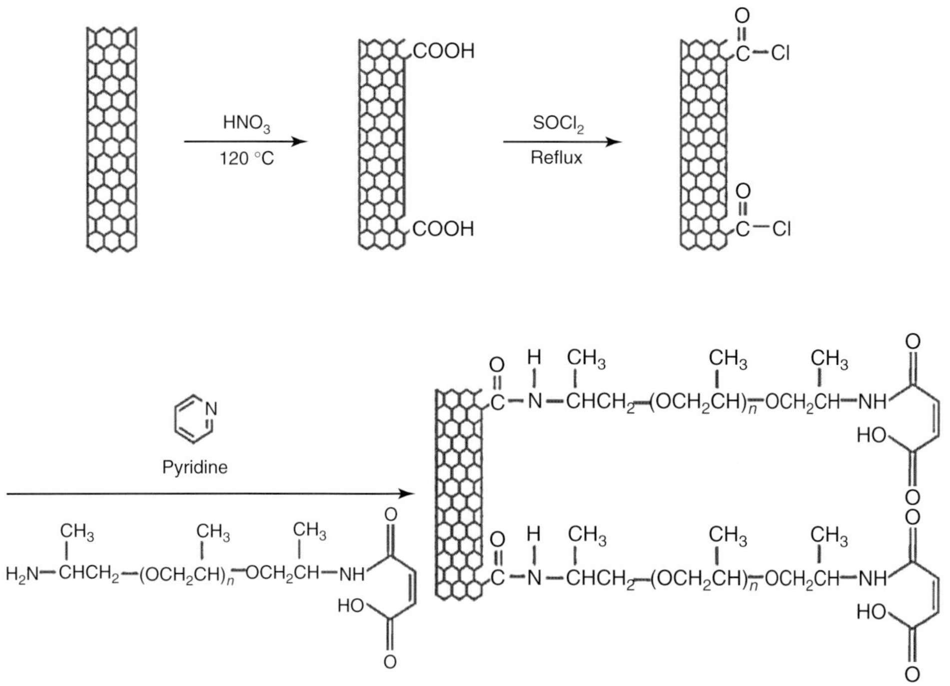

Figure 6.4 Synthesis scheme for MWNTs-POAMA. (Source: Reproduced from Ref. [22d] with permission from Elsevier.)

of colloidal suspensions can reach 118 and 351 S cm^{-1} by heat treating at 220 and 500 °C, respectively (Chapter 1, Ref. 127). With further improvements in the synthesis and processing of rGO, it is anticipated that high-quality rGO papers will be used as the filler components for polymer nanocomposites in near future. These rGO papers can be impregnated with thermosetting resin followed by curing to form bipolar plates.

Compared to CNTs, EGs and GNPs can be homogeneously dispersed in the polymer matrix without forming large agglomerates. Accordingly, the electrical conductivity of polymer/GNP nanocomposites can be greatly enhanced by increasing the EG or GNP content beyond the percolation threshold. For example, Song et al. [23] have developed and patented a process for fabricating polymer/GNP nanocomposites with GNP contents ≤15wt%. The nanocomposites exhibit electrical conductivity higher than 100 S cm^{-1}. Such novel nanocomposites with high GNP contents show potential applications for automotive bipolar plates.

From the literature, very scarce information is available on the fabrication and properties of EG/polymer and GNP/polymer bipolar plates [24–31]. In general, EG/polymer nanocomposite plates can be prepared by means of *in situ* polymerizationpolymerization and compression molding. Recently, Meng and coworkers [26] used *in situ* polymerization to prepare poly(arylenedisulfide)/EG nanocomposites. In this process, cyclic(arylene disulfide) oligomers were blended

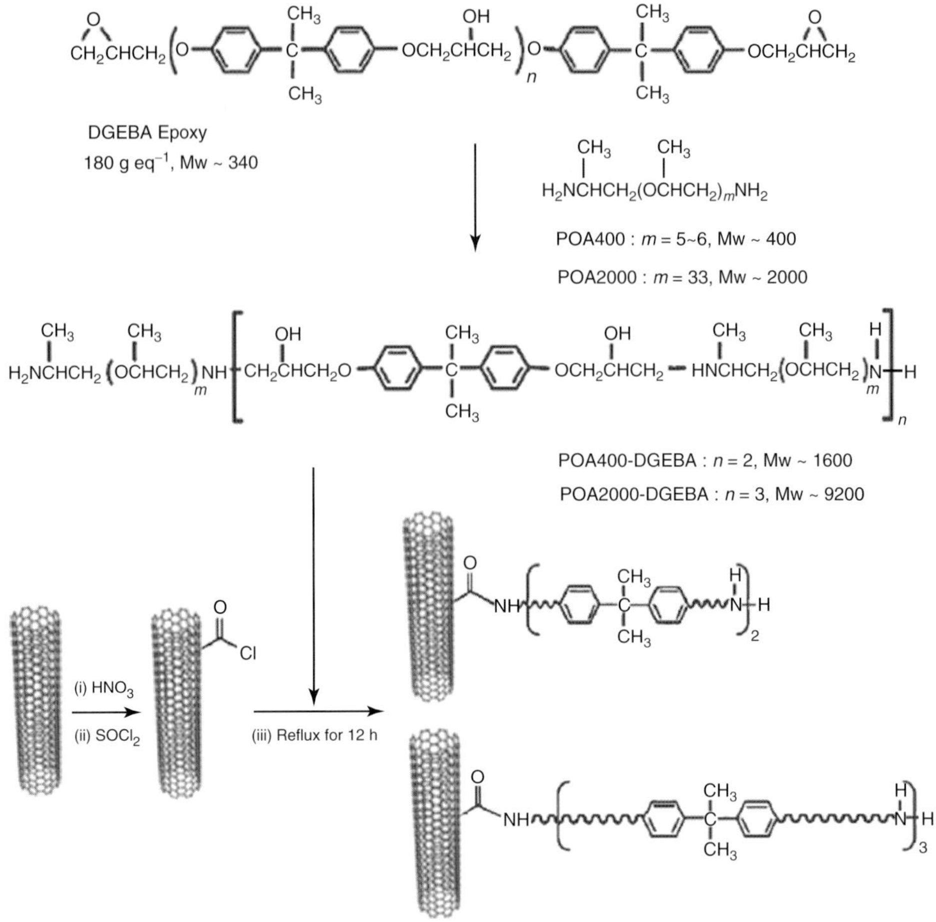

Figure 6.5 Synthesis scheme for MWNTs/POA-DGEBA. (Source: Reproduced from Ref. [22e] with permission from Elsevier.)

with the EG fillers. The oligomers first adsorbed on EG and then penetrated into EG pores. Thus aromatic polydisulfide/EG nanocomposites can be obtained by direct ring-opening polymerization of cyclic oligomer/EG precursor at 200 °C (Figure 6.7). A similar *in situ* polymerization was used to prepare polydisulfide/GNP nanocomposites [27]. Poly(arylene disulfide) exhibits high resistance to environmental degradation, low water vapor absorption, excellent resistance to acids and bases. This polymer is particularly suitable for use in acidic and humid environments of PEMFCs.

Compression molding is an effective process for fabricating epoxy-based composite plates. In this process, EG and epoxy resin were first mixed thoroughly and fed into a mold followed by hot compression [24, 25, 30]. In an alternative approach, EG was first pressed into EG sheets of different densities compressed expanded graphite (CEG). Then the CEG sheet was impregnated with resin in

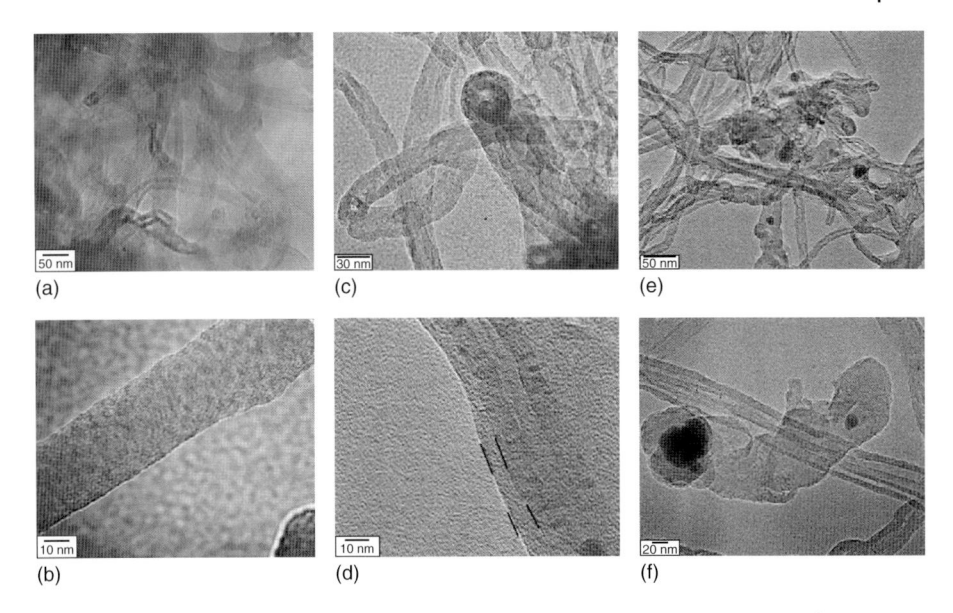

Figure 6.6 TEM images of: (a,b) pristine MWNTs, (c,d) MWNTs/POA400-DGEBA, and (e,f) MWNTs/POA2000-DGEBA. (Source: Reproduced from Ref. [22e] with permission from Elsevier.)

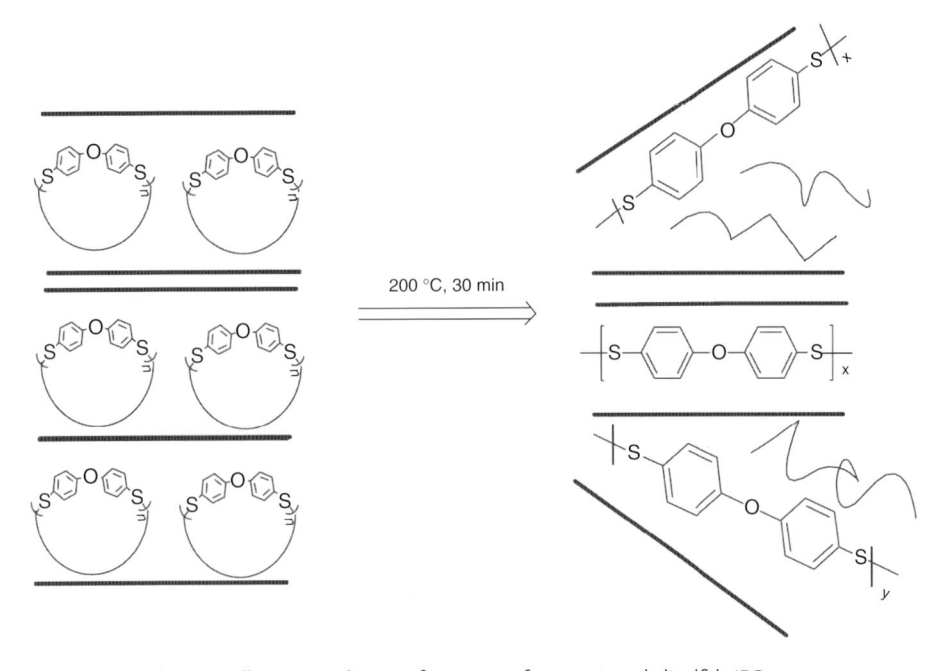

Figure 6.7 Schematic illustration showing formation of aromatic polydisulfide/EG nanocomposite via *in situ* ring-opening polymerization. (Source: Reproduced from Ref. [26] with permission from Elsevier.)

vacuum, compressed, and cured. This process is called *vacuum resin impregnation* [31] and developed independently by Gibb [32] and Yan *et al.* [33]. Very recently, Du *et al.* [31, 34] systematically studied the effect of resin content and different manufacturing process on the performances of epoxy/EG nanocomposites prepared by vacuum resin impregnation. The first process termed as the *impregnation-compression* involves direct impregnation of CEG sheets with epoxy resin solution followed by compression and other subsequent processing, for example, surface postprocessing, drying, and curing. This route allows the resin molecules to enter EG pores easily because of the loose and porous nature of

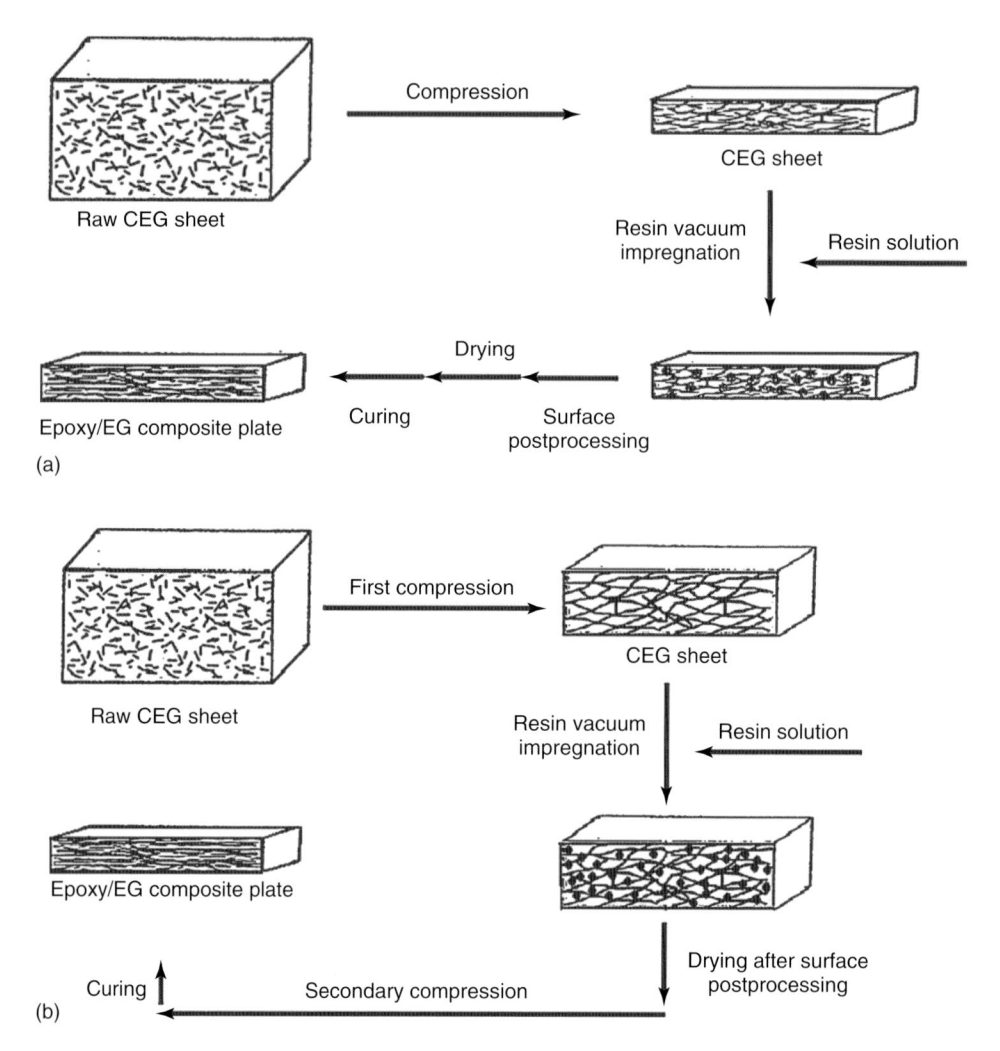

Figure 6.8 Schematic diagrams showing (a) Compression-impregnation and (b) compression-impregnation-compression processes. (Source: Reproduced from Ref. [34] with permission from Elsevier.)

raw CEG sheet. The compression step after the impregnation can further reduce the porosity. The second process is termed as the *compression-impregnation* in which CEG sheets are compressed into desired shape followed by resin solution vacuum impregnation and subsequent processing (Figure 6.8a). The third process is known as the *compression-impregnation-compression* that involves two steps of compression during processing (Figure 6.8b).

6.5
Electrical Characteristics of Nanocomposite Bipolar Plates

6.5.1
Polymer/Graphite/CNT Hybrid Plates

Lee et al. [35] investigated the properties of epoxy/graphite and epoxy/graphite/MWNT composites fabricated by hot pressing techniques. Figure 6.9 shows electrical conductivity versus graphite particle (180 µm) content for the epoxy/graphite particle composites. The conductivity rises with increasing graphite particle content, achieving a maximum value of 124 S cm^{-1} at 75 vol% graphite. Above this concentration, the conductivity decreases markedly because of insufficient amount of polymer resin to bind fillers. By incorporating pristine MWNT, the electrical conductivity of epoxy/graphite composites improves dramatically (Figure 6.10). The conductivity of hybrid composites increases with increasing MWNT content up to 2 vol%, thereafter decreases as a result of aggregation of pristine MWNTs. A maximum conductivity value of 254.7 S cm^{-1} is reached at a total filler loading (graphite and nanotube) of 75 vol%. This conductivity value is 105% higher than

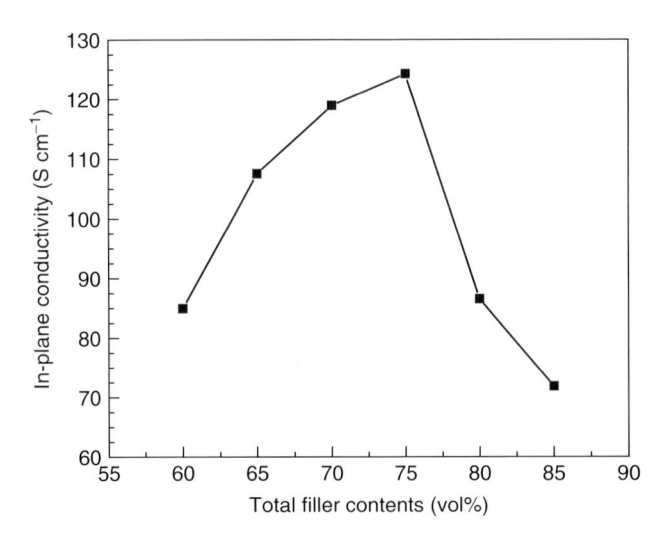

Figure 6.9 Electrical conductivity of epoxy/graphite composites as a function of graphite particle content. (Source: Reproduced from Ref. [35] with permission from Elsevier.)

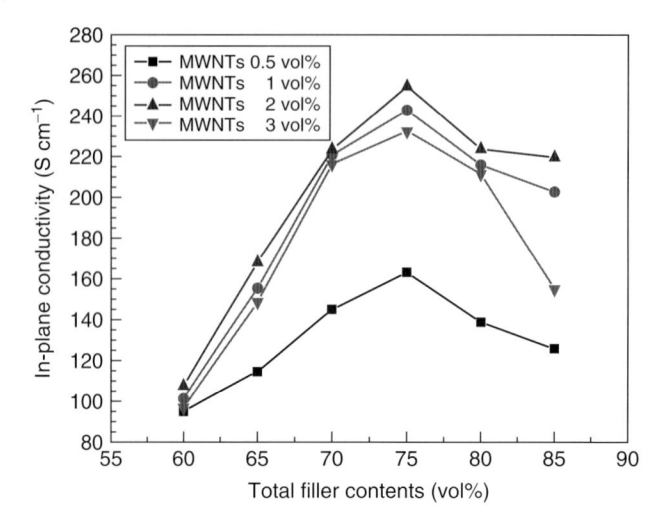

Figure 6.10 Electrical conductivity as a function of total filler content for epoxy/graphite/MWNT composites. (Source: Reproduced from Ref. [35] with permission from Elsevier.)

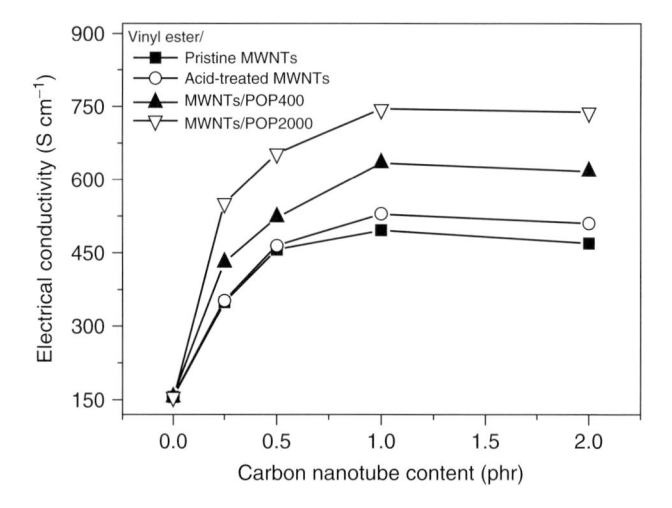

Figure 6.11 Electrical conductivity versus MWNT content for vinyl ester/graphite/MWNT bipolar plates. POP-diamines of different molecular weights were grafted onto MWNTs as shown in Figure 6.3. (Source: Reproduced from Ref. [22a] with permission from Elsevier.)

that of the epoxy/graphite composite with 75 vol% graphite. The enhancement in electrical conductivity of hybrid composites is attributed to the high aspect ratio of MWNTs. This facilitates the formation of additional conductive channels between graphite particles in the epoxy/graphite/MWNT 25/73/2 composite.

Figures 6.11 and 6.12 show the effects of adding MWNTs grafted with different functional groups on electrical conductivity of the VE/graphite 30/70 composite.

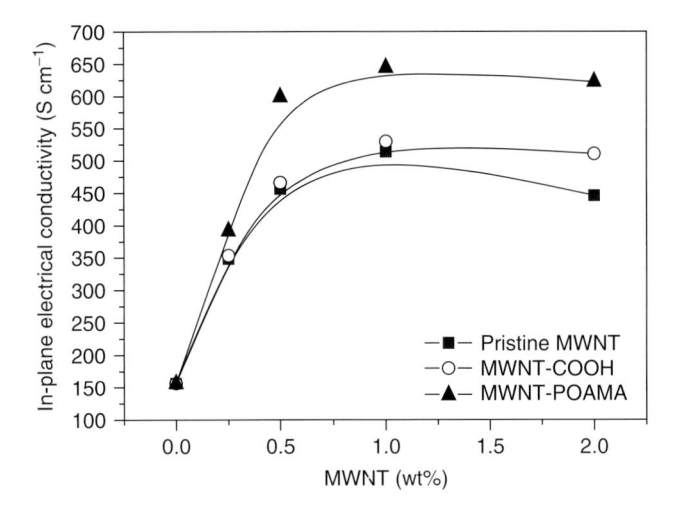

Figure 6.12 Electrical conductivity versus MWNT content for vinyl ester/graphite/MWNT bipolar plates. POA-MA was grafted onto MWNTs as shown in Figure 6.4. (Source: Reproduced from Ref. [22d] with permission from Elsevier.)

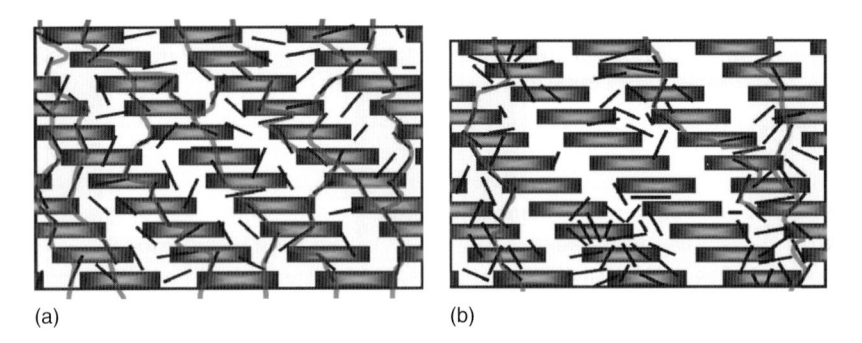

Figure 6.13 Schematic diagrams showing conductive paths in vinyl ester/graphite composite plates with (a) homogeneous MWNTs/POP2000 and (b) MWNTs aggregation. Gray rectangles are graphite microparticles and black rods are carbon nanotubes. (Source: Reproduced from Ref. [22a] with permission from Elsevier.)

The electrical conductivity of VE/graphite 30/70 composite improves considerably by adding pristine MWNTs, acid-treated MWNTs, and POP-grafted MWNTs. At 1 phr filler level, the conductivity values of pristine MWNT, MWNT-COOH, MWNTs/POP400, and MWNTs/POP2000, are found to be 496, 529, 633 and 744 S cm^{-1}, respectively. The electrical conductivity increases markedly in the order MWNTs/POP2000 > MWNTs/POP400 > MWNT-COOH > pristine MWNT composite plate. The high conductivity of the composite plate filled with MWNTs/POP2000 can be attributed to the uniform dispersion of MWNTs in the matrix, favoring formation of a large number of conducting channels between graphite microparticles (Figure 6.13a). In contrast, fewer conducting channels are established between graphite microparticles by adding pristine MWNTs

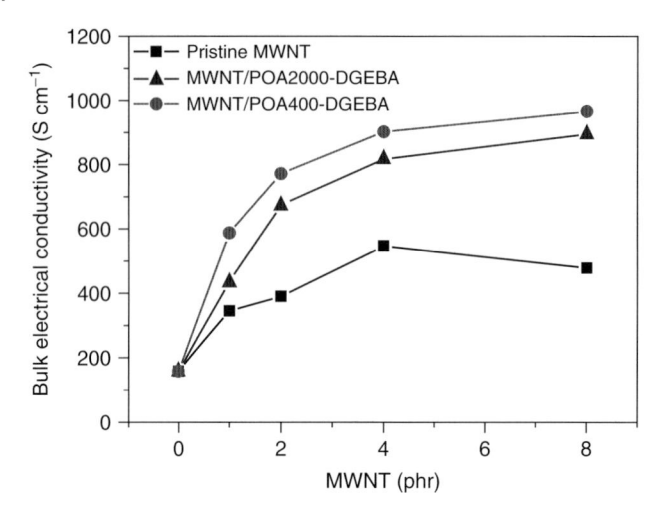

Figure 6.14 Electrical conductivity versus MWNT content for PP/graphite/MWNT bipolar plates with 80 wt% graphite. POA400-DGEBA and POA2000-DGEBA were grafted onto MWNTs as shown in Figure 6.5. (Source: Reproduced from Ref. [22e] with permission from Elsevier.)

(Figure 6.13b). The conductivity versus MWNT content for the PP/graphite 20/80 plates is shown in Figure 6.14. The beneficial effect of MWNTs/POA-DGEBA400 additions on enhancing the electrical conductivity can be readily observed.

6.5.2
Polymer/EG and Polymer/GNP Composite Plates

Figure 6.15 shows electrical conductivity versus EG content for *in situ* polymerized poly(arylenedisulfide)/EG nanocomposites. EGs of different expansion ratios were used as nanofillers. The electrical conductivity of nanocomposites increases with increasing EG content and EG expansion ratio. The conductivity of nanocomposite with 50 wt% EG having an expansion ratio of 3 is 129 S cm^{-1}, that is, higher than the DOE target value. For the purpose of comparison, the conductivity behavior of poly(arylenedisulfide)/GNP nanocomposites is also shown in this figure. Obviously, the poly(arylenedisulfide)/GNP nanocomposites display higher conductivity than poly(arylenedisulfide)/EG with the same filler content because of the high aspect ratio of GNPs.

Figure 6.16 shows the variation of in-plane electrical conductivity with resin content for phenolic resin/EG nanocomposites prepared by compression molding [25]. The conductivity decreases with increasing resin content. This is due to an increase of insulating resin amount in the composites. Nanocomposites with resin content 40–45 wt% are sufficient to obtain conductivity value >100 S cm^{-1}.

Figure 6.17 shows the variation of in-plane electrical conductivity with resin solution concentration for the epoxy/EG nanocomposites prepared from a series combination of vacuum resin impregnation and compression molding [34]. For the

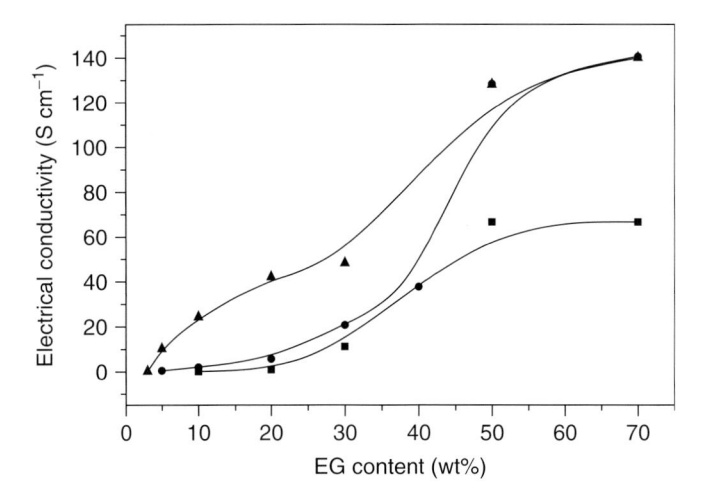

Figure 6.15 Electrical conductivity versus EG content for poly(arylenedisulfide)/EG nanocomposite. ■: EG expansion ratio of 35; ●: EG expansion ratio of 100; ▲: EG expansion ratio of 200. △: GNP. (Source: Reproduced from Ref. [26] with permission from Elsevier.)

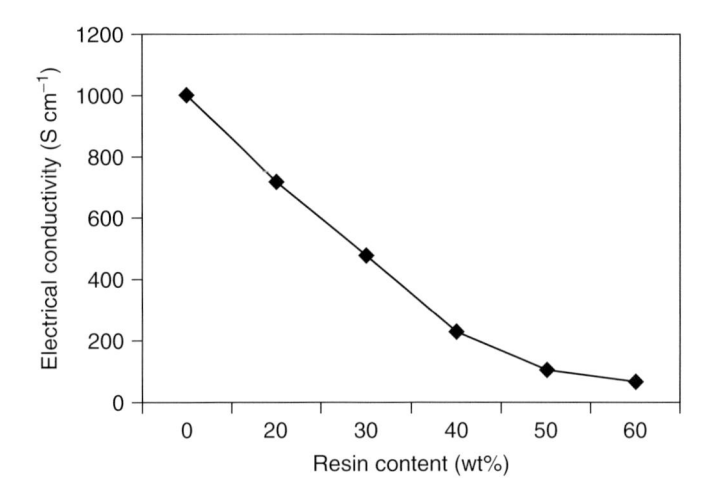

Figure 6.16 Electrical conductivity as a function of resin content for EG/phenolic resin composites. (Source: Reproduced from Ref. [25] with permission from The American Chemical Society.)

impregnation-compression route, the conductivity decreases with increasing resin solution concentration. Because of the loose porous structure of CEG sheet, resin molecules enter into the CEG pores readily, forming dense materials and disturbing the graphite conductive network structure. For the compression-impregnation route, the conductivity shows little changes with resin solution concentration ranging from 5 to 40%. The resin molecules enter into the interlayer pores

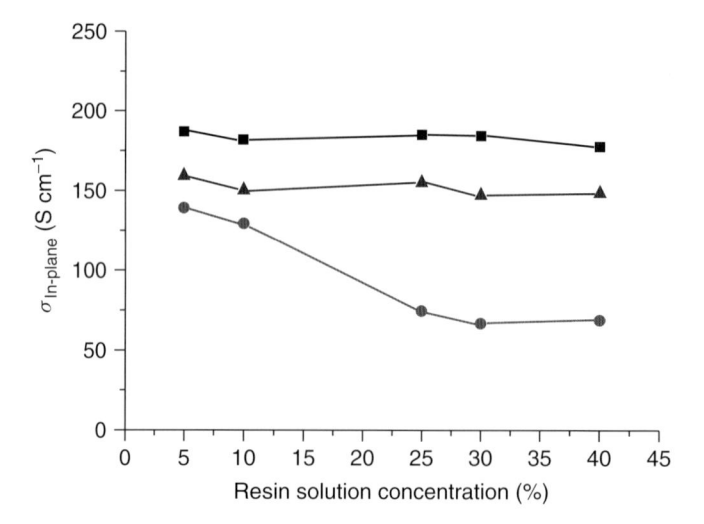

Figure 6.17 Variation of in-plane electrical conductivity with resin solution concentration content for epoxy/EG nanocomposites prepared by different processes. ●: Impregnation-compression; ■: compression-impregnation; and ▲: compression-impregnation-compression. (Source: Reproduced from Ref. [34] with permission from Elsevier.)

that cannot destroy the graphite conductive network. The conductivity of the composite system prepared from the "compression-impregnation-compression" system is between those of other two systems. Table 6.4 summarizes binary polymer nanocomposites and ternary graphite/polymer composites containing low loading levels of MWNTs or GNPs.

6.5.3
Polymer/CB Composite Plates

CB-filled polymer nanocomposite plates generally have very low electrical conductivity. Thus they are unsuitable for use as the filler materials in bipolar plates. Del Rio et al. [36] prepared PVDF/CB nanocomposites filled with 20, 30, and 40 wt% CB particles (20 nm) by means of injection molding. The highest conductivity obtained is 2.36 S cm^{-1} for the PVDF/CB 60/40 nanocomposite at 30 °C. This value is far smaller than that of the DOE target requirement. Recently, Chen and Kuo [37] used injection molding to fabricate the PA6/CB (100–600 nm) composite containing 10 wt% carbon nanoparticles. The measured electrical conductivity is 11.67 S cm^{-1}. Moreover, the resistivity of such a nanocomposite increases markedly at temperatures exceeding 90 °C. These results demonstrate that the electrical performance of the CB-filled composite system is unsatisfactory for use as a single filler component for bipolar plate applications. They can only employ as a hybrid filler for the graphite/polymer composite bipolar plates.

Table 6.4 Properties of binary polymer nanocomposites and ternary graphite/polymer composites containing MWNTs or GNPs.

Polymer matrix	Binary composite	Ternary hybrid composite	Processing method	Thermal conductivity (W m^{-1} K^{-1})	Electrical conductivity (S cm^{-1})		Mechanical strength (MPa)	
	Nanofiller (%)	Graphite/filler (%/%)			In-plane	Through-plane	Tensile	Flexural
Epoxy (Chapter 5, Ref. [65])	16.5 CNT	–	Resin transfer molding	–	130.8	–	231.5	–
Arylenedisulfide Ref. [27]	30 GNP	–	In situ polymerization	–	162	–	–	32
Arylenedisulfide Ref. [27]	30 EG	–	In situ polymerization	–	125	–	–	28.5
Phenol formaldehyde [25]	55 EG	–	Compression	–	250	–	–	56
Epoxy Ref. [28]	50 EG	–	Compression	–	248	77	45	72
Epoxy Ref. [31]	70 EG	–	Vacuum resin impregnation	–	119.8	–	–	45.8
Phenolic resin Ref. [21]	–	64 graphite/1 MWNT	Compression	50	178	30	–	56
Vinyl ester Ref. [22d]	–	69 graphite/1 FMWNT	Bulk-molding compound	–	634	–	–	39.4
Vinyl ester Ref. [22f]	–	69 graphite/1 MWNT	Bulk-molding compound	25.0	233.7	–	–	47.6
Vinyl ester Ref. [22f]	–	69.8 graphite/0.2 GNP	Bulk-molding compound	27.2	286.4	–	–	49.2
Low crystalline polypropylene Ref. [22c]	–	79 graphite/1 MWNT	Compression	–	346	–	–	23
Low crystalline polypropylene Ref. [22c]	–	76 graphite/4 MWNT	Compression	–	548	–	–	32.6

FMWNTs: functionalized multiwalled carbon nanotubes.

6.5.4
Fuel Cell Performance

6.5.4.1 Voltage Loss

Voltage loss or overvoltage occurs at the electrodes during the operating period of PEMFCs, leading to a marked reduction in the cell efficiency and performance. Three major types of voltage losses are found in fuel cells including: (i) activation overpotential associated with slow reaction kinetics in the cathode catalyst layer, (ii) mass transport losses due to failure of transport of sufficient reactant to the electrodes and (iii) ohmic losses resulting from low electronic and protonic conduction through the bipolar plate and the polymer electrolyte, respectively [38]. A polarization curve is used to describe the performance of a fuel cell by plotting the cell voltage behavior against operating current density. Figure 6.18 is a schematic diagram showing polarization curve and voltage losses of the PEMFC [39]. This figure shows that the reversible cell voltage (1.23 V) is above the actual open circuit voltage (OCV) of the curve. This is the theoretical cell voltage determined from the Gibbs free energy of the fuel cell reactions and assuming no energy loss during the energy conversion process. Apparently, reducing the activation loss of the cathode catalyst is the most effective way for improving cell efficiency. Furthermore, the GDL must be optimized to reduce mass transport loss. And the reduction of membrane resistance is necessary to reduce the ohmic loss and contact resistance between each component [39].

The activation overvoltage (ΔV_{act}) can be well described by the Tafel equation given as Refs. [40, 41]

$$\Delta V_{act} = \beta \ln\left(\frac{i}{i_0}\right) \tag{6.3}$$

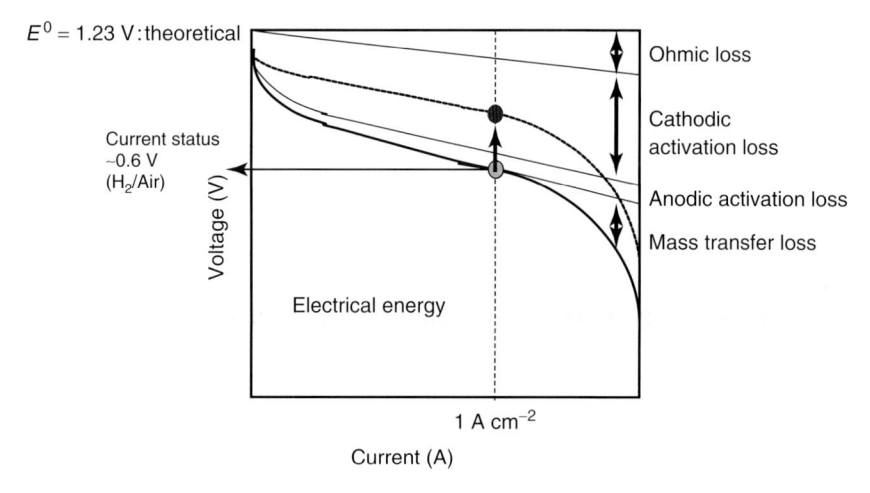

Figure 6.18 I–V curve and losses for PEMFC. (Source: Reproduced from Ref. [39] with permission from Wiley Interscience.)

where β is the Tafel slope and i_0 the exchange current density for an electrochemical reaction. From Eq. (6.3), ΔV_{act} can be suppressed by increasing i_0. Platinum group metals are well recognized to be very effective to increase i_0 value of the hydrogen electrode reaction (Equation (6.1)). Accordingly, the overvoltage at the anode is rather small compared to that of the cathode. The i_0 value for oxygen reduction at the cathode is much smaller that at the hydrogen anode, typically 10^5 times smaller [38].

The overvoltage caused by the change in concentration of the reactants at the surface of the electrodes as the fuel is used is termed *concentration overpotential*. The reduction in concentration is the result of the concentration gradient required to transport sufficient reactant to the electrode surface. Mathematically, it can be expressed as

$$\Delta V_{con} = \frac{RT}{nF} \ln\left(1 - \frac{i}{i_L}\right) \tag{6.4}$$

where R and T are the gas constant and absolute temperature, respectively, F the Faraday constant ($=96500\ C\ mol^{-1}$), n the number of electrons involved in an electrochemical reaction ($n = 4$ for the ORR) and i_L the limiting current density. This type of loss usually occurs at high-current-density region because of the high demand of the reactants and only limited amount of reactants can be transported to electrode surfaces.

The overvoltage from the ohmic losses is resulted from the resistance to the ionic and electronic flows. The resistance to the flow of ions through the electrolyte is termed as *ionic resistance*. The electronic resistance includes the resistance to the flow of electrons through the cell interconnects or bipolar plates. The overpotential of ohmic loss can be expressed in terms of Ohm's Law

$$\Delta V_{ohm} = ir \tag{6.5}$$

where i is the current density of the cell and r the overall area specific contact resistance (Ωcm^2) within the cell. Thus good design and proper use of appropriate materials for bipolar plates can reduce the internal resistance of the cell.

6.5.4.2 Polarization Curves

Figure 6.19 shows polarization curves and the performance of single cells assembled from pristine MWCNTs, MWNTs-COOH and MWNTs-POA-MA filled VE composite plates, and VE/graphite particle 30/70 (wt%) composite plate, respectively. The OCV of single cells for all systems is located at 1 V. The overvoltages due to cathodic activation and mass transfer losses can be readily seen in the voltage versus current density plots. The POA-MA group is grafted onto MWNTs via acid oxidation, acylation, and amidation processes. Bipolar plates with both pristine MWNT and FMWNT generally outperform the plate without nanotubes. The MWNTs/POA-MA plate exhibits a maximum current density of 1230 mA cm^{-2} and a maximum power density of 0.518 W cm^{-2}. The power density of this MWNTs/POA-MA plate is 32% higher than that of the cell with VE/graphite 30/70 (wt%) composite (i.e., 0.392 W cm^{-2}). The higher performance of the cell with MWNTs/POA-MA plate is attributed to better dispersion of nanotubes in the VE matrix due to the tube

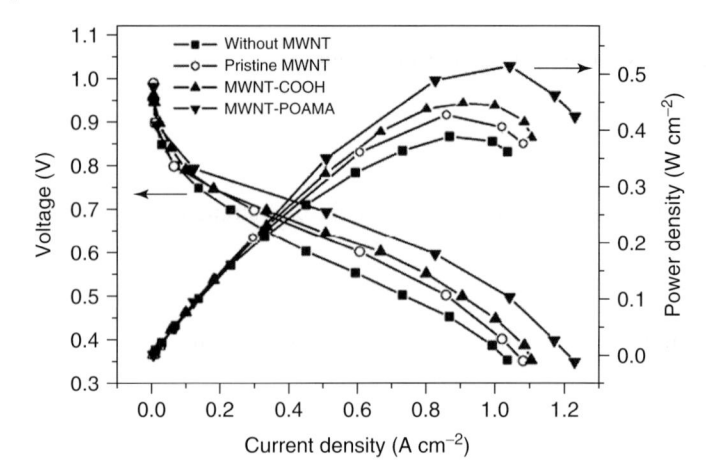

Figure 6.19 Performance of the single cells assembled with pristine MWNTs, MWNTs-COOH, MWNTs-POAMA, and without MWNT composite bipolar plates. POA-MA was grafted onto MWNTs via a series of acid oxidation, acylation, and amidation process. (Source: Reproduced from Ref. [22d] with permission from Elsevier.)

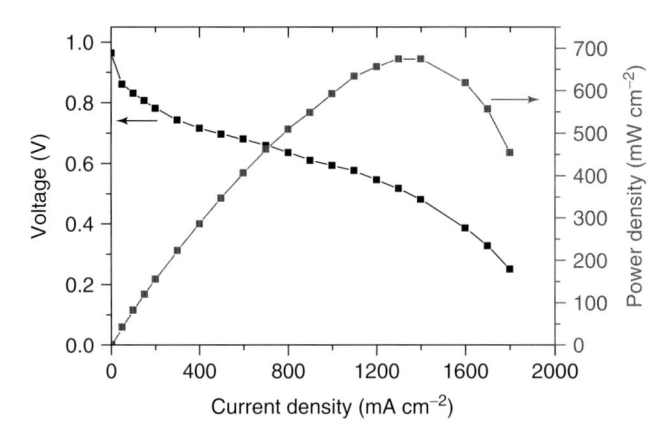

Figure 6.20 Polarization and performance curves of the single cell with epoxy/EG 30/70 (wt%) bipolar plates. (Source: Reproduced from Ref. [31] with permission from Elsevier.)

functionalization. A similar beneficial effect in the cell performance is found by incorporating MWNT-POP2000 into VE matrix of bipolar plates [22a].

The polarization and power density curves of PEMFC single cell assembled with epoxy/EG 30/70 composite bipolar plates are shown in Figure 6.20. From the voltage–current responses, a maximum power density value of 0.674 W cm^{-2} with a current density of 1400 mA cm^{-2} can be achieved. It is apparent that the cell performance of the epoxy/EG 30/70 composite bipolar plate is even higher than that of the VE/graphite/MWNTs-POAMA and VE/graphite/MWNTs-POP2000 hybrid plates. Thus low-cost, two-dimensional EGs with ease of fabrication can replace expensive pristine or FMWNTs for enhancing performance of PEMFCs.

6.6
Mechanical Properties of Nanocomposite Bipolar Plates

6.6.1
Hybrid Composite Bipolar Plates

Figure 6.21a,b shows the respective plots of the flexural strength and modulus versus graphite particle (180 μm) content for conventional epoxy/graphite particle composites prepared by hot pressing. The flexural strength and modulus of the composites decrease with increasing graphite particle content. Thus graphite microparticles mainly act as conducting materials rather than reinforcing fillers

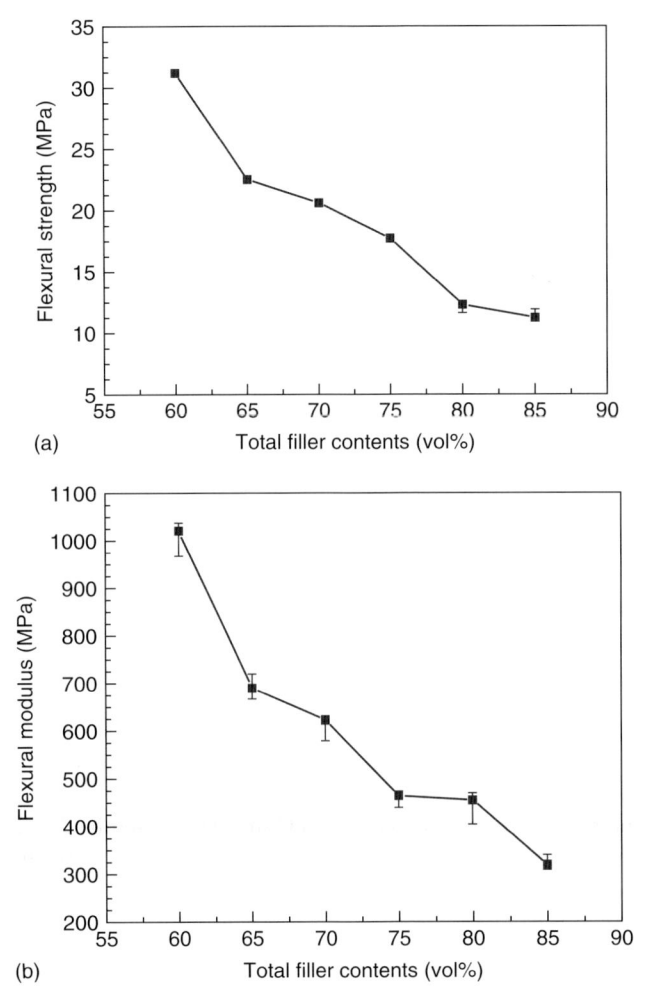

Figure 6.21 Flexural (a) strength and (b) modulus of epoxy/graphite composites. (Source: Reproduced from Ref. [35] with permission from Elsevier.)

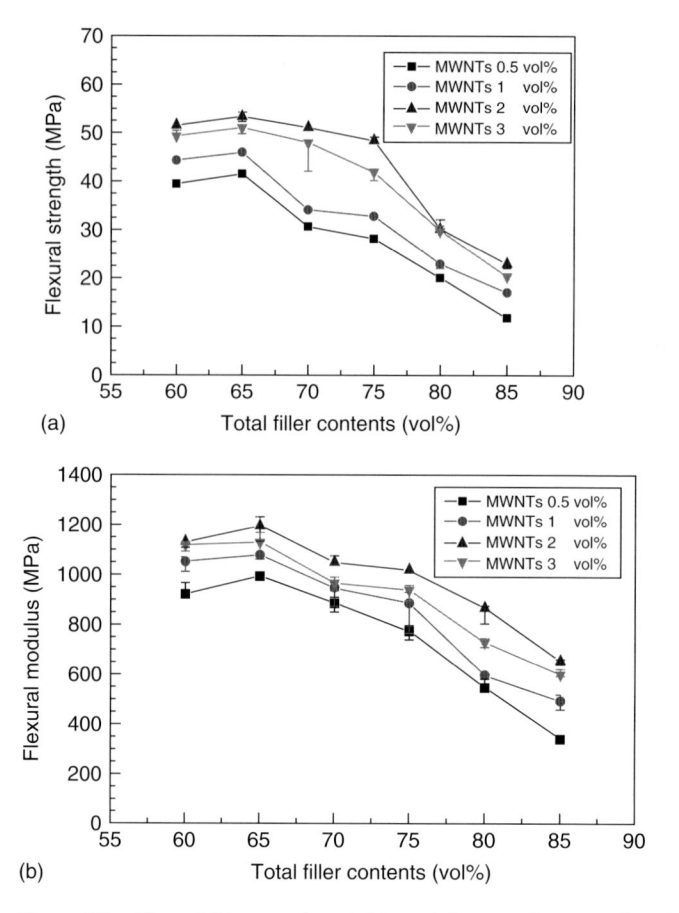

Figure 6.22 Flexural (a) strength and (b) modulus as a function of total filler content for epoxy/graphite/pure MWNT composites. (Source: Reproduced from Ref. [35] with permission from Elsevier.)

because of the low aspect ratios. However, MWNTs with extraordinary high stiffness and strength can reinforce epoxy/graphite composites effectively (Figure 6.22a,b). These figures reveal that the flexural strength and modulus of bipolar plates increases with increasing MWNT up to 2 vol%, thereafter decreases as the MWNT content is further increased because of the agglomeration of pristine nanotubes.

Functionalization of CNTs can improve their dispersion in the polymer matrix and enhance the interfacial bonding between the matrix and fillers, thereby promoting effective stress transfer across the matrix–filler interface. Figure 6.23 shows the variation of flexural strength with MWNT content for the VE/graphite 30/70 composite. The MWNTs used include pristine, acid-oxidized, and POAMA-grafted nanotubes prepared via acidification, acylation, and amidation. The nanotube additions are effective to improve the flexural strength of VE/graphite composite bipolar plate. By increasing the nanotube content to 2 wt%, the flexural

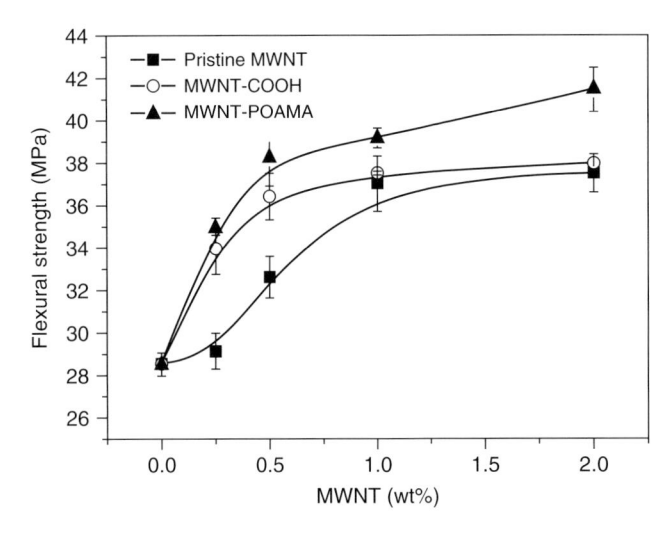

Figure 6.23 Flexural strength versus MWNT content for vinyl ester/graphite/MWNT particle bipolar plates. (Source: Reproduced from Ref. [22d] with permission from Elsevier.)

strengths of VE nanocomposite bipolar plates reinforced with pristine MWNTs, MWNTs-COOH, and MWNTs-POAMA are determined to be 37.51, 37.96, and 41.44 MPa, respectively. The flexural strength of the composite bipolar plate with 2 wt% MWNTs-POAMA improves by more than 40% over pure VE/graphite 30/70 composite. The increase in flexural strength derives from the high aspect ratio and functionalization of MWNTs. The latter aspect facilitates stronger MWNT–polymer interaction, enabling load transfer from the host polymer matrix to the MWNTs more efficiently during mechanical testing.

For thermoplastic PP matrix, functionalized nanotubes also exhibit high reinforcing effect. Figure 6.24 shows the effect of pristine and FMWNT additions on the flexural strength of PP/graphite 20/80 composite. MWNTs were grafted with POA400-DGEBA and POA2000-DGEBA as mentioned previously. All the MWNT-filled PP/graphite bipolar plates are stiffer than pure composite bipolar plate without MWNTs. Furthermore, FMWNTs, particularly those grafted with POA400-DGEBA are found to be very effective to reinforce PP/graphite composite bipolar plates. When the MWNT content reached 8 phr, the flexural strengths of PP/graphite composite bipolar plates filled with pristine MWNTs, MWNTs/POA400-DGEBA, and MWNTs/POA2000-DGEBA were 29.49, 34.18, and 31.81 MPa, respectively. The flexural strength of the PP/graphite bipolar plate with 8 phr loading of MWNTs/POA400-DGEBA improves by more than 59% than that of the pure composite bipolar plate without MWNTs. From Figure 6.24, the flexural strength of PP/graphite 20/80 (wt%) composite without nanotubes is only 21.44 MPa, smaller than that of the DOE target value (25 MPa). It is quite obvious that the flexural strength of PP/graphite 20/80 (wt%) composite can be raised to the DOE target requirement by adding low loading levels of MWNTs.

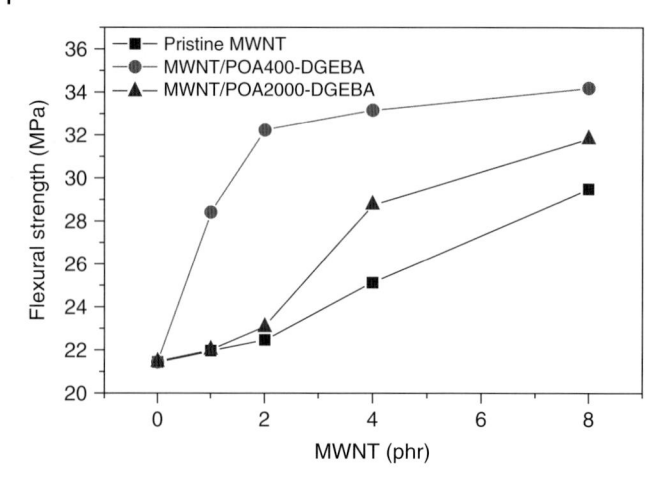

Figure 6.24 Flexural strength of PP/graphite/MWNT composite bipolar plates with 80% graphite and various MWNTs. (Source: Reproduced from Ref. [22e] with permission from Elsevier.)

6.6.2
Polymer/EG Composites

EGs of different expansion ratios are incorporated into high-performance poly(arylenedisulfide) matrix. The flexural strength versus EG content is shown in Figure 6.25. Additions of EGs with expansion ratios of 35 and 200 to neat aromatic polydisulfide result in a decrease in the flexural strength. However, the flexural strength of poly(arylenedisulfide) remains nearly unchanged by adding EGs with

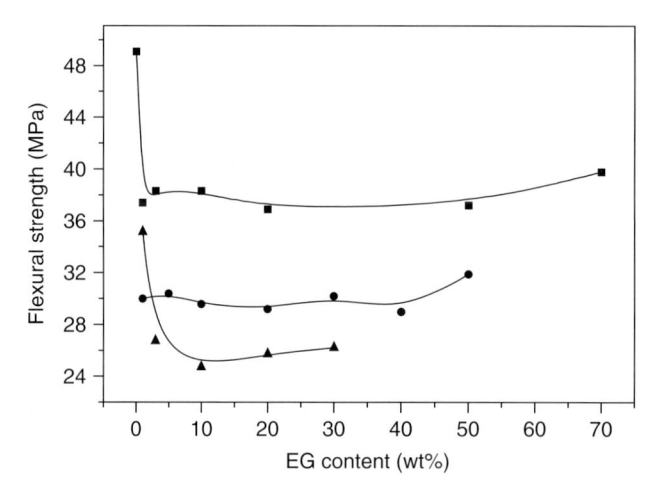

Figure 6.25 Flexural strength versus EG content for poly(arylenedisulfide)/EG nanocomposite. ■: EG expansion ratio of 35; ●: EG expansion ratio of 100; and ▲: EG expansion ratio of 200. (Source: Reproduced from Ref. [26] with permission from Elsevier.)

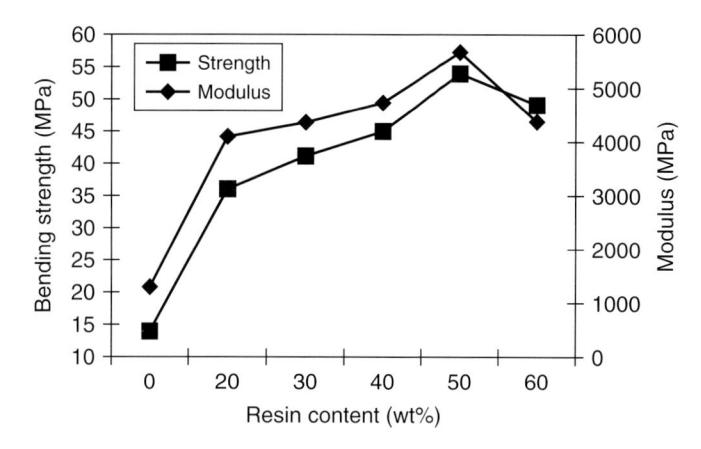

Figure 6.26 Variations of bending strength and modulus with resin content for EG/phenolic resin nanocomposites. (Source: Reproduced from Ref. [25] with permission from the American Chemical Society.)

an expansion ratio of 100. In general, the flexural strength of all composites meets the DOE target value of 25 MPa.

Figure 6.26 shows the variations of flexural modulus and strength with resin content for phenolic resin/EG composites prepared by hot molding [25]. The flexural (bending) strength and modulus of the nanocomposites increase with increasing resin content up to 50 wt%, thereafter decrease substantially. For the nanocomposites with lower resin content, the properties of materials are mostly dominated by EG, resulting in fragile and brittle nature. When the resin content approaches 40–50 wt%, the mechanical properties of nanocomposites improve considerably.

Figure 6.27 shows the flexural modulus versus resin solution concentration for EG/epoxy nanocomposite systems prepared from a series of combination of vacuum resin impregnation and compression molding. The flexural strength of high density composites prepared from the impregnation-compression route decreases with increasing resin concentration. On the contrary, the flexural strength of the composites prepared from other two routes increases with increasing resin solution concentration. More resin content tends to reduce the porosity of composites, thereby improving the flexural strength. For the impregnation-compression route, the low resin solution concentration ($\leq 10\%$) is the optimum range for obtaining composites with outstanding gas impermeability, flexural strength, and electrical conductivity. For the compression-impregnation route, the high resin solution concentration (40%) is the optimum condition for preparation. The optimum range of the resin solution concentration is between 25 and 40% for the "compression-impregnation-compression" route. It can be concluded that the epoxy/EG composites prepared by different routes can meet the DOE target values for bipolar plate applications provided the optimum conditions are selected.

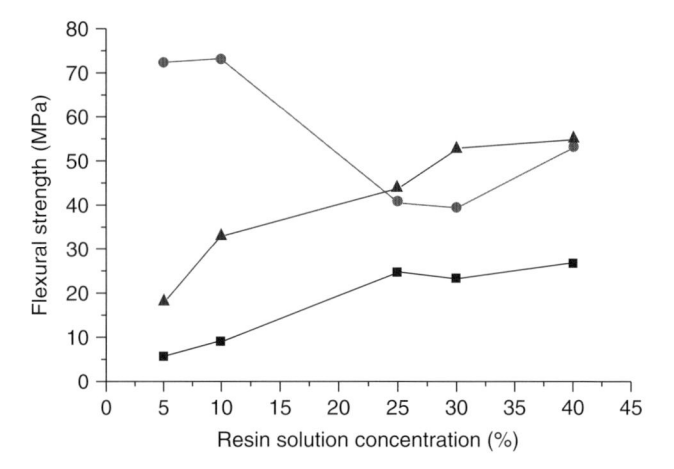

Figure 6.27 Variation of flexural strength with resin solution concentration content for epoxy/EG nanocomposites prepared by different processes. ●: Impregnation-compression; ■: compression-impregnation; and ▲: compression-impregnation-compression. (Source: Reproduced from Ref. [34] with permission from Elsevier.)

6.7
Electrocatalyst Supports

6.7.1
Carbon-Nanotube-Supported Platinum Electrocatalysts

The exchange current density for hydrogen oxidation reaction (HOR; reaction (6.1)) of a platinum electrode at 20–25 °C is 10^{-2} Acm^{-2}, nearly eight orders of magnitude larger than that of ORR on a Pt electrode having a value of 9×10^{-11} Acm^{-2} [42]. The activation overvoltage of HOR is considerably small compared with that of ORR according to Eq. (6.3). The overall electrochemical kinetics of PEMFCs is therefore limited by the relatively slow ORR. The insufficient activity and durability of electrocatalysts are the main limiting factors for achieving a high power output of PEMFCs.

The support for Pt catalysts practically should have several functions including: (i) high surface area and good electronic property, (ii) good reactant gas access to the electrocatalysts, and (iii) high electrochemical stability under fuel cell operating conditions [43]. Platinum is generally supported by electrically conductive carbon particles with high surface areas to form supported catalysts. CBs such as Vulcan XC-72, Black Pearls 2000, Ketjen black, and so on are commonly used for this purpose [44]. However, CBs undergo electrochemical oxidation to surface oxides, and eventually to carbon dioxide at the cathode of a fuel cell, where low pH, high humidity, and high potential conditions prevail [45]. Oxidation of CB support can lead to the loss of Pt catalyst particles from the electrode. In this regard, CNTs are attractive catalyst supports because of their large specific surface areas, high chemical and thermal stability, exceptionally high strength, and superior electrical conductivity. CNT-supported platinum catalysts offer a dramatic improvement in the ORR activity when compared with conventional carbon supports [43, 46–52]. For example, Matsumoto et al. [46] reported that the 12 wt% Pt-deposited CNT electrode gives 10% higher voltage than the 29 wt% Pt-deposited CB. Such nanotube electrode also reduces the Pt usage by 60% in PEMFCs. Wang et al. [52] demonstrated that MWNTs exhibit higher electrochemical stability, better durability, and improved oxidation resistance over Vulcan XC-72. MWNT-catalyst supports have lower loss of Pt and higher ORR activity during the measurements. In spite of this, the synthesis of highly dispersed Pt nanoparticles supported on CNTs with high particle loadings still remains a big challenge.

Since CNTs are relatively inert, functionalization of nanotubes is needed to facilitate the deposition of Pt particles on their surfaces. Deposition of Pt nanoparticles on acid FMWNTs can be achieved by reacting Pt salt precursors such as K_2PtCl_4 and H_2PtCl_6 with ethylene glycol (EG) or formaldehyde (HCHO). In the process, EG acts as a reducing agent by supplying electrons to the reaction. The carboxyl, carbonyl, and hydroxyl molecules of functionalized nanotubes with negatively charged groups act as anchors for Pt^{2+} ions via electrostatic interaction. Heating the solution facilitates the oxidation of EG to form oxalic acid and glycolic acid, thereby producing sufficient electrons to reduce the Pt^{2+} ions to the Pt^0 atoms [53]. The nucleation of Pt particles starts only when a critical concentration of zerovalent atoms is reached [54]. Microwave irradiation through a dielectric medium enables rapid conversion of EG and Pt salt precursor into Pt metal catalysts, thus reducing the synthesis time significantly [55–57].

Figure 6.28 shows schematic diagrams for depositing Pt nanoparticles onto MWNTs using polyol synthesis method [55]. In general, Pt nanoparticles tend to aggregate into clusters owing to less surface functional groups being available for anchor sites (Figure 6.29). To improve the dispersion of Pt nanoparticles on MWNTs, Xing et al. [58, 59] employed a sonochemical method to treat MWNTs in nitric and sulfuric acids in which more surface functional groups are created for Pt nanoparticle deposition. Figure 6.30a–c shows TEM images of Pt nanoparticles supported on sonochemically treated MWNTs. Dense deposition of Pt nanoparticles up to 30 wt% loading can be achieved. These nanoparticles are dispersed uniformly on the nanotube surfaces, particularly for low Pt loading catalyst. The mean particle size of Pt nanoparticles determined from image analysis is 2.78, 3.57, and 4.46 nm for the 10, 20, and 30 wt% catalysts, respectively. Cyclic voltammograms (CV) of three catalysts supported on MWNTs in 1 M H_2SO_4 acid is shown in Figure 6.31. During CV testing, a cyclic linear potential is imposed onto the electrode immersed in an electrolyte, and the resulting current is recorded. Cyclic voltammetry can give information on the kinetics of the electrode reaction and mass transport associated with it. From Figure 6.31, the cathodic and anodic peaks appearing at ~0.05–0.3 V originate from the adsorption and desorption of atomic hydrogen in an acid medium. Both the peak current densities of hydrogen adsorption/desorption increase in the following order: 10 wt% Pt catalyst > 20 wt% Pt catalyst > 30 wt% Pt catalyst. The 10% catalyst exhibits the largest current density in the hydrogen

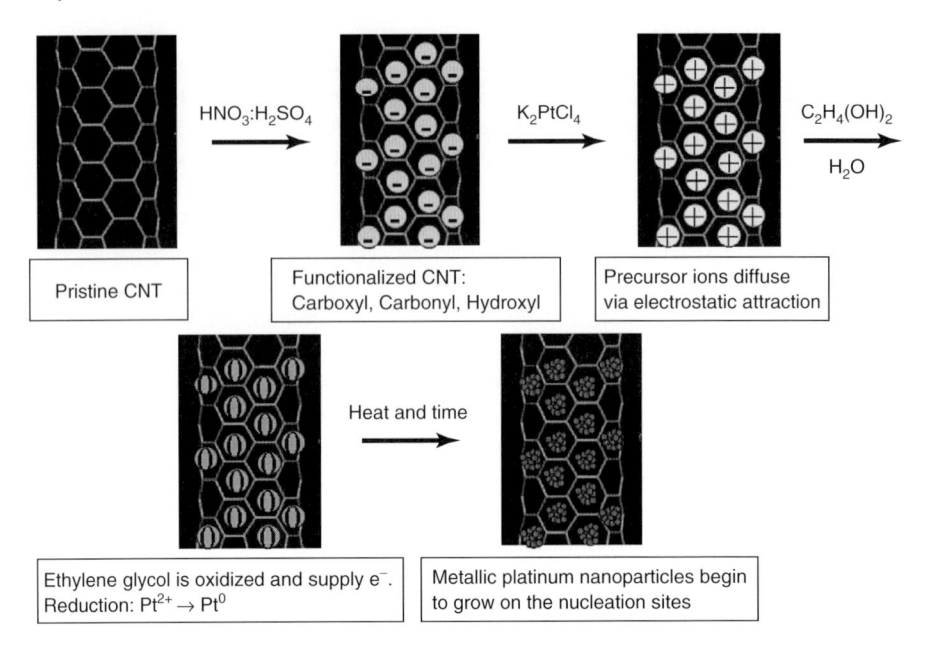

Figure 6.28 Schematic representation of the working mechanism behind the polyol synthesis method. (Source: Reproduced from Ref. [55] with permission from Elsevier.)

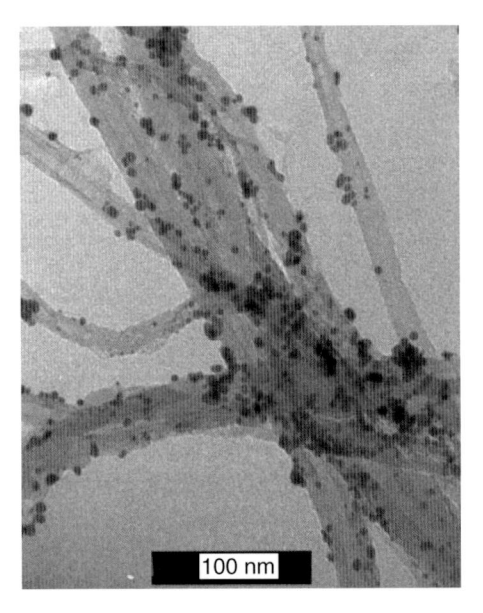

Figure 6.29 TEM micrograph showing deposition of 20 wt% Pt nanoparticles on refluxed carbon nanotubes. (Source: Reproduced from Ref. [58] with permission from the American Chemical Society.)

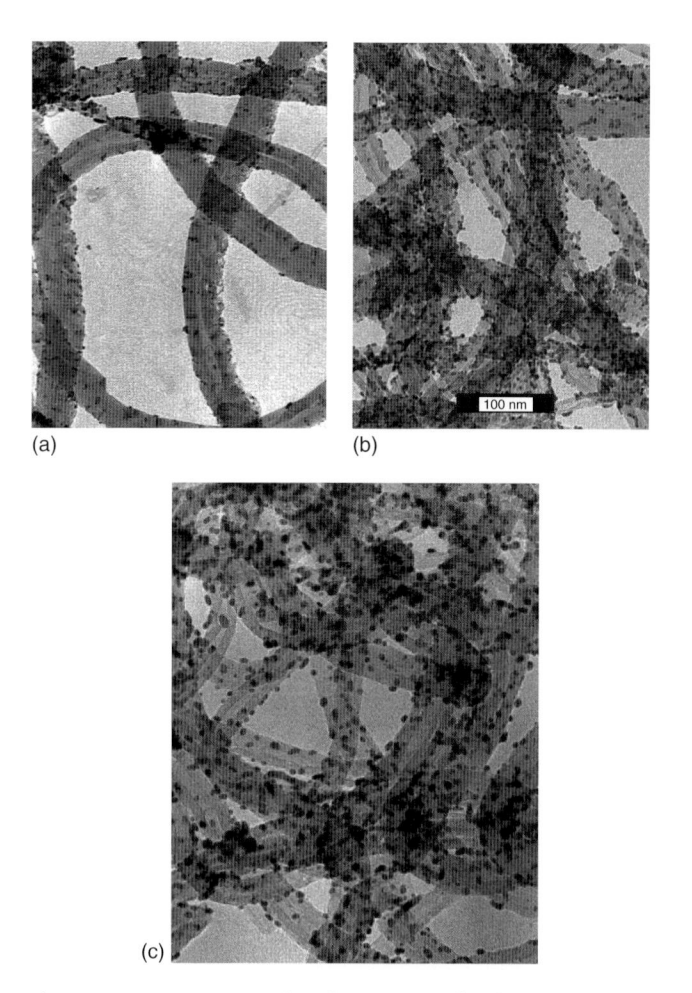

Figure 6.30 TEM micrographs of Pt nanoparticles deposited on sonochemically treated carbon nanotubes with Pt loading of (a) 10 wt%, (b) 20 wt%, and (c) 30 wt%, respectively. (Source: Reproduced from Ref. [58] with permission from the American Chemical Society.)

region because of its finest particle size and better dispersion on the nanotube surfaces. The area of H-adsorption or H-desorption on the CV curve can be used to estimate the catalyst activity. Figure 6.32 is the plot of catalyst activity versus Pt loading showing the 10 wt% catalyst exhibits the highest activity.

The PEMFC performance can be further improved by supporting Pt nanoparticles on well-aligned nanotubes. Recently, Li et al. [60] reported that PEMFC with oriented CNT film as the cathode exhibits higher single cell performance than those with CB or disordered CNT-based cathode. Figure 6.33 shows schematic representation for the fabrication process of oriented Pt/CNT-film-based MEA. In this process, acid-oxidized MWNTs were suspended in EG solution and mixed with H_2PtCl_6. The Pt/CNT suspension was filtered through hydrophilic nylon filter paper. Pt/CNT

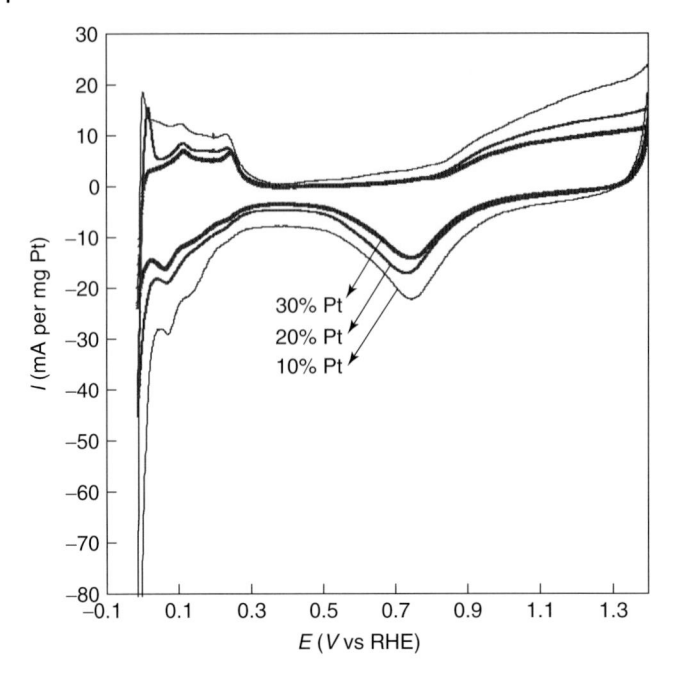

Figure 6.31 Cyclic voltammograms of the three catalysts of Pt nanoparticles supported on MWNTs. Measurements were performed using a rotating disk electrode in 1.0 M H_2SO_4 with a scan rate of 20 mV s^{-1} and rotating speed at 1000 rpm. (Source: Reproduced from Ref. [58] with permission from the American Chemical Society.)

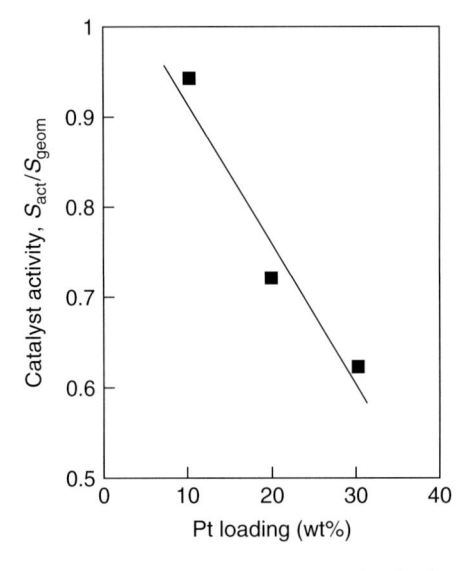

Figure 6.32 Catalyst activity versus of Pt loading. The lowest loading catalyst has the highest activity (∼0.96). Catalyst activity decreases with increase in Pt loading for the catalysts. (Source: Reproduced from Ref. [58] with permission from the American Chemical Society.)

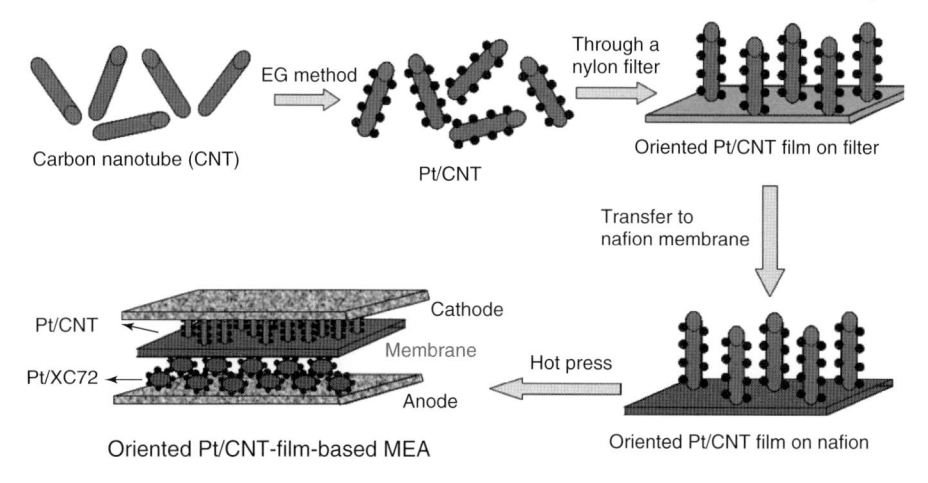

Figure 6.33 Schematic representation of the fabrication process for oriented Pt/CNT-film-based MEA. (Source: Reproduced from Ref. [60] with permission from the American Chemical Society.)

catalyst with a metal loading of 30 wt% was obtained. Since the nylon filter is hydrophilic and the nanotubes are hydrophobic, the MWNTs tend to self-assemble into an oriented film. The I–V responses of PEMFCs with oriented Pt/MWNT film, nonoriented Pt/MWNT film, and Pt/CB (Pt/C; ETEK, a registered trademark pf pt/c catalyst from De Nora N.A, Inc., USA) as the cathode catalyst layers and an ETEK electrode as anode are shown in Figure 6.34. It can be seen that the oriented Pt/MWNT film exhibits higher performance than other cathode catalyst layers. This can be attributed to the enhanced electrocatalytic activity of Pt/MWNT film and improved mass transport within the oriented film.

6.7.2
Graphene-Supported Platinum Electrocatalysts

Two-dimensional graphene sheets can also act as effective anchoring sites for Pt nanoparticles. Seger and Kamat [61] deposited Pt nanoparticles onto graphene sheets by means of borohydride reduction of H_2PtCl_6 in a graphene oxide suspension (Figure 6.35). Graphene oxide (GO) is an insulator and undergoes partial reduction during the $NaBH_4$ treatment. The partially reduced GO-Pt catalyst was then deposited as films on glassy carbon and carbon Toray paper using a drop cast method. In general, the presence of functional groups on GO resulting from acid oxidation decreases its conductivity as a result of a loss in the conjugated sp^2 network. Typical reductants such as hydrazine and $NaBH_4$ are used to reduce graphene oxide and increase the conductivity by restoring the sp^2-hybridized network [62, 63]. Figure 6.36 shows CV of glassy carbon electrode coated with different electrocatalyst films immersed in 0.1 M H_2SO_4 solution. The area under the desorption peak in the voltammogram represents the total charge relating to H^+ desorption (Q_H), which in turn provides an estimate of electrochemically

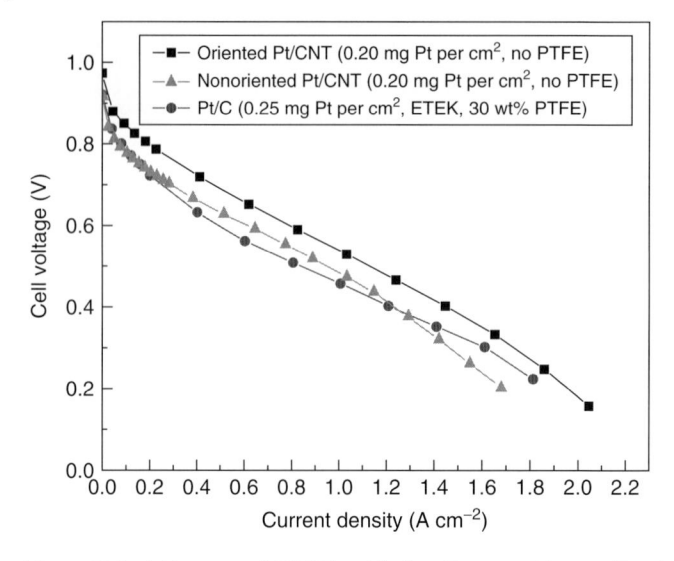

Figure 6.34 I–V curves of PEMFCs with the oriented Pt/CNT film, nonoriented Pt/CNT film, Pt/C with 30 wt% polytetrafluoroethylene (PTFE) as the cathode catalyst layers. 0.2 mg of Pt per cm^2 (Pt/C, 20 wt%, ETEK) in the anode and Nafion 112 as the membrane. Test conditions: cell and O$_2$ humidification at 70 °C; H$_2$ humidification at 85 °C; H$_2$ and O$_2$ pressure and flow rate at 0.2 MPa and 0.2 l min^{-1}, respectively. (Source: Reproduced from Ref. [60] with permission from the American Chemical Society.)

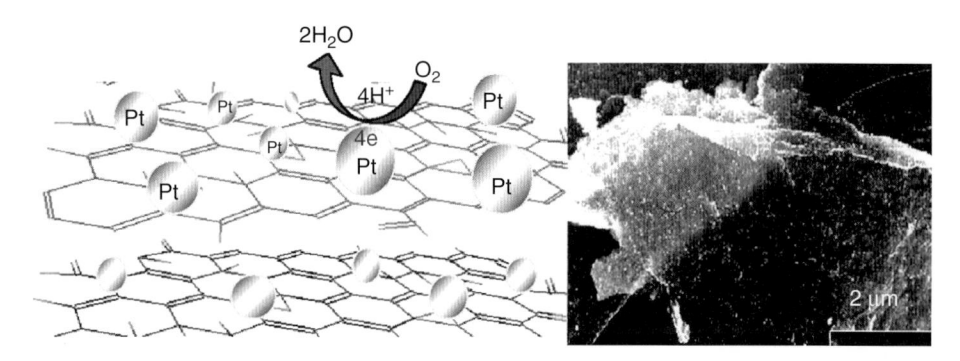

Figure 6.35 Schematic diagram and SEM image showing dispersion of Pt nanoparticles on a 2D carbon sheet (graphene) to facilitate an electrocatalytic reaction. (Source: Reproduced from Ref. [61] with permission from the American Chemical Society.)

active surface area (ECSA). Apparently, the partially reduced GO-Pt film exhibits ECSA higher value than the unsupported Pt but lightly lower than CB-Pt film. To enhance ECSA values, the electrodes with GO-Pt catalyst were treated with hydrazine followed by annealing at 300 °C for different time periods. It appears that the hydrazine electrode subjected to 8 h annealing exhibits the highest ECSA value of ∼20 m^2 g^{-1}. The performance of a cell assembly using different electrocatalyst deposited on Toray paper as the cathode and Pt-dispersed on CB (Pt/C; ETEK) as

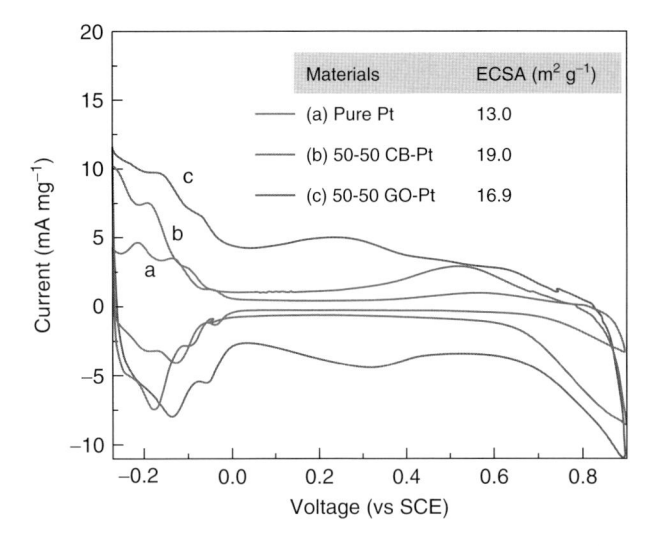

Figure 6.36 Cyclic voltammetry graphs of Pt, 50–50 (by weight) carbon black-Pt, and 50–50 (by weight) partially reduced graphene oxide-Pt. 20 μg cm^{-2} of Pt was deposited on each electrode. 0.1 M H$_2$SO$_4$ was used as the electrolyte and the scan rate was 20 mV s^{-1}. (Source: Reproduced from Ref. [61] with permission from the American Chemical Society.)

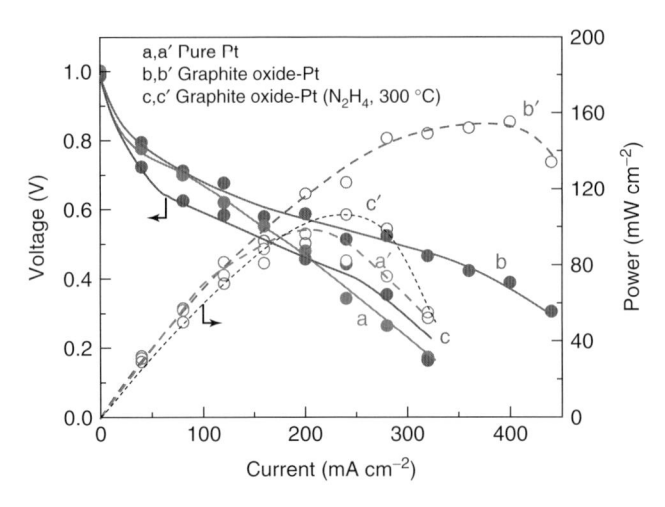

Figure 6.37 Polarization (I–V) curves (a,b, and c) and power characteristics (a′, b′, and c′) of a fuel cell at 60 °C and 1 atm of back pressure. The cathode was composed of (a, a′) Pt, (b, b′) 1 : 1 GO-Pt and (c, c′) 1 : 1 GO-Pt (hydrazine, 300 °C treated). Electrocatalyst concentration was maintained at 0.2 mg cm^{-2} Pt. The anode consisted of 0.5 mg cm^{-2} Pt of ETEK CB-Pt. The anode and cathode catalyst films were heat pressed on Nafion 115 membrane. (Source: Reproduced from Ref. [61] with permission from the American Chemical Society.)

the anode is shown in Figure 6.37. For the purpose of comparison, an unsupported Pt cathode is also used. The partially reduced GO-Pt based fuel cell delivers a maximum power of 161 mW cm^{-2} compared to 96 mW cm^{-2} for an unsupported Pt based fuel cell. Hydrazine and thermal treated GO-Pt-based fuel cell shows a major drop in power output possibly due to the loss of proton conductivity of Nafion ionomer during the deposition of electrocatalysts via heat pressing.

The cost of fuel cells can be markedly reduced by decreasing the amount of Pt catalyst needed for the application. But this can affect the overall ORR activity of the electrodes. The manufacture cost can be further reduced by using Pt-free catalyst. In this regard, scientists are continuously engaged in a search for next-generation catalysts free from noble metals. Dai and coworkers [64] reported that vertically aligned (VA) nitrogen-doped CNT can act as a metal-free electrocatalyst for oxygen reduction in 0.1 M KOH solution. The N-doped CNT exhibits higher electrocatalytic activity, long-term operation stability than pure platinum for ORR in alkaline fuel cells. In a subsequent study, they also found that nitrogen-doped graphene synthesized by chemical vapor deposition of methane in the presence of ammonia also acts as a metal-free catalyst and exhibits similar electrocatalytic behaviors [65]. In a recent study, Geng et al. [66] synthesized N-graphene on a large scale using ammonia and heat treatment. A similar beneficial effect of N-graphene on the ORR activity in oxygen-saturated 0.1 M KOH solution has also been observed.

Nomenclature

β	Tafel slope
F	Faraday constant
I	Current
i_0	Exchange current density
i_L	Limiting current density
N	Number of electrons involved in an electrochemical reaction
R	Specific contact resistance
R	Gas constant
T	Temperature
ΔV_{act}	Activation overvoltage
ΔV_{con}	Concentration overvoltage
ΔV_{ohm}	Ohmic overvoltage

References

1. Jacobson, M.Z. (2011) *Energy Environ. Sci.*, **2**, 148–173.
2. http://www.fuelcells.org.
3. Ling, P.C., Park, B.Y., and Madou, M.J. (2008) *J. Power Sources*, **176**, 207–214.
4. http://www1.eere.energy.gov/hydrogenandfuelcells/mypp/pdfs/fuel_cells.pdf.
5. Mahreni, A., Mohamad, A.B., Khadhum, A.A., and Daud, W.R. (2011) in *Advances in Nanocomposites-Synthesis, Characterization and Industrial Applications*, Chapter 12 (ed. B.S. Reddy), Intech, Croatia, pp. 263–288.
6. Hermann, A., Chaudhuri, T., and Spagnol, P. (2005) *Int. J. Hydrogen Energy*, **30**, 1297–1302.

7. Huang, J., Baird, D.G., and McGrath, J.E. (2005) *J. Power Sources*, **150**, 110–119.
8. Cele, N.P., Sinha Ray, S., Pillai, S.K., Ndwandwe, M., Nonjola, S., Sikhwivhilu, L., and Mathe, M.K. (2010) *Fuel Cells*, **10**, 64–71.
9. Cooper, J.S. (2004) *J. Power Sources*, **129**, 152–169.
10. Cunningham, B.D., Huang, J., and Baird, D.G. (2007) *Int. Mater. Rev.*, **52**, 1–13.
11. Clulow, J.G., Zappitelli, F.E., Carlstrom, C.M., Zemsky, J.I., Busick, D.N., and Wilson, M.S. (2002) Fuel cell technology: opportunities and challenges. AIChE Spring National Meeting, New Orleans, LA, March 10–14, 2002, pp. 417–425.
12. Wissler, M. (2006) *J. Power Sources*, **156**, 142–150.
13. Tawfik, H., Hung, Y., and Mahajan, D. (2007) *J. Power Sources*, **163**, 755–767.
14. de Bruijn, F.A., Dam, V.A., and Janssen, G.J. (2008) *Fuel Cells*, **8**, 3–22.
15. Radhakrishnan, S., Ramanujam, B.T., and Sivaram, A.A. (2007) *J. Power Sources*, **163**, 702–707.
16. Xia, L.G., Li, A.J., Wang, W.Q., Yin, Q., Lin, II., and Zhao, Y.B. (2008) *J. Power Sources*, **178**, 363–367.
17. Kuan, H.C., Ma, C.C.M., Chen, K.H., and Chen, S.M. (2004) *J. Power Sources*, **134**, 7–17.
18. (a) Blunk, R., Elhamid, M.H., Lisi, D., and Mikhail, Y. (2006) *J. Power Sources*, **156**, 151–157; (b) Blunk, R., Zhong, F., and Owens, J. (2006) *J. Power Sources*, **159**, 533–542.
19. Sichel, E.K. (1982) *Carbon Black-Polymer Composites*, Marcel Dekker, New York.
20. Tjong, S.C. (2011) *Energy Environ. Sci.*, **4**, 605–626.
21. Dhakate, S.R., Sharma, S., Chauhan, N., Seth, R.K., and Mathur, R.B. (2010) *Int. J. Hydrogen Energy*, **35**, 4195–4200.
22. (a) Liao, H., Hung, C.H., Ma, C.C.M., Yen, C.Y., Lin, Y.F., and Weng, C.C. (2008) *J. Power Sources*, **176**, 175–182; (b) Liao, S.H., Yen, C.Y., Hung, C.H., Weng, C.C., Tsai, M.C., Lin, Y.F., Ma, C.C., Pan, C., and Su, A. (2008) *J. Mater. Chem.*, **18**, 3993–4002; (c) Liao, S.H., Yen, C.Y., Weng, C.C., Lin, Y.F., Ma, C.C., Yang, C.H., Tsai, M.C., Yen, M.Y., Hsiao, M.C., Lee, S.J., Xie, X.F., and Hsiao, Y.H. (2008) *J. Power Sources*, **185**, 1225–1232; (d) Liao, S.H., Hsiao, M.H., Yen, M.H., Ma, C.C., Lee, S.J., Su, A., Tsai, M.C., Yen, M.Y., and Liu, P.L. (2010) *J. Power Sources*, **195**, 7804–7817; (e) Liao, S.H., Weng, C.C., Yen, C.Y., Hsiao, M.H., Ma, C.C., Tsai, M.C., Su, A., Yen, M.Y., Lin, F.Y., and Liu, P.L. (2010) *J. Power Sources*, **195**, 263–270; (f) Hsiao, M.H., Liao, S.H., Yen, M.Y., Teng, C.C., Lee, S.H., Pu, N.W., Wang, C.A., Sung, Y., Ger, M.D., Ma, C.C., and Hsiao, M.H. (2010) *J. Mater. Chem.*, **20**, 8496–8505.
23. Song, L., Guo, J., Zhamu, A., and Jang, B.Z. (2009) Highly conductive nano-scaled graphene plate nanocomposites. US Patent 7,566,410.
24. Dhakate, S.R., Sharma, S., Borah, M., Mathur, R.B., and Dhami, T.L. (2008) *Int. J. Hydrogen Energy*, **33**, 7146–7152.
25. Dhakate, S.R., Sharma, S., Borah, M., Mathur, R.B., and Dhami, T.L. (2008) *Energy Fuels*, **22**, 3329–3334.
26. Song, L.N., Xiao, M., and Meng, Y.Z. (2006) *Compos. Sci. Technol.*, **66**, 2156–2162.
27. Xiao, M., Lu, Y., Wang, S.J., Zhao, Y.F., and Meng, Y.Z. (2006) *J. Power Sources*, **160**, 165–174.
28. Du, L. and Jana, S.C. (2007) *J. Power Sources*, **172**, 734–741.
29. Du, L. and Jana, S.C. (2008) *J. Power Sources*, **182**, 223–229.
30. Heo, S.I., Oh, K.S., Yun, J.C., Jung, S.H., Yang, Y.C., and Han, K.S. (2007) *J. Power Sources*, **171**, 396–403.
31. Du, C., Ming, P., Hou, M., Fu, J., Fu, Y., Luo, X., Shen, Q., Shao, Z., and Yi, B. (2010) *J. Power Sources*, **195**, 794–800.
32. Gibb, P.R. (1999) Europe Patent. WO 0041260.
33. Yan, X., Hou, M., Zhang, H., Jing, F., Ming, P., and Yao, B. (2006) *J. Power Sources*, **160**, 252–257.
34. Du, C., Ming, P., Hou, M., Fu, J., Fu, Y., Luo, X., Shen, Q., Shao, Z., and Yi, B. (2010) *J. Power Sources*, **195**, 5312–5319.

35. Lee, J.H., Jang, Y.K., Hong, C.E., Kim, N.H., Li, P., and Lee, H.K. (2009) *J. Power Sources*, **193**, 523–529.
36. Del Rio, C., Ojeda, M.C., Acosta, J.L., Escudero, M.J., Hontanon, E., and Daza, L. (2002) *J. Appl. Polym. Sci.*, **83**, 2817–2822.
37. Chen, C.K. and Kuo, J.K. (2006) *J. Appl. Polym. Sci.*, **101**, 3415–3421.
38. Larminie, J. and Dicks, A. (2000) *Fuel Cell Systems Explained*, Chapter 3, John Wiley & Sons, Inc., New York, pp. 45–60.
39. Arita, M. (2002) *Fuel Cells*, **2**, 10–14.
40. Burstein, G.T. (2005) *Corros. Sci.*, **47**, 2858–2870.
41. Bard, A.J. and Faulkner, L.R. (2000) *Electrochemical Methods: Fundamentals and Applications*, 2nd edn, John Wiley & Sons, Inc., New York.
42. Uhlig, H.H. and Revie, R.W. (1985) *Corrosion and Corrosion Control*, 3rd edn, John Wiley & Sons, Inc., New York, p. 44.
43. Saha, M.S., Li, R., and Sun, X.L. (2008) *J. Power Sources*, **177**, 314–322.
44. Tada, T. (2003) in *Handbook of Fuel Cells-Fundamentals, Technology and Applications*, vol. 3, Chapter 38 (eds W. Vielstich and H.A. Gasteiger), John Wiley, & Sons, Inc., New York, pp. 481–488.
45. Kangasniemi, K.H., Condit, D.A., and Jarvi, T.D. (2004) *J. Electrochem. Soc.*, **151**, E125–E132.
46. Matsumoto, T., Komatsu, T., Arai, K., Yamazaki, T., Kijima, M.M., Shimizu, H., Takasawa, Y., and Nakamura, J. (2004) *Chem. Commun.*, **7**, 840–841.
47. Lee, K., Zhang, J., Wang, H., and Wilkinson, D.P. (2006) *J. Appl. Electrochem.*, **36**, 507–522.
48. Yuan, F. and Ryu, H. (2004) *Nanotechnology*, **15**, S596–S602.
49. Wang, X., Waje, M., and Yan, Y. (2005) *Electrochem. Solid State Lett.*, **8**, A42–A44.
50. Li, W., Wang, X., Chen, Z., Waje, M., and Yan, Y. (2005) *Langmuir*, **21**, 9386–9389.
51. Waje, M., Wang, X., Li, W., and Yan, Y. (2005) *Nanotechnology*, **16**, S395–S400.
52. Wang, X., Li, W., Chen, Z., Waje, M., and Yan, Y. (2006) *J. Power Sources*, **158**, 154–159.
53. Bock, C., Paquet, C., Couillard, M., Botton, G.A., and MacDougall, B.R. (2004) *J. Am. Chem. Soc.*, **126**, 8028–8037.
54. Ciacchi, L.C., Pompe, W., and De Vita, A. (2001) *J. Am. Chem. Soc.*, **123**, 7371–7380.
55. Knupp, S.L., Li, W.Z., Paschos, O., Murray, T.M., Snyder, J., and Haldar, P. (2008) *Carbon*, **46**, 1276–1284.
56. Yu, W., Tu, W., and Liu, H.F. (1999) *Langmuir*, **15**, 6–9.
57. Tian, Z.Q., Jiang, S.P., Liang, Y.M., and Shen, P.K. (2006) *J. Phys. Chem. B*, **110**, 5343–5350.
58. Xing, Y. (2004) *J. Phys. Chem. B*, **108**, 19255–19259.
59. Xing, Y., Li, L., Chusuei, C.C., and Hull, R.V. (2005) *Langmuir*, **21**, 4185–4190.
60. Li, W.Z., Wang, X., Chen, Z.W., Waje, M., and Yan, Y. (2005) *Langmuir*, **21**, 9386–9389.
61. Seger, B. and Kamat, P.V. (2009) *J. Phys. Chem. C*, **113**, 7990–7995.
62. Si, Y. and Samulski, E.T. (2008) *Nano Lett.*, **8**, 1679–1682.
63. Bourlinos, A.B., Gournis, D., Petridis, D., Szabo, T., Szeri, A., and Dekany, I. (2003) *Langmuir*, **19**, 6050–6055.
64. Gong, K., Du, F., Xia, Z., Durstock, M., and Dai, L.M. (2009) *Science*, **323**, 760–764.
65. Qu, L., Liu, Y., Baek, J.B., and Dai, L.M. (2010) *ACS Nano*, **4**, 1321–1326.
66. Geng, D.S., Chen, Y., Chen, Y.G., Li, Y.L., Li, R.Y., Sun, X.L., Ye, S.Y., and Knights, S. (2011) *Energy Environ. Sci.*, **4**, 760–764.

7
Polymer Nanocomposites for Biomedical Applications

7.1
Overview

Recent scientific advances in nanoscience and nanotechnology enable the rapid development of nanomedicine, which concerns the applications of nanomaterials for diagnostic and therapeutic purposes of human diseases at the molecular level. The integration of nanotechnology and medicine has innovated new approaches for the treatment of human diseases and injuries. Nanotechnology is capable of producing a wide variety of novel biomaterials, probes, and sensors having improved interactions between the material surfaces and biological entities. It offers great opportunities to produce materials with nanoscale features that can mimic the structures of human tissues. It aids in the design for the next generation of tissue engineering scaffolds.

Carbon nanotubes (CNTs) with unique chemical, electrical, and mechanical characteristics are particularly attractive for biomedical applications. Their large aspect ratio, hollow interior, and cagelike structures allow biomolecules to be loaded onto them for penetrating into cells and tissues. In this regard, CNTs act as nanocarriers for a variety of biomolecules such as proteins, peptides, genes, and carbohydrates [1]. CNTs are found to be an effective drug delivery system showing advantages over conventional drug therapies with many shortcomings such as limited drug solubility, poor biodistribution, and lack of selectivity [2]. CNTs can also be utilized as biomaterials for bone tissue scaffolding and regeneration, load-bearing implants, and neuron engineering because of their remarkable mechanical and electrical properties [3]. The nanoscale features of nanotubes greatly enhance their interactions with human cells, fostering cell adhesion, differentiation and growth. Figure 7.1 is a schematic diagram illustrating potential biomedical applications of CNTs. The main concern of using CNTs for biomedical applications is the possible health hazard caused by implanting these materials into human body. Despite the fact that numerous studies of nanotube hazard and cytotoxicity have been conducted by the researchers, it still remains unclear which attributes of nanotubes contribute to specific toxicity. As recognized, CNTs synthesized from various techniques and batches have different morphologies and impurity levels. These can produce inconsistent results in toxicity testing and uncertainty about

Polymer Composites with Carbonaceous Nanofillers: Properties and Applications, First Edition. Sie Chin Tjong.
© 2012 Wiley-VCH Verlag GmbH & Co. KGaA. Published 2012 by Wiley-VCH Verlag GmbH & Co. KGaA.

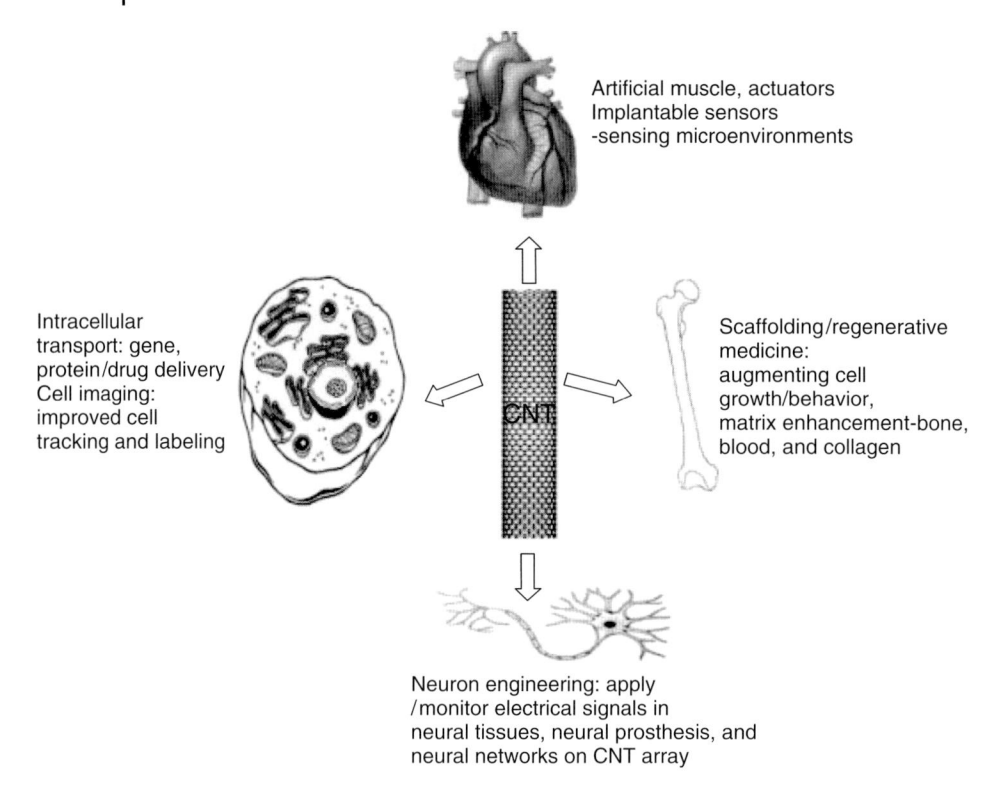

Figure 7.1 Schematic diagrams showing potential applications of carbon nanotubes in biomedical engineering. (Source: Reproduced with permission from [3], Wiley (2009).)

Surrounding labels:
- Artificial muscle, actuators; Implantable sensors -sensing microenvironments
- Intracellular transport: gene, protein/drug delivery; Cell imaging: improved cell tracking and labeling
- Scaffolding/regenerative medicine: augmenting cell growth/behavior, matrix enhancement-bone, blood, and collagen
- Neuron engineering: apply/monitor electrical signals in neural tissues, neural prosthesis, and neural networks on CNT array

the cause of cytotoxicity. For most cases, the impurities are the main sources responsible for biological response and inflammation. Furthermore, *in vitro* testing of nanotube-induced toxicity is closely related to the types of biological cells used. A wide variety of cell lines including fibroblasts, keratinocytes, macrophages, osteoblasts, and so on have been used for cell viability tests, and no consensus has been gained on the appropriate test assays to evaluate cytotoxicity. Accordingly, it is rather difficult to draw conclusive remarks about biological impacts from many studies on CNT-induced toxicity.

7.2
Bone Implants

Recently, there has been an increasing need for bone tissue replacements owing to a large increase in the number of aging populations and patients suffering from trauma and bone cancer. The defective or damaged bone can be replaced with autogenous bone (autograft) taken from the patient by the surgeon. However, the grafting method has several drawbacks such as limited supply of autograft tissue, donor site morbidity, graft rejection, and inadequate bone formation and quality [4]. To overcome these limitations, artificial materials or implants have been developed for tissue reconstruction or replacement in the past decades. Metallic alloys (e.g., austenitic stainless steel, cobalt-chrome-based and titanium alloys) with excellent mechanical strength/stiffness and good ductility are the first implant materials successfully used for load-bearing and bone fixation applications during the twentieth century [5]. However, the use of metallic materials stiffer than bone tissue can lead to a mechanical mismatch problem commonly known as *stress shielding*. This is because bone is insufficiently loaded compared to the metallic implants, resulting in a reduction in bone density [6]. Further, austenitic stainless steel (AISI 316 L) and cobalt-chrome implants have poor corrosion resistance when exposed to human body fluid containing chloride ions. The wear resistance of austenitic stainless steel is very poor owing to its low hardness and high friction characteristics. In view of the health hazards caused by metallic implants, considerable attention is paid to the use of polymers with smaller stiffness [7]. Polymers can be easily manufactured into desired products at low processing temperatures because of their lower melting temperatures. The mechanical stiffness and strength, wear resistance, and biocompatibility of polymers can be tailored by adding selected fillers of appropriate weight fractions.

Bone tissue is a natural biocomposite with a complex hierarchical structure as shown in Figure 7.2 [8]. It comprises living cells (osteoblasts and osteoclasts), organic polymer (collagen), and inorganic (hydroxyapatite (HA)) nanocrystals with a chemical composition of $Ca_{10}(PO_4)_6(OH)_2$. Collagen fibers serve as templates for HA nanoplatelets to nucleate and grow. The toughness of bone is derived from its organic component, while its stiffness comes from the HA nanoplatelets. Osteoblasts or bone-forming cells locate near the surface of bone. They are then matured to form osteocytes and entrapped in the deposited bone matrix. Osteoclasts are the cells that break down the bone matrix and are responsible for bone resorption. Therefore, bone experiences continuous processes of removal and replacement throughout the entire life of humans. Its metabolic function is controlled by the interplay between the processes of bone formation and resorption. Bone can be either compact (cortical) or cancellous (spongy). Cortical bone makes up a large portion of skeletal mass. The mechanical properties of cortical bone are listed in Table 7.1.

The main strategy in the design and fabrication of orthopedic implants is to simulate the properties of bone in these biomaterials. A proper understanding of the biomimetic principle of replicating formation and structure of bone tissues facilitates the development of implant materials with good biocompatibility. The bone formation involves several steps including the assembly of extracellular matrix (ECM), the transportation of inorganic ions to discrete organized structures, and subsequent mineral nucleation and growth. Communication between the cell and the ECM influences cellular activities including adhesion, differentiation, proliferation, and apoptosis. Polymer composites for bone tissue replacement must promote adhesion and proliferation of osteoblasts. Both nondegradable and

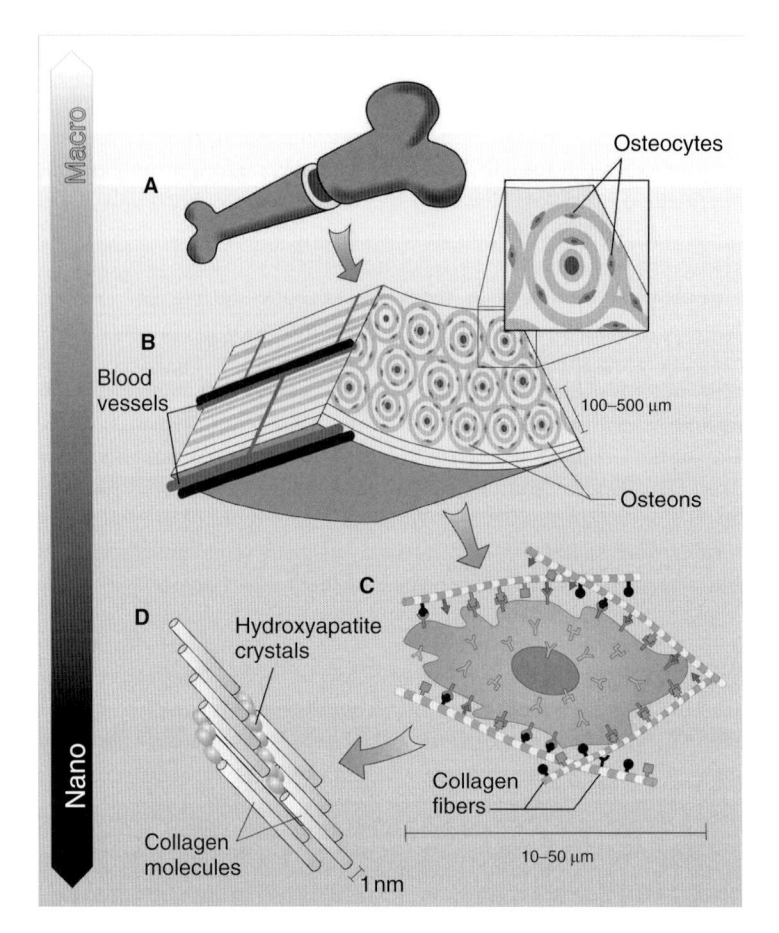

Figure 7.2 Hierarchical organization of bone over different length scales. Bone has a strong calcified outer compact layer (A), which comprises many osteons (B). The resident cells are coated in a forest of cell membrane receptors that respond to specific binding sites (C) and the well-defined surrounding extracellular matrix (D). (Source: Reproduced with permission from [8], American Association for the Advancement of Science (2005).)

Table 7.1 Tensile properties of cortical bone and HAPEX™.

Material	Young's modulus (GPa)	Tensile strength (MPa)	Fracture strain (%)	References
Cortical bone	7–30	50–150	1–3	[9]
HAPEX™	4.29	20.67	2.6	[10]

degradable polymers have been selected as the composite matrix materials. Nondegradable polymers are targeted to fabricate load-bearing implants for replacing defective bones, while degradable polymers are mainly designed for forming scaffolds and therapeutic devices [7, 11–13].

Bonfield and coworkers of Queen Mary College (UK) were the first research group to develop the HAPEX™ polymer composite for bone replacements. The composite consists of the dispersion of 40 vol% HA microparticles in nondegradable, high-density polyethylene (HDPE) matrix [9, 10]. The mechanical properties of HAPEX are also listed in Table 7.1. Apparently, the Young's modulus and strength of HAPEX are considerably lower than those of cortical bone. As a result, HAPEX can only be used for low-strength orbital floor prosthesis, middle ear implant, and maxillofacial surgery [14, 15]. This composite incorporates a large volume fraction of HA particles of several micrometers into HDPE matrix in order to reinforce the polymer and to promote the growth of osteoblasts. However, high HA content can lead to poor processability of the composite. Further, large HA particles fracture readily during tensile deformation. To further improve elastic modulus and strength of biocomposites, high-performance poly(etheretherketone) (PEEK) is employed as the matrix of composites. The Young's modulus of PEEK is ~3.8 GPa and can be increased to 12.38 GPa by adding 30 wt% carbon fiber (CF). PEEK reinforced with CFs find increasing use as biomaterials for load-bearing applications [16–18].

As recognized, polymer nanocomposites only require low ceramic nanofiller loadings to reinforce polymers. In addition, the responses of biological cells to nanocomposite surfaces are different from those of their microcomposite counterparts. Nanoceramics with larger surface areas foster adsorption of greater amounts of protein, thereby promoting osteoblast adhesion and proliferation (Figure 7.3) [19, 20]. Recently, extensive studies have been carried out by many

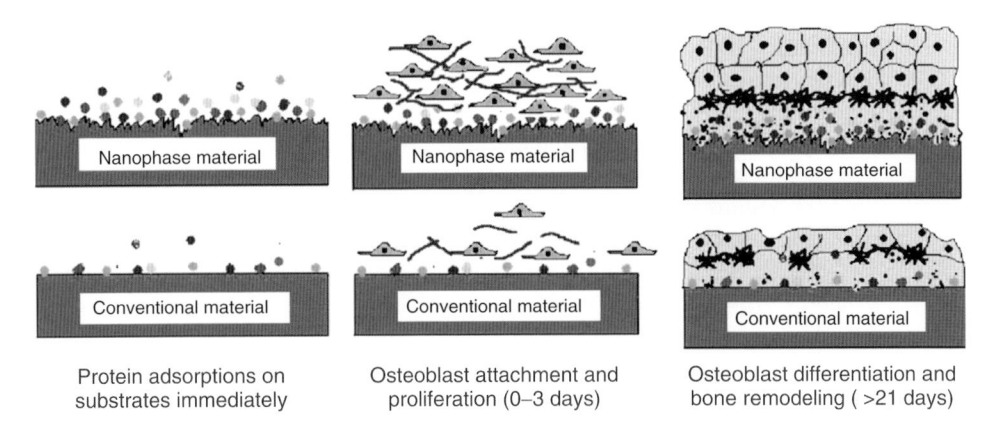

Figure 7.3 The biomimetic advantages of nanomaterials. The bioactive surfaces of nanomaterials mimic those of natural bones to promote greater amounts of protein adsorption and efficiently stimulate more new bone formation than conventional materials. (Source: Reproduced with permission from [20], Elsevier (2009).)

researchers to synthesis HA nanomaterials and their polymer composites [21–25]. In spite of these efforts, the elastic modulus of most HA-polymer nanocomposites remains unsatisfactory [26] and only reaches the lower limit of cortical bone by using PA6 or PEEK as the matrix of composites [25, 27]. In this respect, CNTs, carbon nanofibers (CNFs), and graphene oxides (GOs) with high elastic modulus and tensile strength are ideal nanofillers to reinforce polymers for biomedical applications. Webster and coworkers [28, 29] reported that CNFs are attractive nanomaterials for orthopedic implant applications since they promote osteoblast adhesion and proliferation.

7.3
Biocompatibility of Carbon Nanotubes

The performance of medical implants depends greatly on the interactions between living cells and their biological environment. The structure of living cells is well described in the booklets issued by the National Institute of General Medical Sciences (NIGMS), United States. More information relating the structure and function of human cells can be found online at the NIGMS web site [30]. The surfaces of biomaterials play a crucial role in the interactions with cells because they serve as the active sites for cell adhesion and proliferation. The adhesion of osteoblasts to biomaterial substrates is mediated through proteins, providing integrins on the cell membrane. Integrins are the primary surface receptors that are responsible for cell–substrate adhesion. They also involve in cell signaling, thereby regulating cellular behaviors such as shape, differentiation, mobility, and growth [31]. Understanding the mechanisms of cell responses to chemical, physical, and biological signals from the material surfaces can foster the development of novel biocomposites with desired biological properties. As CNTs have widespread applications in many technological sectors and biomedical fields, the exposure of technical staff to these materials is inevitable, posing some health safety concerns. Before realizing the biomedical applications of CNTs and their composites, it deems necessary to explore the interactions between CNTs and relevant biological cell systems. The bioactivity and biocompatibility of the CNT/polymer composites can be assessed using *in vitro* cell culture, cell viability, and *in vivo* animal tests. *In vitro* cell culture in controlled conditions is commonly used for assessing bioactivity of medical implants. However, it is rather difficult to simulate the real and complex conditions existing within living organisms using laboratory *in vitro* tests. Thus there is a need for accurate determination of *in vivo* cell responses to medical implants in animals.

7.3.1
Potential Health Hazards

Pure carbon is generally accepted as a relatively biocompatible material. Exposure of biological cells to nanostructured carbon may result in toxicity. The most attractive properties of CNTs for biomedical and industrial applications, such as small size, large surface area, and high reactivity, are also the main concerns for their potential toxicity. CNTs would have higher pulmonary deposition and biological activity compared to larger microparticles [32]. CNTs are fiber shaped and so may behave like asbestos and other pathogenic fibers to cause pulmonary fibrosis [33–35]. Thus inhalation and dermal absorption are potential exposure routes to the workers because of the increased use and production of CNTs. CNTs can enter living organisms through inhalation, ingestion, and skin penetration. CNTs can easily enter into the lungs through the respiratory tract with air inhalation. They then translocate rapidly in the nervous system, lymph, and blood. This is followed by rapid migration to heart, spleen, kidney, bone marrow, and liver as shown in Figure 7.4 [35]. After inhalation, inorganic particles can either be cleared from the respiratory tract or be translocated to pulmonary organs across epithelial layers (Figure 7.5). In the latter case, alveolar macrophages and epithelial cells are activated by the particles, thereby initiating alveolar inflammation (alveolitis).

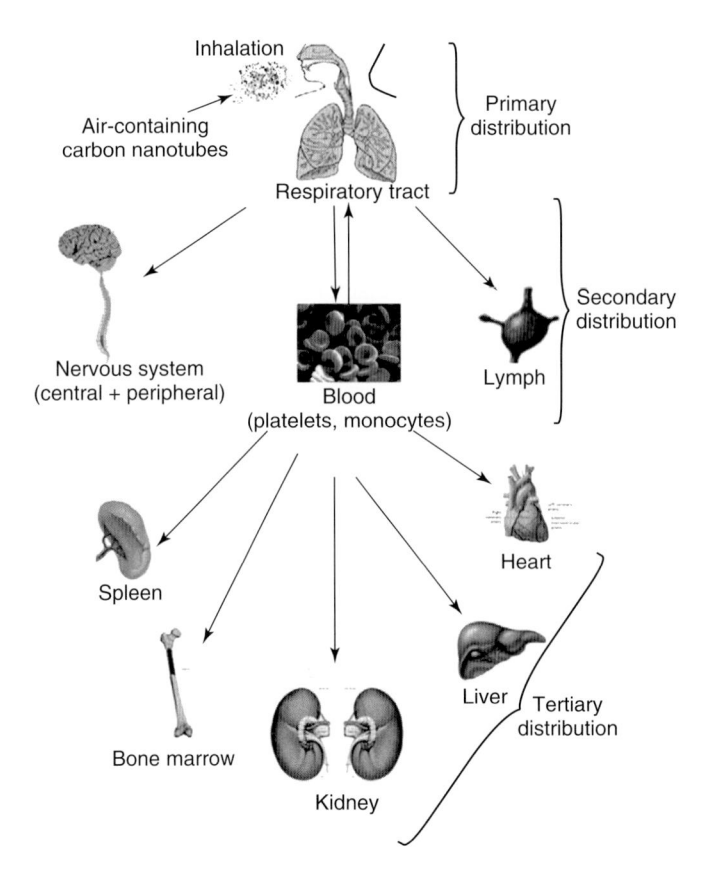

Figure 7.4 Distribution of CNTs in the body. (Source: Reproduced with permission from [35], Elsevier (2011).)

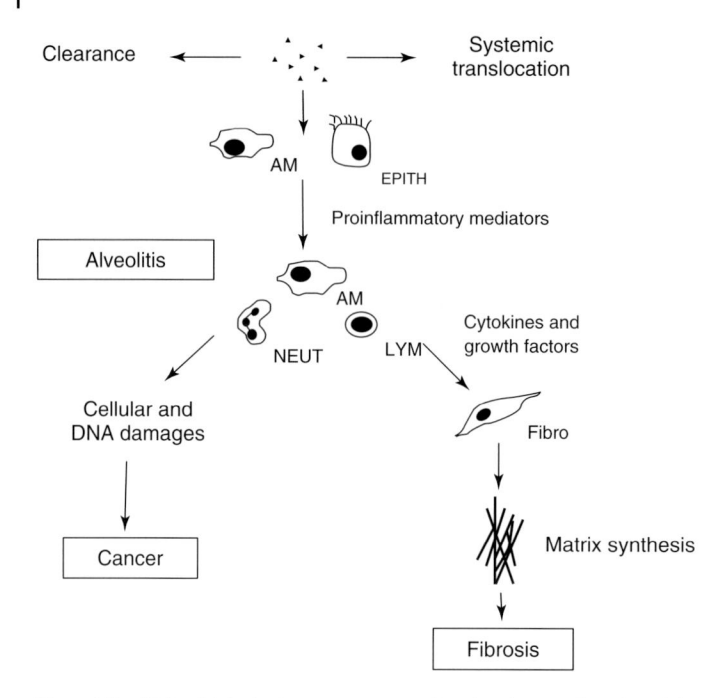

Figure 7.5 Major biological processes associated with particle-induced lung injuries. AM, alveolar macrophages; EPITH, epithelial cells; NEUT, neutrophils; and LYM, lymphocytes. (Source: Reproduced with permission from Ref. [36], Elsevier (2006).)

Macrophages are cells of the immune system that form the initial defense system against foreign objects in many tissues including the lung. The engulfment and ingestion of macrophages through phagocytosis lead to the release of inflammatory mediators such as cytokines (tumor necrosis factor-alpha (TNF-α)), chemokines (interleukin-8 (IL-8), macrophage inflammatory protein-2 (MIP-2)), reactive oxygen species (ROS), nitrogen species, and so on (Table 7.2). All these factors facilitate the recruitment of neutrophils and lymphocytes into the lung. The ROS or oxidant species can also induce cellular and DNA damages leading to cancer [36].

Transmission electron microscopy (TEM) is a powerful technique to observe changes in cell morphology on exposure to CNTs. Direct imaging of nanotubes in macrophage cells was carried out by Porter *et al.* [37] using TEM. They reported that single-walled carbon nanotubes (SWNTs) are first localized mainly within lysosomes, suggesting uptake by phagocytosis. The nanotubes then translocate across the membrane into the neighboring cytoplasm and the nucleus, implying passive uptake through the lipid layer. The cellular uptake of SWNTs in these sites eventually leads to cell mortality in a dose-dependent manner. More recently, Cheng *et al.* exposed human macrophage cells to unpurified and purified multiwalled carbon nanotubes (MWNTs). Scanning transmission electron microscopy (STEM), confocal microscopy, and scanning electron microscopy (SEM) were used in their study [38]. Unpurified MWNTs entered the cell through the plasma membrane

Table 7.2 Experimental approaches for characterizing the toxicity of inhaled particles.

Physicochemical investigations
Morphology and surface properties (size, length, diameter, specific surface charge and area, and hydrophobicity)
Purity and contaminants
Presence of transition metals
Solubility and durability

Capacity to form reactive species *in vitro*
Interaction with biological macromolecules (lipids, proteins, and DNA)
In vitro studies: macrophage and/or epithelial cell cultures
Cytotoxicity and apoptosis
Oxidative stress and responses: 4-hydroxynonenal, intracellular glutathione, calcium, and NF_κ-B activation
Production of proinflammatory mediators: TNF-α IL-1, MIP-2 α, ROS, nitric oxide, and derivatives
Production of pro- and antifibrotic mediators: TGF-β, PGE2, IL-10, and IFN-γ
Genotoxicity: DNA breaks, 8-hydroxydeoxyguanosine, chromosomal aberrations, micronuclei, and gene mutations

In vivo studies (intratracheal or inhalation in rats, mice, or hamsters)
Biopersistence and clearance kinetics
Inflammation: histology, cell proliferation, apoptosis, and broncho-alveolar lavage (BAL) analysis
Fibrosis: histology, biochemical measurements (e.g., hydroxyproline content)
Genotoxicity: mutations in epithelial cells (e.g., micronuclei)
Cancer: long-term bioassays, intrapleural injection

Abbreviations: TGF, transforming growth factor; PGE, prostaglandin; IFN, interferone.
Source: Reproduced with permission from [36], Elsevier (2006).

and then penetrated into the cytoplasm and the nucleus (Figures 7.6a,b and 7.7). Pulskamp *et al.* [39] conducted a series of experiments such as cell culture, cell viability assays, TEM, inflammatory cyto-/chemokines (TNF-α and IL-8), and nitride determination for testing toxicological potential of CNTs on pulmonary epithelial cells and macrophages. Both rat alveolar macrophage cell line (NR 8383) and human alveolar epithelial cell line (A549) obtained from American Type Culture Collection (Rockville, MD) were used. A549 is a widely used model cell line for the toxicity study. Cell lines are often used for cell cultivation and cytotoxicity studies because of their simplicity, good repeatability, and available at relatively low cost. Primary cells have disadvantages of poor reproducibility and the difficulty to correlate the findings of different research groups. Pulskamp *et al.* reported that CNT can penetrate the cell membrane of rat macrophages and reside in the cytoplasm. However, CNTs showed no sign of acute toxicity and induced no inflammatory responses on the basis of cell viability and inflammatory mediator tests. Metal impurities in the nanotubes were responsible for the release of ROS

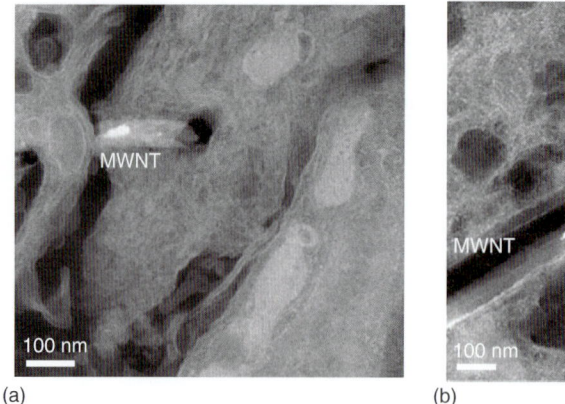

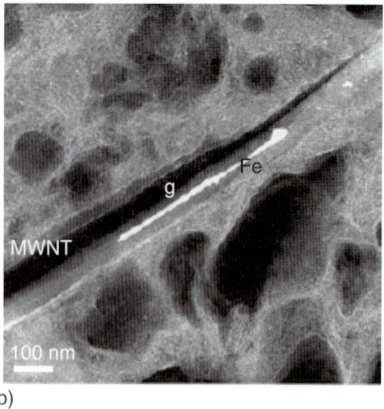

Figure 7.6 Dark-field STEM images of MWNTs within cells. (a) MWNTs penetrating a healthy cell and (b) MWNT traversing the cytoplasm of a healthy cell with evidence for vacuolation of the cytoplasm (g = gap in the microtomed thin section, Fe = internal iron). (Source: Reproduced with permission from [38], Elsevier (2009).)

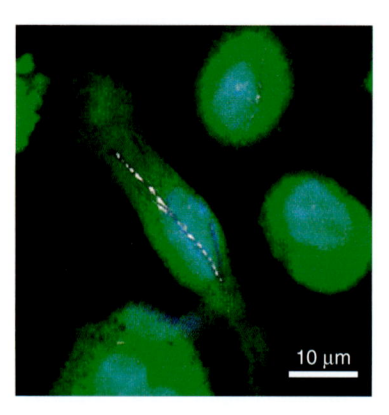

Figure 7.7 Confocal image of 3D averaged stack of MWNT penetrating the cytoplasm and nucleus after four days at 37 °C (green + cytoplasm, blue = nucleus). (Source: Reproduced with permission from [38], Elsevier (2009).)

that may induce oxidative stress. Thus pure CNTs without metal catalysts are attractive materials for biomedical applications.

In the case of *in vivo* animal tests, Chou *et al.* [40] demonstrated that intratracheal instillation of 0.5 mg of SWNTs into male mice (eight weeks old) induces oxidative stress, alveolar macrophage activation, chronic inflammatory responses, and severe pulmonary granuloma formation. Warheit *et al.* [41] also studied acute lung toxicity through intratracheally instilled SWNTs containing trace amount of Co, Ni, and amorphous carbon in rats. SWNTs produced a transient inflammation and a non-dose-dependent accumulation of multifocal granulomas in rats. The formation of granulomas is debatable because large aggregates can be produced during instillation exposure procedures. Muller *et al.* [42] exposed rats intratracheally to

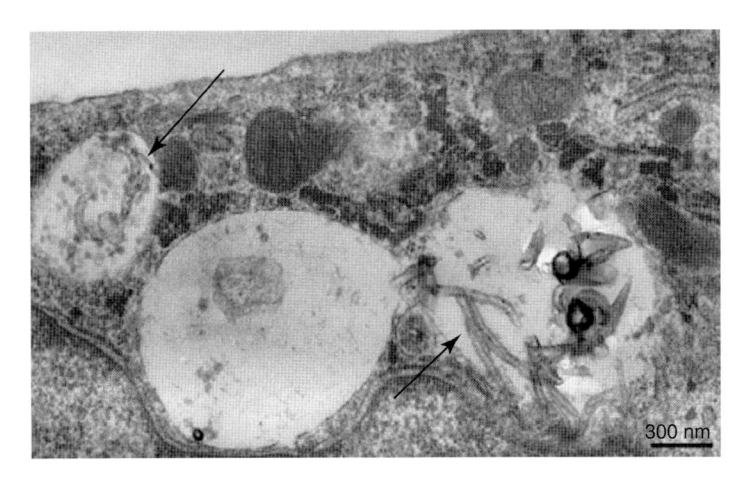

Figure 7.8 Transmission electron micrograph of human epidermal keratinocytes with intracellular localization of MWNTs. Arrows depict the nanotubes present within the cytoplasmic vacuoles of HEKs. (Source: Reproduced with permission from [44], Elsevier (2005).)

pristine and ground MWNTs of 0.7 and 5.9 µm, respectively. Both pristine and ground CNTs induced inflammatory and fibrotic reactions in the lungs after intratracheal administration (0.5, 2, or 5 mg) for 60 days. In other words, the short, ground MWNTs are more toxic than the long, unground nanotubes.

The dermal exposure to CNTs can be effectively studied using mammalian cell lines such as epidermal cells. Exposure of human epidermal keratinocytes (HEKs) to SWNTs (high-pressure carbon oxide (HiPco)) disproportionation containing 30% iron results in oxidative stress and cellular toxicity [43]. This is associated with the formation of free radicals, accumulation of peroxidative products, antioxidant depletion, and loss of cell viability after 18 h of nanotube exposure. The formation of free radicals and peroxide is catalyzed by iron particles within the SWNTs. The transition metal catalysts used to synthesize CNT constitute about 25–40% of the nanotube material by weight. The oxidative stress induced by Fe catalyst elicits inflammatory responses in the HEKs. Excessive oxidative stress may also modify proteins, lipids, and nucleic acids, which further stimulates the antioxidant defense system or even leads to cell death. In another study, Monteiro-Riviere *et al.* indicated that chemically unmodified MWNTs can enter HEKs easily and reside within cytoplasmic vacuoles of the cells (Figure 7.8). SWNTs also induced the release of the proinflammatory mediator, that is, IL-8 from HEKs in a time-dependent mode [44].

The toxicity of CNTs is affected by their degree of dispersion and functionalization. CNTs are generally oxidized in strong acids to generate hydroxyl and carboxyl groups. The carboxylic groups can be used as precursors for further chemical reactions such as amidation in which biomolecules can be attached. Alternatively, a 1,3-dipolar cycloaddition reaction can be used to functionalize CNTs covalently using the process developed by Tagmatarchis and Prato [45]. Azomethine ylides generated by condensation of the α-amino acid and aldehyde are added to a

CNT/dimethylformamide (DMF) suspension, forming pyrrolidine rings. Azomethine ylides are very reactive intermediates and attack the π-system of the CNT effectively, both at the tips and the sidewalls. The reaction takes place as follows:

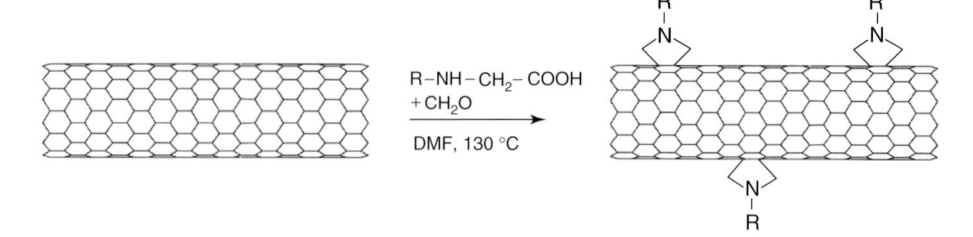

$$\text{CNT} + \text{R-NH-CH}_2\text{-COOH} + \text{CH}_2\text{O} \xrightarrow{\text{DMF, 130 °C}} \text{functionalized CNT}$$

This approach allows the insertion of biofunctional groups at the sidewalls of CNTs by covalently linking a wide variety of bioactive molecular species [46a,b]. The cytotoxic response of cells in culture is dependent on the functional groups attached to the nanotubes. Cytotoxicity is enhanced when the surface of CNTs containing hydroxyl and carboxyl groups after an acid treatment [47]. However, water-soluble CNTs modified with the 1,3-dipolar cycloaddition are not cytotoxic [48]. Functionalization of CNTs with biomolecules generally leads to a marked reduction in toxicity. Accordingly, biofunctionalized CNTs show potential application as effective carriers for the drug delivery system for the diagnosis and treatment of cancer. This potential application is particularly encouraged by their ability to penetrate cell membranes and relatively low toxicity [46].

Till present, there are no conclusive remarks for the CNT cytotoxicity. Most aspects of nanotube toxicity remain uncertain and debatable. This is because toxicity of CNTs is closely related to the types of biological cells used and the properties of nanotube material, including type (SWNTs or MWNTs), length, catalytic impurity, bound functional group, concentration, and dose [35]. As aforementioned, pure CNTs free from catalytic impurities do not induce inflammatory responses and acute cytotoxicity. However, CNTs synthesized from worldwide research institutions and industries have a wide range of metal catalyst contents and lengths. There is a need for reliable techniques for synthesizing and purifying CNTs with consistent properties. Further, several experimental protocols or cell viability assays for assessing nanotube cytotoxicity also produce inconsistency in the test results. It remains a big challenge for the establishment of standard protocols for assessing cytotoxicity of CNTs. Of the *in vivo* studies, no study has conducted inhalation, and all employ instillation in which the high dose and dose rate raise questions about physiological relevance [49]. Nevertheless, the number of studies reporting CNTs to be nontoxic *in vivo* exceeds those proposing otherwise [50].

7.3.2
Cell Viability Assays

Several *in vitro* toxicity tests and assays have been used for determining the effects of CNTs on cell proliferation and cell viability quantitatively. Performing *in vitro* cytotoxicity assay before animal testing is a fast, simple alternative, and essential for the safety use of CNTs for biomedical applications. Cell proliferation is determined by a homogeneous, vital dye process in which a selected dye is added to cells in a 96-well plate and cultured for a period of time. The dye is reduced mainly by the enzymes in living cells such that development of color is measured at a different wavelength than the unaltered dye. Visible wavelength absorbance data are collected and read on a spectrophotometer, showing a high degree of precision. Many dyes commonly used for assessing cytotoxicity, including 3-(4,5-dimethylthiazol-2-yl)-2,5-diphenyltetrazolium bromide (MTT), 3-(4, 5-dimethylthiazol-2-yl)-5(3-carboxymethonyphenol)-2- (4-sulfophenyl)- 2H-tetrazolium (MTS), 2-(4-iodophenyl)-3-(4-nitrophenyl)-5-(2,4-disulfophenyl)-2H-tetrazolium (WST-1), 7-7-hydroxy-10-oxidophenoxazin-10-ium-3-one (Alamar Blue), and 3-amino-*m*-dimethylamino-2-methyl-phenazine hydrochloride (Neutral Red (NR)).

The MTT assay is a calorimetric assay based on reduction of the yellowish, water-soluble tetrazolium salt into water-insoluble formazan product by the mitochondrial enzyme succinate dehydrogenase of living cells in the culture. Tetrazolium ring is cleaved in active mitochondria, indicating the presence of living cells. The amount of formazan formed is directly proportional to the degree of activation of the cells, that is, the number of metabolically active cell. The precipitated formazan is dissolved in an organic solvent such as ethanol and propanol/HCl [51]. The absorbance of the resulting solution is measured photometrically at a wavelength of 550 nm. The MTS calorimetric assay is replacing the MTT assay as a fast, one-step assay of cell viability [52]. The tetrazolium salt aided by an electron coupling reagent (phenazine ethosulfate) can yield a colored formazan product that is soluble in the tissue culture medium. The amount of formazan formed can be quantified by the absorbance at 490 nm. For the WST-1 assay, tetrazolium salt is reduced by cellular mitochondrial dehydrogenases, but the resulting product is water soluble and quantified at 450 nm. Alamar Blue or resazurin is mainly used as an oxidation–reduction indicator in cell viability assays for mammalian cells. It is reduced by mitochondria of viable cells, leading to a color change from oxidized (nonfluorescent, blue) to reduced (fluorescent, red) form. The main advantage of this dye lies in its water solubility and therefore requires no solvent extraction step [53]. NR is cheaper and more sensitive than other cytotoxicity tests based on tetrazolium salts for measuring cell viability. Living cells take up the NR, which is concentrated within lysosomes of the cells [54].

Krug and coworkers [55] exposed human alveolar epithelial cell line (A549) to the SWNTs containing Ni and Co elements in the MTT test. They reported that the MTT assay can produce false cytotoxicity results due to the MTT-formazan crystals clump together with CNTs. Figure 7.9 shows the viability test results from MTT assay for the as-synthesized and acid-purified SWNTs. The nanotubes were initially dispersed in a 1% sodium dodecyl sulfate (SDS) solution. A significant loss in cell viability can be readily in the MTT test results for the as-synthesized SWNTs (about 70% viability reduction) and purified SWNTs (~58%) cultured with A549 for 24 h. This is because the lumped formazan-CNT mixture cannot be dissolved in a propanol/HCl solution. In other words, the crystals cannot be

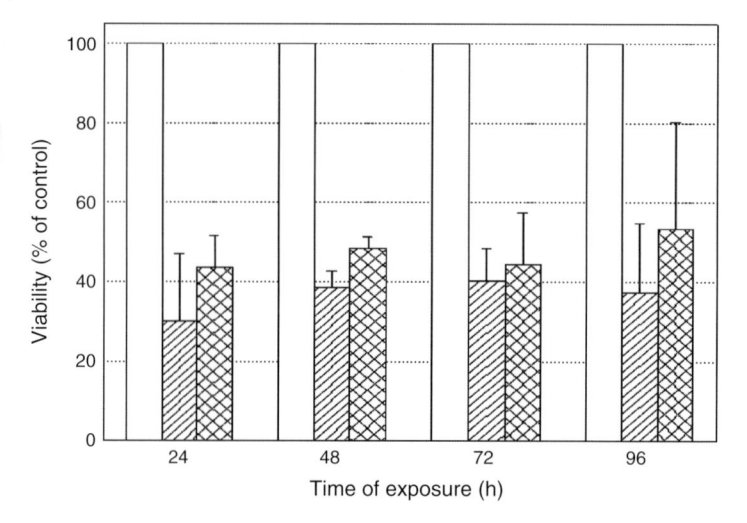

Figure 7.9 MTT assay after incubation with carbon nanotubes in A549 cells at a concentration of 50 µg ml^{-1} for indicated times. Viability results from cells treated with SWNTs without and with catalyst metals are shown in hatched and cross-hatched bars, respectively. The untreated cells are used as positive control, that is, 100% viable (open bars). (Source: Reproduced with permission from [55], The American Chemical Society (2006).)

separated from the SWNTs, preventing dissolution of the MTT-formazan into solution. A reduction in the MTT-formazan content in the assay is related to the loss of crystals binding to nanotubes. However, the WST-1 assay reveals no significant loss of viability for epithelial, endothelial (ECV), and rat alveolar macrophage (NR8383) cells (Figure 7.10). The reduced tetrazolium salt is water soluble in the WST assay; thus, no extraction in an organic solvent is needed. This implies that CNTs interact with tetrazolium salt of the MTT assay but not with that of the WST-1 assay. Consequently, MTT assay produces false negative results. This interference does not affect the enzymatic reaction but relates to insoluble nature of the MTT-formazan. It is noteworthy to mention that the surfactants used to suspend CNTs can also influence the MTT assay results. To study nanotube–cell interaction, the CNTs must be homogeneously suspended in aqueous culture medium by adding surfactants with good biocompatibility and low toxicity [56]. Belyanskaya et al. [57] indicated that SDS-suspended SWNTs interact with MTT-formazan crystals, thereby affecting cytotoxicity results.

As aforementioned, Porter et al. [37] used TEM for imaging the distribution of SWNTs in macrophage cells. TEM can also provide direct estimates of cell viability by treating macrophages with SWNTs at 5 µg ml^{-1}. The localized effect of the uptake of SWNTs on cell death is quantified by indexing 100 cells in TEM. Figure 7.11a,b shows the cytotoxicity of SWNTs in macrophages determined by cell viability assays (NR, MTT) and TEM, respectively. The NR assay shows no significant toxicity after two days but displays a decrease in viability after three days, particularly at the highest SWNT concentration of 10 µg ml^{-1}. For the MTT assay, cell viability decreases markedly when the SWNT concentrations are higher

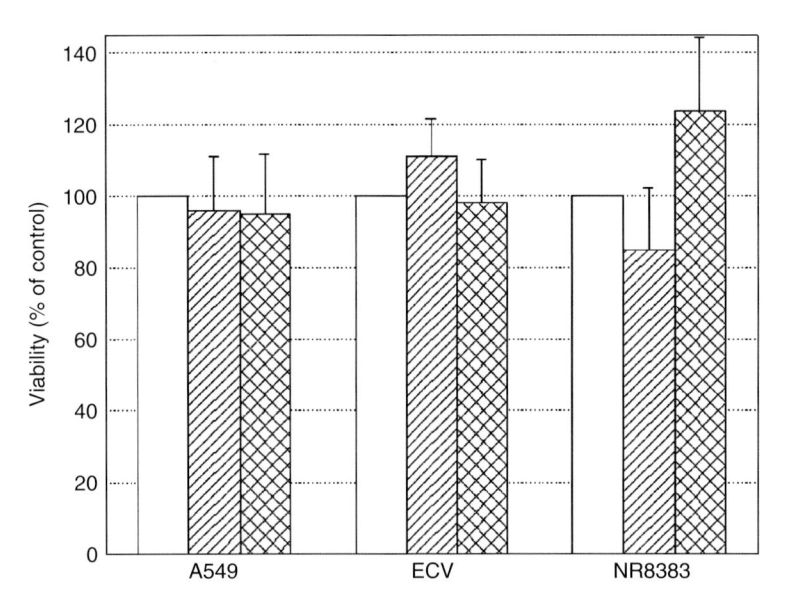

Figure 7.10 WST-1 assay on A549, ECV, and NR8383 cells incubated for 24 h with carbon nanotubes (50 µg ml^{-1}). The reduction of WST-1 is unaffected by SWNTs, and cells remain viable to ~100% compared to the control (open bars). No significant changes in viability between pristine (hatched) and purified SWNT (cross-hatched) can be observed after 24 h ($n \geq 4$) in either cell lines exposed. (Source: Reproduced with permission from [55], The American Chemical Society (2006).)

than 0.3 µg ml^{-1}. The NR assay detects uptake of the dye by lysosomes and does not discriminate the routes of cell death. The MTT assay gives lower cell viability since the SWNTs bind to the insoluble formazan, preventing its dissolution for subsequent photometric analysis. On the contrary, TEM determines the mode of cell death in terms of necrosis and apoptosis. Necrosis is a disorganized and unregulated process of traumatic cell destruction. The process completes with the release of intracellular components. A distinctive set of morphological features are observed, including membrane distortion, organelle degradation, and cellular swelling [24]. Necrosis is caused by factors external to the cell or tissue, such as infection or inflammation. Apoptosis is a programmed cell death used by the body to remove damaged or unwanted cells, that is, a naturally occurring cause of cellular death. Figure 7.11b reveals that the observed increase in cell death by SWNTs is mainly due to necrosis. Similarly, Cheng et al. [38] combined cell viability assays (NR, MTT, Live-Dead) and TEM imaging for assessing the toxicity of pristine MWNTs containing 6.4 wt% Fe and purified MWNTs (0.0005 wt% Fe) cultured with human macrophage cells (Figure 7.12a,b). A concentration-dependent decrease in viability is found by exposing macrophages to both pristine and purified MWNTs. Distinguishing between live and dead cells is crucial for assessment of growth control and cell death. The LIVE-DEAD viability assay applies two-color fluorescence to differentiate these cells. The viable cells are stained with a cell-permeable green

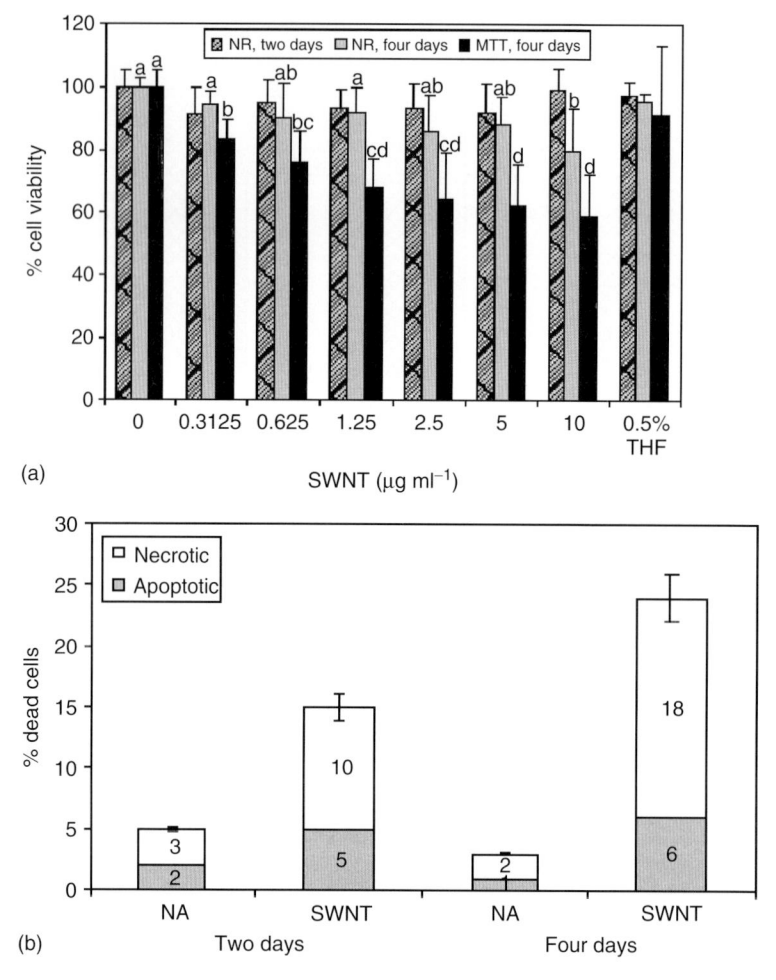

Figure 7.11 (a) SWNT-treated cells show decreased viability in the neutral red assay at four days when compared to control cells (NA) (ANOVA with, $F = 4.406$ and $p < 0.001$). With the MTT assay, at all concentrations, SWNT-treated cells show significantly decreased viability when compared to control cells (NA) at four days (ANOVA, $F = 16.67$, $p < 0.0001$). (b) TEM index of the analysis of cell death (apoptosis/necrosis: ±statistical error %) between control (NA) and SWNT-treated cells (5 µg ml^{-1}). Numbers indicate % dead cells. (Source: Reproduced with permission from [37], Nature Publishing (2007).)

fluorescent dye. The nucleus of the dead cells can be easily stained by propidium iodide, a cell-non-permeable red fluorescent dye. Stained live and dead cells can be observed by means of fluorescence microscopy. From Figure 7.12, cytotoxicity is significant for all three assays at 20 µg ml^{-1} MWNT, with NR showing 26%, MTT 43%, and live-dead 21% toxicity. For purified MWNTs, significant toxicity occurs when the nanotube content reaches 10 µg ml^{-1} and above. No significant differences exist between the viability results of pristine and purified MWNTs. The

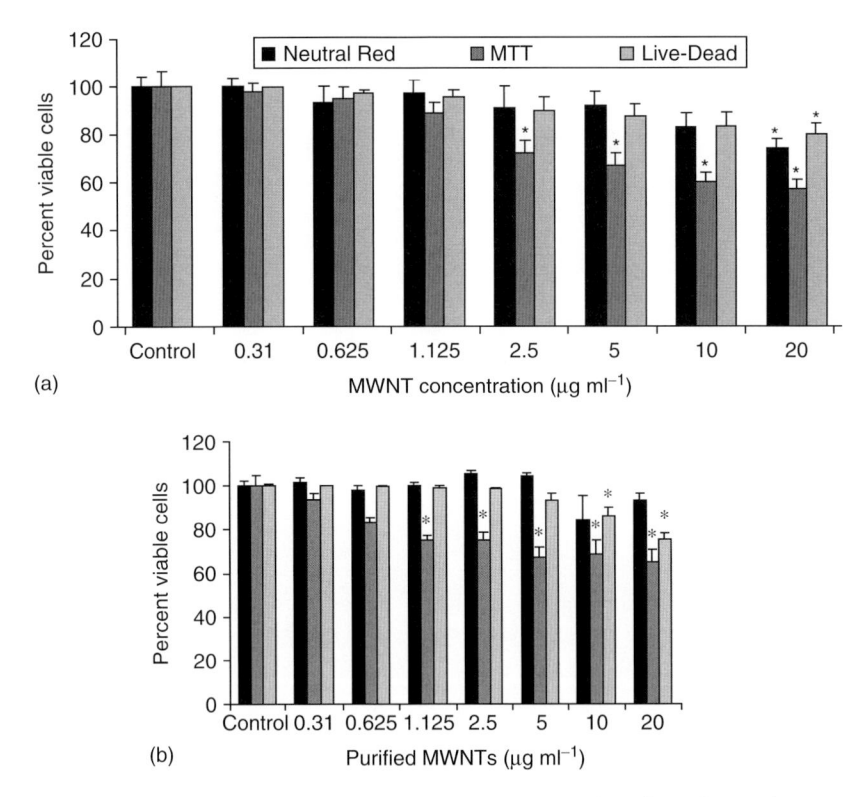

Figure 7.12 Toxicity of (a) pristine MWNTs and (b) purified MWNTs cultivated with human macrophage cells for four days at 37 °C. Cell viability was measured using NR, MTT, and live-dead assays. Cells treated under similar conditions but without carbon nanotubes were used as controls. NR and MTT values represent the mean ± SE of three experiments each performed in triplicates. Live-dead values are the mean ± SE of five individual experiments (ANOVA, $p < 0.01$). Statistical significant differences from the control are indicated with (*). (Source: Reproduced with permission from [38], Elsevier (2009).)

Fe catalyst content in purified MWNTs is extremely low, that is, 0.0005 wt%; thus, CNTs are considered to be the primary cause of toxicity. MWNTs can pass through the plasma membrane (Figures 7.6a,b and 7.7), thereby resulting in oxidative stress and cell death.

7.3.3
Tissue Cell Responses

CNFs and CNTs show promise for orthopedic implant applications because they promote osteoblast adhesion and proliferation as well as sustain bone formation [28, 29, 58–64]. Osteointegration plays a key role in the development of bone implants. These implant materials must promote adhesion and proliferation of osteoblasts and foster biomineralization. Adhesion of osteoblasts to biomaterial

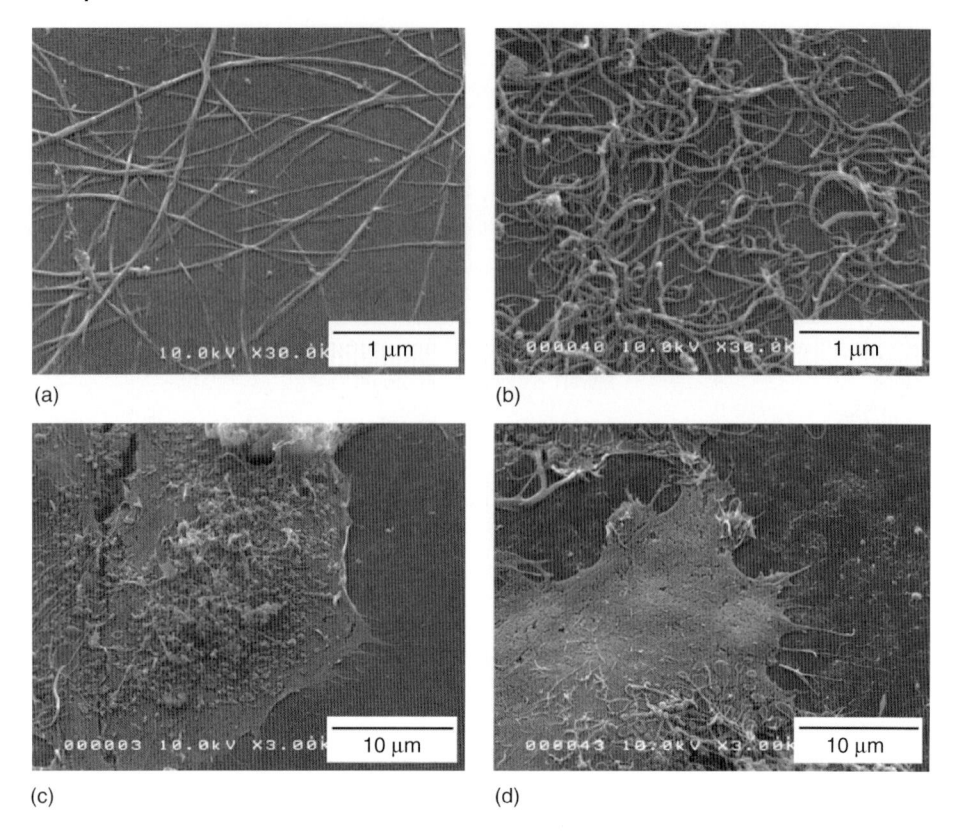

Figure 7.13 SEM images showing morphologies of (a) SWNT-coated dish, (b) MWNT-coated dish, (c) Saos-2 cells cultured on SWNTs, and (d) Saos-2 cells cultured on MWNTs. (Source: Reproduced with permission from [65], Elsevier (2010).)

substrates is critical for bone generation and involves several cell functions including interactions with neighboring cells and the ECM. In living tissues, cells are surrounded by ECM that composes of hierarchically arranged of collagen, laminin, and glycosaminoglycan in a complex topography in the nanometer scale. ECM plays an important role in several cellular activities such as providing structural cell support and guiding cell behavior and signaling. Tissue cells can attach to ligands present in the ECM via integrin receptors that are connected to the cytoskeleton in the interior of the cell [65]. Within the cell, the intracellular domain of integrin binds to the cytoskeleton via adapter proteins such as actin and vinculin. Understanding the signals that guide cell behavior is crucial for developing strategies to control tissue formation and function. The exceptional topography of the CNT surfaces provides enhanced osteoblast cell adhesion, which is confirmed by SEM and atomic force microscopy observations [64, 65]. Very recently, Matsuoka *et al.* [65] coated culture dishes with SWNT and MWNT suspensions and then incubated with human osteoblast-like cell line (sarcoma osteogenic, SAOS-2) for three days (Figure 7.13a–d). SEM images show that osteoblast cells attach and adhere firmly

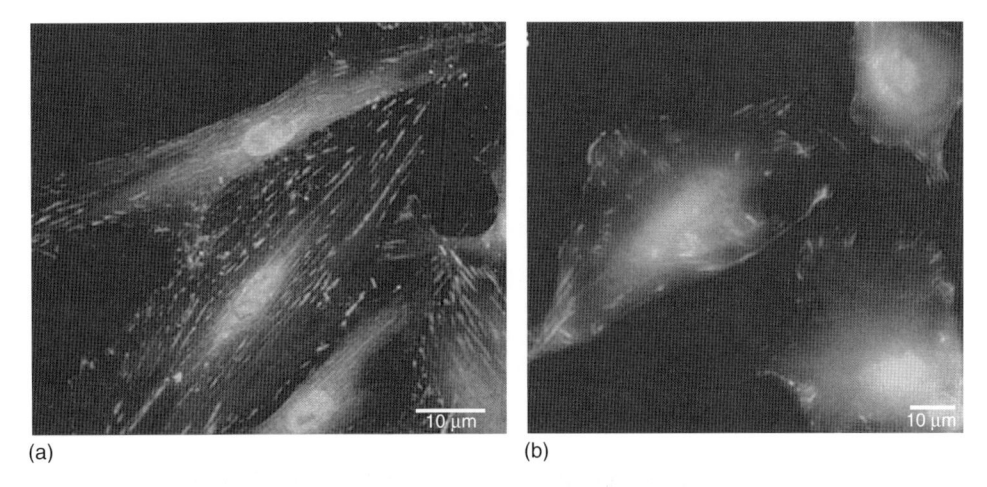

Figure 7.14 Typical vinculin expression of human osteoblast cells cultured on (a) randomly distributed CNTs and (b) glass substrate. (Source: Reproduced with permission from [64], Wiley-VCH (2008).)

on the nanotube surfaces. The cells then spread and anchor on the nanotube surfaces through fine filopodia (Figure 7.13d). For a more detailed analysis of cell adhesion to the nanotube surfaces, immunofluorescence microscopy can be used for observing the distribution of stained vinculin, a structural protein involved in the formation of focal adhesions. Osteoblasts adhered to CNTs exhibit well-developed vinculin clusters visible throughout the entire cell body (Figure 7.14a). For comparison, osteoblasts attached to the glass surface are shown in Figure 7.14b. Apparently, the focal contacts are mostly limited to the periphery of osteoblasts [64].

Mikos and coworkers [58] employed the Live-Dead fluorescent assay to assess fibroblast cell viability exposed to HiPco SWNTs of different forms, that is, purified nanotubes, ultrashort (US), and functionalized ultrashort (F-US) tubes. The US tubes with a length of 80 to 20 nm were prepared by fluorination of pristine nanotubes (~1 μm length) followed by pyrolysis. Fibroblasts are cell functions for the synthesis of ECM and collagen. Figure 7.15a–f shows fluorescent microscopy images of SWNT samples after seeding with fibroblasts (American type culture collection (ATCC), CRL-1764) for 24 h and treating with live-dead dyes. A high density of live cells (denoted with green color) with normal morphology and very few dead cells (red color) comparable with that of the live control can be readily seen in these images. Because of the hydrophobic nature of nanotubes, they tend to form aggregates within confluent cell monolayer (Figure 7.14c–f). For the SWNT and US tubes, fibroblasts can surround and attach to these aggregates without showing any signs of cell death. However, few dead cells are observed on the F-US tube aggregates owing to hindered nutrient transport in the nanotube agglomerates.

Figure 7.16 shows the MTT assay results of mouse fibroblast cells treated with acid-purified MWNTs and CNFs for 24 h. Cytotoxicity is expressed as a mean percentage decrease relative to the unexposed control. It is evident that the nanotubes do not significantly affect cell viability with the exception of the highest

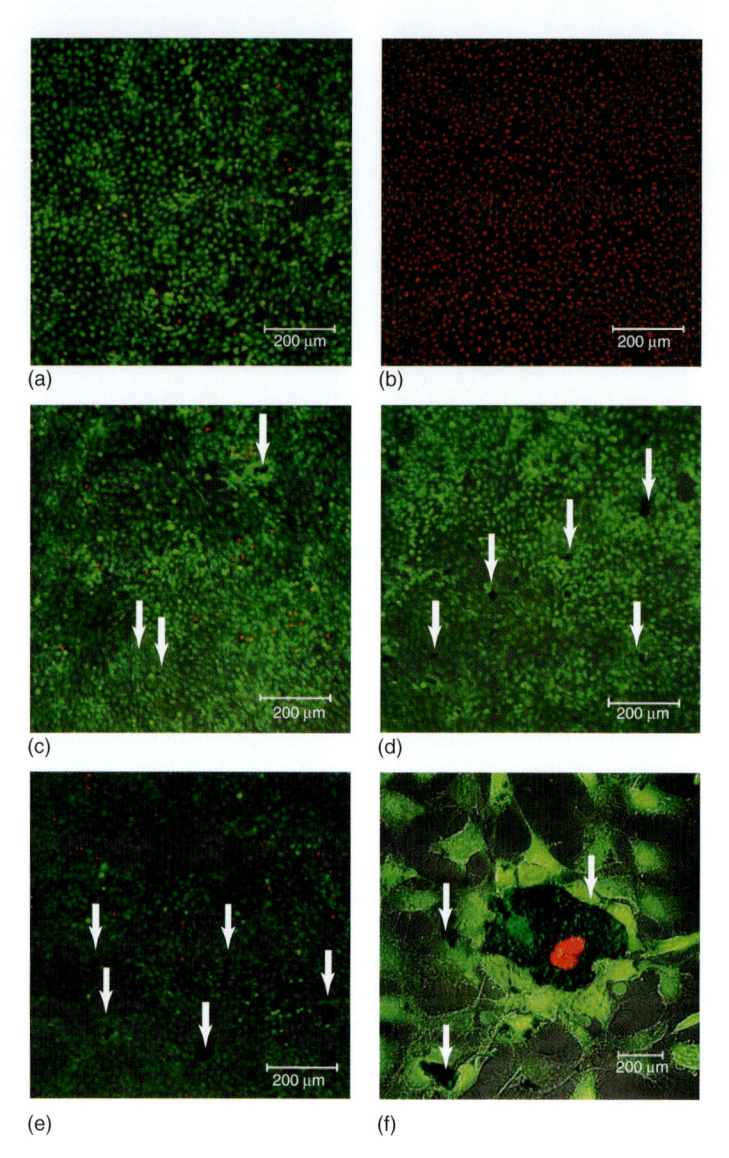

Figure 7.15 Fluorescent microscopy images of fibroblasts after 24 h exposure to experiment media and treated with LIVE-DEAD reagent. (a) Live (positive control), (b) dead (negative) control, (c) SWNT media, (d) functionalized SWNT media, and (e,f) ultrashort SWNT media (arrows point to nanotube aggregates). (Source: Reproduced with permission from [58], John Wiley & Sons (2008).)

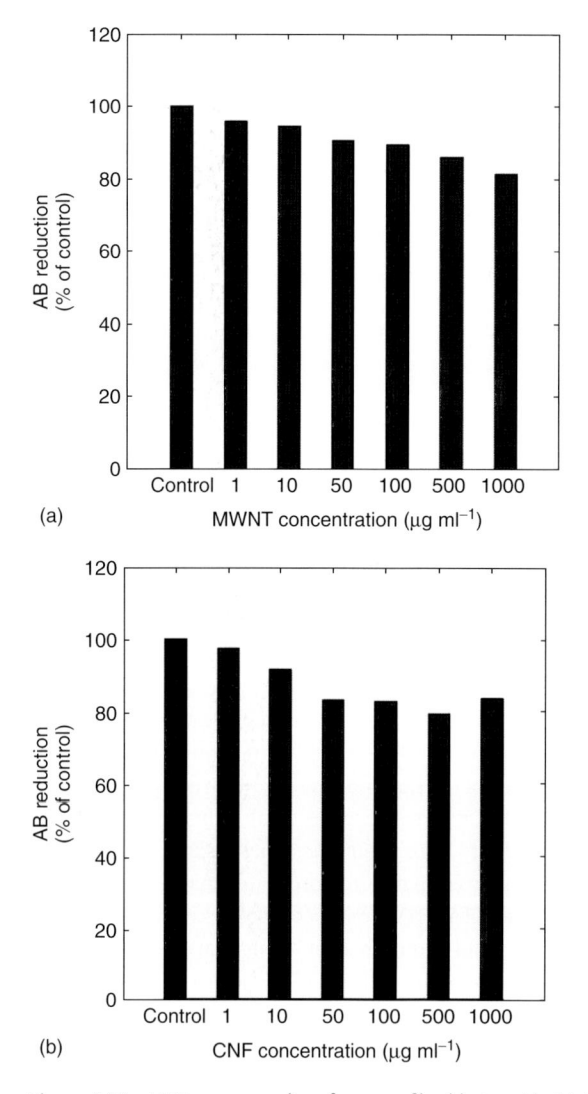

Figure 7.16 MTT assay results of mouse fibroblasts with (a) MWNTs and (b) CNFs after 24 h incubation. Data were averaged and expressed as percentage of control. (Source: Reproduced with permission from [61], Elsevier (2009).)

concentration of nanotubes. At 1000 µg ml^{-1}, a 20% inhibition effect is observed. There is no significant difference between fibroblast cell adhesion and growth for MWNTs and CNFs [61]. The effect of longer cell culture time on the viability of osteoblasts exposed to MWNTs is shown in Figure 7.17.

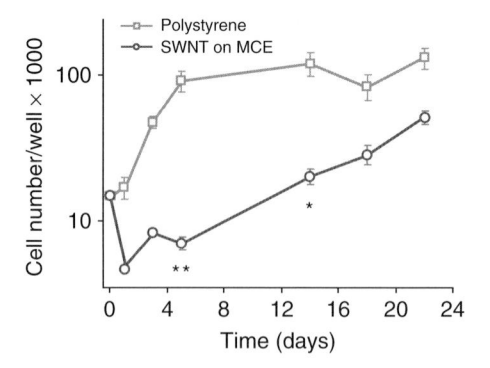

Figure 7.17 MTT calorimetric assay of primary osteoblastic rat cells with acid-purified SWNTs deposited onto mutticellulose ester multicellulose ester (MCE) membrane and polystyrene (control). Values are ±SD of three independent cell cultures. Statistical significant differences from control are indicated with $*p < 0.05$ and $**p < 0.001$. (Source: Reproduced with permission from [62], IOP Publishing Ltd (2009).)

7.4
CNT/Polymer Nanocomposites for Load-Bearing Implants

HAPEX polymer composite developed by Bonfied and coworkers [9, 10] in the 1980s was aimed for replacing defective bones. The Young's modulus and strength of HDPE filled with 40 vol% HA microparticles are much lower than those of cortical bone (Table 7.1). The low mechanical strength of the composite limits its use in orthopedic surgery as the implant material for load-bearing purposes. As recognized, nanocrystalline HA with unique structures exhibits higher mechanical strength and superior biocompatibility than their micrometer-sized counterparts [24]. Tjong and coworkers used hydroxyapatite nanorods (nHAs) as reinforcing fillers to strengthen HDPE [26]. The nHA/HDPE nanocomposites exhibit improved hardness, Young's modulus, and yield strength over their neat HDPE polymer (Table 7.3). The improved hardness of the nHA/HDPE nanocomposites over pure HDPE by adding nHA results in an increase in wear resistance of the composite materials. However, the mechanical stiffness values of the nHA/HDPE nanocomposites are still unsatisfactory for load-bearing applications.

To address low stiffness problem, high-performance PEEK and CF with high stiffness are selected as the composite materials for bone implants [17, 66, 67]. PEEK is a rigid, semicrystalline thermoplastic with a high melting temperature of 343 °C. PEEK possesses good mechanical properties, superior chemical resistance, high-temperature durability, and radiation stability [68]. Such characteristics render it an attractive biomaterial for trauma, orthopedic, and spinal implants [69]. The wear rate of the 30 wt% CF/PEEK composite is nearly 2 orders of magnitude smaller than that of conventional ultrahigh-molecular-weight polyethylene (UHMWPE)/metal and UHMWPE/ceramic couples [66]. The Young's modulus of PEEK is ∼3.9 GPa and can be increased to 18 GPa by adding 30 wt% CF. The modulus of the composite lies in the midrange of the cortical bone's stiffness.

Table 7.3 Mechanical properties of HDPE/nHA nanocomposites [26].

Material	Young's modulus (MPa)	Yield strength (MPa)	Vickers hardness number (VHN)	Izod impact strength (kJ m^{-2})
HDPE	1028.4 ± 10.2	22.56 ± 0.29	4.5 ± 0.2	3.18 ± 0.05
HDPE/2 wt% nHA	1139.5 ± 4.6	23.98 ± 0.26	4.8 ± 0.1	3.01 ± 0.04
HDPE/4 wt% nHA	1163.7 ± 6.8	24.23 ± 0.15	5.1 ± 0.1	2.85 ± 0.03
HDPE/6 wt% nHA	1194.1 ± 11.6	24.98 ± 0.26	5.6 ± 0.3	2.62 ± 0.05
HDPE/8 wt% nHA	1286.8 ± 8.1	25.21 ± 0.27	6.1 ± 0.2	2.41 ± 0.07
HDPE/15 wt% nHA	1430.6 ± 13.7	25.02 ± 0.31	6.6 ± 0.3	2.28 ± 0.14
HDPE/20 wt% nHA	1580.2 ± 14.0	25.09 ± 0.13	7.2 ± 0.2	2.07 ± 0.08

The CF/PEEK composite developed by Invibio Company (USA) offers enhanced properties for load-bearing implants, including significantly increased strength and stiffness and a modulus similar to that of cortical bone. Such a composite receives regulatory approval from the Japanese Ministry of Health, Labor, and Welfare for use as the load-bearing implant material [70].

Compared with rigid CFs, CNTs with large aspect ratios, extraordinary high tensile stiffness, and strength as well as large fracture strain can be incorporated into polymers in low loading levels. The unique properties of CNTs offer tremendous potentials for tailoring polymer composites into lightweight, high strength, and biocompatible materials. Furthermore, CNFs are very effective to promote adhesion and proliferation of osteoblasts [28]. The mechanical strength of the CNT–polymer and CNF–polymer nanocomposites depends greatly on the efficient load transfer from the polymer matrix to nanotubes and homogeneous dispersion of nanofillers in the polymer matrix. Strong interfacial bonding facilitates an effective stress transfer across the filler–matrix interface during tensile testing. The interfacial bonding between the polymer matrix and fillers can be enhanced considerably by adding coupling agents. In general, no coupling agent is added to avoid cytotoxicity in the CNT/polymer and CNF–polymer nanocomposites during cell culture and viability tests.

7.4.1
Mechanical Properties

7.4.1.1 MWNT/PE Nanocomposites

McNally et al. fabricated MWNT/PE (polyethylene) nanocomposites by direct melt mixing of MWNTs and PE pellets [Chapter 2, Ref. 93]. Figure 7.18 shows the yield stress, break stress, and failure strain versus CNT content for MWNT/PE nanocomposites. The yield stress increases slightly with increasing filler content. However, the ultimate tensile strength and elongation at break decrease significantly with increasing nanotube content. These results indicate that the filler–matrix interfacial bonding is very poor, leading to ineffective load transfer

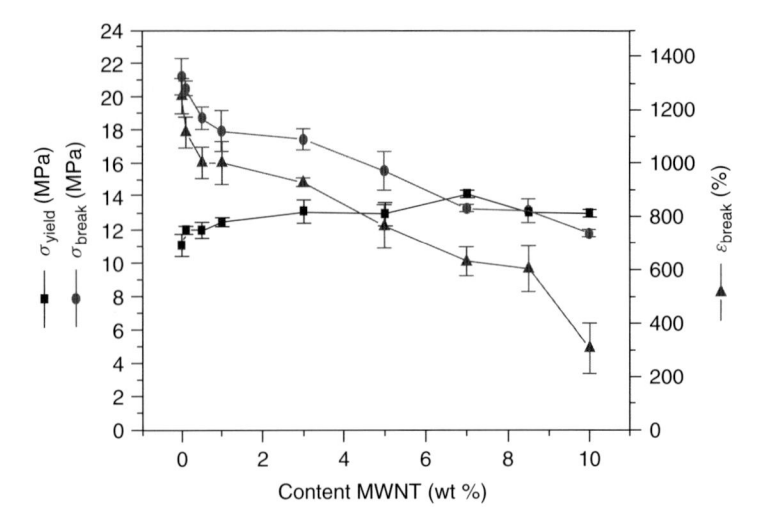

Figure 7.18 Tensile properties of melt-compounded MWNT/PE nanocomposites. (Source: Reproduced with permission from [Chapter 2, Ref. 93], Elsevier (2005).)

Table 7.4 Tensile properties of functionalized MWNT/HDPE nanocomposites.

MWNT content (vol%)	Young's modulus (GPa)	Tensile strength (MPa)	Elongation at fracture (%)	Tensile toughness (J)
0	1.095	105.80	863.4	634.53
0.11	1.169	105.51	948.5	743.35
0.22	1.228	106.67	978.5	756.24
0.33	1.287	106.38	1020.4	776.27
0.44	1.338	109.86	1069	842.27

Source: Reproduced with permission from [71], Elsevier (2007).

across the nanotube–matrix interface. To improve the filler–matrix bonding, Kanagaraj et al. [71] used acid-purified MWNTs to reinforce HDPE. They reported that the composite specimens undergo full yielding, necking, and ductile failure. The tensile properties of HDPE and its nanocomposites are listed in Table 7.4. Considerable enhancements in Young's modulus, ultimate tensile stress, and fracture strain of the MWNT/HDPE nanocomposites can be realized by using functional MWNTs containing carboxylic groups. Further, the nanotube additions do not impair the ductility of HDPE but rather improve it. The stiffness of resulting nanocomposites is still quite low for load-bearing implant applications because of the small amount of MWNT additions.

Recently, Park et al. [72] demonstrated that the elastic modulus of MWNT/HDPE nanocomposites can be increased markedly by *in situ* polymerization of ethylene with a zirconocene-based catalyst, Cp_2ZrCl_2. The direct interactions between nanotubes and the cyclopentadiene (Cp) ring lead to the adsorption of catalyst onto the sidewalls of MWNTs. Consequently, MWNTs can be well dispersed in the polymer matrix. Table 7.5 lists the tensile properties of MWNT/HDPE nanocomposites. The tensile properties of HDPE can be enhanced by adding 2.22 wt%

Table 7.5 Tensile properties of *in situ* polymerized MWNT/HDPE nanocomposites.

MWNT content (wt%)	Young's modulus (GPa)	Yield strength (MPa)	Flexural modulus (GPa)	Elongation at fracture (%)
0	1.47 ± 0.31	22.3 ± 0.2	1.12 ± 0.01	709 ± 45
1.25	4.61 ± 1.44	23.3 ± 0.4	1.27 ± 0.06	320 ± 39
2.22	6.75 ± 1.51	24.6 ± 0.3	1.42 ± 0.07	77 ± 31
4.11	2.58 ± 1.16	23.3 ± 0.4	1.13 ± 0.05	44 ± 22

Source: Reproduced with permission from [72], The American Chemical Society (2008).

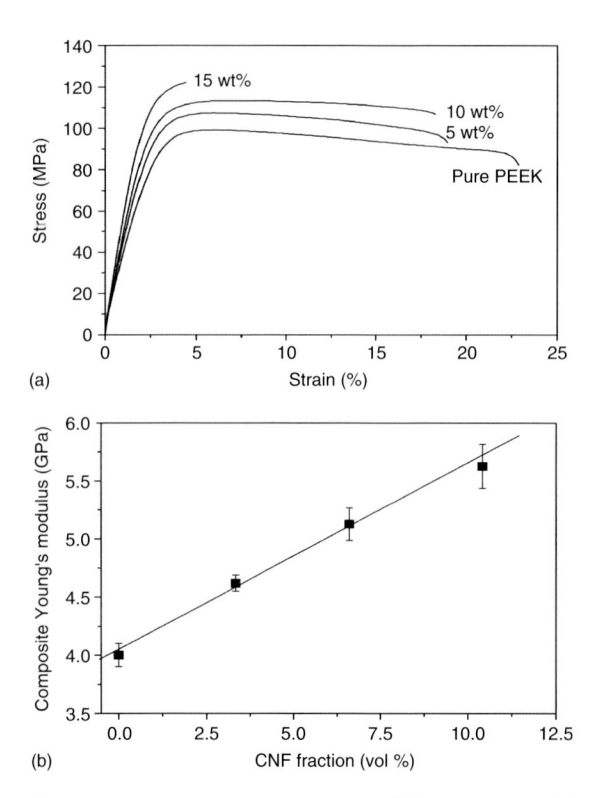

Figure 7.19 (a) Stress–strain curves and (b) Young's modulus versus carbon nanofiber content of CNF/PEEK nanocomposites. (Source: Reproduced with permission from [73], Elsevier (2002).)

Table 7.6 Tensile properties of injection-molded MWNT/PEEK nanocomposites.

MWNT content (wt%)	Young's modulus (GPa)	Tensile strength (MPa)	Elongation at fracture (%)	References
0	4.0	93.55	>20	[74]
6.5	5.32	102.15	12.49	[74]
9	6.0	104.44	10.01	[74]
12	6.35	107.14	8.28	[74]
15	7.55	110.90	6.28	[74]
0	4.2	106	–	[75]
9	6.3	114	–	[75]
15	8.3	121	–	[75]

MWNT. The Young's modulus of the 2.22 wt% MWNT/HDPE nanocomposite is 6.75 GPa, being 359% improvement over that of pure HDPE. The stiffness value of this composite is higher than that of HAPEX and approaches the lower limit of cortical bone.

7.4.1.2 MWNT/PEEK Nanocomposites

Sandler et al. [73] studied the tensile properties of injection-molded CNF/PEEK nanocomposites. They reported a linear increase in tensile strength and stiffness of the composites with increasing CNF content up to 15 wt%, while matrix ductility was maintained up to 10 wt% (Figure 7.19a,b). Deng et al. [74] then investigated the tensile behavior of MWNT/PEEK nanocomposites at room and high temperatures. They reported that the MWNT additions increase the Young's modulus and tensile strength o f PEEK but decrease the fracture strain at room temperature (Table 7.6). The Young's modulus of pure PEEK is 4 GPa and increased to 7.55 GPa by adding 15 wt% MWNT, being 89% improvement. The tensile strength and ductility of this composite also meet the requirement of cortical bone. Thus this nanocomposite material shows promise for use as the bone replacement in orthopedics. Very recently, this research group also presented similar results for the tensile properties of injection-molded MWNT/PEEK nanocomposites as listed in Table 7.6 [75].

7.5
CNT/Polymer Nanocomposite Scaffolds

Cells *in vivo* are embedded in a three-dimensional (3D) environment, surrounded by other cells and the ECM. Most of the cell cultivation experiments are conducted on 2D flat substrates such as glass slides and multiwell plates. The 2D cell cultures are different from *in vivo* environments in which cellular attachment and growth, transport of nutrient, waste removal, and metabolism occur in a 3D mode. Scaffold is an artificial structure for ECM that supports 3D tissue formation, acting as a template for cell attachment, migration, and differentiation. Scaffolds can mimic the structure and biological functions of ECM. To achieve the goal of tissue reconstruction, scaffolds must meet several specific requirements including [76]

a) good compatibility with local cell environment;
b) high porosity with an adequate pore size and pore interconnectivity;
c) excellent cell adhesion and growth;
d) controllable degradation rate;
e) good mechanical strength to support regeneration of new tissues.

In general, scaffolds must exhibit large pores for effective transfer of cells and nutrients and metabolic waste removal [77]. Different kinds of cells and bone tissue have different pore size ranges in the scaffolds for effective cell growth and tissue regeneration. Oh et al. fabricated scaffolds with a gradient of pore diameters ranging from ~88 to 405 μm and examined the interaction of osteoblasts, fibroblasts, and chondrocytes with the scaffolds in vitro [78]. The scaffolds with pore sizes of 380–405 μm show better cell growth for chondrocytes and osteoblasts, while the scaffolds with 186–200 μm pore size exhibit better fibroblasts growth. The scaffolds with 290–310 μm pore size encourage faster tissue penetration, resulting in new bone formation in vivo.

7.5.1
Types and Structures of Scaffolds

Polymers used for scaffold fabrication are biodegradable in nature, including natural polysaccharides such as starch, chitosan (CS), and alginate; proteins such as collagen, silk, and soya; and synthetic polymers such as poly(lactic acid) (PLA), poly(glycolic acid) (PGA), poly(lactic-co-glycolic acid) (PLGA), poly(ε-caprolactone) (PCL), and so on. Among them, natural polymers are particularly attractive, mainly because of their similarities with the ECM, chemical versatility, and good biological performance [79]. However, naturally derived polymers have poor mechanical properties. Synthetic biomaterials such as PCL and PLA are hydrophobic and lack of functional groups, thereby showing low efficiency for cell adhesion. To improve the hydrophilicity and biological properties of PCL nanofibrous scaffolds, gelatine and collagen are incorporated into the PCL scaffolds [80]. There are a number of excellent reviews that discuss the syntheses and properties of natural and synthetic biodegradable polymers for tissue engineering applications [79, 81–85]. In general, CNTs are potential nanofillers for polymer scaffolds in order to improve cellular activities, particularly promoting cell attachment and proliferation for tissue engineering applications.

Polymeric scaffolds with different porosities can be prepared by several techniques including freeze drying, porogen leaching, gas foaming, phase separation, microsphere sintering, and electrospinning [24]. Freeze drying is a rather simple method based on the formation of ice crystals for inducing fine pores through ice sublimation. Porosity levels up to 90% with different interconnectivities can be produced [86]. The pore size and structure of scaffolds depend mainly on the

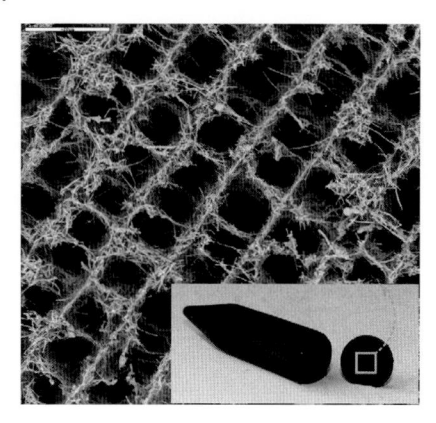

Figure 7.20 Scanning electron micrograph of cross-sectioned MWNT/CS scaffold. Bar is 20 μm. Inset is a picture of monolithic scaffold prepared from freeze-drying process. (Source: Reproduced with permission from [87], Elsevier (2008).)

nucleation and growth rate of ice crystals during freezing process. The porosity level of scaffolds can be tailored by monitoring the freeze time [87]. This technique is difficult to produce scaffolds with better structural integrity and adequate mechanical properties. Particle leaching requires the use of a porogen agent such as salt or sugar to induce pores formation. The porogen is first dispersed in a solvent, and the suspension is then blended with polymer solution followed by solvent casting. Finally, the porogen is leached away using water to form pores. The pore size and porosity level of scaffolds can be controlled by the size and volume fraction of the porogen. However, the use of organic solvent may cause cytotoxicity.

The mechanical properties of polymer scaffolds generally have insufficient stiffness and compressive strength compared to human cancellous bone. Therefore, CNTs with high stiffness and strength offer a tremendous opportunity for materials scientists and biomedical engineers to develop novel scaffolds with enhanced biological and mechanical performances. For example, Abarrategi et al. [88] used freeze-drying method for making porous MWNT/CS scaffold (Figure 7.20). The process involved an initial dispersion of purified nanotubes in a CS solution. The mixed suspension was quickly frozen by dipping into liquid nitrogen bath followed by freeze drying. The scaffold was subsequently cultured with myoblastic mouse cells (C2C12 cell line). Scanning electron micrographs show that the myoblastic cells (indicated by arrows) spread uniformly over the scaffold surface (Figure 7.21a). The cells cover an entire scaffold surface after four days incubation (Figure 7.21b). This implies that the scaffold is nontoxic and capable of supporting the attachment of myoblastic cells.

7.5.2
Electrospinning: Principle and Applications

Electrospinning is a simple and versatile process to produce fibers of nanometric size. Electrospun nanofibrous scaffolds with large surface area, high porosity, and

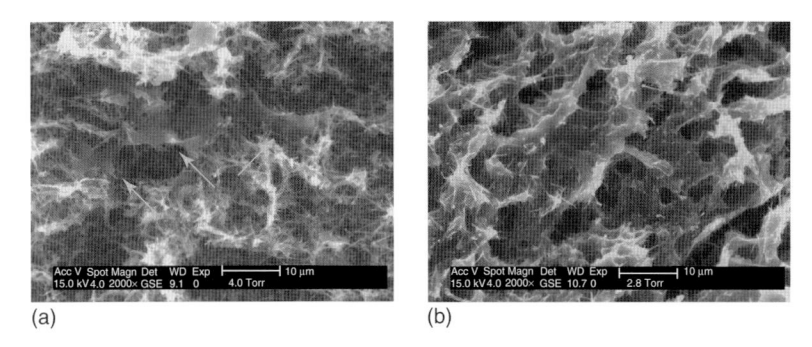

Figure 7.21 Scanning electron micrographs of MWNT/CS scaffold seeded with C2C12 cells after (a) one and (b) four days culturing. (Source: Reproduced with permission from [88], Elsevier (2008).)

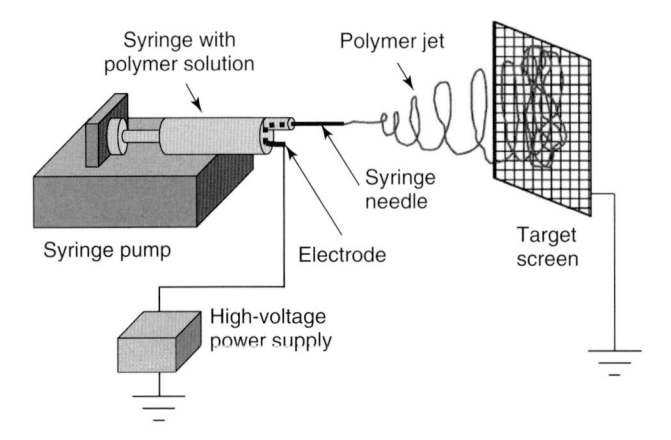

Figure 7.22 Electrospinning facility. (Source: Reproduced with permission from [94], Elsevier (2006).)

well-interconnected pore network mimic the architecture and biological functions of ECM. A typical electrospinning facility consists of a syringe, syringe pump, spinneret, collector (stationary or rotating), and high-voltage power supply (Figure 7.22). The diameter, morphology, and property of electrospun fibers are controlled by the material components and spinning parameters, including polymer concentration, solvent volatility, solution conductivity, applied voltage, and distance between the spinneret and substrate [89–94]. Improper selection of those parameters can lead to the formation of beads in nanofibers. In a typical electrospinning process, a high voltage is applied between the spinneret and the collector to create an electrically charged polymer solution. As the polymer flows to the syringe needle at a constant rate by a syringe pump, it forms a droplet because of the surface tension of the solution. With the increase of electric field, the induced charges on the liquid surface repel each other. At this stage, the electrostatic repulsion force counteracts the surface tension, and droplet is stretched and deformed into a conical shape known as *Taylor cone*. When the electrical potential at the surface of a polymer overcomes

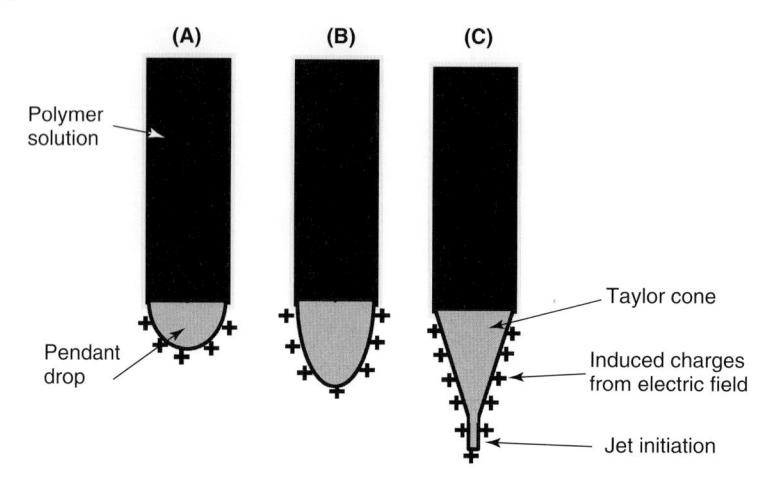

Figure 7.23 Schematic diagrams showing the formation of Taylor cone: (a) surface charges induce in the polymer solution due to the electric field. (b) Elongation of the droplet. (c) Liquid jet initiation from the Taylor cone. (Source: Reproduced with permission from [92], Elsevier (2010).)

the surface tension, a liquid jet is ejected from the Taylor cone (Figure 7.23). The charged jet then undergoes solvent evaporation and a series of electrically driven bending instabilities in the form of looping and spiraling motions and finally deposits on a grounded metal screen/collector as ultrafine porous membranes (mats). These instabilities cause the deposited fibers to be randomly oriented and nonwoven. By monitoring the process parameters, mats with fully interconnected open structures can be obtained [91]. The nanofibers can also be collected on a rotating drum, rotating disk, or static parallel electrodes to obtain aligned fibers or arrays of fibers [92]. Figure 7.24 shows the formation of aligned nanofibers using a rotating drum. The diameter of the fiber can be monitored based on the rotational speed of the drum. Uniaxially aligned nanofibers can be collected by using static parallel electrodes (Figure 7.25). Two nonconductive material strips are placed along a straight line, and a metal foil is placed on each of the strips and then grounded. This technique enables fibers to be deposited at the end of the strips such that the fibers adhere to the strips in an alternate manner and collected as aligned arrays [93]. Uniaxially aligned nanofibers generally exhibit better mechanical and electrical properties than randomly oriented fibers.

The continuous nature of the fibers within the mats is particularly useful for fabricating 3D scaffolds with high spatial connectivity. Electrospun nanofiber scaffolds show morphological similarities to the ECM and can induce favorable interactions between cells and scaffolds. Accordingly, those scaffolds exhibit good biocompatibility by promoting cell adhesion and proliferation. Several natural and synthetic polymers including collagen, PLA, PGA, PCL, and their copolymers can be electrospun into nanofiber scaffolds [94–96]. The main challenge of using electrospun scaffolds for tissue engineering applications is a lack of cell infiltration into fibrous mats owing to the inherent small pore sizes between fibers. As

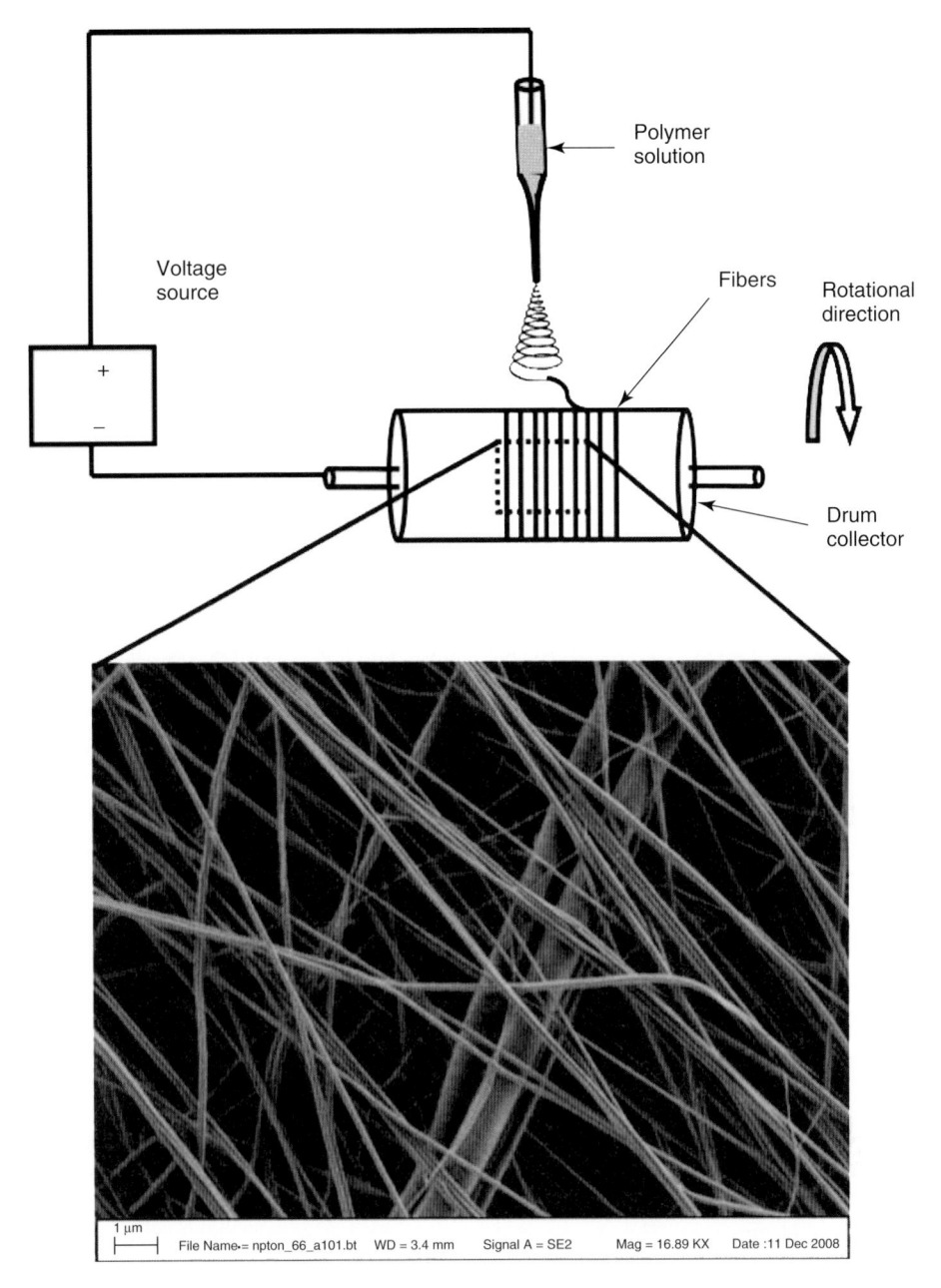

Figure 7.24 Schematic of the rotating drum used for fiber collection. The inset Scanning electron micrograph shows the aligned fibers obtained using the rotating drum. (Source: Reproduced with permission from [92], Elsevier (2010).)

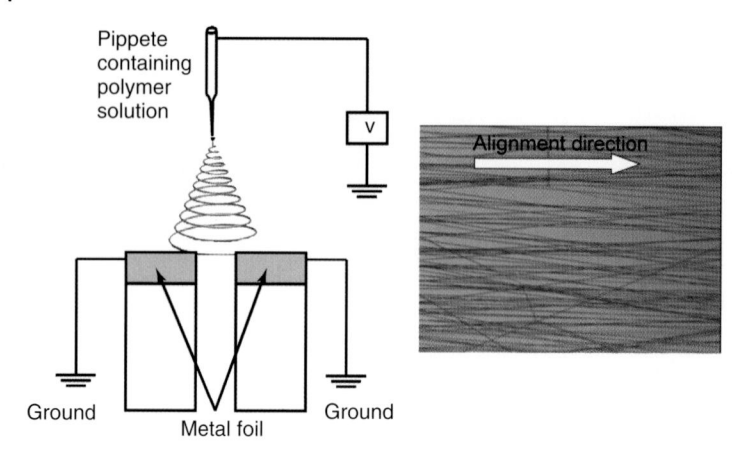

Figure 7.25 Schematic of static electrodes used for collecting aligned fiber arrays. (Source: Reproduced with permission from [92], Elsevier (2010).)

mentioned above, pore size of the scaffolds must be sufficient large enough to allow for cell penetration [78]. For electrospun mats, pore size and fiber diameter are highly correlated. Reducing the fiber diameter generally results in a decrease of the pore size of fibrous mats [97]. Collectors of desired patterns have been found to improve cellular colonization in electrospun scaffolds effectively [98, 99]. Vaquette and Cooper-White demonstrated that fibroblasts can penetrate up to 250 µm into the patterned PCL scaffolds, compared with 30 µm for the standard PCL scaffolds. However, the tensile strength and modulus of patterned scaffolds are lower than those of standard electrospun PCL scaffolds because of an enlargement in the pore size [99]. It is considered that the [97] mechanical performance of electrospun polymer nanofibers can be improved by adding CNTs. A number of nondegradable polymer/CNT nanofibers have been electrospun, including poly(vinyl chloride), poly(methyl methacrylate) (PMMA), PA6, polystyrene (PS), and poly(trimethylene terephthalate) [100–104]. In contrast, there have been few studies reporting the effect of CNTs on the fabrication and properties of electrospun scaffolds based on degradable polymers for tissue engineering applications [105–107].

From the literature, electrical signals are very effective to regulate cell functions by stimulating the growth of neuronal and osteoblastic cells on conducting CNT substrates [3, 108, 109]. Very recently, Shao et al. [106] carried out a systematic study on the effect of electrical stimulation on the function of osteoblasts on electrospun MWNT/PLA nanofiber mats. Both randomly oriented (denoted as R-type) and aligned nanofiber (denoted as A-type) mats with 1–5 wt% MWNT were electrospun. The A-type mats were collected with a rotating drum at a speed of 2000 rpm. Figure 7.26 shows SEM images and fiber diameter histograms for both R- and A-type mats. R0, R1, R2, R3, R4, and R5 correspond to the randomly oriented fibers with 0, 1, 2, 3, 4, and 5 wt% MWNT, respectively, and A0, A1, A2, A3, A4, and A5 stand for aligned composite with 0, 1, 2, 3, 4, and 5 wt% MWNT, respectively. Apparently, the diameter of R- and A-type nanofibers

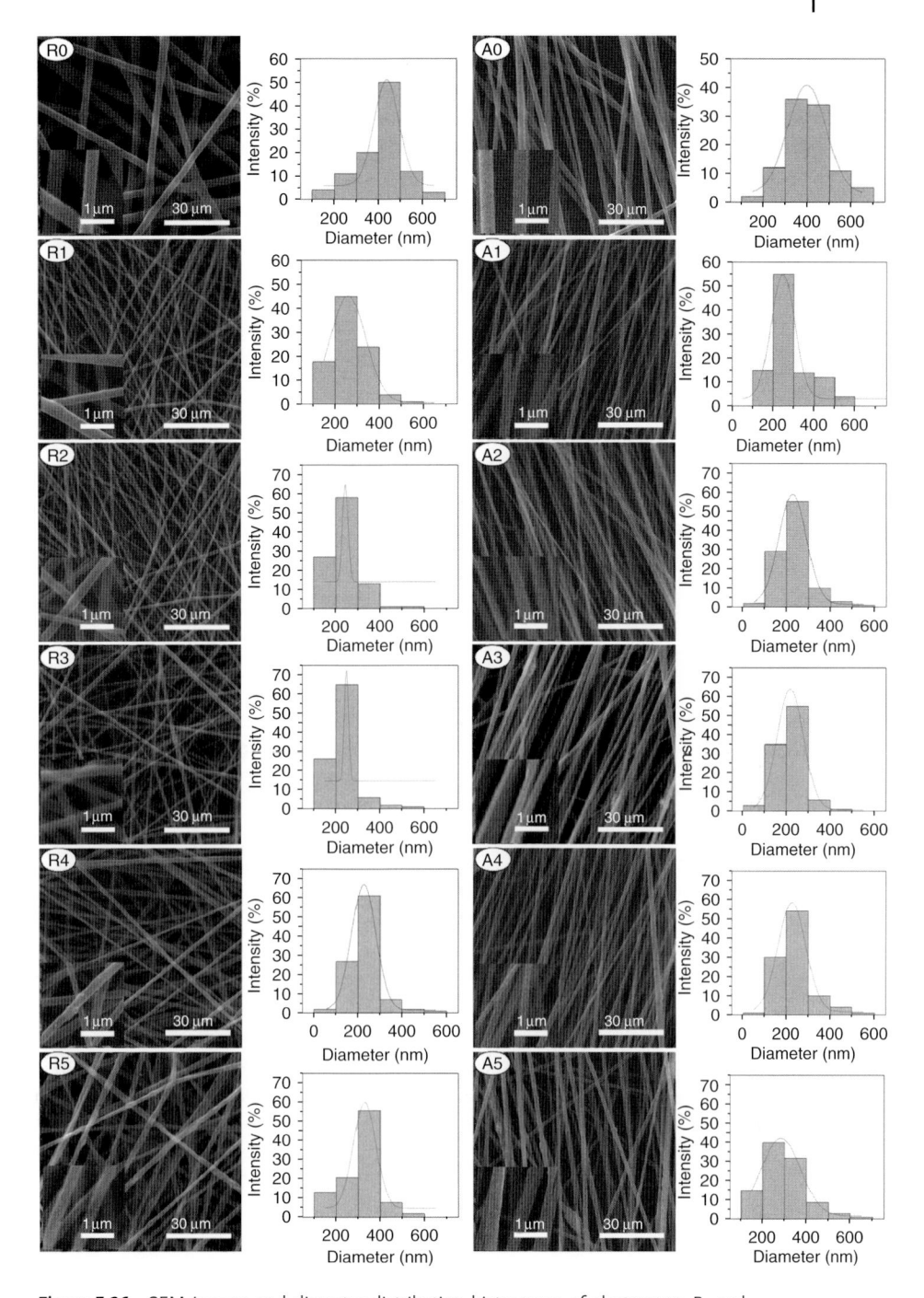

Figure 7.26 SEM images and diameter distribution histograms of electrospun R- and A-type nanofiber mats. (Source: Reproduced with permission from [106], Elsevier (2011).)

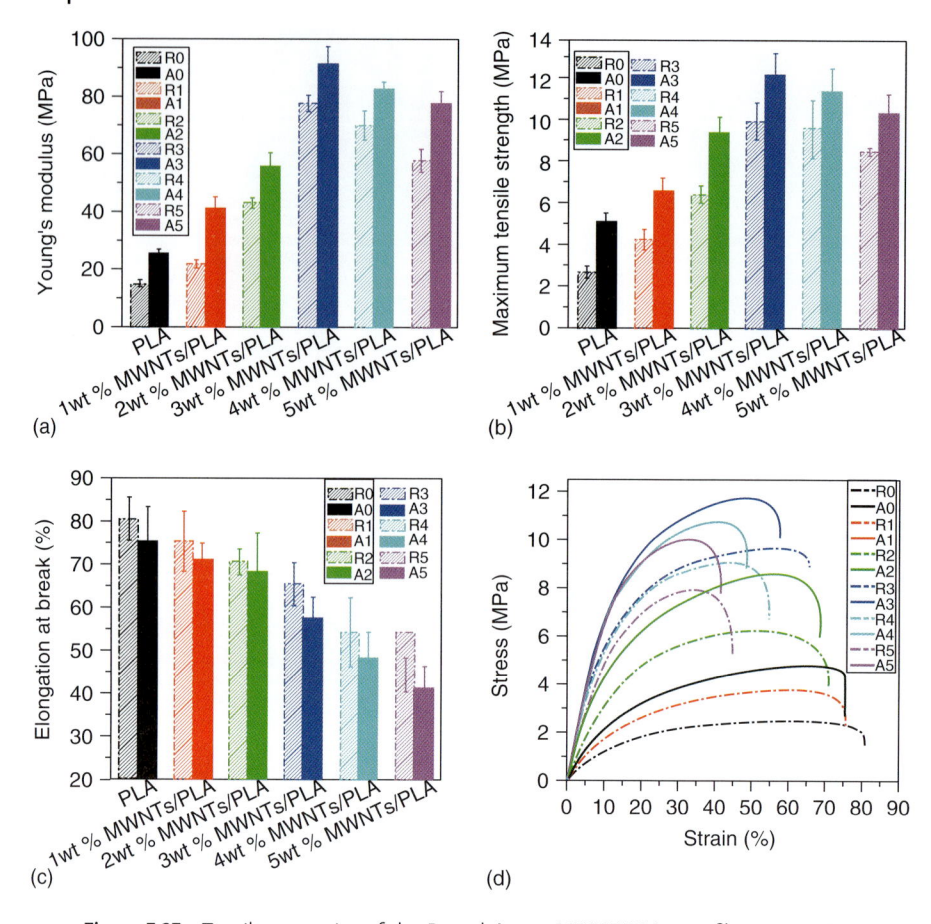

Figure 7.27 Tensile properties of the R- and A-type MWNT/PLA nanofiber mats. (Source: Reproduced with permission from [106], Elsevier (2011).)

decreases with increasing nanotube content up to 3 wt% MWNT. Above 3 wt%, the diameter increases slightly but is still smaller than that of pure PLA nanofibers. The average diameter of A-type fibers is somewhat smaller than that of R-type nanofibers with the same nanotube content due to additional stretching effect during high-speed drum collection. Furthermore, the surface of MWNT/PLA nanofibers turns rougher as the nanotube content increases. The tensile properties of electrospun nanofiber mats are shown in Figure 7.27a–d. The Young's modulus and tensile strength of the R- and A-type nanofiber mats reach their peak values by increasing MWNT content to 3 wt%. The A-type mats generally exhibit higher reinforcing effect than the R-type mats. This can be attributed to the A-type mats possess finer fiber diameter and a higher degree of fiber alignment. The modulus and tensile strength of the fiber mats appear to decrease when the MWNT content reaches 4 wt% and above because of the agglomeration of CNTs. The elongation at break of the R- and A-type nanofiber mats decreases with increasing nanotube

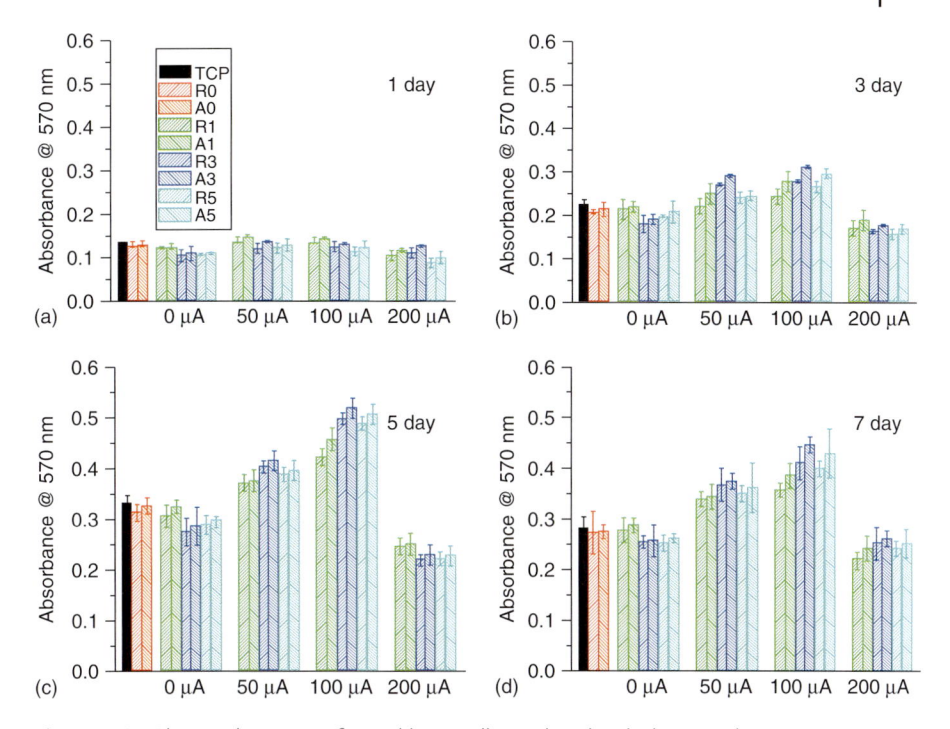

Figure 7.28 Alamar Blue assay of osteoblastic cells incubated with the R- and A-type MWNT/PLA nanofiber mats under DC electrical stimulation for (a) one day, (b) three days, (c) five days, and (d) seven days. (Source: Reproduced with permission from [106], Elsevier (2011).)

content (Figure 7.27c). Figure 7.28 shows the cytotoxicity results determined from Alamar Blue assay for osteoblasts cultured on the tissue culture plate (TLP) and nanofiber mats with different direct current (DC) electrical signals of 50, 100, and 200 µA. The viability of osteoblasts as determined from the absorbance at 570 nm reveals that the electrical signals have little effect after one day cultivation. Without electrical stimulation, cell viability is slightly higher for the aligned A-type mats than that for randomly oriented R-type mats after culturing for three, five, and seven days. However, electrical signals of 50 and 100 µA exert a beneficial effect for the proliferation of osteoblasts on the A-3 and R-3 mats after three day cultivation. This beneficial effect is more pronounced after culturing for five and seven days. The application of DC signal of 200 µA leads to a drastic reduction in cell viability after culturing for three, five, and seven days. Excessive electrical stimulation exceeds the tolerance of osteoblasts, leading to the death of some cells. Figure 7.29a shows SEM images of osteoblasts cultured on the R3 and A3 mats for three and seven days at different electrical signals. Without electrical stimulation, osteoblasts appear to be more elongated on aligned A3 mat. The morphology of osteoblasts becomes narrower and longer in the presence of electrical stimulation. The aspect ratios of osteoblasts cultured on the A3 and R3 mats for three and seven days are shown in Figure 7.29b,c, respectively.

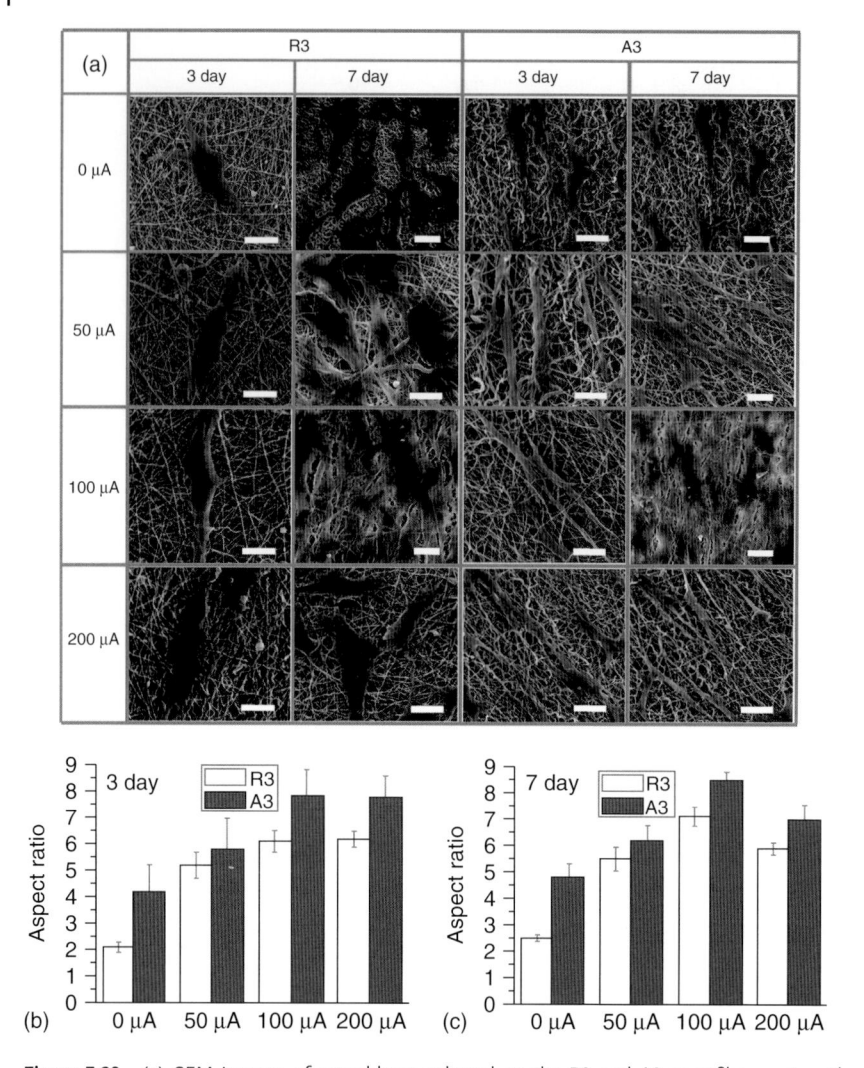

Figure 7.29 (a) SEM images of osteoblasts cultured on the R3 and A3 nanofiber mats with and without electrical stimulation for three and seven days. The scale bar is 30 μm. Elongation of osteoblasts as measured by the aspect ratio after cultivation for (b) three and (c) seven days. (Source: Reproduced with permission from [106], Elsevier (2011).)

7.6
Nervous System Remedial Applications

The nervous system embedded with neurons is a complex network that transmits information by electrical and chemical signaling. The nervous system can be classified into the central nervous system including the brain and the spinal cord and the peripheral nervous system including the spinal and autonomic nerves [20]. Damage to the nervous system can range from loss or impairment of sensory to severe cognitive disruption. Nerve repair and regeneration are key challenge issues for medical practitioners because of the large number of patients suffering from nerve injury globally [110]. In this case, nanotechnology plays an important role in the design and fabrication of biocompatible materials for neural probes and scaffolds for nervous tissue engineering. The ideal materials for neural tissue engineering applications should possess excellent cytocompatible, mechanical, and electrical properties [20].

Cell attachment, proliferation, and differentiation are primarily affected by the substrate surface onto which the cells reside. CNTs have been reported to serve as active sites for supporting neuronal growth and differentiation (Figure 7.1) [3, 111, 112]. Mattson *et al.* reported that neurons grow preferentially on the MWNTs coated with 4-hydroxynonenal compared to pristine MWNTs. A 200% increase in total neurite length and nearly a 300% increase in the number of branches and neurites were found in coated nanotubes [113]. CNTs enhance neural performance by delivering electrical signals to guide neuron growth, favoring the formation of electrical shortcuts [109, 114]. Accordingly, CNTs are promising materials to make neural probes for the treatment of neural degenerative diseases. Neural electrodes are mainly made from metallic materials such as platinum, gold, titanium, and stainless steels. However, these metallic electrodes show long-term instability of electrical stimulation because of their poor contact with tissues. In this context, conducting polymers can be deposited on neural electrodes to enhance stability and conductivity of the electrodes [115]. The electrical stimulus to neuroendocrine PC-12 cells cultured on polypyrrole surfaces leads to a significant increase in neurite length compared to the same material surface without electrical stimulation. This cell line is popular for examining neurite growth and spreading

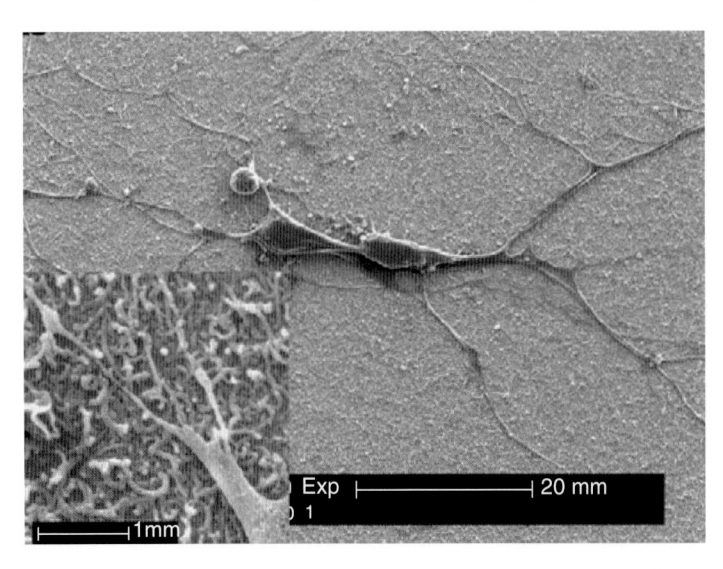

Figure 7.30 SEM image of neurons cultivated on MWNT/PEDOT surface. (Source: Reproduced with permission from [117], Elsevier (2011).)

[116]. To further improve the performance of neural electrodes, CNTs are incorporated into conducting polymers as the additives [116–118]. For example, Luo et al. [117] electrochemically deposited MWNT/poly(3,4-ethylenedioxythiophene) (PEDOT) films on the platinum and gold/plastic substrates and examined their physical and cellular behavior. They reported that the MWNT/PEDOT film is nontoxic and support the growth of neurons. Figure 7.30 shows typical SEM image of neurons cultured on the MWNT/PEDOT surface. The neurons attach firmly to the surface and possess long neurite extension (inset figure).

The nervous network transmits information by electrical and chemical signaling, thus electrical stimulation may affect the growth of neural cells on CNT/polymer nanocomposites. Cho and Borgens prepared MWNT/collagen composites containing 0.1–90 wt% MWNT using solution-casting process. The effect of electrical stimulation (i.e., 100 mV for 6 h) on the growth of PC12 cells on such composites was examined [119]. Figure 7.31a,b depict SEM images showing the growth of neurites on the 5 wt% MWNT/collagen composite without and with electrical stimulation, respectively. It is obvious that electrical stimulation facilitates neurite outgrowth on the composite as evidenced by appreciable filopodia and microspike extensions. In contrast, neurites without filopodia and microspikes are observed on unstimulated composite surface. The MTT assay test reveals that the metabolic activity of the cells gradually decreases with increasing nanotube content in the composites. The cell viability decreases markedly at high MWNT contents, that is, $\geq 10\%$, regardless the presence of electrical stimulation (Figure 7.32). There exists dose-dependent relationship between cell viability and MWNT content. The metabolic activity of PC12 cells is significantly affected by the loading amount of MWNTs, rather than the presence of electrical stimulation.

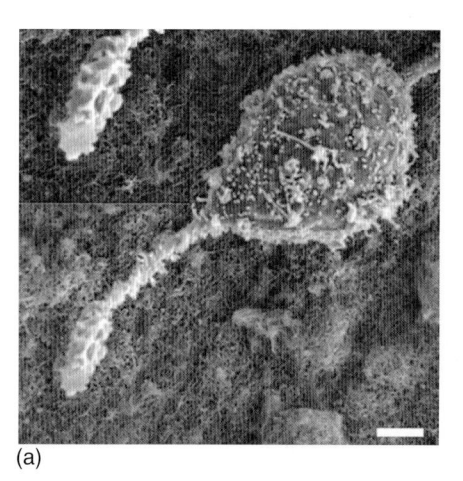

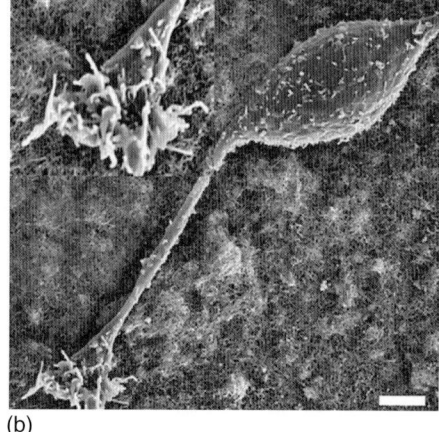

Figure 7.31 SEM images of PC Cells grown on (a) unstimulated and (b) electrically stimulated 5% MWNT/collagen composite. The cells were exposed to 100 mV for 6 h. Inset shows the initiation of microspike and filopodia extension on electrically stimulated composite surface. Scale bar = 5 μm. (Source: Reproduced with permission from [119], Wiley (2010).)

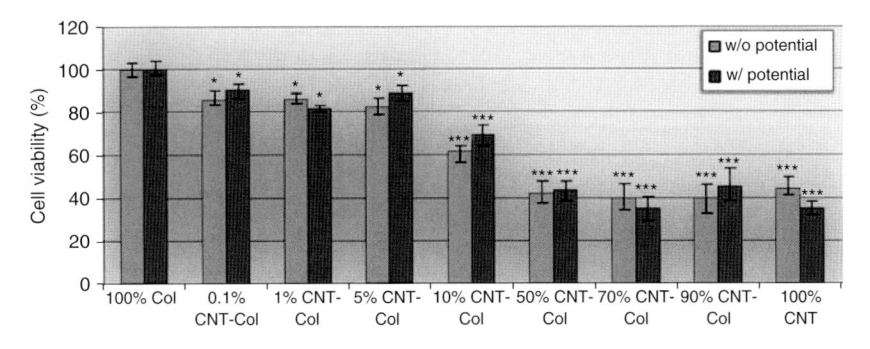

Figure 7.32 MTT cell viability assessment of PC12 cells in the absence and presence of electrical stimulation. (Source: Reproduced with permission from [119], Wiley (2010).)

7.7
Biocompatibility of Graphene Oxide and Its Nanocomposites

As aforementioned, CNTs containing metallic impurities may cause cytotoxicity and inflammation. In contrast to CNTs, graphene can be prepared in a relatively pure form, thus minimizing toxic effect of impurities on biological cells.

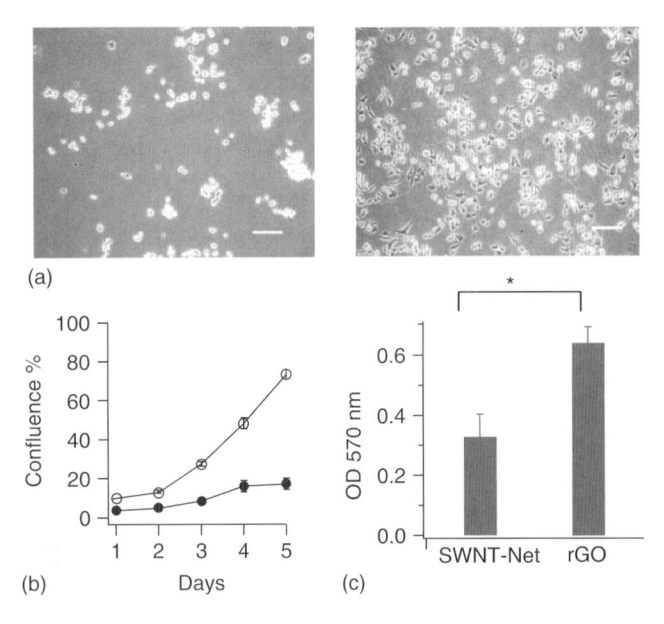

Figure 7.33 (a) Optical micrographs of PC12 cells grown on SWNT (left) and GO (right) surfaces for five days. Scale bar = 100 μm. (b) Proliferation curves of PC12 cell on SWNT (filled circles) and GO (open circles). Error bars indicate the standard error (SE). (c) MTT assay using optical absorbance at 570 nm as the indicator of cell metabolic activities. Data are shown as mean ± SE from three independent experiments. (Source: Reproduced with permission from [120], The American Chemical Society (2010).)

Accordingly, biocompatibility and cell viability of GOs have received increasing attention in recent years. Agarwal et al. [120] compared the viability of SWNT and GO specimens cultured with several types of cells, including neuroendocrine PC12 cells, oligodendroglia cells, and osteoblasts. They reported that GO is biocompatible with all these cell types and shows higher cell viability than the SWNT specimen. Typical cultivation and MTT assay results for PC12 cells grown on the SWNT and GO surfaces are shown in Figure 7.33a–c.

GO exhibits good biocompatibility as reported by some researchers [120; Chapter 1, Ref. 127], thus can be used as a nanocarrier for the loading and targeted delivery of anticancer drugs [121, 122]. However, some studies reported that GOs may induce cell toxicity, especially at high doses. Chang et al. studied the effects of GO additions on the morphology, viability, mortality, and membrane integrity of A549 cells. They reported that GOs do not enter A549 cell and have no obvious cytotoxicity. However, GOs can cause a dose-dependent oxidative stress

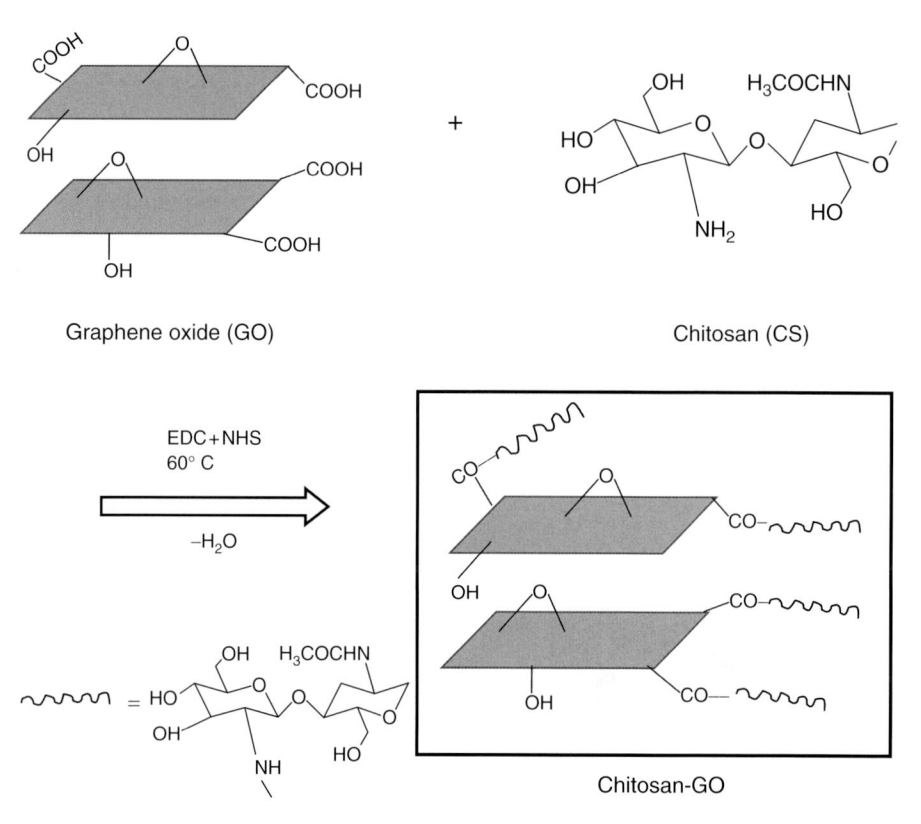

Figure 7.34 The reaction scheme shows covalent linking of carboxyl groups of GO with amine groups of CS to yield a CS–GO network structure in the presence of 1-ethyl-3-(3-dimethylaminoprophyl) carbondiimide hydrochloride (EDC) and N-hydroxyl succinimide (NHS) reagents. (Source: Reproduced with permission from [126], Elsevier (2011).)

in cells and induce a slight loss of cell viability at high concentrations [123]. Wang et al. [124] reported that GOs exhibit dose-dependent toxicity to human fibroblast cells and mice, such as inducing cell apoptosis and lung granuloma formation. The inconsistency derived may come from the use of different GO synthesis processes and cell testing procedures/approaches.

Considering functional hydroxyl and carboxyl groups bound to the surface of GO can react with a polymer matrix for improving interfacial bonding, some researchers used this strategy to prepare GO/CS, GO/PCL, and GO/gelatine nanocomposite films and nanofibers very recently [125–129]. CS is a linear polysaccharide, composed of glucosamine and N-acetyl glucosamine linked in a β(1–4) manner. CS exhibits good biodegradability but very poor mechanical strength; thus, GO additions are beneficial to enhance its strength. Figure 7.34 shows the reaction between the carboxyl groups of GO and the amine groups of CS to form a CS–GO network structure. Such a network effectively modulates the biological response of osteoblasts, such that cell attachment, proliferation, and growth are significantly enhanced [126]. Fan et al. [125] demonstrated that the elastic modulus of CS increases over ∼200% by adding 0.1–0.3 wt% GO. The solution-cast GO/CS nanocomposite films exhibit good biocompatibility for L929 mouse fibroblast cells as shown in Figure 7.35. This figure reveals that no apparent reduction in viability between the negative control and composite specimens cultivated for 24 and 48 h. Finally, Wan et al. [128] prepared GO/PCL membranes using electrospinning techniques and found that the incorporation of 0.3 wt% GO increases the tensile strength, modulus, and ductility (energy at break) of the PCL membrane by 95, 66, and

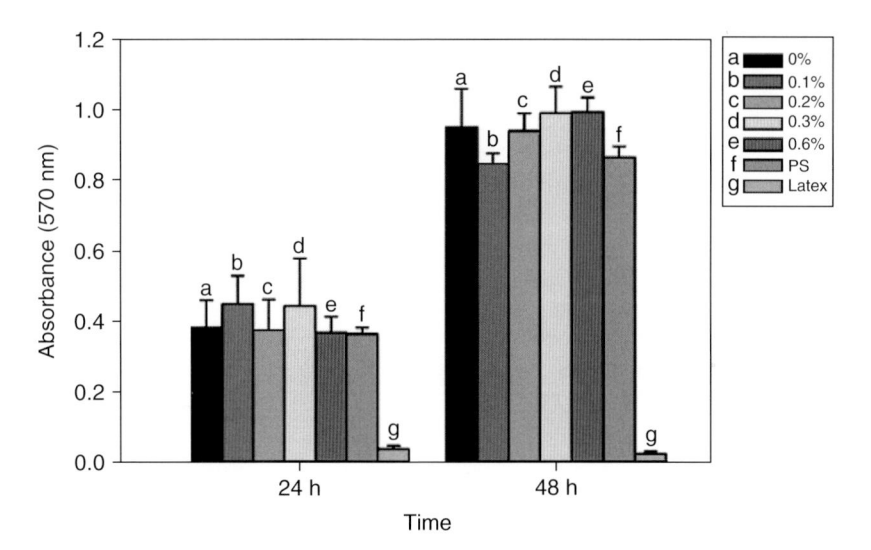

Figure 7.35 MTT assay results of L929 cells incubated with GO/CS composite films containing 0.1–0.6 wt% GO for 24 and 48 h. The bar represents standard deviation of three replicates ($p < 0.05$). (Source: Reproduced with permission from [125], The American Chemical Society (2010).)

416%, respectively. The reinforcing effect derived from the intermolecular hydrogen bonding between the carbonyl groups of PCL and hydrogen-donating groups of GO. The 0.3 wt% GO/PCL membrane exhibits good bioactivity as revealed by biomineralization tests using simulated body fluid solution.

References

1. Foldvari, M. and Bagonluri, M. (2008) Nanomed.: Nanotechnol., Biol. Med., 4, 183–200.
2. Jafar, E.N., Yadollah, O., and Dusan, L. (2011) Curr. Nanosci., 7, 297–314.
3. Veetil, J.V. and Ye, K.M. (2009) Biotechnol. Prog., 25, 709–721.
4. Yaszemski, M.J., Payne, R.G., Hayes, W.C., Langer, R., and Mikos, A.G. (1996) Biomaterials, 17, 175–185.
5. Navarro, M., Michiardi, A., Castano, O., and planell, J.A. (2008) J. R. Soc. Interface, 5, 1137–1158.
6. Uhthoff, H.K. and Finnegan, M. (1983) J. Bone Joint Surg., 65B, 66–71.
7. Vert, M. (2007) Prog. Polym. Sci., 32, 755–761.
8. Stevens, M.M. and George, J.H. (2005) Science, 310, 1135–1138.
9. Bonfield, W., Wang, M., and Tanner, K.E. (1998) Acta Mater.. 46, 2509.
10. Wang, M., Joseph. R., and Bonfield, W. (1998) Biomaterials, 19, 2357–2366.
11. Seal, B.L., Otero, T.C.N.D., and Panitch, A. (2001) Mater. Sci. Eng., R, 34, 147–230.
12. Middleton, J.C. and Tipton, A.J. (2000) Biomaterials, 21, 2335–2346.
13. Nair, L.S. and Laurencin, C.T. (2007) Prog. Polym. Sci., 32, 762–798.
14. Downes, R.N., Vardy, S., Tanner, K.E., and Bonfield, W. (1991) Bioceramics, 4, 239–246.
15. Bonfield, W., Grynpas, M.D., Tully, A.E., Bowman, J., and Abram, J. (1998) Biomaterials, 2, 185–185.
16. Wang, A., Lin, R., Polineri, V.K., Essner, A., Stark, C., and Dumbleton, J.H. (1998) Tribol. Int., 31, 661–667.
17. Godara, A., Raabe, D., and Green, S. (2007) Acta Biomater., 3, 209–220.
18. Schambron, T., Lowe, A., and McGrehor, H.V. (2008) Compos. Part B, 39, 1216–1229.
19. Webster, T.J., Ergun, C., Doremus, R.H., Siegel, R.W., and Bizios, R. (2000) Biomaterials, 21, 1803–1810.
20. Zhang, J. and Webster, T.J. (2009) Nano Today, 4, 66–80.
21. Kumta, P.N., Sfeir, C., Lee, D.H., Olton, D., and Choi, D. (2005) Acta Biomater., 1, 65–83.
22. Kannan, S., Lemos, A.F., and Ferreira, J.M. (2006) Chem. Mater., 18, 2181–2186.
23. Guo, X., Gough, J.E., Xiao, P., Liu, J., and Shen, J. (2007) J. Biomed. Mater. Res. Part A, 82, 1022–1032.
24. Tjong, S.C. (2010) Advances in Biomedical Sciences and Engineering, Chapter 4, Bentham Science, San Francisco.
25. Li, K. and Tjong, S.C. (2011) J. Nanosci. Nanotechnol., 12, 10644–10648.
26. Li, K. and Tjong, S.C. (2011) J. Macromol. Sci., Part B: Phys., 50, 1325–1337.
27. Li, K., Yuen, C.Y., Yeung, K.W., and Tjong, S.C. (2012) Adv. Biomater., 14, B155–B165.
28. Elias, K.L., Price, R.L., and Webster, T.J. (2002) Biomaterials, 23, 3279–3287.
29. Price, R.L., Waid, M.C., Haberstroh, K.M., and Webster, T.J. (2003) Biomaterials, 24, 1877–1887.
30. http://publications.nigms.nih.gov/insidethecell/.
31. Sibers, M.C., ter Brugge, P.J., Walboomers, X.F., and Jansen, J.A. (2005) Biomaterials, 26, 137–146.
32. Li, Z., Salmen, R., Hulderman, T., and Simeonova, P. (2004) Free Radical Biol. Med., 37, S142–S142.
33. Shvedova, A.A., Kisin, E.R., Porter, D., Schulte, P., Kagan, V.E., and Fadeel, B. (2009) Pharmacology, 211, 192–204.
34. Shvedova, A.A., Kisin, E.R., Mercer, R., Murray, A.R., Johnson, V.J., and Potapovich, A.I. (2005) Am. J.

Physiol. Lung Cell Mol. Physiol., **289**, L698–L708.
35. Kayat, J., Gajbhiye, V., Tekade, R.K., and Jain, N.K. (2011) *Nanomed.: Nanotechnol., Biol. Med.*, **7**, 40–49.
36. Muller, J., Huaux, F., and Lison, D. (2006) *Carbon*, **44**, 1048–1056.
37. Porter, A., Gass, M., Muller, K., Skepper, J.N., Midgley, P.A., and Welland, M. (2007) *Nat. Nanotechnol.*, **2**, 713–717.
38. Cheng, C., Muller, K.H., Koziol, K.K., Skepper, J.N., Midgley, P.A., Welland, M.E., and Porter, A.E. (2009) *Biomaterials*, **30**, 4152–4160.
39. Pulskamp, K., Diabate, S., and Krug, H.F. (2007) *Toxicol. Lett.*, **168**, 58–74.
40. Chou, C.C., Hsiao, H.Y., Hong, Q.S., Chen, C.H., Peng, Y.W., Chen, H.W., and Yang, P.C. (2008) *Nano Lett.*, **8**, 437–445.
41. Warheit, D.B., Laurence, B.R., Reed, K.L., Roach, D.H., Reynolds, G.A., and Webb, T.R. (2004) *Toxicol. Sci.*, **77**, 117–125.
42. Muller, J., Huaux, F., Moreau, N., Mison, P., Heilier, J.F., Delos, M., Arras, M., Fonseca, A., Nagy, J.B., and Lison, D. (2005) *Toxicol. Appl. Pharmacol.*, **207**, 221–231.
43. Shvedova, A.A., Castranova, V., Kisin, E.R., Schwegler-Berry, D., Murray, A.R., Gandelsman, V.Z., Maynard, A., Baron, P., and Part, A. (2003) *J. Toxicol. Environ. Health*, **66**, 1909–1926.
44. Monteiro-Riviere, N.A., Nemanich, R.J., Inman, A.O., Wang, Y.Y., and Riviere, J. (2005) *Toxicol. Lett.*, **155**, 377–384.
45. Tagmatarchis, N. and Prato, M. (2004) *J. Mater. Chem.*, **14**, 437–439.
46. (a) Bianco, A., Kostarelos, K., Partidos, C.D., and Prato, M. (2005) *Chem. Commun.*, 571–577; (b) Kostarelos, K., Lacerda, L., Pastorin, G., Wu, W., Wieckowski, S., Luansgsivilay, J., Godefroy, S., Pantarotto, D., Briand, J.P., Muller, S., Prato, M., and Bianco, A. (2007) *Nat. Nanotechnol.*, **2**, 108–113.
47. Magrez, A., Kasas, S., Salicio, V., Pasquier, N., Seo, J.W., Celio, M., Catsicas, S., Schwaller, B., and Forro, L. (2006) *Nano Lett.*, **6**, 1121–1125.
48. Dumortier, H., Lacotte, S., Pastorin, G., Marega, R., Wu, W., Bonifazi, D., Briand, J.P., Prato, M., Muller, S., and Bianco, A. (2006) *Nano Lett.*, **6**, 1522–1528.
49. Donaldson, K., Aitken, R., Tran, L., Stone, V., Duffin, R., Forrest, G., and Alexander, A. (2006) *Toxicol. Sci.*, **92**, 5–22.
50. Firme, C.P. and Bandaru, P.R. (2010) *Nanomed.: Nanotechnol., Biol. Med.*, **6**, 245–256.
51. Mosmann, T. (1983) *J. Immunol. Methods*, **65**, 55–63.
52. Patel, M.I., Tuckerman, R., and Dong, Q. (2005) *Biotechnol. Lett.*, **27**, 805–808.
53. Casey, A., Herzog, E., Davoren, M., Lyng, F.M., Byrne, H.J., and Chambers, G. (2007) *Carbon*, **45**, 1425–1432.
54. Repetto, G., del Peso, A., and Zurita, J.L. (2008) *Nat. Protoc.*, **3**, 1125–1131.
55. Worle-knirsch, J.M., Pulskamp, K., and Krug, H.F. (2006) *Nano Lett.*, **6**, 1261–1268.
56. Monteiro-Riviere, Inman, A.O., Wang, Y.Y., and Nemanich, R.J. (2005) *Nanomed.: Nanotechnol., Biol. Med.*, **1**, 293–299.
57. Belyanskaya, L., Manser, P., Spohn, P., Bruinink, A., and Wick, P. (2007) *Carbon*, **45**, 2643–2648.
58. Shi, X., Sitharaman, B., Pham, Q.P., Spicer, P.P., Hudson, J.L., Wilson, L.J., Tour, J.M., Raphael, R.M., and Mikos, A.G. (2008) *J. Biomed. Mater. Res.*, Part A, **86**, 813–823.
59. Zanello, L., Zhao, B., Hu, H., and Haddon, R.C. (2006) *Nano Lett.*, **6**, 562–567.
60. Chlopec, J., Czajkowska, B., Szaraniec, B., Frackowiak, E., Szostak, K., and Beguin, F. (2006) *Carbon*, **44**, 1106–1111.
61. Yun, Y., Dong, Z., Tan, Z., Schulz, M., and Shanov, V. (2009) *Mater. Sci. Eng., C*, **29**, 719–725.
62. Tutak, W., Park, K.H., Vasilov, A., Starovoytov, V., Fanchini, G., Cai, S.Q., Partridge, N.C., Sesti, F., and Chhowalla, M. (2009) *Nanotechnology*, **20**, 255101 - 1–255101-8.
63. Li, X., Gao, H., Uo, M., Sato, Y., Akasaka, T., Abe, S., Feng, Q., Cui, F.,

and Watari, F. (2009) *Biomed. Mater.*, **4**, 015005-1–015005-8.
64. Firkowska, I., Godehardt, E., and Giersig, M. (2008) *Adv. Funct. Mater.*, **18**, 3765–3771.
65. Matsuoka, M., Akasaka, T., Totsuka, Y., and Watari, F. (2010) *Mater. Sci. Eng., B*, **173**, 182–186.
66. Wang, A., Lin, R., Polineri, K., Essner, A., Stark, C., and Dumbleton, J.H. (1998) *Tribol. Int.*, **31**, 661–667.
67. Howling, G.I., Sakoda, H., Antonarulrajah, A., Marrs, H., Stewart, T.D., Appleyard, S., Rand, B., Fisher, J., and Ingham, E. (2003) *J. Biomed. Mater. Res. Part B*, **67**, 758–764.
68. Rae, P.J., Brown, E.N., and Orler, E.B. (2007) *Polymer*, **48**, 598–615.
69. Kurtz, S.M. and Devine, J.N. (2007) *Biomaterials*, **28**, 4845–4869.
70. http://www.invibio.com.
71. Kanagaraj, S., Varanda, F.R., Zhil'tsova, T.V., Oliveira, M.S., and Simoes, J.A. (2007) *Compos. Sci. Technol.*, **67**, 3071–3077.
72. Park, S., Yoon, S.W., Choi, H., Lee, J.S., Cho, W.K., Kim, J., Park, H.J., Yun, W.S., Choi, C.H., Do, Y.K., and Choi, I.S. (2008) *Chem. Mater.*, **20**, 4588–4594.
73. Sandler, J., Werner, P., Shaffer, M.S., Demchuk, V., Altstadt, V., and Windle, A.H. (2002) *Compos. Part A*, **33**, 1033–1039.
74. Deng, F., Ogasawara, T., and Takeda, N. (2007) *Compos. Sci. Technol.*, **67**, 2959–2964.
75. Ogasawara, T., Tsuda, T., and Takeda, N. (2011) *Compos. Sci. Technol.*, **71**, 73–78.
76. Jones, J.R. (2009) *J. Eur. Ceram. Soc.*, **29**, 1275–1281.
77. Karageorgiou, V. and Kaplan, D. (2005) *Biomaterials*, **26**, 5474–5491.
78. Oh, S.H., Park, I.K., Kim, J.M., and Lee, J.H. (2007) *Biomaterials*, **28**, 1664–1671.
79. Mano, J.F., Silva, G.A., Azevedo, H.S., Malafaya, P.B., Sousa, R.A., Silva, S.S., Boesel, L.F., Oliveira, J.M., Santos, T.C., Marques, A.P., Neveis, N.M., and Reis, R.L. (2007) *J. R. Soc. Interface*, **4**, 999–1030.
80. Ghasemi-Mobarakeh, L., Prabahkaran, M.P., Morshed, M., and Ramakrishna, S. (2010) *Mater. Sci. Eng., C*, **30**, 1129–1136.
81. Seal, B.L., Otero, T.C., and Panitch, A. (2001) *Mater. Sci. Eng., R*, **34**, 147–230.
82. Unatilake, P., Mayadunne, R., and Adhikari, R. (2006) *Biotechnol. Annu. Rev.*, **12**, 301–347.
83. Nair, L.S. and Laurencin, C.T. (2007) *Prog. Polym. Sci.*, **32**, 762–798.
84. Armentano, I., Dottori, M.F., Fortunati, E., Mattioli, S., and Kenny, J.M. (2010) *Polym. Degrad. Stab.*, **95**, 2126–2146.
85. Puppi, D., Chiellini, F., Piras, A.M., and Chiellini, E. (2010) *Prog. Polym. Sci.*, **35**, 403–440.
86. Whang, K., Thomas, C.H., Healy, K.E., and Nuber, G. (1995) *Polymer*, **36**, 837–842.
87. Hottot, A., Vessot, S., and Andrieu, J. (2004) *Dry Technol.*, **22**, 2009–2021.
88. Abarrategi, A., Gutierrez, M.C., Moreno-Vicente, C., Hortiguela, M.J., Ramos, V., Lopez-Lacomba, J.L., Ferrer, M.L., and del Monte, F. (2008) *Biomaterials*, **29**, 94–102.
89. Sill, T.J. and von Recum, H.A. (2008) *Biomaterials*, **29**, 1989–2006.
90. Greiner, A. and Wendorff, J.H. (2007) *Angew. Chem. Int. Ed.*, **46**, 5670.
91. Huang, Z.M., Zhang, Y.Z., Kotaki, M., and Ramakrishna, S. (2003) *Compos. Sci. Technol.*, **63**, 2223–2253.
92. Baji, A., Mai, Y.W., Wong, S.C., Abtahi, M., and Chen, P. (2010) *Compos. Sci. Technol.*, **70**, 703–718.
93. Wong, S.C., Baji, A., and Leng, S.W. (2008) *Polymer*, **21**, 4713–4722.
94. Badami, A.S., Kreke, M.R., Thompson, M.S., Riffle, J.S., and Golstein, A.S. (2006) *Biomaterials*, **27**, 596–606.
95. Mo, X.M., Xu, C.Y., Kotaki, M., and Ramakrishna, S. (2004) *Biomaterials*, **25**, 1883–1890.
96. Mathews, J.A., Wnek, G.E., Simpson, D.G., and Bowlin, G.L. (2002) *Biomacromolecules*, **3**, 232–238.
97. Eichhorn, S.J. and Sampson, W.W. (2005) *J. R. Soc. Interface*, **2**, 309–318.
98. Zhang, D.M. and Chang, J. (2007) *Adv. Mater.*, **19**, 3664–3667.

99. Vaquette, C. and Cooper-White, J. (2011) *Acta Biomater.*, **7**, 2544–2557.
100. Lee, K.H., Kim, H.Y., La, Y.M., Lee, D.R., and Sung, N.H. (2002) *J. Polym. Sci., Part B: Polym. Phys.*, **40**, 2259–2268.
101. Sung, J.H., Kim, H.S., Jin, H.J., Choi, H.J., and Chin, I.J. (2004) *Macromolecules*, **37**, 9899–9902.
102. Jose, M.V., Steinert, B.W., Thomas, V., Dean, D.R., Abdalla, M.A., Price, G., and Janowski, G.M. (2007) *Polymer*, **48**, 1096–1104.
103. Mazinani, S., Ajji, A., and Dubois, C. (2009) *Polymer*, **50**, 3329–3342.
104. Wu, T., Shi, T., Yang, T., Sun, Y., Zhai, L., Zhou, W., Zhang, M., and Zhang, J. (2011) *Eur. Polym. J.*, **47**, 284–293.
105. Meng, Z.X., Zheng, W., Li, L., and Zheng, Y.F. (2010) *Mater. Sci. Eng., C*, **30**, 1014–1021.
106. Shao, S., Zhou, S., Li, L., Li, J., Luo, C., Wang, J., Li, X., and Weng, J. (2011) *Biomaterials*, **32**, 2821–2833.
107. Liao, H., Qi, R., Shen, M., Cao, X., Guo, R., Zhang, Y., and Shi, X. (2011) *Colloids Surf. B: Biointerfaces*, **84**, 528–535.
108. Supronowicz, P.R., Ajayan, P.M., Ullmann, K.R., Arulanandam, B.P., Metzger, D.W., and Bizios, R. (2002) *J. Biomed. Mater. Res.*, **59**, 499–506.
109. Lovat, V., Pantarotto, D., Lagostena, L., Cacciari, B., Grandolfo, M., Righi, M., Spalluto, G., Prato, M., and Ballerini, L. (2005) *Nano Lett.*, **5**, 1107–1110.
110. Roach, P., Parker, T., Gadegaard, N., and Alexander, M.R. (2010) *Surf. Sci. Rep.*, **65**, 145–173.
111. Zhang, X., Prasad, S., Niyogi, S., Morgan, A., Ozkan, M., and Ozkan, C.S. (2005) *Sens. Actuator B: Chem.*, **106**, 843–850.
112. Ni, Y., Hu, H., Malarkey, E.B., Zhao, B., Montana, V., Haddon, R.C., and Parpura, V. (2005) *J. Nanosci. Nanotechnol.*, **5**, 1707–1712.
113. Mattson, P., Haddon, R.C., and Rao, A.M. (2000) *J. Mol. Neurosci.*, **14**, 175–182.
114. Cilia, E., Cellot, G., Cipollone, S., Rancic, V., Sucapane, A., Giordani, S., Gambazzi, L., Markram, H., Grandolfo, M., Scaini, D., Gelain, F., Casalis, L., Prato, M., Giugliano, M., and Ballerini, L. (2009) *Nat. Nanotechnol.*, **4**, 126–133.
115. Guimard, N.K., Gomez, N., and Schmidt, C.E. (2007) *Prog. Polym. Sci.*, **32**, 876–921.
116. Schmidt, C.E., Shastri, V.R., Vacanti, J.P., and Langer, R. (1997) *Proc. Natl. Acad. Sci. U.S.A.*, **94**, 8948–8953.
117. Luo, X., Weaver, C.L., Zhou, D.D., Greenberg, R., and Cui, X.T. (2011) *Biomaterials*, **32**, 5551–5557.
118. Chen, H., Guo, L., Ferhan, A.R., and Kim, D.H. (2011) *J. Phys. Chem. C*, **115**, 5492–5499.
119. Cho, Y. and Borgens, R.B. (2010) *J. Biomed. Mater. Res. Part A*, **95**, 510–517.
120. Agarwal, S., Zhou, X., Ye, F., He, Q., Chen, G.C., Soo, J., Boey, F., Zhang, H., and Chen, P. (2010) *Langmuir*, **26**, 2244–2247.
121. Zhang, L., Xia, J., Zhao, Q., Liu, L., and Zhang, Z. (2010) *Small*, **6**, 537–544.
122. Liu, Z., Robinson, J.T., Sun, X., and Dai, H. (2008) *J. Am. Chem. Soc.*, **130**, 10876–10877.
123. Chang, Y., Yang, S.T., Liu, J.H., Dong, E., Wang, Y., Cao, A., Liu, Y., and Wang, H. (2011) *Toxicol. Lett.*, **200**, 201–210.
124. Wang, K., Ruan, J., Song, H., Zhang, J., Wo, Y., Guo, S., and Cui, D. (2011) *Nanoscale Res. Lett.*, **6**, 1–8.
125. Fan, H., Wang, L., Zhao, K., Li, N., Shi, Z., Ge, Z., and Jin, Z. (2010) *Biomacromolecules*, **11**, 2345–2351.
126. Depan, D., Girase, B., Shah, J.S., and Misra, R.D. (2011) *Acta Biomater.*, **7**, 3432–3445.
127. Lim, H.N., Huang, N.M., and Loo, C.H. (2012) *J. Non-Cryst. Solids*, **358**, 525–530.
128. Wan, C.Y. and Chen, B.Q. (2011) *Biomed. Mater.*, **6**, 055010 (8 pp).
129. Wan, C.Y., Frydrych, M., and Chen, B.Q. (2011) *Soft Matter*, **7**, 6159–6166.

8
Polymer Nanocomposites for Electromagnetic Interference (EMI) Shielding

8.1
Introduction

Electrical sources emitting electromagnetic radiation at microwave and radio frequencies have received considerable attention in recent years. This radiation causes interference in the electrical equipments and electronic devices, leading to degradation and failure in their electrical performances. There is an increasing demand in industrial and military sectors for lightweight electromagnetic interference (EMI) shielding materials owing to the rapid increase in electromagnetic radiation sources such as cell phones, computers, power lines, and telecommunication equipments in our surroundings. EMI shielding in the range of 8.2–12.4 GHz (X-band) is very important for military and commercial applications. The weather radar, TV picture transmission, and telephone microwave relay systems lie in the X-band range [1, 2]. EMI shielding of a material refers to its capability to reflect and absorb electromagnetic radiation at high frequencies, acting as a shield against the penetration of the radiation through it. The shield must possess mobile charge carriers for interacting with electromagnetic field in the radiation [3]. This implies that the shield is an electrically conducting material. Conventional EMI shielding materials are made from metals such as copper, aluminum, nickel, stainless steel, and so on. Most of these metals are heavy and susceptible to degradation on exposure to corrosive environments. Conductive coatings are the most widely used EMI shielding materials. Electrically conductive coatings are prepared by incorporating metallic pigment or graphite into the polymer binder [4]. These metal coated polymers have some limitations. They can delaminate easily and require secondary operations that are expensive. Polymer microcomposites filled with carbon fibers and carbon blacks are alternative EMI shielding materials owing to their unique combination of electrical conduction, flexibility, light weight, moldability, and processability [5–8]. However, large filler loadings are needed in polymer composites to achieve desired EMI shielding levels. Therefore, carbonaceous nanofillers of high aspect ratios offer distinct advantages over carbon fibers because they can percolate at very low filler contents. Furthermore, carbon nanotubes (CNTs) are lightweight conducting materials having strong microwave absorption properties in the gigahertz frequency

Polymer Composites with Carbonaceous Nanofillers: Properties and Applications, First Edition. Sie Chin Tjong.
© 2012 Wiley-VCH Verlag GmbH & Co. KGaA. Published 2012 by Wiley-VCH Verlag GmbH & Co. KGaA.

range [9]. It is considered that the incorporation of carbonaceous nanofillers into insulating polymers can increase their EMI effectiveness significantly.

8.2
EMI Shielding Efficiency

An electromagnetic shielding material can attenuate incident electromagnetic energy effectively. For an incident electromagnetic wave (E_0) that penetrates through a shield with a thickness t, reflection (E_R), the transmission (E_1) and absorption of the wave can take place as shown in Figure 8.1 [10]. The corresponding reflectivity (R), absorptivity (A), and transmissivity (T) follow the relation:

$$A + R + T = 1 \tag{8.1}$$

The shielding efficiency (SE) of a material expressed in decibels (dB) is defined as the ratio of transmitted power (P_1) to incident power (P_0) of an electromagnetic wave [10–12]

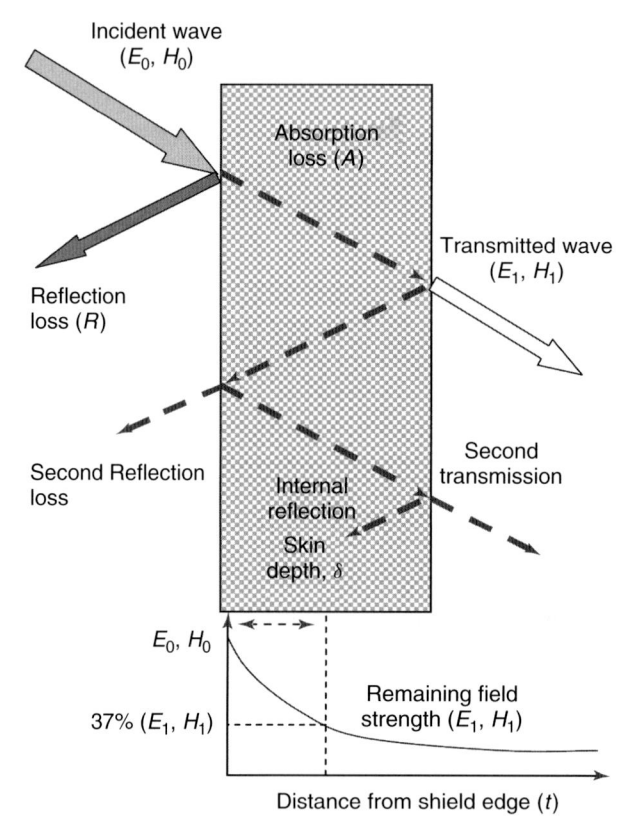

Figure 8.1 Schematic representation of EMI shielding in a conductive plate. (Source: Reproduced with permission from Ref. [10], Intech (2011).)

$$SE = -10\log\left(\frac{P_1}{P_0}\right) = -20\log\left(\frac{E_1}{E_0}\right) \tag{8.2}$$

To shield by reflection, the charge carrier of a material should interact with electromagnetic radiation. Shielding by absorption is enhanced when the electric and/or magnetic dipoles of a material interact with the electromagnetic field. The electric dipoles may derive from the materials having high dielectric constants or magnetic permeability. The multiple reflection mechanism is associated with the reflections at various surfaces or interfaces in the shield. This requires the presence of large surface or interface areas in the shield such as a porous foam material.

The total EMI SE of a shielding material with a thickness t is the summation of the SE due to absorption (SE_A), reflection (SE_R) and multiple reflection (SE_M), that is

$$EMI\ SE = SE_A + SE_R + SE_M \tag{8.3}$$

The contribution by multiple reflections can be ignored when $SE_A \geq 8$ dB. The reflection loss is related to the relative mismatch between the incident wave and the surface impedance of a shield. From the literature [13, 14], SE_R (dB) is given by the following expression

$$SE_R = 20\log\left|\frac{(Z_i - Z_0)}{Z_i + Z_0}\right| \tag{8.4}$$

where Z_i is the impedance at the air-absorber interface, Z_0 the free space impedance (~ 377 Ω). The SE_R becomes infinity when $Z_i = Z_0$. In this case, the shielding material is impedance matched to the incident medium.

The absorption loss, SE_A, can be determined from the following relation

$$SE_A = 20\log\left[\exp\left(\frac{t}{\delta}\right)\right] \tag{8.5}$$

where δ is the skin depth defined by

$$\delta = \frac{1}{\sqrt{\pi f \mu \sigma}} \tag{8.6}$$

where $\mu (= \mu_0 \mu_r)$ is the magnetic permeability of a shielding material, μ_r the relative permeability, μ_0 the magnetic permeability of free space ($=4\pi \times 10^{-7}$ H m^{-1}), σ the conductivity of a shield with unit in S m^{-1}, and f the frequency in hertz. It is generally known that the strength of electromagnetic radiation decays exponentially as it penetrates through a conducting material. At a skin depth distance δ, the electric field reduces to $(1/e)$ or 37% of its incident strength. Thus the skin depth decreases with increasing frequency and with increasing conductivity or permeability.

Assuming the sample thickness is considerably greater than the skin depth, SE_R for plane electromagnetic wave at higher frequencies can be expressed by [13]:

$$SE_R \approx 10\log\left[\frac{\sigma}{16\omega\varepsilon_0\mu_r}\right] \tag{8.7}$$

where $\omega(=2\pi f)$ is angular frequency of the radiation and ε_0 the free space permittivity. Therefore, reflection loss increases with an increase in the shield conductivity and is inversely proportional to the logarithm frequency of the incident electromagnetic radiation. From Eq. (8.5), SE_A can be rewritten as:

$$SE_A = 20 \left(\frac{t}{\delta}\right) \log e \tag{8.8}$$

Combining Eqs. (8.6) and (8.8), SE_A takes the following form:

$$SE_A = 8.7t\sqrt{\pi f \mu \sigma} \tag{8.9}$$

It appears that the absorption loss is directly proportional to the thickness of shielding material and the square root of frequency, conductivity, and permeability. The standard procedures for measuring EMI SE of conducting materials are specified in detail in the ASTM D4935-99 [15]. An electromagnetic wave is incident directly onto the sample using a waveguide setup. A two port network analyzer equipped with a scattering parameter (S parameter) test set and the waveguide at 8.2–12.4 GHz are commonly used for the EMI SE measurements of the composite materials (Figure 8.2) [16–20].

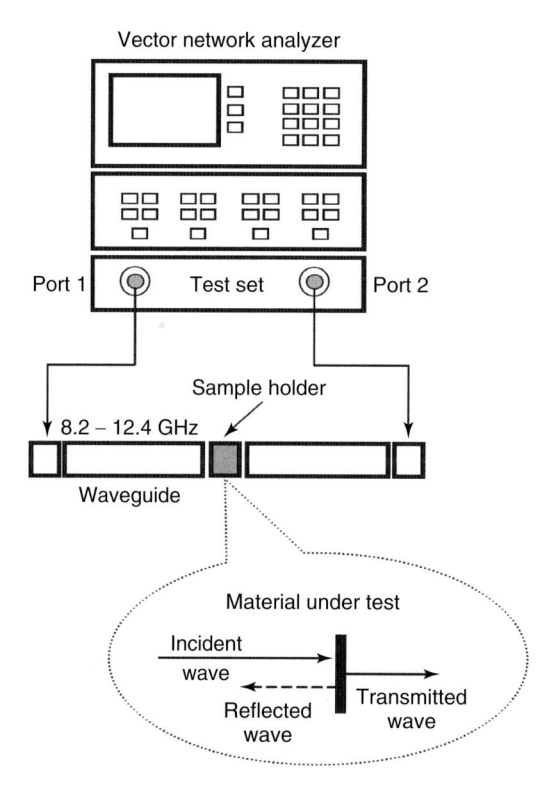

Figure 8.2 Schematic representation of the setup for EMI SE measurements. (Source: Reproduced with permission from Ref. [16], The American Chemical Society (2005).)

8.3
Nanocomposites with Graphene Fillers

The EMI SE of a nanocomposite material is rather complicated owing to the presence of conducting nanofillers of large surface areas for reflection and multiple reflections. Therefore, EMI SE depends on several factors, including the filler's intrinsic conductivity, dielectric constant, and aspect ratio. In practice, the EMI SE target value needed for commercial applications is ~20 dB. Bigg [21] demonstrated that a shield material must have a volume resistivity of 2 Ω cm or less to provide a minimum of 30 dB attenuation. An SE value of 30 dB attenuation blocks nearly 99.9% of an impinging signal, but 20–30 dB of attenuation is considered acceptable for most industrial and consumer applications.

Figure 8.3 shows the variation of EMI SE with frequency of the (reduced graphene oxide) rGO/epoxy nanocomposites in the X-band (Chapter 2, Ref. 56). The SE value increases with increasing filler loading over entire frequency range studied. This can be attributed to the electrical conductivity of rGO/epoxy nanocomposites increases with increasing rGO content. Thus the 15 wt% rGO/epoxy nanocomposite with an SE value of ~21 dB can meet the commercial application requirement.

8.4
Nanocomposites with GNPs

Park and coworkers [22] studied dielectric and EMI properties of styrene-acrylonitrile (SAN) filled with graphite nanoplatelet (GNPs). The dielectric permittivity and dissipation factor of these composites were determined with a

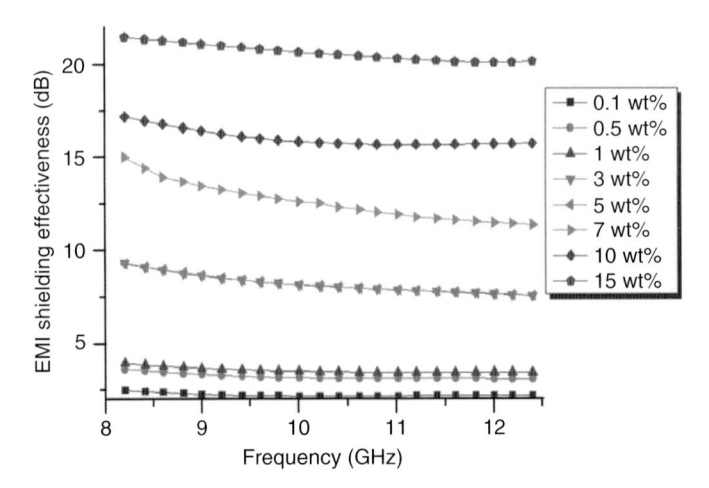

Figure 8.3 EMI SE versus frequency of solvent-cast rGO/epoxy composites of different rGO contents. (Source: Reproduced with permission from (Chapter 2, Ref. 56), Elsevier (2009).)

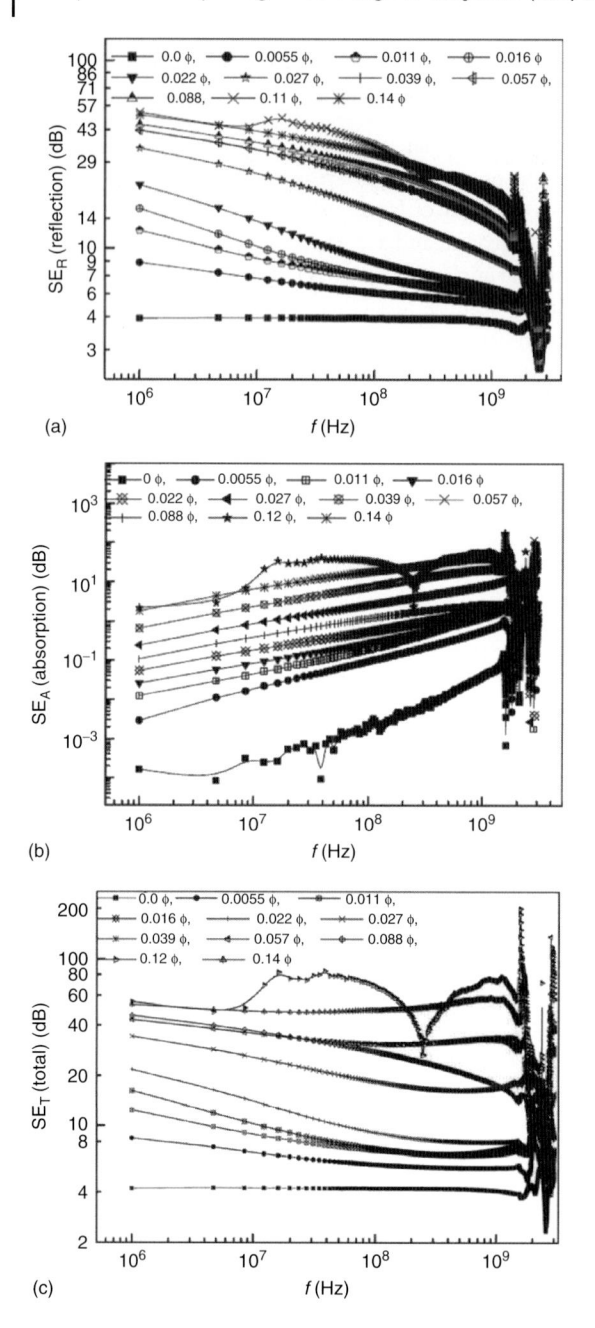

Figure 8.4 Variations of (a) SE_R, (b) SE_A, and (c) total SE with frequency for GNP/SAN nanocomposites. The GNP volume fractions of the composites are given in upper part of figures. (Source: Reproduced with permission from Ref. [22], Elsevier (2009).)

conventional impedance analyzer. They deduced empirical relations for SE_R and SE_A based on the measured dielectric permittivity of the nanocomposites. Figure 8.4a,b shows respective plots of SE_R or SE_A versus frequency for the GNP/SAN nanocomposites. It can be seen that the SE_R decreases while SE_A increases with increasing frequency. Thus the frequency dependence of SE_R and SE_A of the nanocomposites follows Eqs. (8.7) and (8.9). The total SE is the summation of SE_R and SE_A factors, and its variation with frequency is depicted in Figure 8.4c. Obviously, the total SE value of the composites increases with the increasing filler content and decreases with increasing frequency. From Figure 8.4a–c, at frequencies of 1 GHz and above, severe oscillations occur and the results become unreliable. This is because conventional impedance analyzer is effective to measure dielectric behavior of the composite material below 1 GHz only. Obviously, the waveguide setup as shown in Figure 8.2 is more appropriate for measuring EMI shielding effectiveness of a composite material in the 8–12 GHz frequency range (X-band).

More recently, Bellis et al. [23] employed two different sets of waveguides in the X-band and Ku-band (12.4–18 GHz), respectively, to measure shielding effectiveness of GNP/epoxy and CNT/epoxy nanocomposites. They reported that the GNP/epoxy composites exhibit good processability, thereby allowing the incorporation of higher GNP content into the epoxy resin. Figure 8.5 shows effective electrical conductivity versus filler content for the GNP/epoxy and CNT/epoxy nanocomposites at different frequencies. The conductivity of both composite systems increases with increasing frequency. It reaches a value of $4\,S\,m^{-1}$ for the 2 wt% GNP/epoxy composite at 18 GHz. In contrast, a lower value of $\sim 3S\,m^{-1}$ is found for the 0.5 wt% CNT/epoxy composite at 18 GHz. Figure 8.6 shows the shielding effectiveness versus frequency of the 2 wt% GNP/epoxy and 0.5 wt% CNT/epoxy nanocomposites having thicknesses of 0.5 and 1 cm. Apparently, the 2 wt% GNP/epoxy composite exhibits higher SE performance than the 0.5 wt% CNT/epoxy composite over the whole frequency range, particularly at low frequencies. This is attributed to the

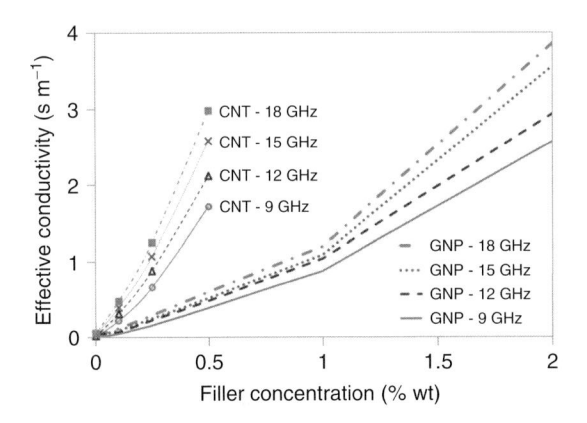

Figure 8.5 Variation of effective electrical conductivity with filler content at different frequencies for GNP/epoxy and CNT/epoxy nanocomposites. (Source: Reproduced with permission from Ref. [23], Elsevier (2011).)

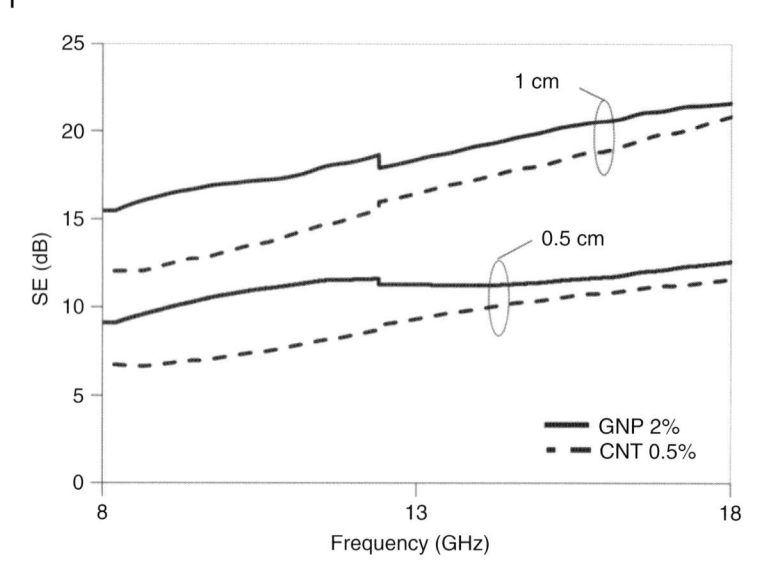

Figure 8.6 Comparison of the shielding effectiveness of the 2 wt% GNP/epoxy and 0.5 wt% CNT/epoxy nanocomposites having thicknesses of 0.5 and 1 cm, respectively. (Source: Reproduced with permission from Ref. [23], Elsevier (2011).)

conductivity of the 2 wt% GNP/epoxy composite is nearly twice higher than that of the 0.5 wt% CNT/epoxy composite at 9 GHz but is only about 30% higher than that of 0.5 wt% CNT/epoxy composite at 18 GHz (Figure 8.5). Moreover, the 2 wt% GNP/epoxy composite with 1 cm thick panel exhibits shielding effectiveness higher than 15 dB over the whole frequency range.

8.5
Nanocomposites with CNTs and CNFs

Free-standing CNT papers or films with high electrical conductivity tend to show large EMI SE values. Wu et al. [24] studied EMI behavior of CNT macrofilms of large area (>30 × 30 cm^2) prepared by the chemical vapor deposition (CVD) process. Figure 8.7 shows the variation of EMI SE with frequency for the CNT films with thicknesses of ∼1, 2, and 4 μm. Apparently, the SE value of the CNTs increases with increasing film thickness. The CNT film with a thickness of ∼4 μm exhibits high EMI SE values of 61–67 dB in the X-band.

The EMI SE values of the (multiwalled carbon nanotube) MWNT/polymer and (carbon nanofiber) CNF/polymer nanocomposites are well documented and generally increase with increasing filler content [12, 16, 20, 25–27]. At the same filler loading, MWNT-filled polystyrene (PS) composites exhibit higher shielding effectiveness compared to those filled with CNFs. In other words, CNTs are more effective than nanofibers to yield high EMI shielding at low filler loadings [25].

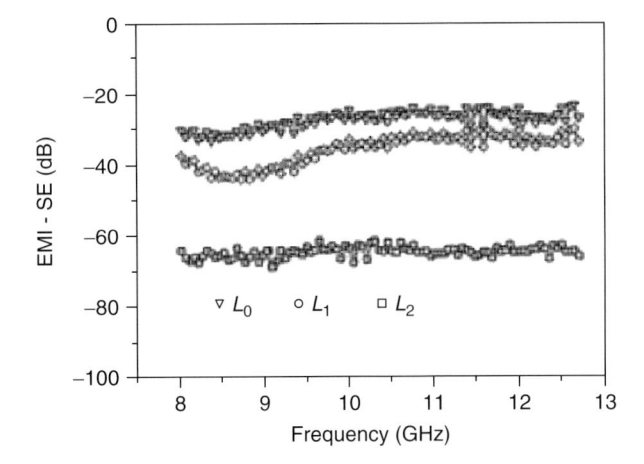

Figure 8.7 EMI SE versus frequency for CNT films of different thicknesses ($L_0 = 1$ μm, $L_1 = 2$ μm and $L_2 = 4$ μm). (Source: Reproduced with permission from Ref. [24], Elsevier (2011).)

Figure 8.8a shows a typical plot of EMI SE versus frequency of melt-compounded MWNT/polytrimethylene terephthalate (PTT) nanocomposites [20]. The EMI SE of the nanocomposites is independent of frequency with the exception at 10 wt% filler loading. At 5 wt% MWNT, the composite exhibits an EMI SE value >23 dB, higher than the required value for commercial applications. Figure 8.8b shows the effect of filler content on SE_R and SE_A. Both SE_R and SE_A values increase with increasing nanotube content, but the increase rate of SE_A is much higher than SE_R. This demonstrates that the absorption loss is the main contributor to the total EMI SE for the composites with filler content above the percolation threshold (i.e., 1 wt% MWNT) as shown in Figure 8.9. Similarly, Kim *et al.* also reported that the contribution of SE_A to the total EMI SE is much larger than the SE_R in solvent-cast MWNT/PMMA (poly(methyl methacrylate)) films [17].

As discussed in Chapter 5, dielectric constant and loss tangent of the CNT/polymer nanocomposites increase sharply at the percolation threshold. At this stage, many conducting particles are isolated by thin insulating layers, resulting in the formation of a network of minicapacitors. The tan δ value is related to the ability of a material to convert applied energy into heat, while the imaginary part of the complex permittivity (ε''), also known as *loss factor*, is related to the ability of the materials to absorb electromagnetic wave. Therefore, materials with high ε'' and loss tangent values are needed for electromagnetic shielding applications in the high-frequency region. In other words, they can be used as electromagnetic absorbing materials and in stealth technology. Figure 8.10a,b shows real and imaginary parts of permittivity as a function of frequency of MWNT/PTT nanocomposites. It is obvious that the real part of permittivity (ε') increases as the filler content exceeds the percolation threshold. Its value decreases in the high-frequency region owing to the dipole moment and cannot follow the reversing electric field in the high-frequency regime of the Ku-band. It is observed that the nanocomposite with 10 wt% MWNT exhibits

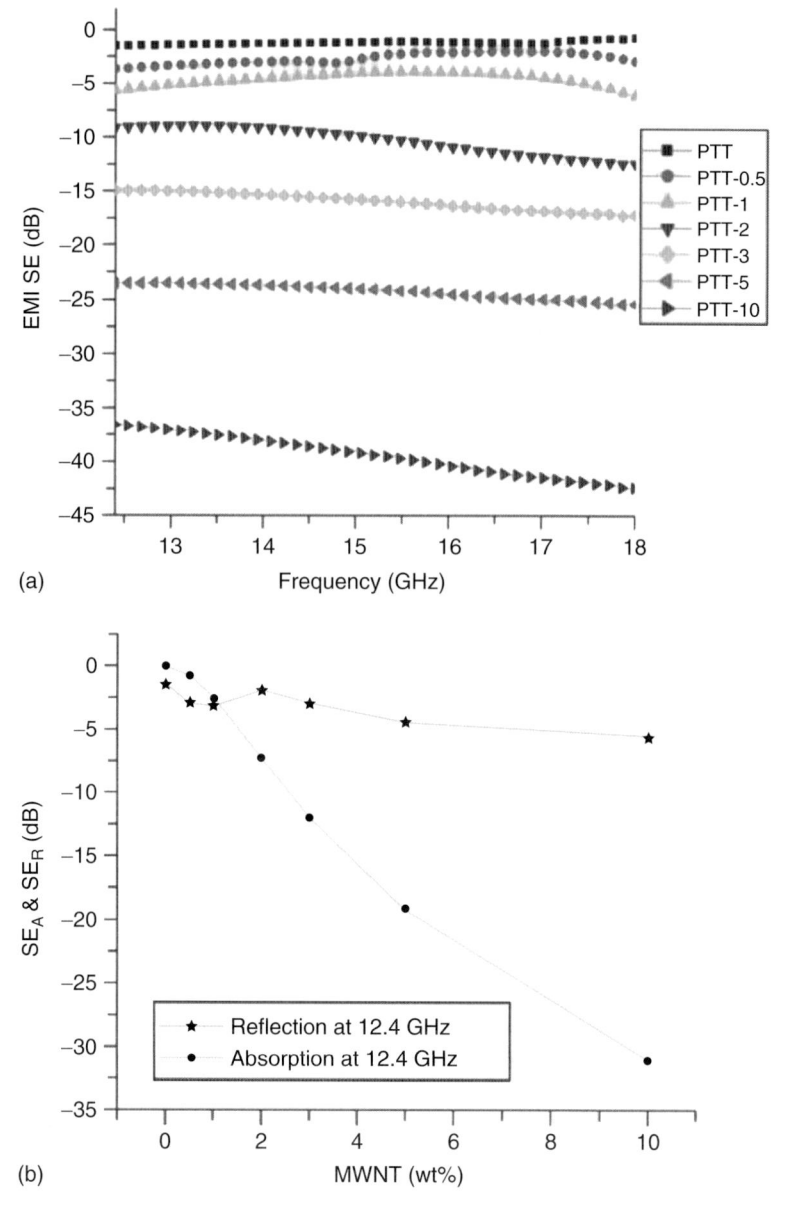

Figure 8.8 (a) EMI SE versus frequency of melt-compounded MWNT/PTT nanocomposites from 12.4 to 18 GHz. The composition of composites is designated as PTT-x, where x is the MWNT content in weight percentage. (b) Contribution of reflection and absorption to EMI SE of MWNT/PTT nanocomposites. (Source: Reproduced with permission from Ref. [20], Elsevier (2011).)

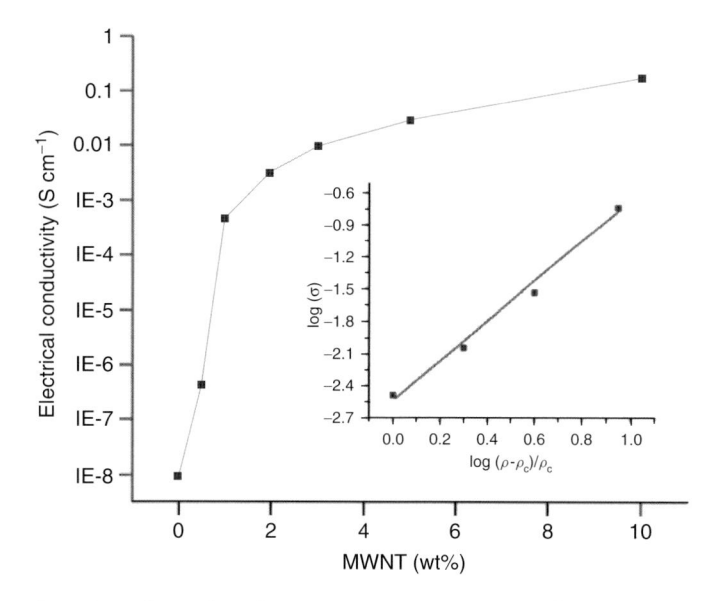

Figure 8.9 Electrical conductivity versus MWNT content of melt-compounded MWNT/PTT nanocomposites at 12.4 Hz. (Source: Reproduced with permission from Ref. [20], Elsevier (2011).)

the largest ε'' value. Figure 8.11 shows the variations of tan δ and SE_A with MWNT content for the MWNT/PTT nanocomposites. The tan δ value of neat PTT is almost zero, demonstrating that pure PTT can barely attenuate or absorb electromagnetic radiation. By incorporating 10 wt% MWNT into PTT, the tan δ value increases markedly to 2.7, while the corresponding SEA value increases to 31 dB. Thus the 10 wt% MWNT/PTT nanocomposite can absorb incident electromagnetic wave effectively. These results clearly indicate that the mechanism of wave absorption is mainly due to heat dissipation effects and hence is strongly linked to the electrical conductivity and dielectric properties of the material.

Comparing with the MWNT/polymer composites, the EMI SE study of SWNT/polymer (single-walled carbon nanotube) composites remains largely unexplored [1, 28–31]. Huang et al. [1] investigated the influence of arc-grown SWNT additions on the EMI SE of the epoxy composites in the X-band. They achieved EMI SE values of 20–30 dB for the SWNT/epoxy composites by adding 15 wt% SWNT. Das and Maiti [29] studied the EMI shielding of melt-compounded SWNT/ethylene vinyl acetate (EVA) over microwave (200–2000 MHz) and X-band frequency ranges. The percolation threshold of SWNT/EVA nanocomposites was determined to be 1.8 wt%. The variation in EMI SE with frequency for SWNT/EVA nanocomposites in the X-band is shown in Figure 8.12a. The SE value of nanocomposites increases with increasing SWNT content. The SE values reach up to 22 and 37 dB by adding 15 and 30 wt% SWNT, respectively. A similar behavior is observed in the 200–2000 MHz band but with lower SE values (not shown). The EMI SE of SWNT/EVA nanocomposites is closely related to their

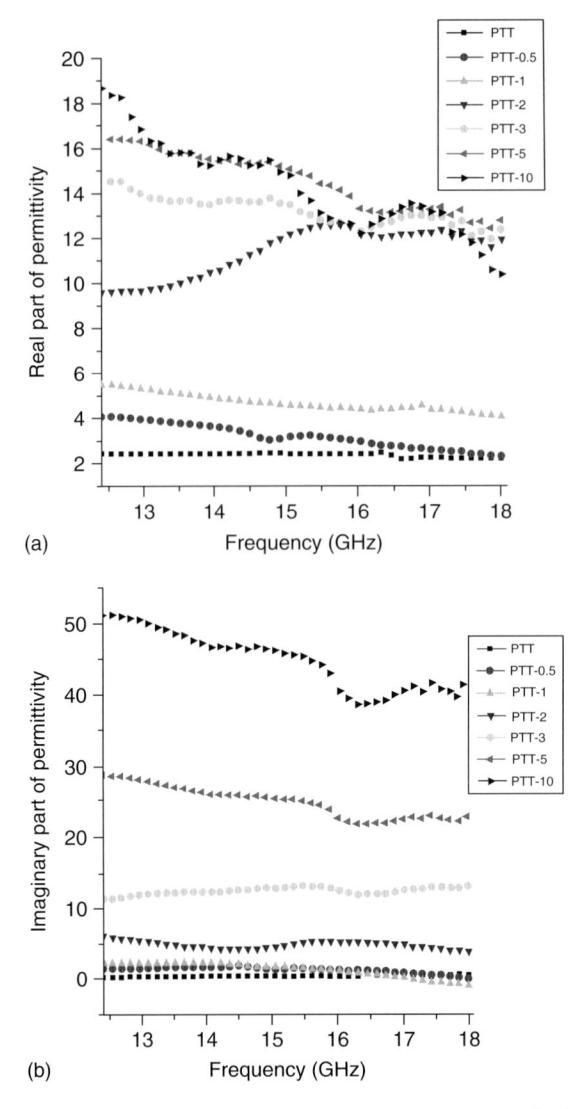

Figure 8.10 (a) Real part (ε') and (b) imaginary part (ε'') of permittivity as a function of frequency of melt-compounded MWNT/PTT nanocomposites. (Source: Reproduced with permission from Ref. [20], Elsevier (2011).)

electrical conductivity, thus depending on the filler content (Figure 8.12b). A sharp increase in SE takes place when the conductivity reaches 10^{-4} S cm^{-1} and higher as a result of the formation of a conductive path network. The EMI SE values of the nanocomposites in the microwave and X-band frequencies also depend greatly on the sample thickness (Figure 8.13). Table 8.1 summarizes EMI SE values of representative polymer composites filled with one-dimensional carbonaceous nanofillers.

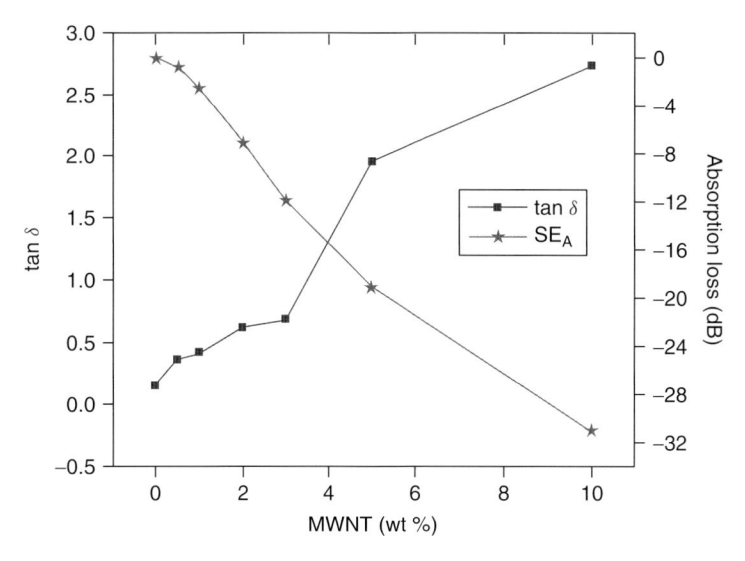

Figure 8.11 Variations of tan δ and SE$_A$ with filler content of melt-compounded MWNT/PTT nanocomposites at 12.4 GHz. (Source: Reproduced with permission from Ref. [20], Elsevier (2011).)

8.6
Foamed Nanocomposites for EMI Applications

8.6.1
CNT/Polymer Foamed Nanocomposites

The contribution from multiple reflections to the total EMI SE is ignored for most CNT/polymer nanocomposites when SE$_A$ values >8 dB. In general, foams can be introduced into the MWNT/polymer composites for increasing their shielding effectiveness via multiple internal reflections as shown in Figure 8.14 [16, 32, 33]. The beneficial effect of multiple reflections is that the material attenuates and absorbs electromagnetic radiation rather than reflecting it. Yang et al. [16] prepared MWNT/PS and CNF/PS nanocomposite foams using a foaming agent, 2,2′-azobisisobutyrronitrile (AIBN), to decompose into nitrogen gas within the composite systems during hot pressing. The foams produced are quite heterogeneous so that the dominant EMI shielding mechanism is reflection rather than absorption. Accordingly, the 7 wt% MWNT/PS foam composite only exhibits an EMI SE value of 18.56 dB (Figure 8.15). The CNF/PS foam composite even requires a high-CNF content of 20 wt% to yield an EMI SE value of 20.51 dB. Thomassin et al. [32] then fabricated MWNT/PCL (polycaprolactone) nanocomposite foams using supercritical CO_2 technique. EMI SE values of 60–80 dB together with a low reflectivity can be achieved in these nanocomposite foams at a very low MWNT content of 0.25 vol%. Such large EMI SE values are comparable to those of the most effective metallic EMI shielding materials. These are resulted from the formation

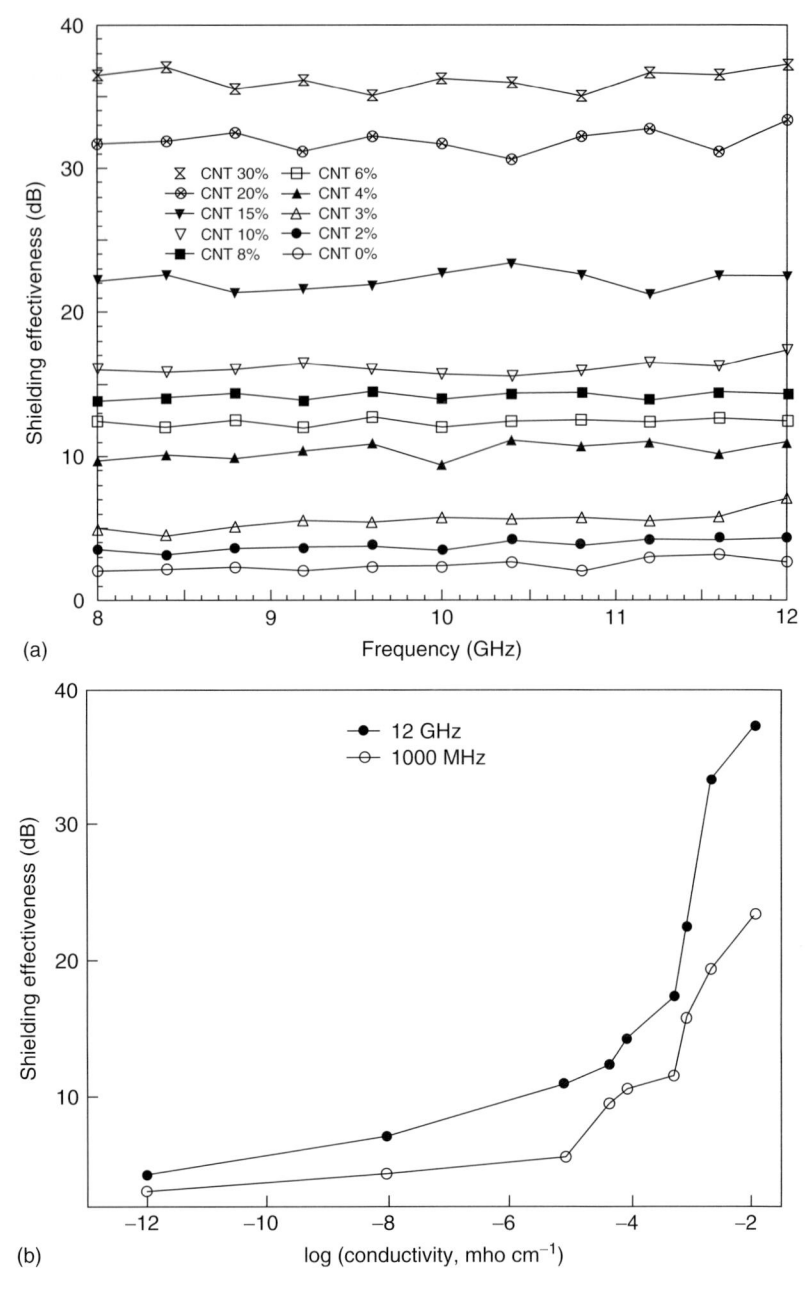

Figure 8.12 EMI SE as a function of (a) frequency and (b) electrical conductivity of SWNT/EVA nanocomposites. (Source: Reproduced with permission from Ref. [29], Springer Science + Business Media (2008).)

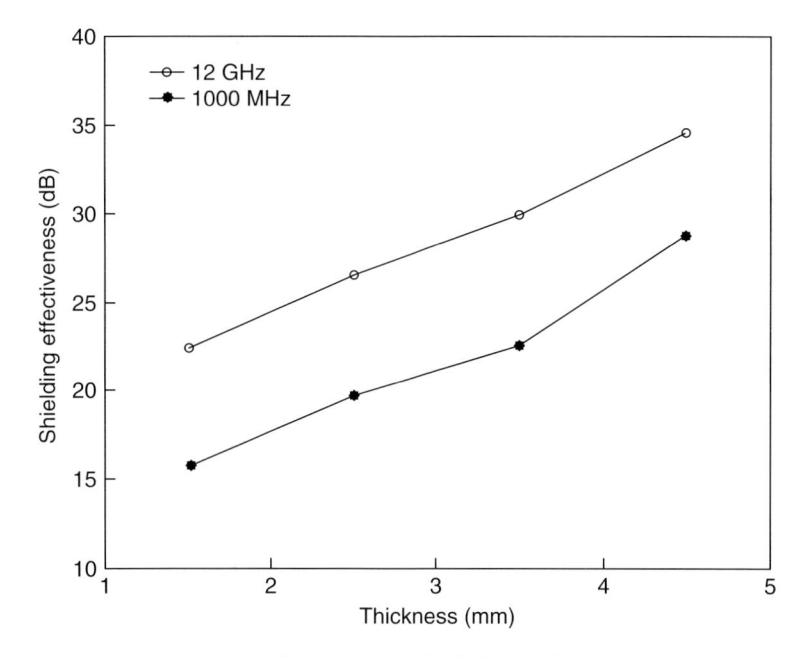

Figure 8.13 EMI SE as a function of sample thickness of SWNT/EVA nanocomposites. (Source: Reproduced with permission from Ref. [29], Springer Science + Business Media (2008).)

Table 8.1 EMI shielding effectiveness of polymer nanocomposites filled with CNTs and CNFs.

Polymer	Filler type and content	Average EMI SE value (dB)	Frequency range (GHz)	References
PTT	5 wt% MWNT	23	12.4–18	[20]
PTT	10 wt% MWNT	39	12.4–18	[20]
PS	7 wt% MWNT	26	8.2–12.4	[25]
PS foam	7 wt% MWNT	18.56	8.2–12.4	[16]
PS foam	7 wt% CNF	8.53	8.2–12.4	[16]
PS foam	20 wt% CNF	20.51	8.2–12.4	[16]
PS	10 wt% CNF	12.9	12.4–18	[26]
PS	10 wt% CNF + 1 wt% MWNT	20.3	12.4–18	[26]
PS	10 wt% CNF + 3 wt% MWNT	21.9	12.4–18	[26]
PS	1 wt% MWNT	7.9	12.4–18	[26]
PMMA	10 wt% SWNT	26	8.2–12.4	[30]
EVA	15 wt% SWNT	22	8.2–12.4	[29]
Epoxy	15 wt% SWNT (arc-grown)	22	8.2–12.4	[1]

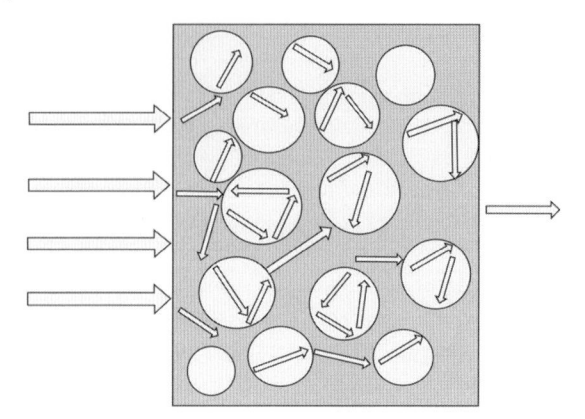

Figure 8.14 Multiple reflections of electromagnetic waves in nanocomposite foams. (Source: Reproduced with permission from Ref. [33], Elsevier (2010).)

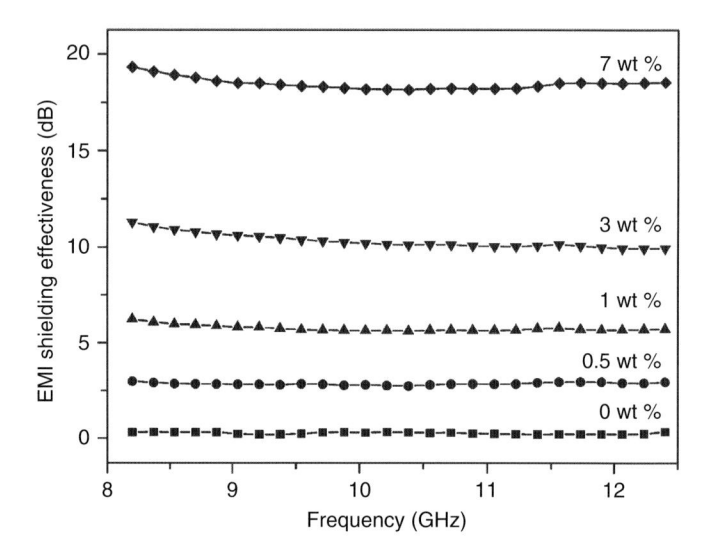

Figure 8.15 EMI shielding effectiveness versus frequency measured in the 8.2–12.4 GHz range of MWNT/PS nanocomposite foams. (Source: Reproduced with permission from Ref. [16], the American Chemical Society (2005).)

of uniformly distributed microcellular foams in the MWNT/PCL nanocomposites. Supercritical CO_2 is considered as a typical supercritical fluid for which both pressure and temperature are above the critical values. Accordingly, supercritical CO_2 does not behave as a gas or liquid but exhibits hybrid properties of these two states. These properties render supercritical CO_2 becoming excellent solvent and foaming agent for polymers and their composites [33–37]. Recent advancements in supercritical-CO_2-mediated material processing have led to development of microcellular foams in both the polymers and polymer composites for biomedical scaffolding and EMI shielding applications.

8.6.2
Graphene/Polymer Foamed Nanocomposites

More recently, Zhang *et al.* [38] prepared (thermally reduced graphene) TRG/PMMA nanocomposite foams by a batch foaming process with the aid of subcritical CO_2. Figure 8.16a,b is the SEM micrographs showing morphology of pure PMMA and 0.1 wt% TRG/PMMA nanocomposite foam. The addition of 0.1 wt% TRG reduces the average cell of PMMA foam from 110 to 30 µm in the nanocomposite foam. The electrical conducting behavior of the solution-mixed TRG/PMMA bulk nanocomposites and microcellular foams are shown in Figure 8.17. It can be seen that the nanocomposite foams exhibit nearly the same percolation behavior but slightly higher conductivity compared with the bulk PMMA nanocomposites.

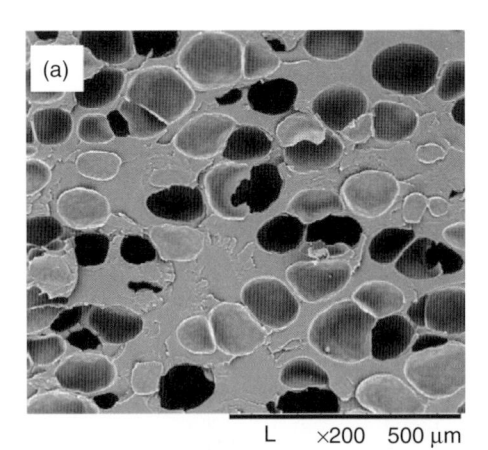

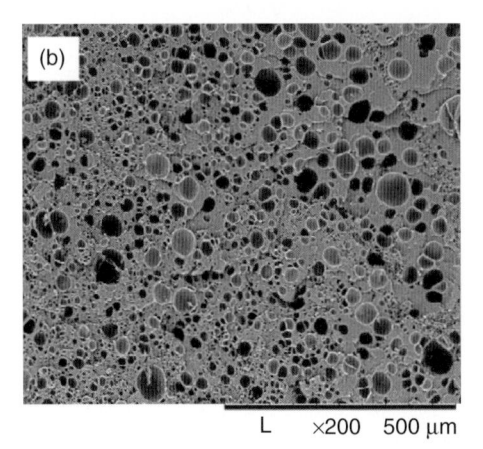

Figure 8.16 SEM images of (a) neat PMMA foam and (b) 0.1 wt% TRG/PMMA nanocomposite foam. (Source: Reproduced with permission from Ref. [38], The American Chemical Society (2011).)

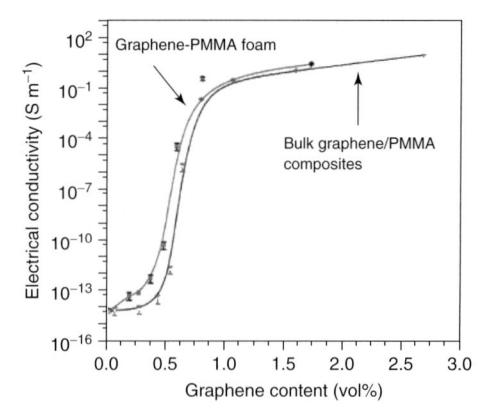

Figure 8.17 Electrical conductivity versus filler content for TRG/PMMA bulk nanocomposites and microcellular foams. (Source: Reproduced with permission from Ref. [38], The American Chemical Society (2011).)

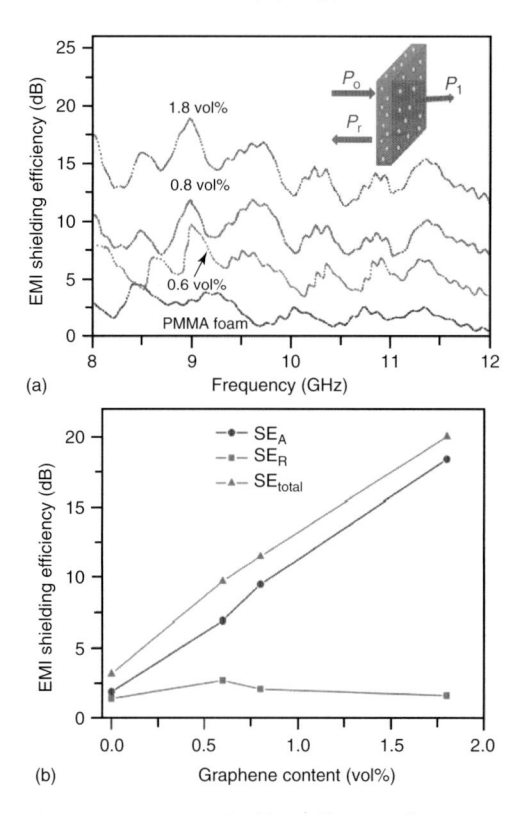

Figure 8.18 (a) EMI shielding efficiency of TRG/PMMA nanocomposite microcellular foams and (b) contribution of microwave absorption (SE_A), and microwave reflection (SE_R) at 9 GHz. (Source: Reproduced with permission from Ref. [38], The American Chemical Society (2011).)

The EMI SE of TRG/PMMA foams with different TRG contents at 8–12 GHz is displayed in Figure 8.18a. Obviously, the 1.8 vol% TRG/PMMA foam exhibits a high conductivity of 3.11 S m^{-1} and good EMI SE of 13–19 dB at 8–12 GHz. The EMI SE is mainly contributed by the absorption rather than the reflection mechanism (Figure 8.18b). From Figure 8.16b, the cell size is nonuniform but ranges from ∼1 to 10 μm. By properly monitoring the foaming process, it is anticipated that the EMI SE value of the TRG/PMMA foams can be further increased at even smaller TRG loadings.

Nomenclature

A	Absorptivity
δ	Skin depth
E_0	Electric field of incident electromagnetic wave
E_1	Electric field of transmitted electromagnetic wave
ε_0	Free space permittivity
ε'	Real permittivity
ε''	Imaginary permittivity
f	Electromagnetic wave frequency
μ	Magnetic permeability of a shield
μ_0	Magnetic permeability of free space
μ_r	Relative magnetic permeability
P_0	Electromagnetic wave incident power
P_1	Electromagnetic wave transmitted power
R	Reflectivity
SE	Shielding effectiveness
SE$_A$	Shielding effectiveness by absorption
SE$_M$	Shielding effectiveness by multiple reflections
SE$_R$	Shielding effectiveness by reflection
σ	Electrical conductivity
T	Transmissivity
t	thickness of a shield
$\tan \delta$	Tangent loss
ω	Angular frequency of electromagnetic wave
Z_0	Free space impedance
Z_i	Impedance at the air-absorber interface

References

1. Huang, Y., Li, N., Ma, Y.F., Du, F., Li, F.F., He, X.B., Lin, X., Gao, H.J., andf Chen, Y.S. (2007) *Carbon*, **45**, 1614–1621.
2. Watts, P.C., Hsu, W.K., Barnes, A., and Chambers, B. (2003) *Adv. Mater.*, **15**, 600–603.
3. Chung, D.D.L. (2001) *Carbon*, **39**, 279–285.
4. Syed Azim, S., Satheesh, A., Ramu, K.K., Ramu, S., and Venkatachari, G. (2006) *Prog. Org. Coat.*, **55**, 1–4.
5. Li, L. and Chung, D.D.L. (1994) *Composites*, **25**, 215–224.
6. Das, N.C., Khastgir, D., Chaki, T.K., and Chakraborty, A. (2000) *Composites Part A*, **31**, 1069–1081.

7. Al-Saleh, M.H. and Sundararaj, U. (2008) *Macromol. Mater. Eng.*, **293**, 621–630.
8. Wong, K.H., Pickering, S.J., and Rudd, C.D. (2010) *Composites Part A*, **41**, 693–702.
9. Che, R.C., Peng, L.M., Duan, X.F., Chen, Q., and Liang, X.L. (2004) *Adv. Mater.*, **16**, 401–405.
10. Dhawan, S.K., Ohlan, A., and Singh, K. (2011) in *Advances in Nanocomposites-Synthesis, Characterization and Industrial Applications*, Chapter 19 (ed. B.S. Reddy), Intech, Croatia, pp. 429–482.
11. Joo, J. and Lee, C.Y. (2000) *J. Appl. Phys.*, **88**, 513–518.
12. Al-Saleh, M.H. and Sundararaj, U. (2009) *Carbon*, **47**, 1738–1746.
13. Colaneri, N.F. and Schacklette, L.W. (1992) *IEEE Trans. Instrum. Meas.*, **41**, 291–297.
14. Ramo, S., Whinnery, J., and Van Duzer, T. (1994) *Fields and Waves in Communications Electronics*, John Wiley & Sons, Inc., New York.
15. ASTM (1999) D4935-99. *Standard Test Method for Measuring the Electromagnetic Shielding Effectiveness of Planar Materials*, American Society for Testing and Materials, West Conshohocken, PA.
16. Yang, Y.L., Gupta, M.C., Dudley, K.L., and Lawrence, R.W. (2005) *Nano Lett.*, **5**, 2131–2134.
17. Kim, H.M., Kim, K., Lee, C.Y., Joo, J., Cho, S.J., Yoon, H.S., Pejakovic, D.A., Yoo, J.W., and Epstein, A.J. (2004) *Appl. Phys. Lett.*, **84**, 589–591.
18. Pande, S., Singh, B.P., Mathur, R.B., Dhami, T.L., Saini, P., and Dhawan, S.K. (2009) *Nanoscale Res. Lett.*, **4**, 327–334.
19. Im, J.S., Park, I.J., In, S.J., Kim, T., and Lee, Y.S. (2009) *J. Fluorine Chem.*, **130**, 1111–1116.
20. Gupta, A. and Choudhary, V. (2011) *Compos. Sci. Technol.*, **71**, 1563–1568.
21. Bigg, D.M. (1987) *Polym. Compos.*, **8**, 1–7.
22. Panwar, V., Kang, B.S., Park, J.O., Park, S., and Mehra, R.M. (2009) *Eur. Polym. J.*, **45**, 1777–1784.
23. Bellis, G., Tamburrano, A., Dinescu, A., Santarelli, M.L., and Sarto, M.S. (2011) *Carbon*, **49**, 4291–4300.
24. Wu, Z.P., Li, M.M., Hu, Y.Y., Li, Y.S., Wang, Z.X., Yin, Y.H., Chen, Y.S., and Zhou, X. (2011) *Scr. Mater.*, **64**, 809–812.
25. Yang, Y.L, Gupta, M.C., Dudley, K.L., and Lawrence, R.W (2005) *J. Nanosci. Nanotechnol.*, **5**, 927–931.
26. Yang, Y.L., Gupta, M.C., and Dudley, K.L. (2007) *Nanotechnology*, **18**, 345701 (4 pp).
27. Mathur, R.B., Pande, S., Singh, B.P., and Dhami, T.L. (2008) *Polym. Compos.*, **29**, 717–727.
28. Liu, Z., Bai, Z., Haung, Y., Ma, Y.F., Du, F., Li, F.F., Guo, T.Y., and Chen, Y.S. (2007) *Carbon*, **45**, 821–827.
29. Das, N.C. and Maiti, S. (2008) *J. Mater. Sci.*, **43**, 1920–1925.
30. Das, N.C., Liu, Y.Y., Yang, K.K., Peng, W.Q., Maiti, S., and Wang, H. (2009) *Polym. Eng. Sci.*, **49**, 1627–1634.
31. Park, S.H., Thielemann, P., Asbeck, P., and Bandaru, P.R. (2009) *Appl. Phys. Lett.*, **94**, 243111 (3 pp).
32. Thomassin, J.M., Pagnoulle, C., Bdnarz, L., Huynen, I., Jerome, R., and Detrembleur, C. (2008) *J. Mater. Chem.*, **18**, 792–796.
33. Chen, L.M., Ozisik, R., and Schadler, L.S. (2010) *Polymer*, **51**, 2368–2375.
34. Nalawade, S.P., Picchioni, F., and Janssen, L.P. (2006) *Prog. Polym. Sci.*, **31**, 19–43.
35. Zhu, Q., Meng, Y.Z., Tjong, S.C., Zhao, X.S., and Chen, Y.L. (2002) *Polym. Int.*, **51**, 1079–1085.
36. Van Ngo, T.T., Duchet-Rumeau, J., Whittaker, A.K., and Gerard, J.F. (2010) *Polymer*, **51**, 3436–3444.
37. Blacher, S., Calberg, C., Kerckhofs, G., Leonard, A., Wevers, M., Jerome, R., and Pirard, J.P. (2007) *Stud. Surf. Sci. Catal.*, **160**, 681–688.
38. Zhang, H.B., Yan, Q., Zheng, W.G., He, Z.X., and Yu, Z.Z. (2011) *ACS Appl. Mater. Interfaces*, **3**, 918–924.

9
Polymer Nanocomposites for Sensor Applications

9.1
Introduction

In general, the resistivity of conductive polymer composites is highly sensitive to applied stimuli including temperature, gaseous environment, pressure, and mechanical deformation. The influence of the temperature factor on the conductivity of polymer nanocomposites is manifested in the positive temperature coefficient effect. The increased emission of carbon dioxide and other greenhouse gases has led to severe global warming in recent years. Moreover, hazardous industrial chemical gases pose a great threat to the environment and to human health. Therefore, there is an urgent need in both scientific and technological communities to develop miniaturized sensors with light weight, high sensitivity, fast response, and low-power consumption for measuring and detecting polluting gases. In this regard, conductive polymer composites are ideal materials for sensing gaseous molecules. In the past two decades, chemical sensors based on the carbon black CB/polymer composites have been widely investigated in relation to their ability to detect various gases and organic vapors [1–3]. The sensor response generally derives from a change in electrical resistance due to the swelling of the polymer matrix. The expansion of the polymer matrix increases the distance between the CB particles, giving rise to an increase in resistance. As the response of this type of device is associated with the change of resistance, it is often termed as *chemiresistor*.

Carbonaceous nanomaterials possess a variety of sizes and shapes, rendering them suitable for a wide range of sensing applications. Recent advances in the synthesis of carbon nanotubes (CNTs) and graphene have large impact on the development of polymer nanocomposites designed for chemical sensor applications. The advantages of CNT/polymer composites over conventional CB/polymer composites for gas sensing applications derive from CNTs that exhibit higher sensitivity and excellent electrical conductivity. CNTs are extremely sensitive to charge transfer and chemical doping effects by gaseous molecules, resulting in an electrical signal that is related to the type and number of gas molecules. These gases include H_2, NO_2, NH_3 and O_2, and organic vapors such as acetone, methanol and propanol, [4–7]. The density of main charge carriers of semiconducting single-walled carbon nanotubes (SWNTs) changes considerably by reacting with electron-withdrawing molecules (e.g., NO_2, O_2) or electron-donating molecules (e.g., NH_3) [5]. Similarly, two-dimensional graphene with high specific area ($2600\,m^2\,g^{-1}$) shows great potential for gas sensing applications [8–13]. This is because its whole area can be exposed to adsorbed gases, which maximize the effect. Ultrahigh sensitivity of order of single-molecule detection has been claimed for NO_2 gas [12a]. The sensitivity can be attributed to electron transfer between gas molecule and the graphene surface. Similarly, reduced graphene oxide (rGO) and thermally reduced graphene (TRG) have been found to be very effective for detecting NO_2 NH_3 and CO_2 gases [12b,c]. In general, gas sensors based on pristine CNTs have certain shortcomings, such as low sensitivity to analytes of low adsorption energy, lack of selectivity to different gaseous molecules, and long recovery time [5]. Conductive polymer nanocomposites appear to be very attractive for gas and vapor sensor applications since they combine selectivity of both carbon nanofillers and polymers. Therefore, polymer nanocomposite films with high sensitivities, improved selectivity, and fast response can be developed and fabricated.

Piezoelectric materials such as barium titanate and lead zirconate titanate are widely used in a number of applications as sensors and actuators. However, brittle piezoceramics exhibit poor mechanical performance in their whole service lives. In contrast, conducting polymer nanocomposites can function efficiently as high sensitivity transducers and actuators on applications of pressure, compressive, or tensile strain deformation. The change in electrical resistance due to the pressure or strain deformation is termed as *piezoresistivity*. Piezoresistivity induced by mechanical deformation in CNTs finds attraction applications in electromechanical devices. Baughman *et al.* [14] reported that electromechanical actuators based on SWNT sheets generate higher stresses than natural muscle and higher strains than high-modulus ferroelectrics. Some other applications of SWNTs in the actuators include switches and nanotweezers [15–17]. Furthermore, *in situ* electrical measurements reveal that the conductance of a nanotube sample can be reduced by a few orders of magnitude on deformation using an atomic force microscopy (AFM) tip [18, 19]. These properties make CNTs attractive fillers for composites designed for electromechanical applications. The piezoresistivity found in the CNT/polymer strain sensors can be attributed to the changes of conductive networks with strain, such as loss of contact between the fillers, tunneling effect in neighboring fillers, and conductivity change in deformed CNTs [20].

9.2
Pressure/Strain Sensors

9.2.1
Piezoresistivity

For percolative nanocomposites, the application of uniaxial pressure reduces the gaps between two adjacent conductive particles smaller, thereby decreasing the electrical resistance of a conducting path network. Therefore, tunneling can take place readily between two neighboring particles of the nanocomposites. On the basis of Simmons's [21] analysis of tunneling current between two electrodes separated by a thin insulating film, Zhang et al. [22] derived the following equation for tunneling resistance between two conducting particles of a composite

$$R = \left(\frac{L}{N}\right)\left(\frac{8\pi h d}{3a^2 \gamma e^2}\right)\exp(\gamma d) \qquad (9.1)$$

where L is the number of particles forming a single conducting path, N the number of conducting paths, a^2 the cross-sectional area where tunneling occurs, d the least distance between conducting particles, e the electron charge, h the Plank's constant, and γ is defined by

$$\gamma = \frac{4\pi\sqrt{2m\phi}}{h} \qquad (9.2)$$

where m is the electron mass and Φ the height of potential barrier between adjacent particles. It is evident from Eq. (9.1) that a reduction of d owing to the application of uniaxial pressure results in a decrease of R.

Considering an initial interparticle distance d_0 of the composite changes to d under an applied stress, the relative resistance (R/R_0) can be expressed by Knite et al. [23]

$$\frac{R}{R_0} = \left(\frac{d}{d_0}\right)\exp[\gamma(d-d_0)] \qquad (9.3)$$

where R_0 is the initial resistance. For a polymer composite of an initial length l_0 subjected to compressive deformation, the particle distance can be written as

$$d = d_0(1+\varepsilon) = d_0\left[1+\left(\frac{\Delta l}{l_0}\right)\right] \qquad (9.4)$$

where ε is the compressive strain and $\Delta l\,(=l-l_0)$ the deformation length of the composite. Substitution of Eq. (9.4) into Eq. (9.3) produces

$$\ln(R/R_0) = \ln\left[1+\left(\frac{\Delta l}{l_0}\right)\right] + A\left(\frac{\Delta l}{l_0}\right) \qquad (9.5)$$

where $A = \gamma d_0$. It is noted that Eq. (9.5) prevails only at small deformations. At lager deformations, a large increase in electrical resistance occurs as a result of destruction of the structure of conducting path network. In this regard, the following equation is more appropriate to describe an abrupt increase of R/R_0

$$\ln(R/R_0) = \ln\left[1+\left(\frac{\Delta l}{l_0}\right)\right] + A\left(\frac{\Delta l}{l_0}\right) + B\left(\frac{\Delta l}{l_0}\right)^2 + C\left(\frac{\Delta l}{l_0}\right)^3 + D\left(\frac{\Delta l}{l_0}\right)^4 \qquad (9.6)$$

where A, B, C, and D are constants [23].

Knite et al. indicated that CB/polyisoprene nanocomposites show reversible resistive change under both tensile and compressive strains. In other words, Eqs. (9.5) and (9.6) can be used to describe piezoresistivity of the polymer nanocomposites subjected to either tensile or compressive deformation [23]. A universal tensile tester combined with an electrometer or impedance analyzer is used to measure the resistance of a specimen subjected to tensile loading (Figure 9.1a). Two copper

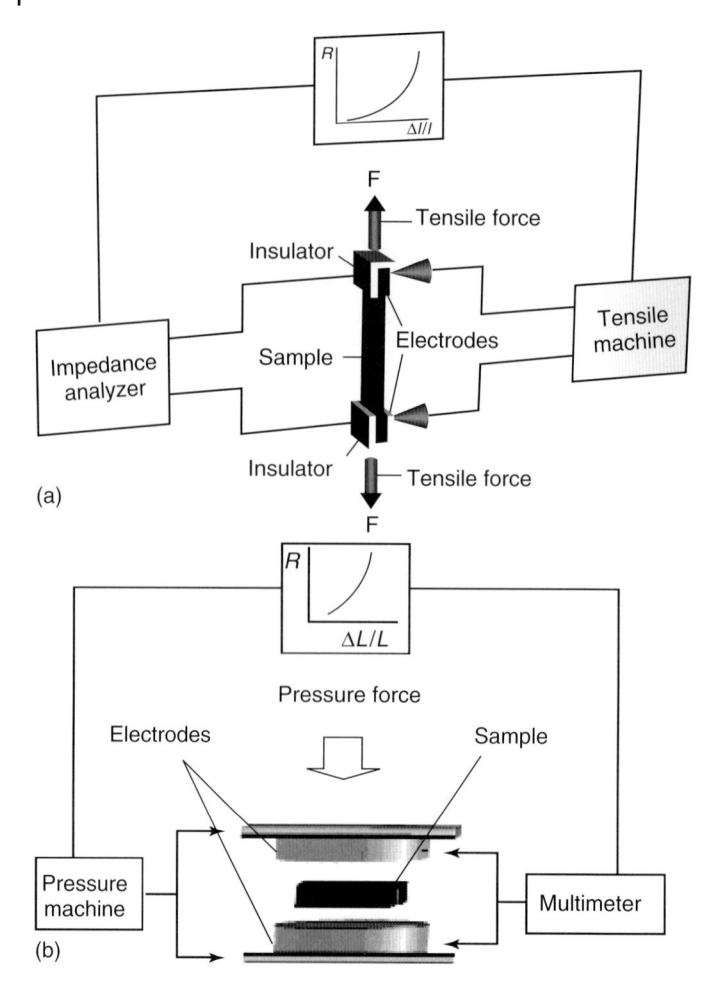

Figure 9.1 (a) Experimental setup for measuring electrical resistance of a specimen as a function of tensile strain or tensile force. (Source: Reproduced with permission from Ref. [23], Elsevier (2004).) (b) The setup for measuring electrical resistance under dynamic pressure and compressive strain. (Source: Reproduced with permission from Ref. [26], Wiley-VCH (2007).)

foils are mounted at both ends of the specimen surface for acting as electrodes. The universal tester can also be used for compressive deformation as shown in Figure 9.1b. Figure 9.2 shows the dependence of relative resistance on tensile strain of the 10 wt% CB/polyisoprene nanocomposite. Resistance values computed using Eqs. (9.5) and (9.6) match well with experimental data at small and large strains, respectively. Fitting Eq. (9.5) to experimental data at low strains yields $A = 6.491$ and $R_0 = 3.770 \times 10^5 \ \Omega$ (dashed line). A good agreement between Eq. (9.6) and experimental data at high strains yields $A = 8.206$, $B = -90.979$, $C = 873.911$, $D = -1333.339$, and $R_0 = 3.835 \times 10^5 \ \Omega$ (solid curve).

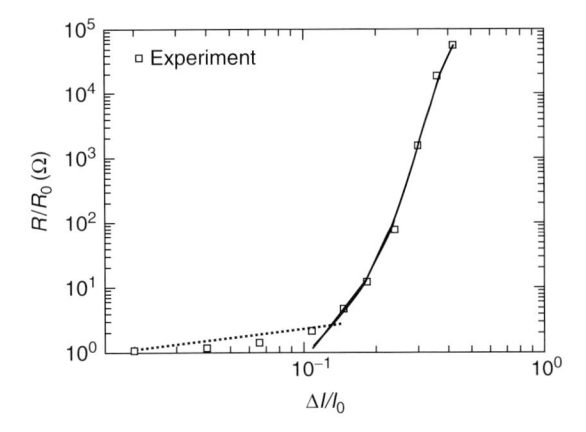

Figure 9.2 Variation of relative resistance with tensile strain for 10 wt% CB/polyisoprene nanocomposite at room temperature. Predictions from Eqs. (9.5) and (9.6) are shown in dashed line and solid curve, respectively. (Source: Reproduced with permission from Ref. [23], Elsevier (2004).)

9.2.2
Nanocomposites with GNPs

Recently, Chen and coworkers [24–26] have studied piezoresistive behavior of graphite nanoplatelet/high-density polyethylene (GNP/HDPE) and GNP/silicone rubber (SR) composites under uniaxial pressure conditions. Figure 9.3 shows the dependence of resistivity on applied pressure of the 6 vol% GNP/HDPE nanocomposite. There exists a critical pressure above which the resistivity increases

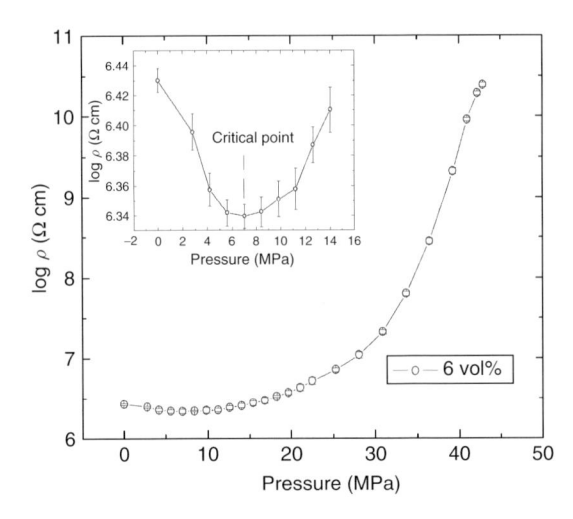

Figure 9.3 Relative resistance as a function of applied pressure for 6 vol% GNP/HDPE nanocomposite. (Source: Reproduced with permission from Ref. [25], Elsevier (2006).)

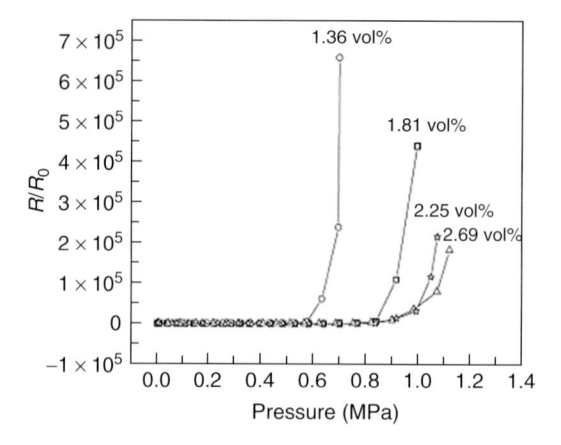

Figure 9.4 The dependence of relative resistance on deformation strain under pressure for GNP/SR nanocomposites. (Source: Reproduced with permission from Ref. [26], Wiley-VCH (2007).)

markedly with increasing pressure (inset). This is due to the breakdown of the conducting path network resulting from deformation of the polymer matrix. Figure 9.4 shows the effect of GNP content on the relative resistance of GNP/SR nanocomposites subjected to dynamic compression. The percolation threshold of these nanocomposites is 0.9 vol% GNP. A sharp increase in relative resistivity is observed for the nanocomposite with 1.36 vol% GNP because its filler content is in the vicinity of the percolation threshold. A quadratic form of Eq. (9.6) fits reasonably well with experimental data of the 1.36 vol% GNP/SR nanocomposite at small deformed strains, that is, region 1 and region 2 of the curve (Figure 9.5). This yields $A = 10.604$ and $B = 42.819$ for region 1 and $A = 3.4298$ and $B = 65.824$ for region 2. The experimental data of this nanocomposite under high deformation strains (region 3) can be best fitted with a cubic polynomial form of Eq. (9.6), yielding $A = -572.896$, $B = 3166.828$, and $C = -4145.811$. These results imply that the tunneling effect is responsible for the transport of charge carriers in the 1.36 vol% GNP/SR nanocomposite subjected to applied pressure. Furthermore, the repeatability of piezoresistivity of this nanocomposite for 10 successive cycles of compression is good as evidenced by very small variations in the relative resistance change (Figure 9.6).

9.2.3
Nanocomposites with CNTs

The piezoresistive behavior of polymers filler with CNTs and carbon nanofibres (CNFs) has been studied under dynamic tensile testing [20, 27–30] and flexural [31] conditions by several researchers. The overall resistance change response versus strain of these nanocomposites consists of linear and nonlinear regions. The linear behavior can be attributed to the tunneling effect between adjacent nanotubes under small strains. Nonlinear piezoresistivity occurs in these nanocomposites

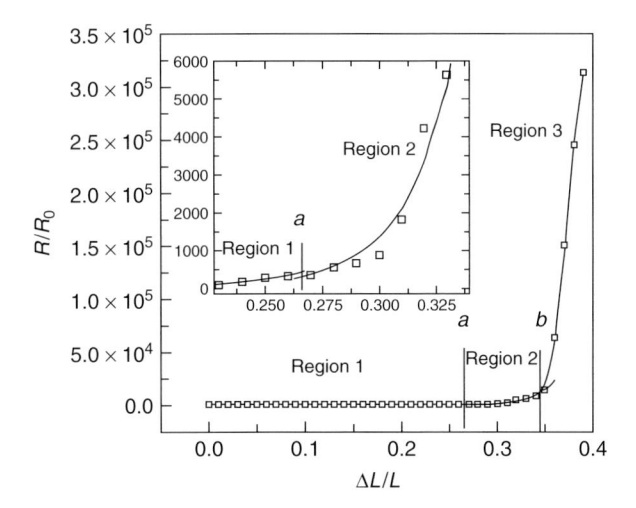

Figure 9.5 The dependence of relative resistance on compressive strain of 1.36 vol% GNP/SR nanocomposite. The $\ln(R/R_0) = \ln\left[1 + (\Delta l/l_0)\right] + A(\Delta l/l_0) + B(\Delta l/l_0)^2$ relation fits reasonably well in region 1 and region 2 of the curve under low pressure. The $\ln(R/R_0) = \ln\left[1 + (\Delta l/l_0)\right] + A(\Delta l/l_0) + B(\Delta l/l_0)^2 + C(\Delta l/l_0)^3$ relation can be used to describe experimental data in region 3 of the curve under high pressure. (Source: Reproduced with permission from Ref. [26], Wiley-VCH (2007).)

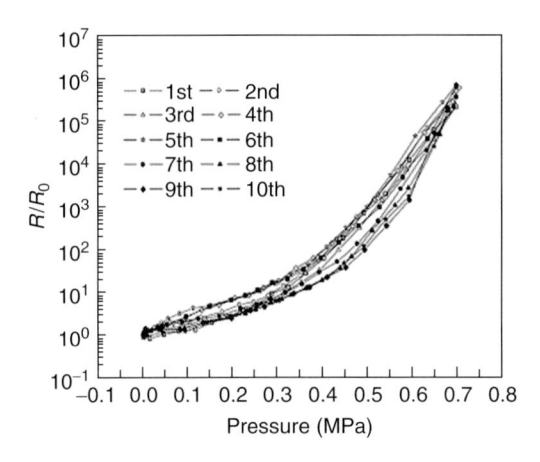

Figure 9.6 Piezoresistive behavior of 1.36 vol% GNP/SR nanocomposite under cyclic compression. (Source: Reproduced with permission from Ref. [26], Wiley-VCH (2007).)

under high strain conditions. As an example, Bautista-Quijano et al. [30] fabricated multiwalled carbon nanotube/polysulfone (MWNT/PSF) films by solvent casting. They bonded the films on aluminum tensile specimens for piezoresistive measurements. Figure 9.7a shows the tensile stress–strain curves of the 0.5 wt% MWNT/PSF film up to final failure. The normalized electrical resistance versus

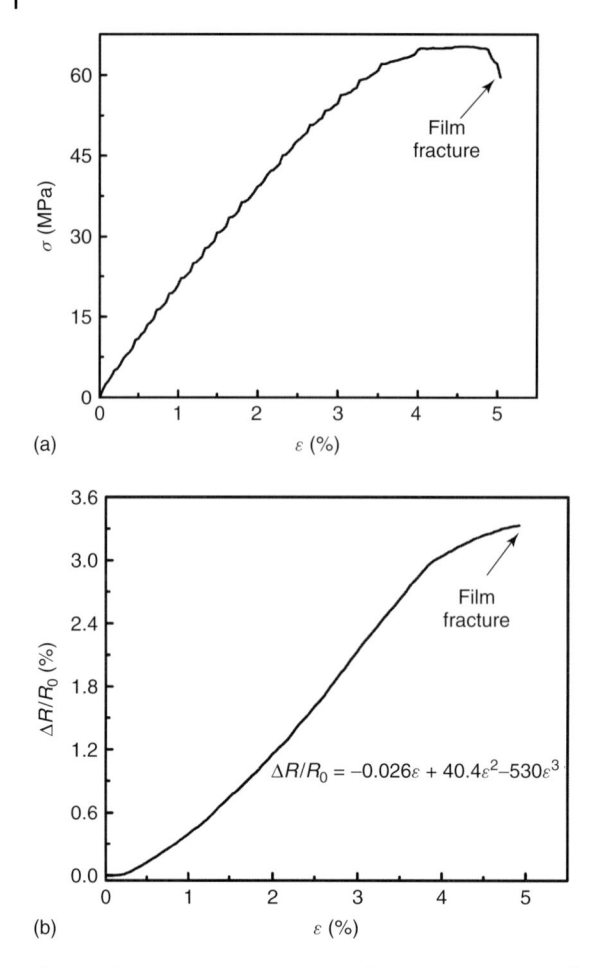

Figure 9.7 Piezoresistive response of 0.5 wt% MWNT/PSF film loaded up to fracture. (a) Stress–strain curve and (b) normalized resistance versus tensile strain. (Source: Reproduced with permission from Ref. [30], Elsevier (2010).)

tensile strain is shown in Figure 9.7b. The filler content of this composite film is well above the percolation threshold of 0.06 wt%. A linear piezoresistivity is found at tensile strain levels up to ∼1.3%, indicating the onset of nonlinearity. The nonlinear resistance change versus strain response for $\varepsilon > 1.3\%$ can be best fitted with a cubic polynomial equation in the form of $\Delta R/R_0 = -0.026\,\varepsilon + 40.4\,\varepsilon^2 - 530\,\varepsilon^3$.

The strain gauge factor (K) or sensitivity factor, defined by $\Delta R/R\varepsilon$, is an important parameter to characterize the performance of strain sensors. This parameter can be determined from the linear slope of a piezoresistive response curve, that is, ($\Delta R/R$) versus ε. In general, MWNT/polymer and CNF/polymer nanocomposites exhibit K values >2, demonstrating their suitability for use as strain sensors of high sensitivity [28, 29, 31]. Figure 9.8 shows the piezoresistive behavior of MWNT/epoxy sheets [29]. Higher sensitivity is observed in these nanocomposite

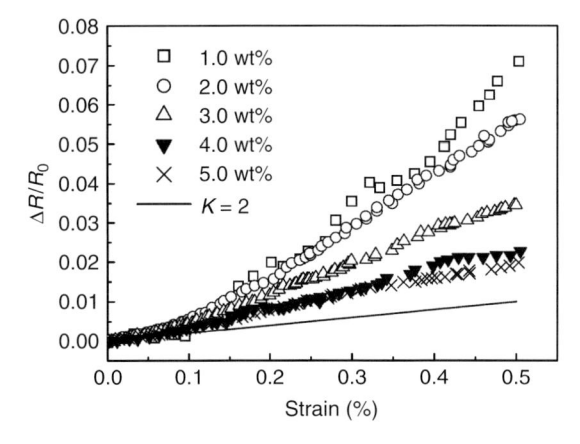

Figure 9.8 Piezoresistive responses of MWNT/epoxy nanocomposites containing different nanotube contents. (Source: Reproduced with permission from Ref. [29], Elsevier (2008).)

sheets, particularly for the specimen with filler content close to the percolation threshold. The gauge factor of the composite sheet with 1 wt% MWNT is about eight times higher than that of conventional strain gauges having values between 2.0 and 3.2 [28]. Therefore, nanocomposite sensors with high sensitivity can be bonded to the surfaces of engineering components for monitoring the macroscopic strain in the structures and for detecting internal cracks or flaws of structural components.

9.3
Gas and Humidity Sensors

9.3.1
Gas Sensitivity

Apart from the greenhouse gases, organic compounds are often utilized in chemical, petrochemical and transportation industries, household chemicals, coatings, adhesives, and so on. Their high volatility renders them potential hazards to the environment and health of mankind. Furthermore, the incinerators of chemical (e.g., garbage treatment and coal) and electric power industries also release toxic organic gases and particles. For example, dioxins and furans are well known halogenated aromatic hydrocarbons of high toxicity and found in air, water, and soil of global environments. Organic pollutants associated with fine particles can affect visibility degradation and radiative transfer through the atmosphere by scattering and absorption [32]. For some cases, organic solvents employed in various industrial processes may be associated with hepatotoxicity. These solvents include carbon tetrachloride, trichloroethylene, tetrachloroethylene, toluene, and 1,1,1-trichloroethene [33]. Benzene is another solvent that is harmful to human blood, causing chronic poisoning and neurasthenia syndrome after long-term contact. Chloroform inhalation not only induces anesthetic effect in our central nervous system but also poses threats to the heart and kidney [32]. Therefore, it is of technological importance to develop gas sensors of high sensitivity to effectively monitor and detect these organic compounds. Conducting polymer composites are well suited for these purposes because of their excellent gas responses compared with existing sensing materials such as semiconductors. In particular, CNTs exhibit higher gas response to organic vapors than CBs [34].

The chemoresistive response of polymer nanocomposite films is based on the modification of electrical conductivity of the films due to adsorbed gas species. The gas response (S) or sensitivity of a "sensitive" film exposed to gas vapor can be defined by

$$S = \frac{R - R_0}{R_0} = \frac{\Delta R}{R_0} \tag{9.7}$$

where R_0 is the initial resistance of a sample in air, R the resistance on exposure to gas/vapor, and ΔR the resistance change. The corresponding maximum responsivity is obtained by replacing R in this equation with a maximum electrical

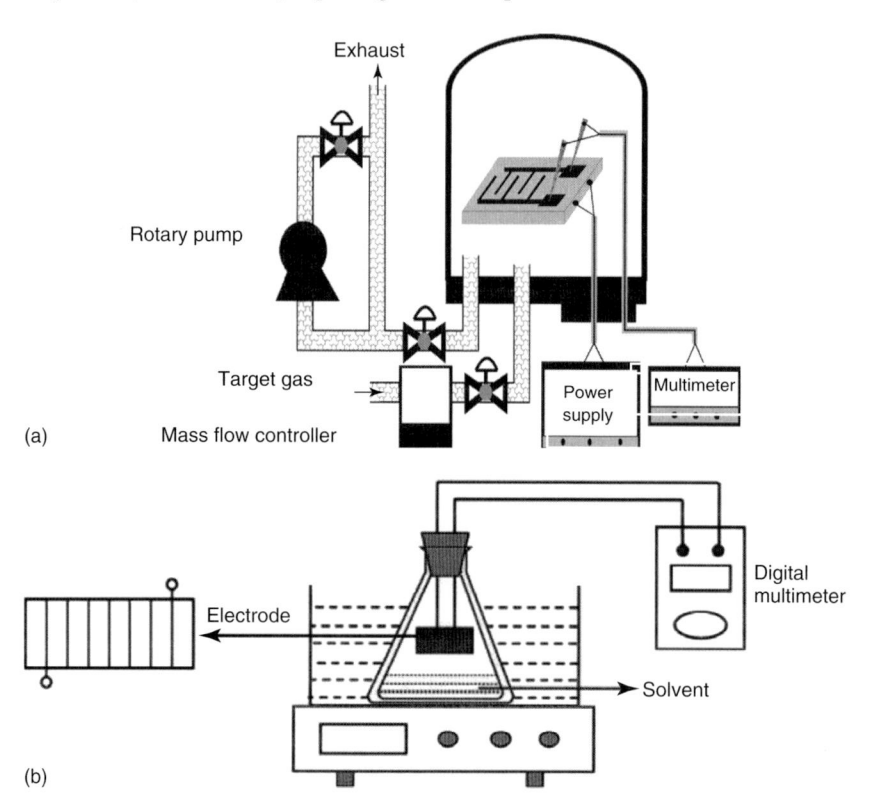

Figure 9.9 Experimental setups for measuring gas sensitivity of conducting polymer composites exposed to (a) gas and (b) organic vapor atmospheres. (Source: Reproduced with permission from (a) [35], Intech (2011), and (b) [36], Elsevier (2003).)

resistance (R_{max}) of the sample exposed to organic vapor. In some cases, maximum responsivity is simply defined as R_{max}/R_0. To fabricate a chemoresistive gas sensor, the nanocomposite film is deposited onto a substrate connected with Au or Pt electrodes. The resistance of the nanocomposite film is monitored by exposing the film to gaseous species [35] (Figure 9.9a). Alternatively, gas sensing measurements can be performed using a simple setup as shown in Figure 9.9b.

9.4
Organic Vapor Sensors

9.4.1
Nanocomposites with CBs

Polymer-CB composite films have been widely used as a sensing material in gas sensors and electronic noses [1–3]. The permeation of gaseous molecules into CB/polymer composites leads to a change in electrical resistivity due to the swelling of the polymer matrix. This response correlates well with their percolation behavior [1, 37]. The detection efficiency of gas sensors can be greatly enhanced by arranging them into sensor arrays [38]. Moreover, the sensitivity of CB/polymer sensors also depends on the nature of polymer matrix, CB content, and the dispersion of CBs in the composites. The diversity of the responses of gas sensors can be achieved by proper selection of polymers with different chemical properties. For instance, poly(N-vinyl-pyrrolidone) is hydrophilic; thus, it swells readily in water vapor but not in toluene vapor. On the contrary, hydrophobic polyisobutylene swells in toluene vapor but not in water vapor [39]. The matrix materials of CB/polymer composite gas sensors generally include amorphous and semicrystalline thermoplastics as well as rubbers [36, 40–47].

Zhang and coworkers [36, 40–43] studied gas sensing behaviors of the CB(50–70 nm)/polymer nanocomposites exposed to different organic vapors systematically. The composite specimen was suspended in a sealed glass flask filled with an organic solvent (Figure 9.9b). It was exposed to a given concentration of organic vapor (parts per thousand by volume; ppt) based on a known quantity of organic solvent inside. The change in electrical resistance of the composite was determined using a digital multimeter. When the electrical resistance of the composite approached its steady-state value, the composite sensor was removed from the vessel and exposed to dry air. Figure 9.10 shows electrical resistance versus time profile of the 14 wt% CB/PBMA composite exposed to 3.7 ppt benzene vapor. The resistance increases quickly to a maximum value in 300 s owing to the swelling of the polymer matrix associated with benzene vapor adsorption. This leads to an increase of interparticle distance between CBs. The removal of the composite from benzene vapor to dry air leads to a large reduction in electrical resistance. This can be attributed to desorption of benzene vapor that yields a contraction of the polymer matrix and reconnection of separated fillers. Therefore, adsorption and desorption of solvent vapors can induce changes in electrical

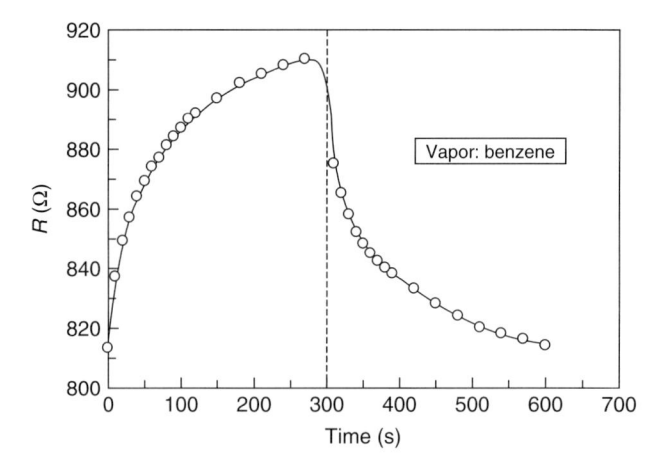

Figure 9.10 Electrical resistance response of 14 wt% CB/PBMA composite in 3.7 ppt benzene vapor followed by air exposure. The dashed line defines the vapor absorption and desorption zone. (Source: Reproduced with permission from Ref. [40], Elsevier (2004).)

resistance of the CB/polymer composites markedly. Figure 9.11a shows electrical resistance versus time profiles of the 14 wt% CB/PBMA composite exposed to toluene vapor of increasing concentrations. Each exposure cycle consists of 300 s in toluene vapor and another 300 s in air. The dependence of maximum responsivity on vapor concentration of this composite is shown in Figure 9.11b. The inset in the upper left shows a linear relationship for maximum responsivity at low vapor pressures. The linear behavior is considered to be of practical importance since gas sensors of high sensitivity are capable of detecting gas molecules of small concentrations.

Polyurethane (PU) is a copolymer that consists of alternating sequence of long nonpolar soft segments (polyol groups) and short polar hard segments (isocyanate groups). Thus PU is characterized by two-phase (soft and hard) morphology. The elastomeric behavior of PU is derived from the phase separation of the hard and soft segments in which the hard segments acting as cross-linkers between soft segment domains. Generally, low polar and nonpolar solvents can swell the mainly nonpolar soft segments of PU, while polar solvents dissolve polar hard segments. Figure 9.12 shows maximum electrical responsivity of the 0.5 wt% CB/PU and 3.5 wt% CB/PU nanocomposites exposed to various organic vapors. Apparently, the 3.5 wt% CB/PU nanocomposite shows broad sensitivities to various organic vapors of different polarities due to the phase separation of hard–soft segments. Polar organic vapors can expand the hard segments of PU, leading to the release of soft segments from cross-linking ties of hard segments of PU. On the other hand, nonpolar and low polar solvents adsorb on nonpolar soft segments of PU. The adsorption of nonpolar solvent vapors leads to the swelling of the composite matrix and increases the distance between CBs as characterized by high responsivity [40]. The CB content of 3.5 wt% is well above the percolation threshold of CB/PU composites, that is, 0.95 wt%. In general, a conducting gas sensor requires high

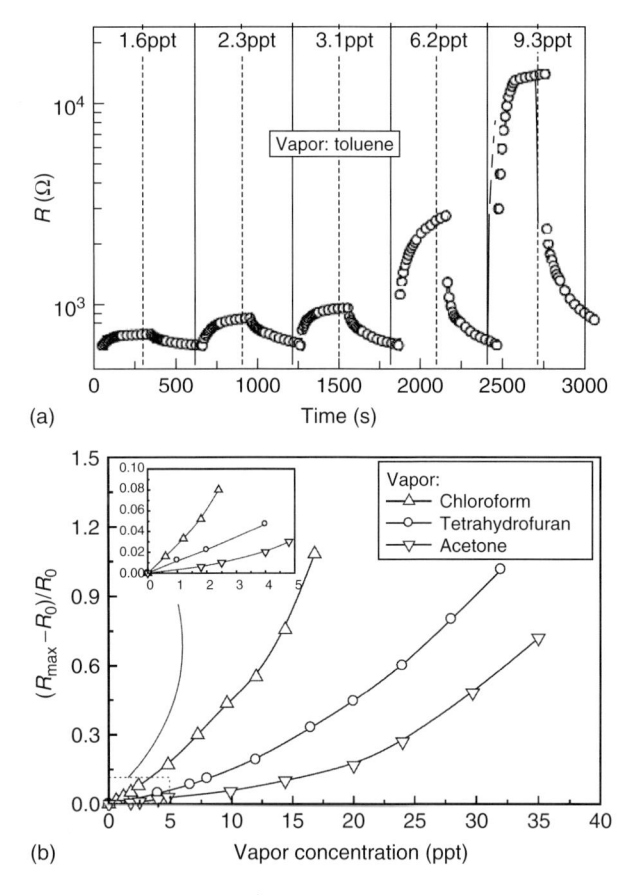

Figure 9.11 (a) Electrical resistance versus time curve of 14 wt% CB/PBMA composite exposed to toluene vapor of different concentrations. The dashed lines define the vapor absorption and desorption zones. (b) Maximum resistance response versus vapor concentration of this composite exposed to acetone, tetrahydrofuran, and chloroform vapors. The inset shows the responses at very low vapor concentrations. (Source: Reproduced with permission from Ref. [40], Elsevier (2004).)

filler content, that is, above its percolation threshold to reach high responsivity. Figure 9.13 shows the vapor pressure dependence of maximum responsivity of the 3.5 wt% CB/PU nanocomposite. The maximum response magnitude of the composites increases with increasing vapor pressure. This demonstrates that the gas response is closely related to the transport of solvent vapor to the polymer matrix. The responsivity exhibits a linear relationship with pressure at low vapor pressures (inset). As aforementioned, linear correlation is considered of technical importance since the gas sensors of high sensitivity are capable of detecting small concentrations of gas molecules. The vapor pressure dependence of maximum responsivity of the 3.5 wt% CB/PU nanocomposite and pure PU at low partial pressures is shown in Figure 9.14.

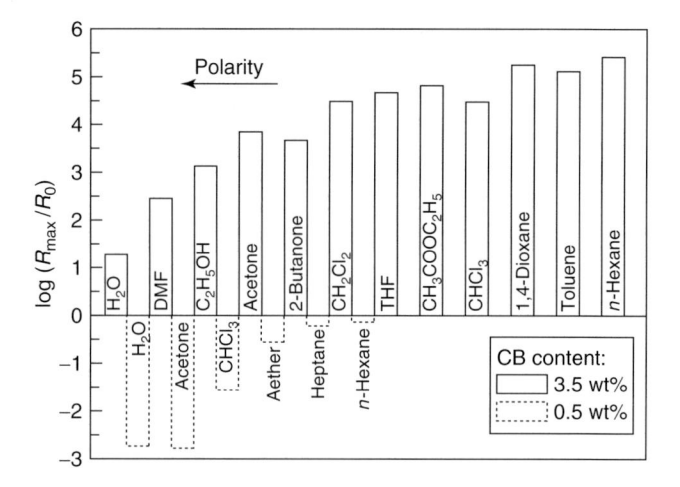

Figure 9.12 Maximum resistance responsivity of CB/PU composites to various saturated organic solvent vapors at 30 °C. (Source: Reproduced with permission from Ref. [42], Elsevier (2005).)

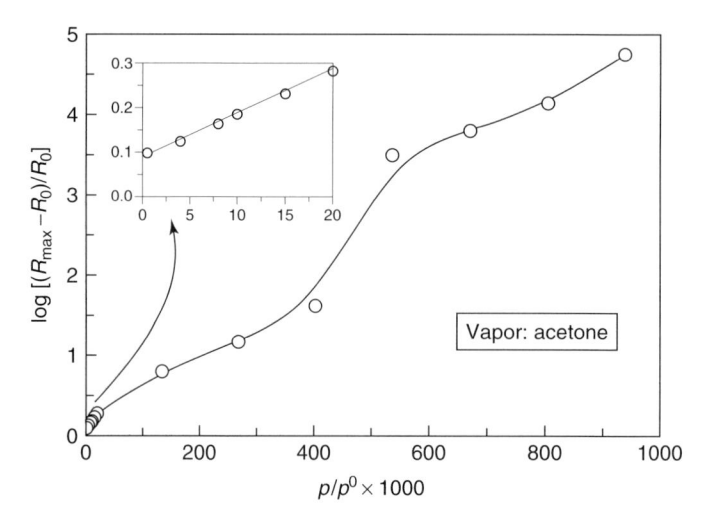

Figure 9.13 Maximum responsivity versus partial pressure of acetone vapor of 3.5 wt% CB/PU nanocomposite. p and p_0 are vapor pressures of the analyte under testing and saturated conditions, respectively. (Source: Reproduced with permission from Ref. [42], Elsevier (2005).)

9.4.1.1 Nanocomposites with CNTs and CNFs

CNTs with remarkable electrical properties show great promise for use as functional materials for fabricating novel gas sensors with high sensitivity and good stability [4–7]. In this regard, considerable attention has been paid to the development and gas sensing properties of the CNT/polymer nanocomposites recently [48–56].

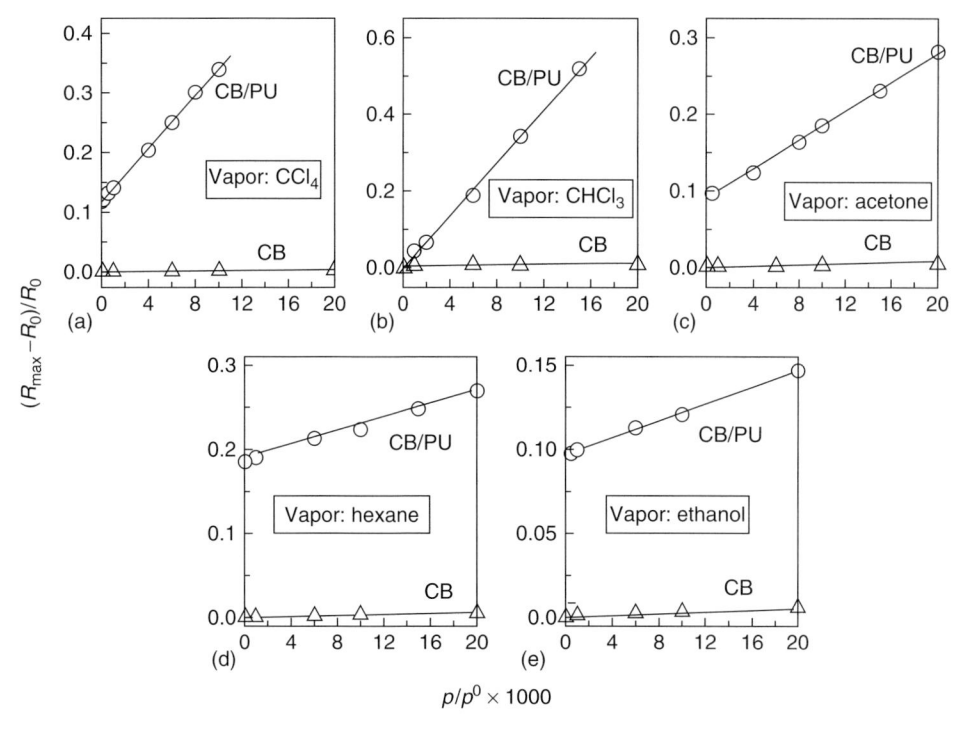

Figure 9.14 Maximum gas response versus partial pressure of pure PU and 3.5 wt% CB/PU nanocomposite exposed to different organic vapors. (Source: Reproduced with permission from Ref. [42], Elsevier (2005).)

Shang et al. [51] prepared MWNT/PMMA (poly(methyl methacrylate)) nanocomposites using both *in situ* microemulsion polymerization and solution-mixing processes. Figure 9.15 shows electrical conductivity versus filler content for *in situ* polymerized and solution-mixed MWNT/PMMA nanocomposites. The percolation thresholds of *in situ* polymerized and solution-mixed MWNT/PMMA nanocomposites determined from the percolation theory are 1 and 4.5 vol%, respectively. Microemulsion polymerization yields better dispersion of nanotubes in the polymer matrix because MMA monomers can penetrate readily into CNTs during polymerization. This leads to *in situ* polymerized nanocomposites that exhibit lower percolation threshold. Figure 9.16 shows maximum responsivity versus filler content for both nanocomposites exposed to tetrahydrofuran (THF) vapor. Maximum responsivity of *in situ* polymerized and solution-mixed MWNT/PMMA nanocomposites peaks at 8 and 10 wt% MWNT. These filler contents are much higher than their respective percolation thresholds of 1 and 4 vol%. Furthermore, the *in situ* polymerized nanocomposites exhibit much higher responsivity at 8 wt% MWNT than solution-mixed counterparts at the same filler loading. This is due to better dispersion of nanotubes in the matrix of *in situ* polymerized nanocomposites. At 10 wt% MWNT and above, both nanocomposite systems show the same responses due to the aggregation of nanotubes at high filler contents. In general,

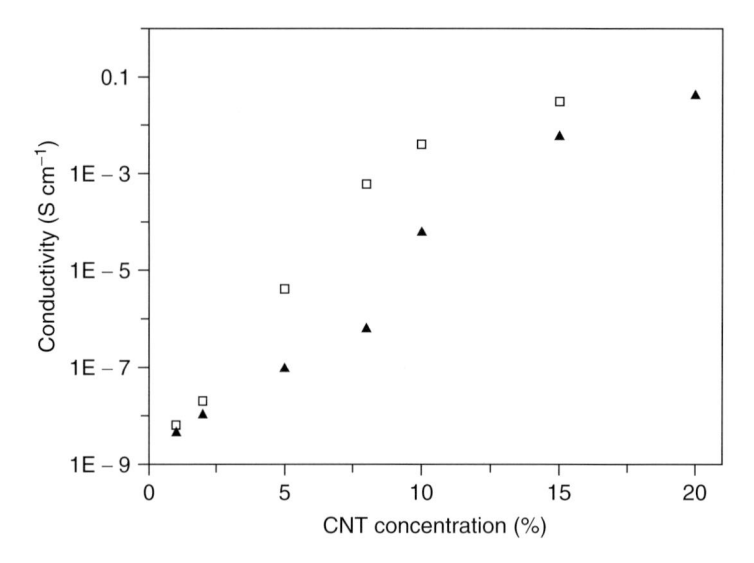

Figure 9.15 Electrical conductivity versus MWNT content (wt%) of *in situ* polymerized and solution-mixed MWNT/PMMA nanocomposites. □: Polymerization; ▲: solution mixing. (Source: Reproduced with permission from Ref. [51], Elsevier (2009).)

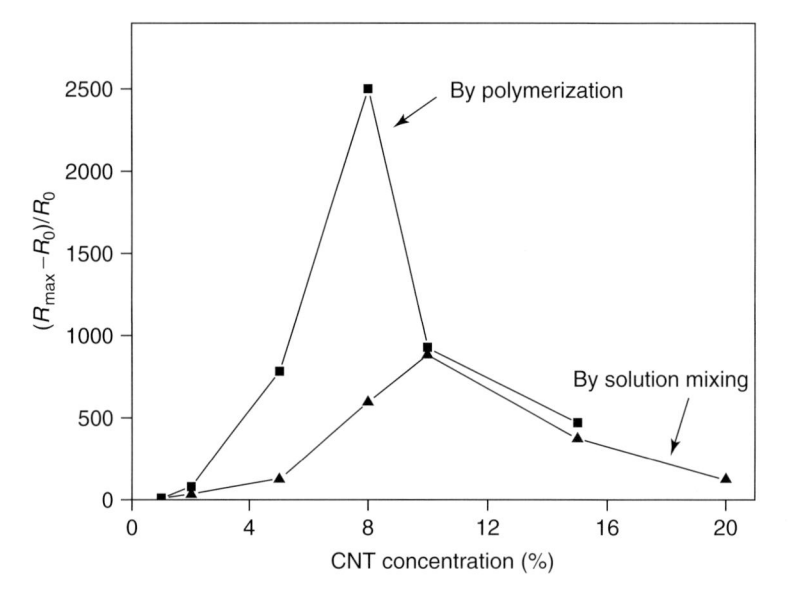

Figure 9.16 Maximum gas response versus MWNT content (wt%) of *in situ* polymerized and solution-mixed MWNT/PMMA nanocomposites exposed to THF vapor. (Source: Reproduced with permission from Ref. [51], Elsevier (2009).)

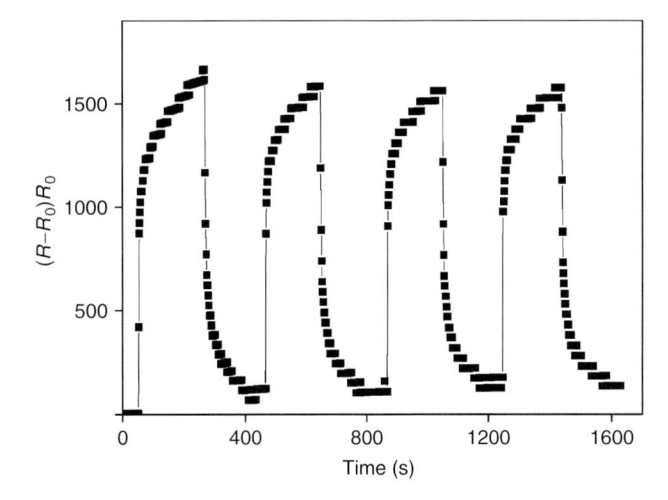

Figure 9.17 Adsorption and desorption response of *in situ* polymerized 5 wt% MWNT/PMMA nanocomposite exposed to acetone vapor and dry air. (Source: Reproduced with permission from Ref. [51], Elsevier (2009).)

good reproducibility or recovery is an important factor of the gas sensor material for practical applications. Good reproducibility can be obtained by exposing *in situ* polymerized 5 wt% MWNT/PMMA nanocomposite in acetone vapor followed by air exposure (Figure 9.17).

In another study, Zhang and coworkers [48] reported that *in situ* polymerized MWNT/PS (polystyrene) nanocomposites also exhibit maximum responsivity at lower filler content than solution-mixed MWNT/PS nanocomposites. They also investigated gas sensing properties of solution-mixed CNF/PS and (CNF + CB)/PS nanocomposites exposed to THF vapor [43, 57]. Figure 9.18 shows the adsorption/desorption behavior of the 6.25 wt% CNF/PS nanocomposite exposed to THF vapor and air for five cycles. The resistance rises sharply to a maximum value in 200 s owing to the adsorption of THF vapor. Good reproducibly of electrical resistance can be seen in this nanocomposite subjected to five adsorption/desorption cycles. The maximum responsivity of CNF/PS nanocomposites exposed to THF vapor as a function of filler content is shown in Figure 9.19. Obviously, the responsivity peaks at 6.25 wt% CNF. For the (CNF + CB)/PS hybrid nanocomposites, the responsivity peaks at 4 wt% total filler content (Figure 9.20). The maximum responsivity of the 4 wt%(CNF + CB)/PS hybrid is considerably higher than that of the 7 wt% CB/PS nanocomposite. The bridging effect of CNFs facilitates the formation of a conductive path network at lower filler content in the hybrid nanocomposites. Thus a total filler content of 4 wt% is needed to achieve maximum responsivity on exposure to THF vapor. In contrast, CNF/PS and CB/PS nanocomposites require the addition of 6.25 wt% CNF and 7 wt% CB, respectively, to obtain maximum responsivity. For a fixed total filler content of 4 wt%, the effect of the CNF/CB ratio on the gas sensitivity of hybrid nanocomposites exposed to various organic vapors is shown in Figure 9.21. The R_{max}/R_0 responses of the (CNF + CB)/PS hybrid nanocomposites

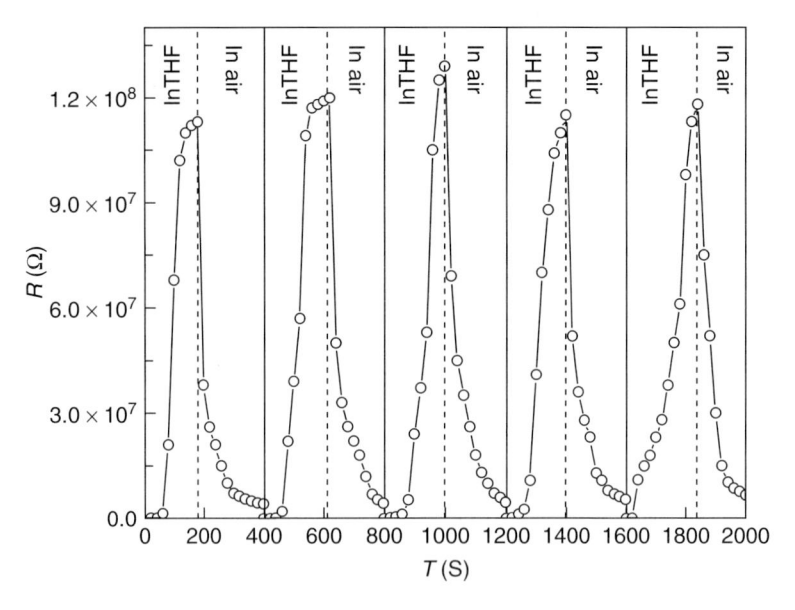

Figure 9.18 Electrical resistance versus time of solution-mixed 6.25 wt% CNF/PS nanocomposite in toluene vapor at 35 °C followed by air exposure. (Source: Reproduced with permission from Ref. [57], Elsevier (2006).)

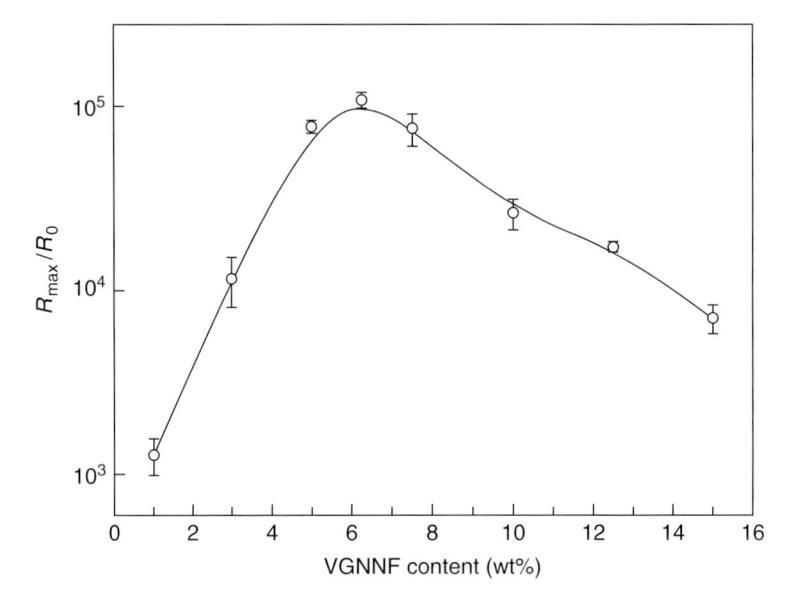

Figure 9.19 Maximum gas response versus CNF content of solution-mixed 6.25 wt% CNF/PS nanocomposite in toluene vapor at 35 °C. (Source: Reproduced with permission from Ref. [57], Elsevier (2006).)

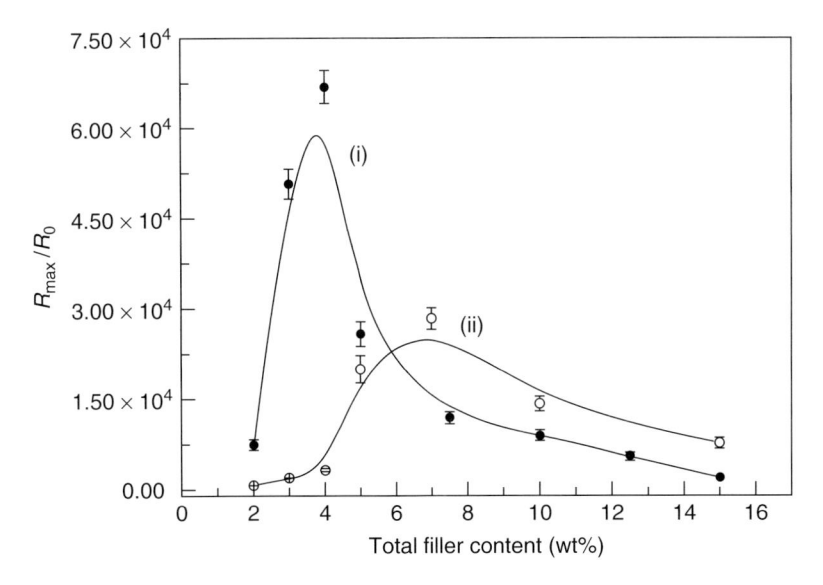

Figure 9.20 Maximum gas response versus total filler content of solution-mixed (i) (CNF + CB) (1 : 1)/PS hybrids and (ii) CB/PS nanocomposites. (Source: Reproduced with permission from Ref. [43], Elsevier (2006).)

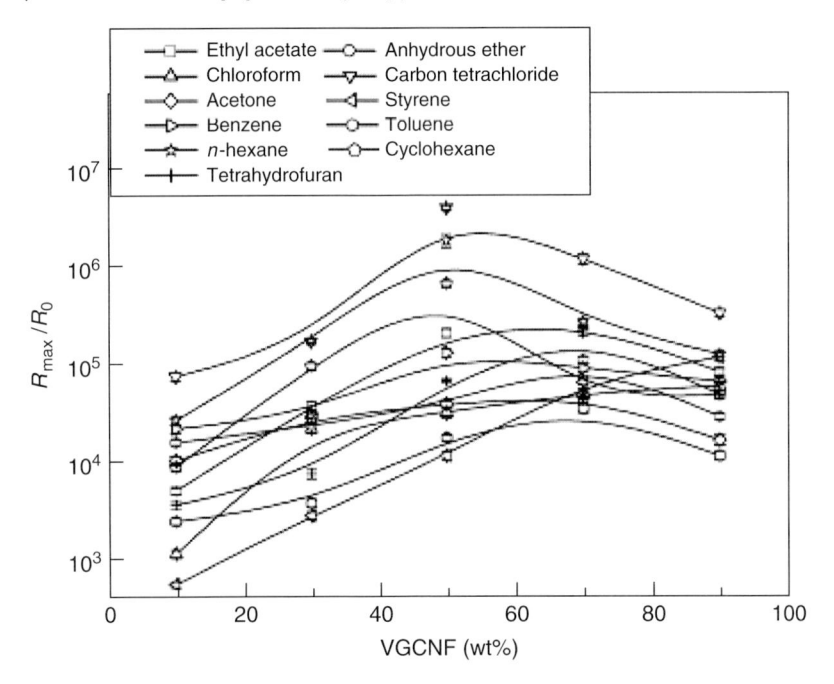

Figure 9.21 Maximum gas response versus CNF content in different vapors for solution-mixed (CNF + CB)/PS hybrids with different CNF/CB content ratios (total filler content: 4 wt%). (Source: Reproduced with permission from Ref. [43], Elsevier (2006).)

peak at different CNF contents on exposure to different vapors, thereafter decreases with increasing filler content. The decrease at higher CNF content is due to the CNF fillers acting as obstacles for the expansion of polymer matrix.

PU generally exhibits excellent gas sensitivity and its selective vapor response can be monitored by incorporating various soft or hard segments with different chain lengths and flexibility. In general, the properties of PU are determined mainly by the choice of polyols. For example, the tensile strength and modulus of PU can be improved by modifying polyols with hydroxyl-terminated acrylonitrile-butadiene copolymer (HTBN) [58]. Recently, Luo et al. [56] investigated the effect of soft–hard segment structures on vapor responsivity of MWNT/PU nanocomposite films. The nanocomposite films were prepared via an *in situ* coupling reaction between linear hydroxyl-terminated (i.e., HTBN and hydrolytically terminated HTBN; h-HTBN) polymer diols, 1,6-hexamethylene diisocyanate (HDI), and various chain extenders such as 1,4-butanediol (BDO), 4,4-methylene-bis(o-chloroaniline) (MOCA), and triethanolamine (TEA). The nanotubes employed were pristine MWNTs, functionalized MWNTs-OH, and MWNTs-COOH. Figure 9.22a shows the reaction scheme of synthesis process of HTBN/BDO/MWNT-PU composites. In the case of MWNTs-OH, its hydroxyl groups involve in the formation of urethane linkages via covalent interactions (Figure 9.22b). The gas sensitivity of h-HTBN-based PU composites filled with pristine MWNTs, MWNTs-OH, and MWNTs-COOH exposed to different organic vapors is shown in Figure 9.23. High responsivity is obtained for the h-HTBN/MWNTs-OH composite film exposed to nonpolar benzene. As

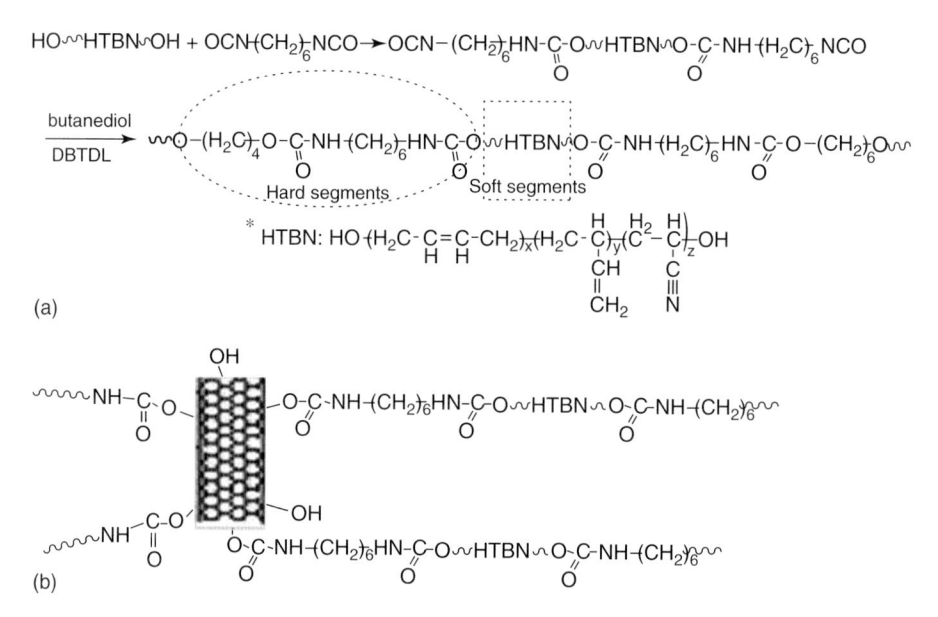

Figure 9.22 (a) Synthesis scheme of HTBN/BDO/MWNTs–PU composites and (b) formation of chemical linkage between hydroxyl groups of MWNTs-OH with HDI during fabrication of PU composites. (Source: Reproduced with permission from Ref. [56], Elsevier (2011).)

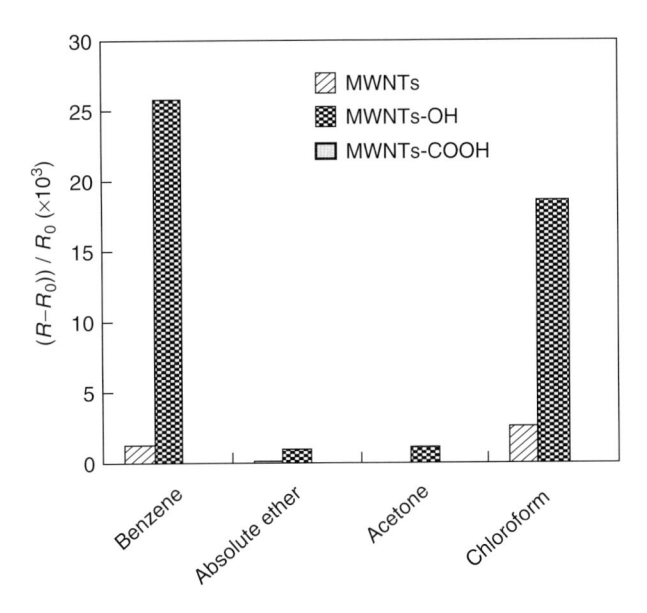

Figure 9.23 Responsivity of PU composite thin films with different types of MWNTs exposed to some solvent vapors (concentration of analytes is 7500 ppm). (Source: Reproduced with permission from Ref. [56], Elsevier (2011).)

pristine MWNTs are encapsulated by the PU material, charge-transfer process between the MWNTs and gas molecules is somewhat obstructed. Therefore, the h-HTBN/MWNTs PU film shows low response. For the MWNTs-OH, a chemical linkage of MWNTs-OH and PU matrix is established as shown in Figure 9.22b. Accordingly, the gas molecules can adsorb onto the surface of h-HTBN/MWNTs-OH film during organic vapor exposure. The swelling of PU matrix is accompanied by the increment of the MWNTs-OH spacing, leading to a substantial increase in electrical resistance. Finally, the gas sensing properties of h-HTBN/BDO-based PU composites filled with pristine MWNTs and functionalized MWNTs-OH exposed to benzene vapor concentrations from 1500 to 7500 ppm are shown in Figure 9.24. It is evident that the h-HTBN/BDO-based PU composite filled with MWNTs-OH displays fast response, good recovery, and response linearity. The composite with pristine MWNTs not only exhibits lower gas sensitivity but also displays a linear relation behavior in the plot of the responsivity versus vapor concentration (inset).

9.4.2
Humidity Sensors

The ability to sense and monitor humidity level is considered of technological importance in biomedical, industrial, and environmental sectors [59–62]. High-performance humidity sensors for these applications must meet several requirements, including linear response, high sensitivity, fast response time,

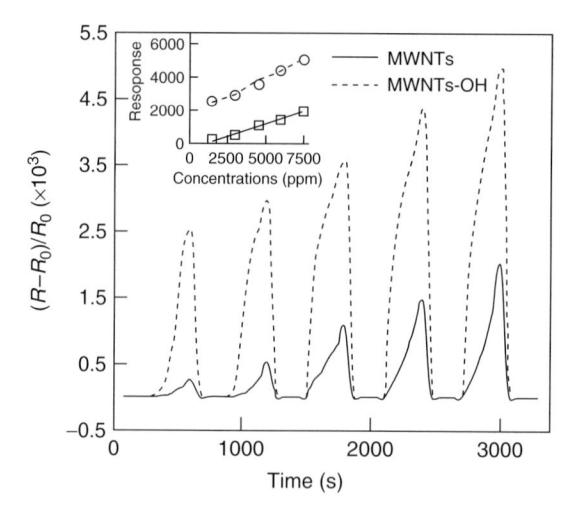

Figure 9.24 Gas sensing properties of h-HTBN/BDO–PU composites with MWNTs and MWNTs-OH exposed to benzene vapor of 1500–7500 ppm. The inset in the upper left shows linear correlativity between responsivity and vapor concentration. (Source: Reproduced with permission from Ref. [56], Elsevier (2011).)

chemical and physical stability, wide operating range of humidity, and low cost [63]. The operating principles of humidity sensors are based on the change in either electrical properties (i.e., resistance or capacitance) or the mass of a thin film due to the adsorption/desorption of water molecules [64–67]. The capacitive type humidity sensors offer advantages of less power consumption and better linear sensitivity with relative humidity (RH). Resistive type sensors are much simpler and more straightforward for sensing humidity than making capacitance measurements [64]. Typical materials such as porous silicon and carbon nitride film have been explored for sensing humidity [68, 69]. However, humidity sensors based on these materials have poor stability and slow response. In this regard, carbonaceous nanomaterials and their composites with high sensitivity and fast response to humidity are ideal materials for fabricating sensors.

9.4.2.1 Graphene Oxide Sensors

Very recently, the properties of humidity sensors made from graphene oxide (GO) were reported by Yao et al. [70]. They deposited GO thin films onto silicon microbridge using spin-coating process. The sensor is then exposed to a humidity test chamber capable of generating required water vapor concentration at a fixed temperature (Figure 9.25). Hydrophilic GO that contains oxygenated functional groups can capture and interact with water molecules from the chamber readily. Consequently, water molecules enter the GO layer, leading to an increase in the GO interlayer distance and swelling of the GO thin film [71]. This causes bending of the silicon membrane. The piezoresistive Wheatstone bridge embedded in silicon microbridge transformed the deformation into a voltage (Figure 9.26a–c). The voltage responses of the sensors exposed to different humidity levels are shown in

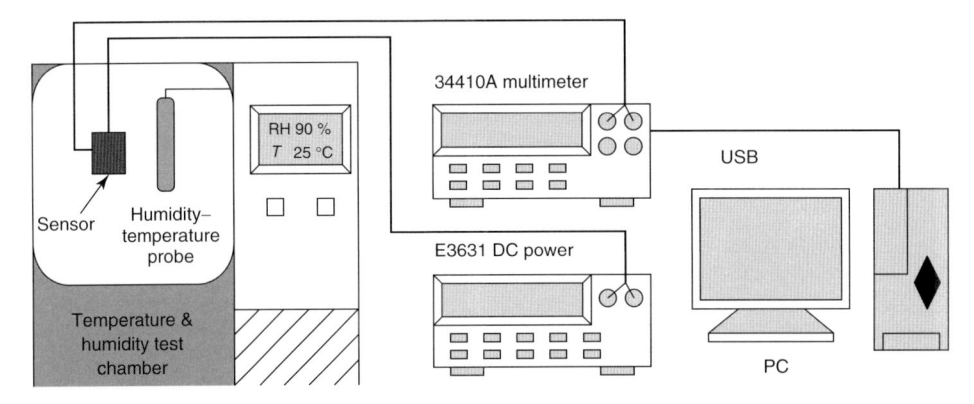

Figure 9.25 Schematic diagram of experimental setup for humidity sensing. (Source: Reproduced with permission from Ref. [70], Elsevier (2012).)

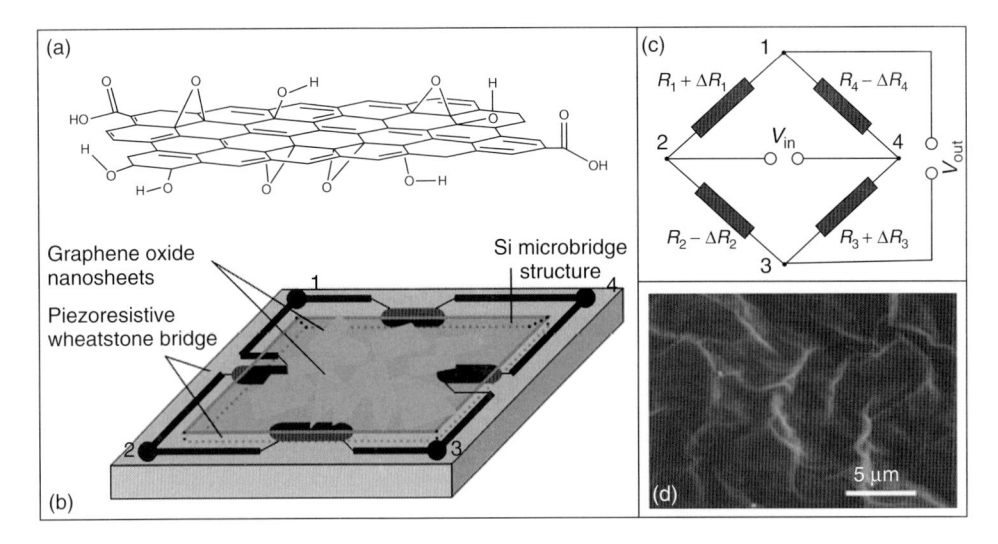

Figure 9.26 (a) Chemical structure of graphene oxide, (b) schematic illustration of graphene oxide-silicon bilayer structure, (c) piezoresistive Wheatstone bridge circuit, and (d) typical scanning electron microscopy (SEM) image of graphene oxide thin film. (Source: Reproduced with permission from Ref. [70], Elsevier (2012).)

Figure 9.27a. The sensor without GO film shows no obvious changes in the voltage response. However, the output voltage of the sensor with GO thin film of 65 nm thickness increases sharply with increasing humidity level. The detection sensitivity can be defined as the ratio of the change in output voltage to the change in RH, that is, $\Delta V/\Delta RH$. The GO sensor exhibits high detection sensitivity of 28.02 µV/%RH. The repeatability of this sensor is reasonably good as shown in Figure 9.27b. The sensitivity can reach 79.3 µV/%RH by increasing GO film thickness to 215 nm (Figure 9.28). This implies that the sensitivity can be tuned by adjusting GO film thickness. The response curves from 10 to 98% RH are best fitted with linear

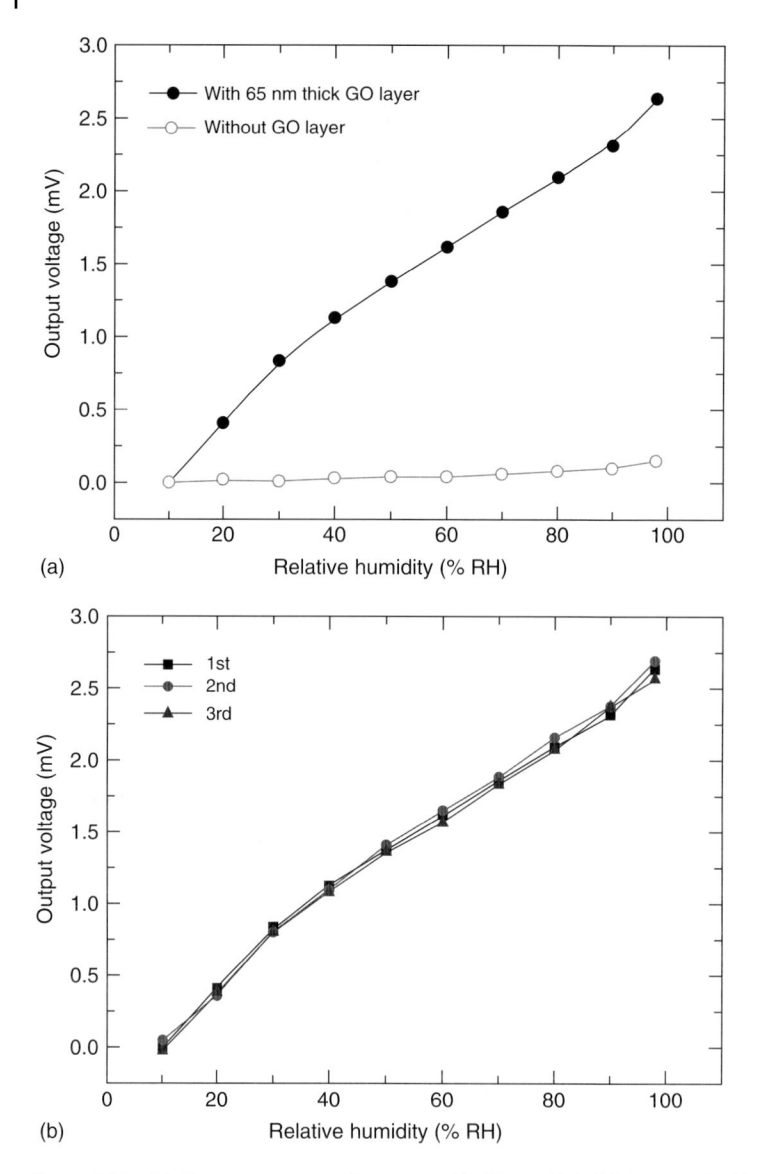

Figure 9.27 (a) Response curves of sensors with 65 nm thick GO layer and without GO layer for various relative humidity levels. (b) Repeatability of the sensor with 65 nm thick GO layer. (Source: Reproduced with permission from Ref. [70], Elsevier (2012).)

regression. The linear regression coefficient shows a slight decrease with increasing GO film thickness. Therefore, a balance between the sensitivity and linearity must be reached in the design of humidity sensors. Till present, there exists no work in the literature reporting the properties of humidity sensors based on graphene/polymer nanocomposites. It is considered that isocyanate-treated graphene oxide/polyimide

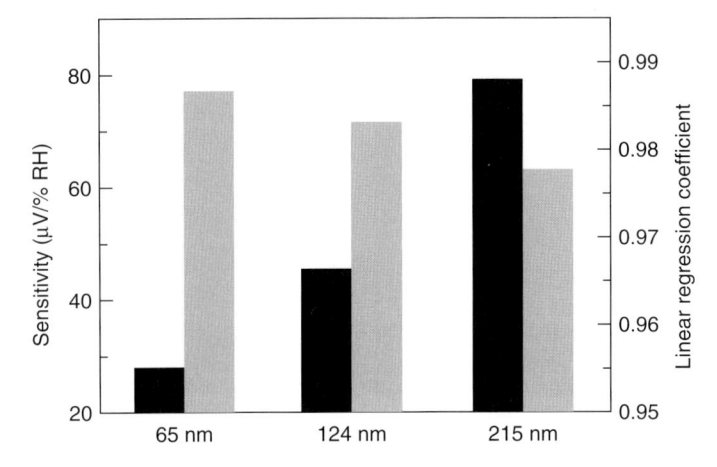

Figure 9.28 Detection sensitivity (black) and linear regression coefficient (gray) of humidity sensors with different GO film thickness. (Source: Reproduced with permission from Ref. [70], Elsevier (2012).)

(iGO/PI) nanocomposites with good electrical and mechanical properties [Chapter 2, Ref. 101] are suitable materials for detecting humidity.

9.4.2.2 Nanocomposite Sensors with CNT Fillers

Water vapor has a great influence on the electrical conductivity of CNTs. Water molecules adsorbed on the nanotube surface reduce the electronic conduction in the tube. The reduction of conductance with water adsorption is due to charge transfer between the adsorbate and the nanotube [72]. In particular, the conductivity type of the SWNT mat can be changed from p-type to n-type by adsorption of water acting as an electron donor [73]. The design of PI resistive type humidity sensors is well documented in the literature [64, 66, 74]. Very recently, the use of the CNT/PI nanocomposites for humidity detection has attracted the attention of researchers [61, 74].

Yoo et al. [53] fabricated resistivity-type humidity sensors using plasma-treated multiwalled carbon nanotube/polyimide (p-MWNT/PI) composite film. The film was deposited onto a suspended silicon nitride membrane. Figure 9.29 shows the room temperature resistance versus filler content of the p-MWNT/PI composite films. The resistance of the composites decreases with increasing nanotube content as expected. The resistance responses of the nanotube sensors exposed to different humidity levels are shown in Figure 9.30a–c. Very good linearity is attained at 0.4 wt% MWNT content. The mechanism responsible for the p-MWCNT/PI composite film is related to the charge transfer between adsorbed water molecules and the p-MWCTs. The humidity sensitivity of the sensors is defined by $\frac{\Delta R/R_0}{\Delta(\%RH)}$. Table 9.1 compares the sensitivity and linearity of humidity sensors made from PI composites filled with both pristine and plasma-treated MWNTs. Among these sensors, it can be seen that the 0.4 wt% p-MWCT/PI composite sensor is capable of working over a wide range of humidity with high sensitivity and linearity.

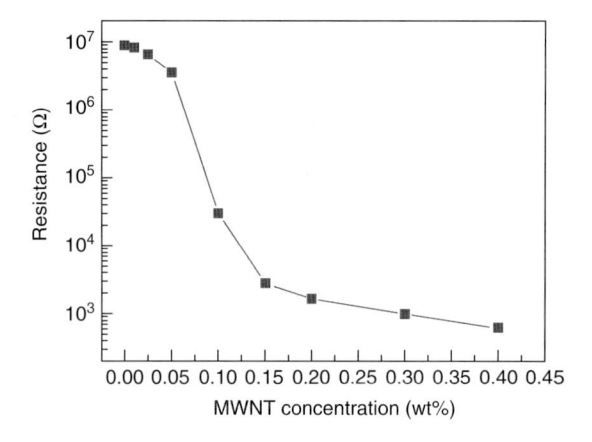

Figure 9.29 Resistance versus filler content of oxygen plasma-treated MWNT/PI composite films. (Source: Reproduced with permission from Ref. [53], Elsevier (2010).)

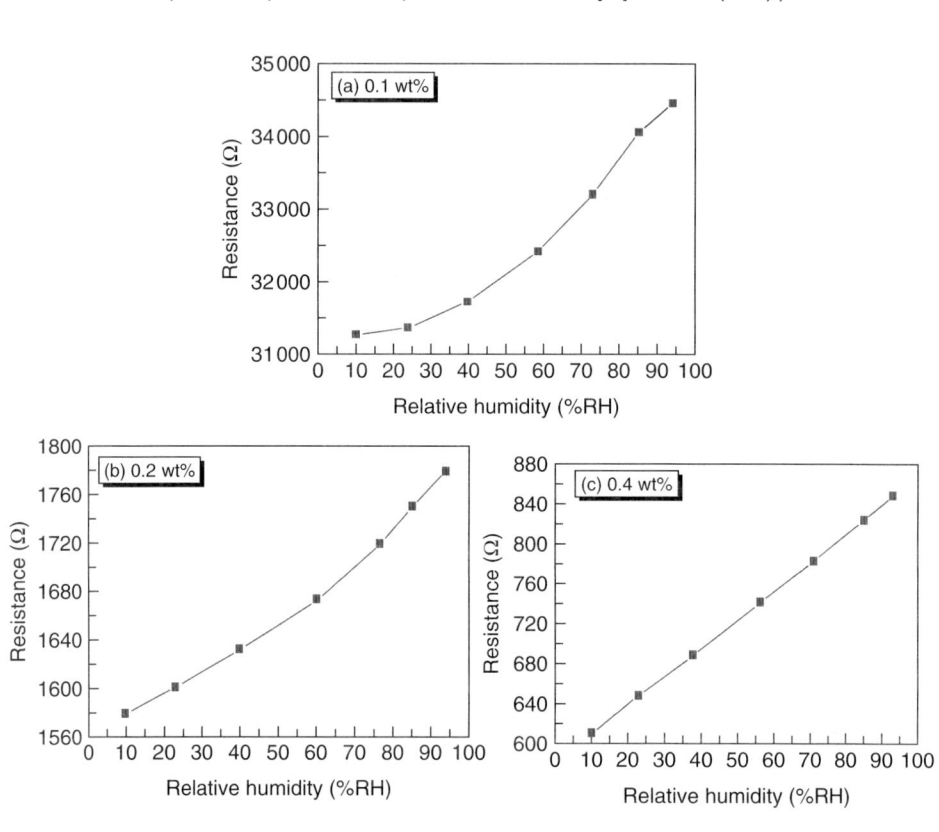

Figure 9.30 Humidity dependence of the resistance of oxygen plasma-treated MWNT/PI composite sensors containing (a) 0.1 wt%, (b) 0.2 wt%, and (c) 0.4 wt% MWNT. (Source: Reproduced with permission from Ref. [53], Elsevier (2010).)

Table 9.1 Sensitivity and linear correlation of p-MWNT/PI and MWNT/PI composite humidity sensors.

CNT content	p-MWNT/PI sensor		MWNT/PI sensor	
	S (/%RH)	R^2	S (/%RH)	R^2
0.1 wt%	0.00127	0.9466	0.00092	0.9379
0.2 wt%	0.00149	0.9822	0.00103	0.9519
0.3 wt%	0.00305	0.9913	0.00182	0.9719
0.4 wt%	0.00466	0.9999	0.00218	0.9823

Source: Reproduced with permission from Ref. [53], Elsevier (2010).

Nomenclature

A	Constant of piezoresistivity
B	Constant of piezoresistivity
C	Constant of piezoresistivity
D	Constant of piezoresistivity
d_0	Initial interparticle distance
d	Final interparticle distance
e	Electronic charge
ε	Strain
h	Plank's constant
K	gauge factor
L	Number of particles forming a single conducting path
l_0	Initial length of a sample before deformation
l	Final length of a sample after deformation
m	Mass of electron
N	Number of conducting paths
Φ	Height of potential barrier between adjacent particles
R_0	Initial resistance of a sample in air
R	Resistance of a sample on exposure to gas/vapor
R^2	Linearity correlation
ΔR	Change in resistance
S	Gas or humidity sensitivity

References

1. Lonergan, M.C., Severin, E.J., Doleman, B.J., Beaber, S.A., Grubb, R.H. and Lewis, N.S. (1996) *Chem. Mater.*, **8**, 2298–2312.
2. Iwata, H., Nakanoya, T., Morohashi, H., Chen, J., Yamauchi, T. and Tsubokawa, N. (2006) *Sens. Actuators, B Chem.*, **113**, 875–892.
3. Wang, H.C., Li, Y., Chen, Y., Yuan, M.Y., Yang, M.J. and Yan, W. (2007) *React. Funct. Polym.*, **67**, 977–985.
4. Collins, P.G., Bradley, K., Ishigami, M. and Zettl, A. (2000) *Science*, **287**, 1801–1804.
5. Zhang, T., Mubeen, S., Myung, N.V. and Deshusses, M.A.

(2008) *Nanotechnology*, **19**, 332001 (14 pp).
6. Li, W., Hoa, N.D. and Kim, D.J. (2010) *Sens. Actuators, B Chem.*, **149**, 184–188.
7. Slobodian, P., Riha, P., Lengalova, A., Svoboda, P. and Saha, P. (2011) *Carbon*, **49**, 2499–2507.
8. Schedin, F., Geim, A.K., Morozov, S.V., Hill, E.W., Blake, P., Katsnelson, M.I. and Novosolov, K.S. (2007) *Nat. Mater.*, **6**, 652–655.
9. Leenaerts, O., Pasrtoens, B. and Peeters, F.M. (2008) *Phys. Rev. B*, **77**, 125416 (6 pp).
10. Zhang, Y.H., Chen, Y.B., Zhou, K.G., Liu, C.H., Zeng, J., Zhang, H.L. and Peng, Y. (2009) *Nanotechnology*, **20**, 185504 (8 pp).
11. Lu, G.H., Ocola, L.E. and Chen, J.H. (2009) *Nanotechnology*, **20**, 445502 (9 pp).
12. (a) Pearce, R., Andersson, T.I., Hultman, L., Spetz, A.L. and Yakimova, R. (2011) *Sens. Actuators, B Chem.*, **155**, 451–455; (b) Lu, H.H., Ocola, L.E. and Chen, J.H. (2009) *Appl. Phys. Lett.*, **94**, 083111 (3 pp); (c) Lu, H.H., Ocola, L.E. and Chen, J.H. (2009) *Nanotechnology*, **20**, 445502 (9 pp).
13. Yoon, H.J., Jun, D.H., Yang, J.H., Zhou, Z.X., Yang, S.S. and Cheng, M.M. (2011) *Sens. Actuators, B Chem.*, **157**, 310–313.
14. Baughman, R.H., Cui, C.X., Zakhidov, A.A., Iqbal, Z., Barisci, J.N., Spinks, G.M., Wallace, G.G., Mazzoldi, A., Rossi, D., Rinzler, A.G., Jaschinski, O., Roth, S. and Kertesz, M. (1999) *Science*, **284**, 1340–1344.
15. Fennimore, A.M., Yuzvinsky, T.D., Han, W.Q., Fuhrer, M.S., Cumings, J. and Zettl, A. (2003) *Nature*, **424**, 408–410.
16. Kaul, A.B., Wong, E.W., Epp, L. and Hunt, B.D. (2006) *Nano Lett.*, **6**, 942–947.
17. Kim, P. and Lieber, C.M. (1999) *Science*, **286**, 2148–2150.
18. Paulson, S., Falvo, M.R., Snider, N., Helser, A., Hudson, T., Seeger, A., Taylor, R.M. and Washburn, S. (1999) *Appl. Phys. Lett.*, **75**, 2936–2938.
19. Tombler, T.W., Zhou, C., Alexseyev, L., Kong, J., Dai, H., Liu, L., Jayanthi, C.S., Tang, M. and Wu, S.Y. (2000) *Nature*, **405**, 769–772.
20. Hu, N., Karube, Y., Arai, M., Watanabe, T., Yan, C., Li, Y., Liu, Y. and Fukunaga, H. (2010) *Carbon*, **48**, 680–687.
21. Simmons, J.G. (1963) *J. Appl. Phys.*, **34**, 1793–1803.
22. Zhang, X.W., Pan, Y., Zheng, Q. and Yi, X.S. (2000) *J. Polym. Sci., Part B: Polym. Phys.*, **38**, 2739–2749.
23. Knite, M., Teteris, V., Kiploka, A. and Kaupuzs, J. (2004) *Sens. Actuators, A*, **110**, 142–149.
24. Lu, J.R., Weng, W.G., Chen, X.F., Wu, D.J., Wu, C.L. and Chen, G.H. (2005) *Adv. Funct. Mater.*, **15**, 1358–1363.
25. Lu, J.R., Chen, X.F., Wei, L. and Chen, G.H. (2006) *Eur. Polym. J.*, **42**, 1015–1021.
26. Chen, L., Chen, G.H. and Lu, L. (2007) *Adv. Funct. Mater.*, **17**, 898–904.
27. Park, M., Kim, H. and Youngblood, J.P. (2008) *Nanotechnology*, **19**, 055705 (7 pp).
28. Pham, G.T., Park, Y.B., Liang, Z.Y., Zhang, C. and Wang, B. (2008) *Composites Part B*, **39**, 209–216.
29. Hu, N., Karube, Y., Yan, C., Masuda, Z. and Fukunaga, H. (2008) *Acta Mater.*, **56**, 2929–2936.
30. Bautista-Quijano, J.R., Aviles, F., Aguilar, J.O. and Tapia, A. (2010) *Sens. Actuators, A Phys.*, **159**, 135–140.
31. Paleo, A.J., van Hattum, F.W., Pereira, J., Rocha, J.G., Silva, J., Sencada, V. and Lanceros-mendez, S. (2010) *Smart Mater. Struct.*, **19**, 065013 (7 pp).
32. Nriagu, J. (2011) *Encyclopedia of Environmental Health*, Elsevier, New York.
33. Brauthar, N. and Williams, I.I.J. (2002) *Int. J. Hyg. Environ. Health*, **205**, 479–491.
34. Parikh, K., Cattanach, K., Rao, R., Suh, D.S., Wu, A.M. and Manohar, S.K. (2006) *Sens. Actuators, B*, **113**, 55–63.
35. Yang, D.F. (2011) in *Advances in Nanocomposites-Synthesis, Characterization and Industrial Applications*, Chapter 37 (ed. B.S. Reddy), Intech, Croatia, pp. 857–882.
36. Li, J.R., Xu, J.R., Zhang, M.Q. and Rong, M.Z. (2003) *Carbon*, **41**, 2353–2360.

37. Sisk, B.C. and Lewis, N.S. (2006) *Langmuir*, **22**, 7928–7935.
38. Xie, H.F., Yang, Q.D., Sun, X.X., Yang, J.J. and Huang, Y.P. (2006) *Sens. Actuators, B*, **113**, 887–891.
39. Lei, H., Pitt, W.G., McGrath, L.K. and Ho, C.K. (2007) *Sens. Actuators, B Chem.*, **125**, 396–407.
40. Dong, X.M., Fu, R.W., Zhang, M.Q., Zhang, B. and Rong, M.Z. (2004) *Carbon*, **42**, 2551–2559.
41. Chen, S.G., Hu, J.W., Zhang, M.Q., Li, M.W. and Rong, M.Z. (2004) *Carbon*, **42**, 645–651.
42. Chen, S.G., Hua, J.W., Zhang, M.Q. and Rong, M.Z. (2005) *Sens. Actuators, B Chem.*, **105**, 187–193.
43. Zhang, B., Fu, R.W., Zhang, M.Q., Dong, X.M., Zhao, B., Wang, L.H. and Pittman, C.U. Jr. (2006) *Composites Part A*, **37**, 1884–1889.
44. Chen, J.H. and Tsubokawa, N. (2000) *Polym. Adv. Technol.*, **11**, 101–107.
45. Feller, J.F. and Grohens, Y. (2005) *Synth. Met.*, **154**, 193–196.
46. Luo, Y.L., Liu, Y.X. and Wang, L. (2006) *Sens. Actuators, B Chem.*, **119**, 516–523.
47. Sau, K.P., Khastgir, D. and Chaki, T.K. (1998) *Die Angewandte Makromolekulare Chemie*, **258**, 11–17.
48. Zhang, B., Fu, R.W., Zhang, M.Q., Dong, X.M., Lan, P.L. and Qiu, J.S. (2005) *Sens. Actuators, B*, **109**, 323–328.
49. Luo, Y.L., Wang, C. and Li, Z.Q. (2007) *Synth. Met.*, **157**, 390–400.
50. Zhang, B., Dong, X.M., Fu, R.W., Zhao, B. and Zhang, M.Q. (2008) *Compos. Sci. Technol.*, **68**, 1357–1362.
51. Shang, A.M., Li, L., Yang, X.M. and Wei, Y.Y. (2009) *Compos. Sci. Technol.*, **69**, 1156–1159.
52. Castro, M., Lu, J.B., Bruzaud, S., Kumar, B. and Feller, J.F. (2009) *Carbon*, **47**, 1930–1942.
53. Yoo, K.P., Lim, L.T., Min, N.K., Lee, M.J., Lee, C.J. and Park, C.W. (2010) *Sens. Actuators, B Chem.*, **145**, 120–125.
54. Lee, J.W., Park, E.J., Choi, J.W., Hong, H.H. and Shim, S.E. (2010) *Synth. Met.*, **160**, 566–574.
55. Yuana, C.L., Chang, C.P. and Song, Y. (2011) *Mater. Sci. Eng. B*, **176**, 821–829.
56. Luo, Y.L., Wang, B.X. and Xu, F. (2011) *Sens. Actuators, B: Chem.*, **156**, 12–22.
57. Zhang, B., Fu, R.W., Zhang, M.Q., Dong, X.M., Wang, L.H. and Pittman, C.U. Jr. (2006) *Mater. Res. Bull.*, **41**, 553–562.
58. Yo, S.S. and Kim, S.C. (2005) *J. Appl. Polym. Sci.*, **95**, 1062–1068.
59. Tetelin, A., Pellet, C., Laville, C. and N'Kaoua, G. (2003) *Sens. Actuators, B Chem.*, **91**, 211–218.
60. Chen, Z. and Lu, C. (2005) *Sens. Lett.*, **3**, 274–295.
61. Tang, Q.Y., Chan, Y.C. and Zhang, K. (2011) *Sens. Actuators, B Chem.*, **152**, 99–106.
62. Liu, H., Agarwal, M., Varahramyan, K., Berney, E.S. and Hodo, W.D. (2008) *Sens. Actuators, B Chem.*, **129**, 599–604.
63. Su, P.G. and Wang, C.S. (2007) *Sens. Actuators, B Chem.*, **124**, 303–308.
64. Packirisamy, M., Stiharu, I., Li, X. and Rinaldi, G. (2005) *Sens. Rev.*, **25**, 271–276.
65. Yao, Z. and Yang, M. (2006) *Sens. Actuators, B Chem.*, **117**, 93–98.
66. Kanamori, Y., Itoh, E. and Miyairi, K. (2007) *Mol. Cryst. Liq. Cryst.*, **472**, 327–335.
67. Shinbo, K., Otuki, S., Kanbayashi, Y., Ohdaira, Y., Baba, A., Kato, K., Kaneko, F. and Miyadera, N. (2009) *Thin Solid Films*, **518**, 629–633.
68. Yarkin, D.G. (2003) *Sens. Actuators, A*, **107**, 1–6.
69. Lee, S.P., Lee, J.G. and Chowdhury, S. (2008) *Sensors*, **8**, 2662–2672.
70. Yao, Y., Chen, X.D., Guo, H.H., Wu, Z.Q. and Li, X.Y. (2012) *Sens. Actuators, B Chem.*, **161**, 1053–1058.
71. Arroso-Bujans, F., Cerveny, S., Alegría, A. and Colmenero, J. (2010) *Carbon*, **48**, 3277–3286.
72. Pati, R., Zhang, Y., Nayak, S.K. and Ajayan, P.M. (2002) *Appl. Phys. Lett.*, **81**, 2638–2640.
73. Zahab, A., Spina, L., Poncharal, P. and Marliere, C. (2000) *Phys. Rev. B*, **62**, 10000–10003.
74. Ueda, M., Nakamura, K., Tanaka, K., Kita, H. and Okamoto, K. (2007) *Sens. Actuators, B: Chem.*, **127**, 463–470.

Index

a
AC universal dynamic response   198
alamar blue assay   319
alamar blue or resazurin   297
alveolar epithelial cell line (A549)   297
annealed paper   33
'as-made' freestanding paper   33
atom transfer radical polymerization (ATRP)   55
atomistic modeling   30
Avrami equation   105–106, 110, 119
2,2-Azobisisobutyrronitrile (AIBN)   343

b
'base' growth model for CNTs   16–18
binary polymer nanocomposites containing MWNTs or GNPs
– electrical characteristics   265
biocompatibility of CNTs   290–306
– cell viability assays   296–301
– – MTT assay   297
– potential health hazards   290–296
– – particle-induced lung injuries   292
– – toxicity of inhaled particles   293–295
– tissue cell responses   301–306
biocompatibility of graphene oxide   323–326
biomedical applications, polymer nanocomposites for   285–326
– biocompatibility of CNTs   290–306
– bone implants   286–290
– CNT/polymer nanocomposite scaffolds   310–320
– load-bearing implants   306–310
– nervous system remedial applications   320–323
bipolar plates (BP), in fuel cells   248–249
– conventional bipolar plates   249–252
– electrical characteristics of   259–268, See also individual entry

– mechanical properties of   269–274, See also individual entry
– in PEMFCs   249–250
– – target values of   250
– polymer nanocomposite bipolar plates   252–259
– – CNT-filled   252–254
– – graphene-sheet-filled composite plates   254–259
– replacement for   250
Boltzmann constant   39
bone implants, polymer nanocomposites for   286–290
– biomimetic advantages of nanomaterials   289
– cortical bone   288–289
– HAPEX^{TM} 288–289
bromoisobutyryl bromide-attached graphene (BIBG)   87
bulk-molding compound (BMC) process   252–254
1,4-butanediol (BDO)   370

c
carbon black (CB) nanoparticles, composites with   44–45
– electrical properties   264–265
carbon nanofibers (CNFs)   28–29, See also under Electrical conductivity
– mechanical properties   172–186
– nanocomposites with
– – from melt mixing   81–84
– – from solution mixing   76–78
– – in situ polymerization   91–98
– properties of   40–41
– vapor-grown carbon nanofibers (VGCNFs)   28

carbon nanotubes (CNTs)   10–27, *See also* electrical conductivity; multiwalled carbon nanotubes (MWNTs); purification of CNTs; single-walled carbon nanotubes (SWNTs)
– 'base' growth model for   16–18
– chemical vapor deposition (CVD)   13–19
– composites with   46
– crystallization, characterization techniques for   111–120
– distribution in the body   291
– mechanical properties   172–186
– nanocomposites with
– – from melt mixing   81–84
– – from solution mixing   76–78
– – *in situ* polymerization   91–98
– patent processes for   43
– properties of   35–40
– – electrical behavior   38–39
– – flexibility   37–38
– – mechanical behavior   35–38
– – thermal behavior   39–40
– structural variety of   11
– synthesis of   11–19
– – direct current (DC) arc discharge   11
– – laser ablation   12–13
– – physical vapor deposition (PVD)   11–13
– 'tip' growth model for   16–17
carbonaceous nanomaterials
– current availability of   41
– manufacturers of   44
cell viability assays   296–301
– MTT assay   297
centrifugation   22
chemical techniques of CNTs purification   19–21
chemical vapor deposition (CVD)
– for CNTs synthesis   13–19
– – DC-plasma-enhanced hot filament CVD   14
– – high-pressure carbon oxide disproportionation (HiPCo)   18
– – plasma-enhanced chemical vapor deposition (PECVD)   14
– – vapor-liquid-solid (VLS) mechanism   16
chemically modified graphene (CMG)   7
coefficient of thermal expansion (CTE)   124
composites
– with carbon nanotubes   46
– with graphene oxide   45
compression-impregnation   258–259
compression-impregnation-compression   259
compression molding   256
concentration overpotential   267

continuum percolation theory   194
covalent functionalization   60–62, 131
Cox model   152
critical aspect ratio   149
critical pigment volume concentrations (CPVCs)   251
crystallization half-life time   107
crystallization, characterization techniques for   104–120
– differential scanning calorimetry (DSC)   105–107
– dynamic mechanical analysis (DMA)   104–105
– nanocomposites with CNTs   111–120
– – ester-functionalized MWNT/PVDF   112
– – melt-cooled samples (MCPs)   113
– – MWNT/PVDF   112
– – nonisothermal crystallization   119
– – transcrystallinity   114
– nanocomposites with graphene nanofillers   107–111
current–voltage relationship   229–234

d

dibutyl phthalate (DBP) absorption test   45
1,2-dichlorobenzene (DCB)   56
differential scanning calorimetry (DSC)   105–107
diglycidyl ether bisphenol A (DGEBA)   253
dimethyl sulfoxide (DMSO)   54–58
dimethylformamide (DMF)   201
3-(4,5-dimethylthiazol-2-yl)-2,5-diphenyltetrazolium bromide (MTT) assay   297
direct current (DC) arc discharge   11
disorder-induced band (D-band)   24–25
dispersion of CNTs   130–133
double-walled carbon nanotube (DWNT), thermosetting matrix   180–182
drawing method   1
dynamic mechanical analysis (DMA)   104–105, 145

e

elastomeric matrices   168–169
– CNT- and CNF-filled polymer composites   228–229
– graphene-filled polymer composites   208–210
electrical behavior
– of CNTs   38–39
– of graphene   31–32
electrical characteristics
– fuel cell performance and   266–268
– – concentration overpotential   267

– – ionic resistance   267
– – polarization curves   267–268
– – voltage loss   266–267
– of nanocomposite bipolar plates   259–268
– – binary polymer nanocomposites containing MWNTs or GNPs   265
– – polymer/CB composite plates   264–265
– – polymer/EG and polymer/GNP composite plates   262–264
– – polymer/graphite/CNT hybrid plates   259–262
– – ternary graphite/polymer composites containing MWNTs or GNPs   265
electrical conductivity   197–229
– CNT- and CNF-filled polymer composites   213–229
– – elastomeric matrices   228–229
– – FMWNT   216
– – thermoplastic matrices   213–222
– – thermosetting matrices   222–228
– – trifluorophenyl (TFP)-functionalized MWNTs   217
– EG- and GNP-filled polymers   210–213
– – thermoplastic matrices   210–213
– – thermosetting matrices   213
– graphene-filled polymer composites   201–210
– – elastomeric matrices   208–210
– – thermoplastic matrices   201–207
– – thermosetting matrices   207–208
electrical properties   193–243, See also Current–voltage relationship; Percolation concentration
– hybrid nanocomposites   238–243
electrocatalyst supports, in fuel cells   274–282
– carbon-nanotube-supported platinum electrocatalysts   274–279
– – carbon-nanotube-supported   274–279
– – graphene-supported   279–282
– – polyol synthesis method   276
electromagnetic interference (EMI) shielding   331–349
– EMI shielding efficiency   332–334
– foamed nanocomposites for EMI applications   343–349
– – graphene/polymer foamed nanocomposites   347–349
– polymer nanocomposites for   331–349
– – with CNTs and CNFs   338–343
– – with GNPs   335–338
– – with graphene fillers   335
– – polytrimethylene terephthalate (PTT) nanocomposites   339

electrospinning
– principle and applications   312–320
– – electrospun R- and A-type nanofiber mats   317–318
– – fiber collection   315–316
– – Taylor cone formation   314
energy dispersive spectroscopy (EDS)   23
epoxy nanocomposites, processing techniques of   72–73
epoxy resin (EP)   163
equivalent-continuum modeling   30
essential work of fracture (EWF)   147
excluded volume   196–197
expanded graphite (EG)   4
– electrical conductivity   210–213
– mechanical properties   169–172
– nanocomposites with
– – from melt mixing   80–81
– – from solution mixing   68–76
– – *in situ* polymerization   89–91
– surface modification of   58–59
Eyring equation   145

f

flexibility of CNTss   37–38
flexural strength   145
foamed nanocomposites for EMI applications   343–349
– CNT/polymer foamed nanocomposites   343–346
– graphene/polymer foamed nanocomposites   347–349
foliated graphene sheets (FGSs)   79
Fowler-Nordheim tunneling   233
fracture toughness   146–148
fuel cells   247–282
– bipolar plates (BP)   248–252, See also *individual entry*
– carbonaceous nanomaterials for   247–282
– electrocatalyst spports   274–282, See also *individual entry*
– polymer exchange membrane fuel cell (PEMFC)   248–249
– polymer nanocomposites for   247–282
functionalization
– of CNFs   59–66
– of CNTs   59–66, 130–133
– covalent functionalization   60–62
– – sidewall functionalization   61
– functionalized graphene sheet (FGS)   9
– functionalized multiwalled carbon nanotube (FMWNT)   252
– functionalized ultrashort (F-US) tubes   303

functionalization (contd.)
- noncovalent functionalization 62–66, See also *individual entry*
- of SWNTs 62–63

g

gas sensitivity, polymer nanocomposites for 359–361
graphene-based nanomaterials 1–10
- 2D graphene 1
- multifunctional composite materials 41–46
- patent processes for 42
- physical properties of graphene 30–35
- – electrical behavior 31–32
- – mechanical behavior 30–31
- – tensile properties 32
- – thermal behavior 32–35
- preparing
- – epitaxial growth of graphene on metal carbide 3
- – graphitization of single-crystal silicon carbide substrate 2
graphene-like fillers
- crystallization, characterization techniques for 107–111
- electrical conductivity 201–210
- mechanical properties 153–169, See also under Mechanical properties
- nanocomposites with 67–68
- – from *in situ* polymerization 84–89
- – from melt mixing 78–80
- – from solution mixing 67–68
- permittivity 201–210
- thermal conductivity of 133–139
graphene nanoplatelets (GNPs) 4, 45
- electrical characteristics 262–264
- electrical conductivity 210–213
- electromagnetic interference (EMI) shielding in 335
- *in situ* polymerization 89–91
- mechanical properties 169–172
- by melt mixing 80–81
- nanocomposites with
- – from melt mixing 80–81
- – from solution mixing 68–76
- – *in situ* polymerization 89–91
- permittivity of 210–213
- in pressure/strain sensors 355–356
- by solution mixing 68–76
- surface modification of 58–59
- thermal conductivity of 133–139
- thermoplastic matrices 153–163, 210–213
- thermosetting matrices 163–168, 213

graphene oxide (GO) 4–10
- chemical structure of 6
- composites with 45
- exfoliation of 5–6
- functionalized graphene sheet (FGS) 9
- humidity sensor applications 372–375
- Hummers process 4
- noncovalent functionalization of 56
- reduced graphene oxide (rGO) 7
- Staudenmaier oxidation method 4
- surface modification of 54–58
- thermally reduced graphene oxide (TRG) 8
graphene-sheet-filled composite plates, in fuel cells 254–259
graphene-supported platinum electrocatalysts in fuel cells 279–282
graphite intercalation compound (GIC) 3–4

h

Halpin-Tsai micromechanical model 152, 155, 160
heat deflection temperature (HDT) 125–127, 174–175
hexamethylphosphoramide (HMPA) 54–58
high-density polyethylene (HDPE) 147
highly oriented pyrolytic graphite (HOPG) 1
High-pressure carbon oxide disproportionation (HiPCo) 18
high-resolution transmission electron microscopy (HRTEM) image
- graphene oxide (GO) 9
humidity sensors 371–377
- graphene oxide sensors 372–375
- nanocomposite sensors with CNT fillers 375–377
- polymer nanocomposites for 359–361
Hummers process 4
hybrid composite bipolar plates, in fuel cells, mechanical properties 269–272
hybrid fillers composites, mechanical properties 186–190
hybrid nanocomposites, electrical properties 238–243
hydrogenated carboxylated nitrile-butadiene rubber (HXNBR) composites 169
hydroxyapatite nanorods (nHAs) 306
hydroxyl-terminated acrylonitrile-butadiene copolymer (HTBN) 370

i

impregnation-compression 258
In situ polymerization 84–98, 257
- *in situ* polymerized MWNT/PS (polystyrene) nanocomposites 367

– for nanocomposites with graphene-like fillers 84–89
– – *in situ* emulsion polymerization 88
– – PMMA-functionalized graphene (MG) 87–88
– – PS-functionalized graphene sheets 88
– – ring-opening polymerization 88
– nanocomposites with CNT and CNF fillers 91–98
– – polyimide (PI) 93
– nanocomposites with EG and GNP fillers 89–91
interfacial interaction 150–151
interfacial shear stress (IFSS) 148–150
ionic resistance 267
isocyanate-treated graphene oxide (iGO) 201
isothermal crystallization 103–105
Izod impact test for polymers 145

j
j-integral measurements 147

l
laser ablation 12–13
linear elastic fracture mechanics (LEFM) 146
linear thermal expansion 124
liquid-phase oxidation for CNTs purification 20
load-bearing implants, CNT/polymer nanocomposites for 306–310
– mechanical properties 307–310
– – MWNT/PE nanocomposites 307–310
– – MWNT/PEEK nanocomposites 310
loss factor 339

m
maleic-anhydride-grafted polypropylene (MA-g-PP) 83
maleic-anhydride-grafted styrene-ethylene/butylene-styrene (MA-g-SEBS) 83
Maxwell-Wagner-Sillars (MWS) polarization 199
mechanical properties 143–190
– of CNTs 35–38
– – experimental measurement 36–37
– – theoretical prediction 35–36
– composites with hybrid fillers 186–190
– – short carbon fibers (SCFs) 186
– flexural strength 145
– fracture toughness 146–148
– of graphene 30–31
– Izod impact test 145
– J-integral measurements 147

– nanocomposites with CNT and CNF fillers 172–186
– – thermoplastic matrix 172–180
– – thermosetting matrix 180–186
– nanocomposites with EG and GNP fillers 169–172
– nanocomposites with graphene fillers 153–169
– – elastomeric matrix 168–169
– – LLDPE 158, 162
– – polyolefins 158
– – rGO/PVA nanocomposites 154–155
– – thermoplastic matrix 153–163
– – thermoplastic matrix 153–163
– – thermosetting matrix 163–168
– of nanocomposite bipolar plates 269–274
– – hybrid composite bipolar plates 269–272
– – polymer/EG composites 272–274
– natural graphite (NG)/PLA composites 169–172
– strengthening and toughening mechanisms 148–153
– – Cox model 152
– – Halpin-Tsai micromechanical model 152
– – interfacial interaction 150–151
– – interfacial shear stress (IFSS) 148–150
– – micromechanical modeling 151–152
– – shear-lag model 151
– – stress transfer across the matrix–nanotube interface 150
– – toughening mechanism 152–153
– tensile yield stress 145
melt-cooled samples (MCPs) 113
melt mixing
– in polymer nanocomposites preparation 78–84
– – nanocomposites with CNT and CNF fillers 81–84
– – nanocomposites with EG and GNP fillers 80–81
– – nanocomposites with graphene-like fillers 78–80
methyl methacrylate (MMA) 85–87
4,4-methylene-bis(o-chloroaniline) (MOCA) 370
micromechanical modeling 151–152
montmorillonite (MMT) clay silicate 242
multifunctional composite materials 41–46
multiwalled carbon nanotubes (MWNTs) 11

n
N,N-dicyclohexylcarbodiimide (DCC) 54–58
N,N-dimethylformamide (DMF) 54–58

Nafion®(Du Pont)   248
nanofillers, dispersion of   54–66
– functionalization of CNTs and CNFs   59–66
– graphene oxide, surface modification of   54–58
natural graphite (NG)/PLA composites   169
negative temperature coefficient (NTC) effect   234
nervous system remedial applications   320–323
N-methylpyrrolidone (NMP)   54–58
noncovalent functionalization
– of CNTs and CNFs   62–66
– of GO sheets   56

o

octadecylamine (ODA)   54–58, 202
organic vapor sensor applications   361–377
– *in situ* polymerized MWNT/PS (polystyrene) nanocomposites   367
– nanocomposites with CBs   361–371
– nanocomposites with CNFs   364–371
– nanocomposites with CNTs   364–371
organoclays   42
Ozawa crystallization   107

p

percolation concentration   193–229, See also Electrical conductivity; Permittivity
– alternating current (AC) field near percolation   196
– continuum percolation theory   194
– direct current (DC) conductivity   195
– excluded volume   196–197
– theoretical modeling   193–196
percolation threshold   194
permittivity   197–229
– CNT- and CNF-filled polymer composites   213–229
– – thermoplastic matrices   213–222
– EG- and GNP-filled polymers   210–213
– – thermoplastic matrices   210–213
– – thermosetting matrices   213
– graphene-filled polymer composites   201–210
– – elastomeric matrices   208–210
– – thermoplastic matrices   201–207
– – thermosetting matrices   207–208
physical techniques for CNTs purification   21–23
physical vapor deposition (PVD)
– in CNTs synthesis   11–13
– direct current (DC) arc discharge   11

– laser ablation   12–13
piezoelectric materials   352
piezoresistivity   352–355
plasma-enhanced chemical vapor deposition (PECVD)   14
– ECR PECVD reactors   15
– microwave PECVD reactors   15
platinum electrocatalysts in fuel cells
– carbon-nanotube-supported   274–279
– graphene-supported   279–282
poly(3,4-ethylenedioxythiophene) (PEDOT) films   322
poly(dimethyl siloxane) (PDMS)   77
poly(ethylene oxide) (PEO)   54–58
poly(oxyalkylene)-amines (POA)   253–255
poly(oxypropylene)-diamines (POP)   252–254
poly(styrene-co-butadiene-co-styrene) (SBS)   56–57
poly(vinyl alcohol) (PVA)   54–58
polyallylamine (PAA)   54–58
polycarbonate (PC)   205–206
polylactic acid (PLA)   205
polymer exchange membrane fuel cell (PEMFC)   248–249
– bipolar plates (BP)   248–249
– gas diffusion layer (GDL)   248
– membrane electrode assembly (MEA)   248
polymer nanocomposites preparation   53–98, See also under *In situ* polymerization; Melt mixing; Solution mixing
– patent processes for   96–98
polymer/EG composites, in fuel cells, mechanical properties   272–274
polyol synthesis method   276
polytrimethylene terephthalate (PTT) nanocomposites   339
polyurethane (PU)   362–364
positive temperature coefficient (PTC) effect   234–238
pressure/strain sensors   352–359
– nanocomposites with CNTs   356–359
– nanocomposites with GNPs   355–356
– piezoresistivity   352–355
purification of CNTs   19–27
– chemical techniques   19–21
– – arc-grown MWNTs   20
– – arc-grown SWNTs   20–21
– – carbonaceous impurities removal   20
– – liquid-phase oxidation   20
– physical techniques   21–23
– – centrifugation   22
– – high-temperature annealing   22
– – selected-area electron diffraction (SAED)   22

– purified CNTs, characterization 23–27
Pyrograf® III nanofibers 29

q
quantum mechanical (QM) modeling 30

r
Raman spectroscopy 23–26
reduced graphene oxide (rGO) 7, 56
resin transfer molding (RTM) technique 227

s
scaffolds 310–320
– CNT/polymer nanocomposite scaffolds 310–320
– – structures 311–312
– – types 311–312
scotch tape 1
selected-area electron diffraction (SAED) 22
sensor applications, polymer nanocomposites for 351–377
– gas sensors 359–361
– humidity sensors 371–377
– organic vapor sensors 361–377
– pressure/strain sensors 352–359
shear-lag model 151
shear mixing 74
short carbon fibers (SCFs), mechanical properties 186
single-edge-notched bending (SENB) 146
single-walled carbon nanotubes (SWNTs) 10–12
– arc-grown SWNTs in H_2 12
– covalent sidewall functionalization of 64
– functionalization strategies for 62
– purified SWNTs 12
– structural defects in 61
– synthesis 11–12
– thermal oxidation for purification of 20
sodium dodecyl sulfate (SDS) 22, 65
solution mixing
– in polymer nanocomposites preparation 66–78
– – nanocomposites with CNT and CNF fillers 76–78
– – nanocomposites with EG and GNP fillers 68–76
– – nanocomposites with graphene-like fillers 67–68
– – shear mixing 74
– – sonication mixing 74
sonication mixing 74
Staudenmaier oxidation method 4
Stone–Wales (S-W) defects 59

strain gauge factor (K) 358
stress shielding 287
styreneacrylonitrile (SAN) 335
surface-initiated polymerization (SIP) 86

t
tangential-mode G-band 24
tensile properties of graphene 32
tensile yield stress 145
ternary graphite/polymer composites containing MWNTs or GNPs, electrical characteristics 265
tetrahydrofuran (THF) 54–58
thermal behavior
– of CNTs 39–40
– of graphene 32–35
thermal black 44
thermal conductivity 127–139
– composites with CNTs 127–133
– – dispersion of 130–133
– – functionalization of 130–133
– – MWNT/epoxy nanocomposites 132
– – thermal interface resistance 127–130
– composites with GNP and graphene nanofillers 133–139
– – GNP–SWNT hybrid 137
– – MWNT/epoxy nanocomposites 136
thermal interface resistance 127–130
thermal oxidation for CNTs purification 20
thermal properties of polymer nanocomposites 103–139, See also Crystallization, characterization techniques for
thermal stability 120–127
– heat deflection temperature 125–127
thermally expanded graphene oxide (TEGO) 56–57, 135
thermally reduced graphene (TRG) 8, 56, 107–108
thermogravimetric analysis (TGA) 24, 27, 121–125
– linear thermal expansion 124
thermoplastic matrices 153–163
– CNT and CNF fillers 172–186, 213–222
– EG and GNP fillers 153–163, 210–213
thermoplastic polyurethane (TPU) 83
thermosetting matrices 163–168
– CNT and CNF fillers 180–186, 222–228
– EG and GNP fillers 163–168, 213
– – electrical conductivity 213
– epoxy resin (EP) 163
– graphene-filled polymer composites
– – electrical conductivity 207–208
– – permittivity 207–208

'tip' growth model for CNTs (*contd.*)
'tip' growth model for CNTs   16–17
tissue cell responses   301–306
toughening mechanism   152–153
transcrystallinity   114
transmission electron microscopy (TEM), graphene oxide (GO)   9–10
triethanolamine (TEA)   370
triethylenetetramine (TETA)   131
trifluorophenyl (TFP)-functionalized MWNTs   217
two-dimensional (2D) graphene   1

v
vapor-grown carbon nanofibers (VGCNFs)   28
vapor-liquid-solid (VLS) mechanism   16
voltage loss, fuel cell performance and   266–267

x
X-ray diffraction (XRD)   23–26
X-ray photoelectron spectroscopy (XPS)   7
– graphene oxide (GO)   7